AF388880

Mining Innovation

Mining Innovation

Editor

Krzysztof Skrzypkowski

MDPI • Basel • Beijing • Wuhan • Barcelona • Belgrade • Manchester • Tokyo • Cluj • Tianjin

Editor
Krzysztof Skrzypkowski
AGH University of Science and
Technology
Poland

Editorial Office
MDPI
St. Alban-Anlage 66
4052 Basel, Switzerland

This is a reprint of articles from the Special Issue published online in the open access journal *Energies* (ISSN 1996-1073) (available at: https://www.mdpi.com/journal/energies/special_issues/ Mining_Innovation).

For citation purposes, cite each article independently as indicated on the article page online and as indicated below:

LastName, A.A.; LastName, B.B.; LastName, C.C. Article Title. *Journal Name* **Year**, *Volume Number*, Page Range.

ISBN 978-3-0365-2219-7 (Hbk)
ISBN 978-3-0365-2220-3 (PDF)

© 2021 by the authors. Articles in this book are Open Access and distributed under the Creative Commons Attribution (CC BY) license, which allows users to download, copy and build upon published articles, as long as the author and publisher are properly credited, which ensures maximum dissemination and a wider impact of our publications.

The book as a whole is distributed by MDPI under the terms and conditions of the Creative Commons license CC BY-NC-ND.

Contents

About the Editor

Krzysztof Skrzypkowski is an associate professor of mining engineering at the AGH University of Science and Technology, Faculty of Civil Engineering and Resource Management in Kraków, Poland. Dr Skrzypkowski obtained his PhD (doctorate) and DSc (habilitation) degrees in AGH in 2014 and 2021, respectively. His scientific research concerns the stability of underground mine workings, particularly the selection of support for room and roadway excavations. In addition, his research interests focus on computer-aided design and the determination of geotechnical parameters of backfill materials. As the head of the rock bolting laboratory, he performs model, numerical and industrial tests on various mechanisms of cooperation between the mining support and the rock mass. He is the author of over 100 scientific publications and several patents and a utility model regarding the monitoring and yielding of the rock bolt and steel arch support. As an academic lecturer, he works closely with scientific institutes, industry and universities related to the exploitation of mineral raw materials.

Preface to "Mining Innovation"

The basic goals of the introduction of new technological solutions in mining are, among others, improvements in work safety, a reduction in production costs and the identification of natural hazards accompanying the exploitation. The Special Issue of "Mining Innovation" includes several articles concerning laboratory, numerical and industrial research into underground, borehole and opencast mining. The issues discussed in the articles include: determining the state of stress and deformation around rooms and longwall workings; monitoring mining machines with the use of specialized programming languages; selection of protection of access, preparatory and operational excavations with the use of rock bolt, arch yielding and lining supports; improving the efficiency of mining with the use of explosives; the use of 3D printing to model the shear mechanism of samples reflecting rocks; determining the most important factors that affect the sealing capacity of the fault; estimation of local anisotropy parameters and detection of voids in the rock mass using geophysical methods; an automatic method of detecting ground subsidence caused by underground mining; a new method for calculating a multi-asset break-even for multi-assortment production; characteristics of the dewatering system in open-cast mining and computer-aided design in order to determine the time of bacfilling for post-mining excavations. The presented solutions are innovative, developmental and, above all, closely related to the geological and mining conditions. The obtained research results clearly confirm that the exploitation of mineral deposits is at a very high level of advancement; however, it still requires further research using modern research techniques.

Krzysztof Skrzypkowski
Editor

Article

An Experimental Investigation into the Stress-Strain Characteristic under Static and Quasi-Static Loading for Partially Embedded Rock Bolts

Krzysztof Skrzypkowski

Faculty of Mining and Geoengineering, AGH University of Science and Technology, Mickiewicza 30 av., 30-059 Kraków, Poland; skrzypko@agh.edu.pl; Tel.: +481-2617-2160

Abstract: This article deals with a static and quasi-static load using the maximum power of a hydraulic pump. Additionally, quasi-static coefficients for the partially embedded rock bolts were determined. The laboratory tests included 2.2 m long bolts, which were embedded segmentally on the lengths of 0.05 m, 0.3 m and 0.9 m and were tested. To fix the ribbed bolt rods in the steel cylinders, resin cartridges with a length of 0.45 m long were used. The main aim of the research was to determine the load-displacement characteristics. Knowing the bolt rod tensile mechanism, the points of failure in the material continuity were identified, on the basis of which stress-strain characteristics are made. Particular attention was paid to the definition of: tensile stress for the yield point (σ_1), maximum stress (σ_2), stress at failure (σ_3), strain in the elastic range (ε_1), strain for maximum stress (ε_2) and strain corresponding to the failure (ε_3).

Keywords: laboratory tensile tests; partially embedded; rock bolt support; load-displacement and stress-strain characteristics

check for
updates

Citation: Skrzypkowski, K. An Experimental Investigation into the Stress-Strain Characteristic under Static and Quasi-Static Loading for Partially Embedded Rock Bolts. *Energies* **2021**, *14*, 1483. https://doi.org/10.3390/en14051483

Academic Editor: Jens Birkholzer

Received: 11 February 2021
Accepted: 5 March 2021
Published: 9 March 2021

Publisher's Note: MDPI stays neutral with regard to jurisdictional claims in published maps and institutional affiliations.

Copyright: © 2021 by the author. Licensee MDPI, Basel, Switzerland. This article is an open access article distributed under the terms and conditions of the Creative Commons Attribution (CC BY) license (https://creativecommons.org/licenses/by/4.0/).

1. Introduction

Rock bolt support as a technological procedure for strengthening excavations is very often tested in terms of optimal cooperation with the rock mass. One of the most important technical parameters for the partially embedded rock bolts is the minimum fixing length at which the support operates in the full load-displacement characteristic. Currently, the bolt support is installed in the rock mass where there are natural hazards of a dynamic nature [1,2]. Masoudi and Sharifzadeh [3] presented the load-displacement characteristics for various bolt support installation mechanisms for dynamic loads, detailing the amount of energy absorbed and the division into load capacity categories. Li et al. [4] stated that in conditions of a rock burst hazard, particular elements of the support are of certain importance, as they must be adapted to deformation in order to avoid premature destruction of the support. Rahimi et al. [5] presented the principles of designing the rock bolt support in conditions of deep mines and the risk of rock bursts, pointing to possible reliability systems. Li [6] stated that under dynamic hazards, the support should be 1 m longer than the range of rocks thrown into the excavation. Kim et al. [7] made a list of parameters influencing the performance of the grouted bolt support, while indicating that the load capacity of the bolt increases with increasing length of the embedment. Ozturk et al. [8] drew attention to the aspect of taking into account micro-seismic events in the support design procedure under dynamic hazard conditions. One of the ways to reflect industrial conditions are laboratory tests in which the support and load conditions are modeled [9]. Cao et al. [10] made a physical model of the stratified rock mass, in which they installed and tested super pre-stressed rock bolts. Feng et al. [11] used two steel cylinders, 0.1 m and 0.3 m long, in which a 22 mm diameter bolt rod was installed in order to model the manners of the bolt support in the stratified rock mass. Feng et al. [12] performed tests of the fixing of the bolt rod in a steel pipe 0.3 m long, which was divided into five different lengths, creating eight variants of installation. Skrzypkowski et al. [13] defined a minimum reinforcement length for a ribbed bolt with a wound bar. Bacić et al. [14] determined experimentally the fixing defects of the 2 m long bolts. The tests used bolt rods made of B500B steel with a diameter of 25 mm, which were fixed in 94 configurations by means of a cement binder. Zhan et al. [15] created a rock mass model using C40 class concrete, in which bolt rods were installed at lengths of 0.15 m, 0.2 m, 0.25 m and 0.3 m. Model tests allow for studying the behavior of the already existing bolt support solution in order to improve it. Moreover, they provide information on the mechanisms of cooperation between the rock mass and the support. First of all, by searching for the optimal solution, the support model can be brought into a state that meets the industrial requirements. Zhao et al. [16] examined the cable anchors on the laboratory stand with bilateral restraint and opposing extension using a hydraulic pump. Rock bolt support installed in a rock mass where dynamic phenomena occur is exposed to additional loads, the intensity of which depends, among others, on the energy and location of the tremors. In the conditions of dynamic hazards, the bolt support should not only have a high load-bearing capacity, but above all, it should be adjusted to absorb the greatest possible kinetic energy. The execution of additional work by the rock bolts can be realized by adding yielding elements, using special geometries of the rock bolt or by partially fixing. As a result of incomplete embedment, the rock bolt may deform more, contributing to the absorption of a greater deformation of the roof rocks. He et al. [17] designed a special rock bolt adapted to large displacements of the rock mass. One of the characteristics of this bolt is the partial embedment at the bottom of the hole. Zhao et al. [18] created a novel J energy-releasing rock bolt support, which can be fixed in the hole with resin cartridges or a cement binder. In the case of using a grouted bolt, the decisive parameter for the effective fixing of the rock bolt is the quality of mixing the adhesive components. One of the most important issues regarding the safe use of construction materials is the measurement of stresses and strains especially in underground mining after tremors. Withers et al. [19] characterized three strain mapping methods: magnetic, synchroton and contour, pointing to the scope and possibilities of its application. Morozov et al. [20] used pulsed eddy current to study residual stress state. Wilson et al. [21] applied the residual magnetic field technique to stress measurement in order to study structures with complex geometries. In the case of bolt support test, Skrzypkowski et al. [22] studied Self-Excited Acoustic System for measuring stress variations in expansion rock bolts. Crompton et al. [23] proved that the cause of poor mixing of the components of the adhesive charge is eccentric bolt location. Feng et al. [24] examined bolts with a diameter of 18 mm, 20 mm and 22 mm, which were fixed in steel cylinders with 450 mm long resin cartridges. Tests found that the addition of fine steel particles to the resin cartridge significantly increases the effectiveness of the embedment. Xu et al. [25] drew attention to the fact that the use of binders for fixing the bolt rods is associated with the formation of defects, which may result from the displacement of rock layers and the existing cracks and fissures in the rock mass. Campbell et al. [26] stated that for a bolt rod installed on a length of 0.5 m with the use of resin cartridge in a glove, about 40% is a good mixing of the components of the load, and connection with the bolt rod.

Rock Mass Stratification at Legnica-Głogów Copper District in Poland

Underground mining of copper ore deposits in Poland is carried out in the Legnica-Głogów Copper District (LGOM) in three mining plants: Rudna, Lubin and Polkowice-Sieroszowice. The dominant natural hazard occurring in all three mines is the release of seismic energy in the form of tremors, stress relief and rock bursts [27]. Occurring stress relief in the excavation, according to Polish regulations [28], is defined as a dynamic phenomenon caused by a rock tremor as a result of the part of the excavation that was damaged to the extent of not causing a loss of its serviceability or deterioration of safety conditions. However, as a result of the occurrence of only a seismic event, the excavation does not lose its functionality. The operation of the bolt support under dynamic loads is associated with the work done over time. By comparing the times of arrival of two different groups of seismic waves generated by the tremor, longitudinal and transverse and their impact on the support, it is possible to determine the path that each of these waves travels from the tremor to the point of their registration [29] (Figure 1a–c).

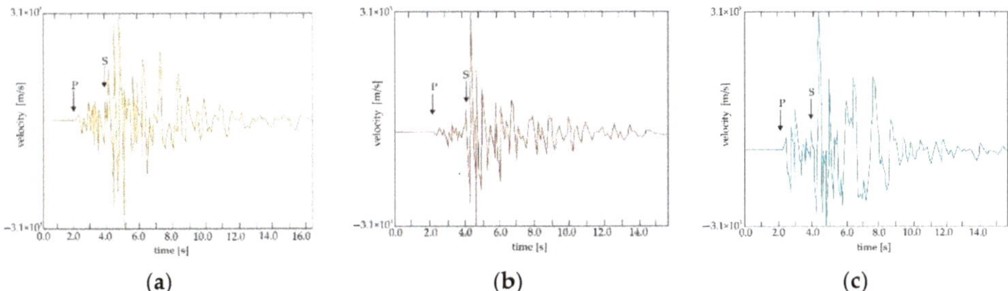

Figure 1. An exemplary seismogram with the P and S waves: (**a**) N-S component; (**b**) the E-W component; (**c**) Vertical component.

The seismogram presented in Figure 1 is intended to show the duration of the additional dynamic load to which the bolt support is exposed as a result of rock mass vibration and longitudinal and transverse wave propagation. Assuming that the duration of the low-energy dynamic load is several seconds, it is necessary to model the behavior of the bolt support that corresponds to the duration of this phenomenon. One of the ways of reflecting industrial conditions are laboratory tests under static and quasi-static loading, in which the stress-strain characteristics of the rock bolt support are obtained. Knowing the values of the maximum stress and strain, it is possible to select a support with greater flexibility and load-bearing capacity, and to modify the embedded length so as to adapt the bolt support to the conditions of rapidly changing loads. For copper ore deposits in the LGOM region, the mineralization zone has an irregular morphology of the roof and floor, which makes it difficult to adjust the geometry of the mining excavations, and to select such a height of the mining gate that the deposit losses and ore dilution are as low as possible. The contractual area determined on the basis of sampling is considered the border. Mining is carried out in three different lithological types of ore with different physical and chemical properties: carbonates, shales and sandstones. The lithological type of copper ores is the carbonate rocks of the Zechstein limestone. It is noteworthy that carbonate rocks show a stratified structure [30]. In the direct roof there are compact limestone dolomites with numerous anhydrite-calcite veins. Anhydrites with a thickness of about 300 m are deposited directly on the carbonate series. A characteristic feature in the vicinity of the deposit is the presence of layers in the roof with significantly higher strength parameters than in the deposit and in the floor. For example, the compressive strength of roof rocks and overburden is more than two and seven times greater than that of deposit and floor rocks, respectively. The type of stratification (thickness of the layers) depends on the intensity of the geological structures that separate individual layers. In the zones of the bolted roof, the intensity of the occurrence of structures emphasizing stratification is varied. This variability can be observed within one mining field, and even in the vicinity of two excavations [31]. Due to the diversity of the stratification structure of the direct roof, seven characteristic types up to a height of about 5 m were distinguished [32]. In particular, very thin-bed, thin-bed, medium-bed and thick-bed layers with a thickness of 0.05 m were separated, up to 0.1 m, 0.1–0.3 m and 0.3–1 m, respectively (Figure 2a–i).

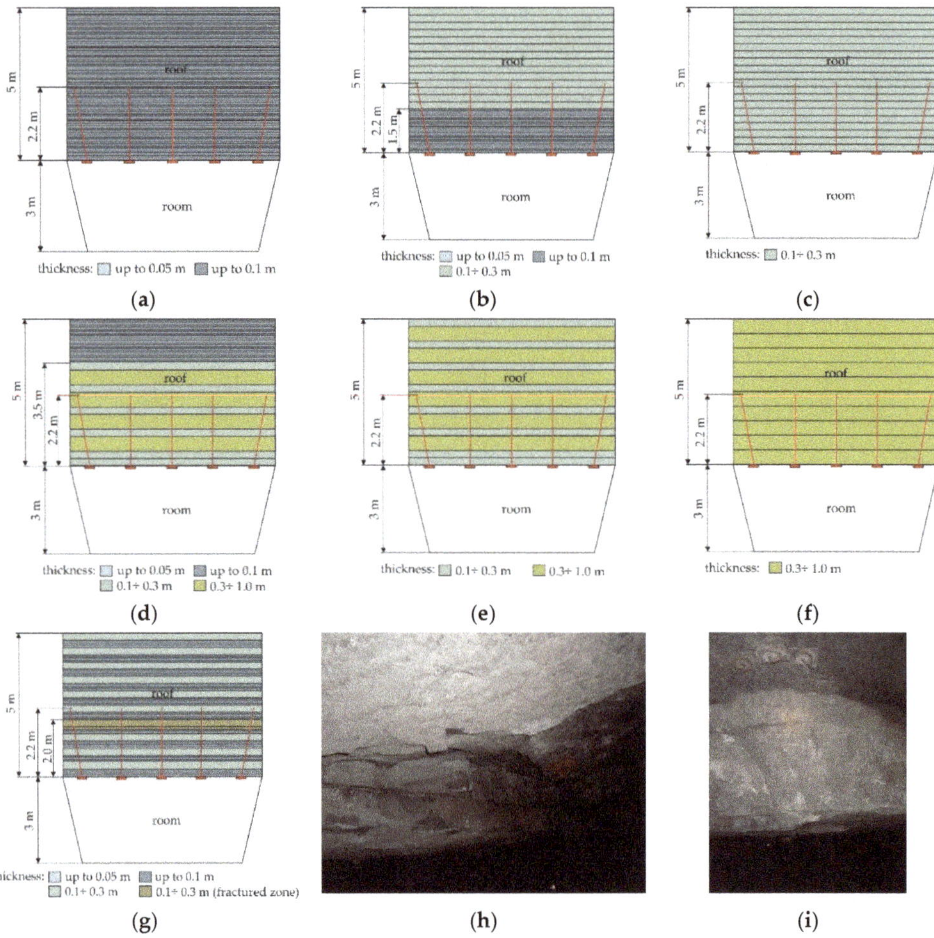

Figure 2. Types of direct roof with stratification structure: (**a**) Type I; (**b**) Type II; (**c**) Type III; (**d**) Type IV; (**e**) Type V; (**f**) Type VI; (**g**) Type VII; (**h**) Mid-bed layers; (**i**) Thick-bed layers.

Taking into account the risk of seismic events and the stratification of the rock mass, laboratory tests were carried out. The purpose of this was to compare the stress-strain characterics for rock bolts embedded at different lengths under static and quasi-static loading. In the research, a short-term quasi-static load was modeled using the maximum power of a hydraulic pump. In contrast, the static load was performed with hold times, modeling a slow load increase. The obtained test results indicate differences in the maximum stress and strain of the bolt for a short-term load of about 3.2 s compared to static tests.

2. Rock Bolt Testing Procedure under Static and Quasi-Static Loading

The tensile tests for the embedded rock bolts were performed according to two procedures. The static tensile tests consisted of a periodic pressure increase of 10 bar, which was performed with the use of the reduction valve knob located on the control panel (Figure 3a,b). Each increase in pressure was preceded by a holding time, which was 10 s. The test duration until the break was on average 450 s and fell within the scope of both the requirements of the Polish standard [33] regarding tensile tests and the conditions corresponding to the increase in static load in the room excavations. The load was measured using four strain gauge force sensors, while the displacement was monitored with the use of an incremental line encoder. The force sensors were connected to the measuring amplifier in a full-bridge strain gauge configuration, while the encoder was set according to the frequency measurement scheme with the signal with the direction of rotation. In quasi-static tests, a sectional anchorage was used, the same as in the static test. In order to obtain the maximum quasi-static load, the control panel (Figure 3c,d) was set to quasi-static load mode, and on the control panel (Figure 3a,b) the reducing valve was opened for the maximum fluid flow range. During the quasi-static load, stop times were not used. The measuring apparatus and the recording of the results did not change, except for the sampling frequency, which was set at 100 Hz for the quasi-static load, and 10 Hz for the static load. The change in sampling frequency resulted from the short loading time. Rock bolts subjected to a quasi-static load were broken by the application of maximum force in the shortest possible time due to the available pump power. A variable displacement multi-piston pump with constant power and constant pressure regulators was used in the tests. The hydraulic unit was characterized by a 18 kW engine and a maximum supply pressure of 31 MPa [34].

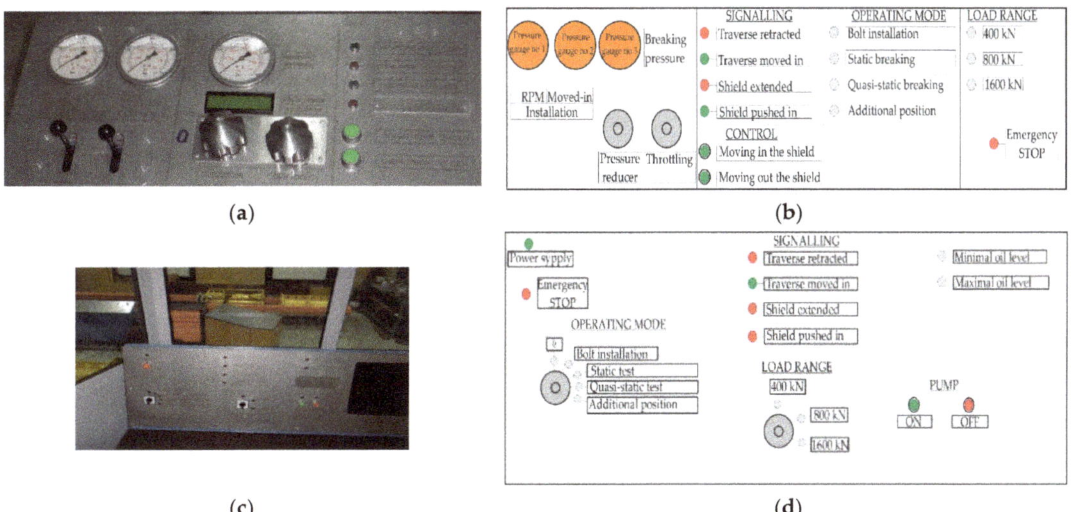

Figure 3. Load settings on the laboratory stand: (**a**) Static (pressure reducer); (**b**) A block diagram of a static setup; (**c**) Quasi-static; (**d**) A block diagram of a quasi-static setup.

Due to the conditions of the underground copper ore mines of the Legnica-Głogów Copper District, the selection of the rock bolt support was made on the basis of determining the class of the roof. The basic support with lengths of 1.2 m, 1.6 m, 1.8 m, 2.2 m and 2.6 m were considered and, taking into account the stratification of the rock mass (Figure 2), a ribbed steel bolt rod 2.2 m long was selected for the tests. The bolt rod cooperated with a dome washer with a diameter of 150 mm, a height of 25.5 mm and a thickness of 6 mm. The threaded end of the rod was secured with a 34 mm high M20 hexagonal nut. The rock bolt was made of B500SP steel [35]. Detailed characteristics of the rods are presented in Table 1 and in Figure 4a–c.

Table 1. Strength and geometrical parameters of the tested rock bolts.

Description	Value	Unit
characteristic yield point	≥500	[MPa]
characteristic tensile strength	≥575	[MPa]
rebar's rib 1 (measurement on the joint along the rod)	22.2	[mm]
rebar's rib 2 (measurement on the rod after turning by 90°)	21.2	[mm]
the thickness of the weld along the rod	3.1	[mm]
weld height	1.1	[mm]
the height of the ribs	1.35	[mm]
rib width	3.6	[mm]
distance between ribs parallel oblique	26.3	[mm]
distance between ribs parallel oblique ribs	10.3; 16	[mm]
rod length without ribs	138.3	[mm]
thread length	111	[mm]
thread diameter	19.2	[mm]
thread core diameter	16.45	[mm]

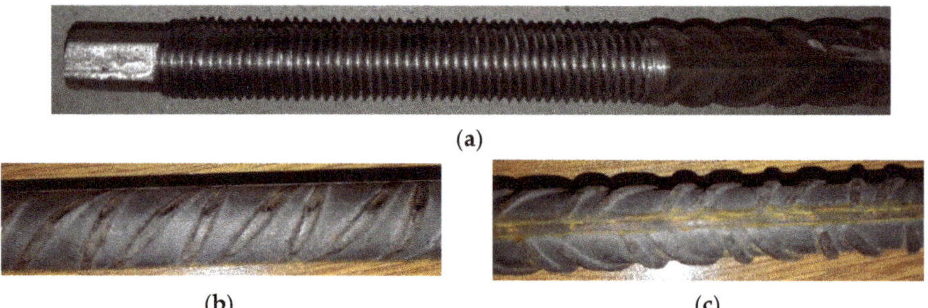

(a)

(b)   (c)

Figure 4. Bolt rod: (**a**) Thread; (**b**) Rebar's rib 1; (**c**) Rebar's rib 2.

The interaction of the bolt support with the rock mass was modeled with a concrete mix filled with specially prepared steel cylinders with a diameter of 0.1 m and lengths of 0.05 m, 0.3 m and 0.9 m. The composition of high-performance concrete consists of the following components: crushed aggregate with a rough surface (49%) and a grain size of up to 2 mm; sand (18%) with a grain size of 0.25 mm to 2 mm with the content of dust fraction to a minimum; and water with the superplasticizer Sika VisciCrete-20HE (14%), Portland cement CEM I 42.5R Górażdże with the addition of SikaFume-HR/-TU microsilica (18%). The strength of concrete, made on cubic samples with a side of 150 mm, ranged from 70 to 75 MPa. In the steel cylinders filled with concrete, holes with a diameter of 28 mm were drilled. Manual hammer drills DD130 and TE70-ATC were used for drilling (Figure 5a,b). The drills were equipped with core drills 0.32 m long and a SDS MAX hammer drill bit with a drilling head made of three parts of high-quality tungsten carbide with four cutting blades 0.92 m long. Resin cartridges with a diameter of 24 mm and a length of 0.45 m were used to fix the rods in the holes. The main purpose of the research was to determine the strength parameters for the embedded length ensuring the achievement of full load capacity (Figure 6b–d). According to earlier studies by Skrzypkowski et al. [36], it was found that the minimum embedded length for the bolt rod with a dome washer was 0.2 m. Nevertheless, in addition, in the research it was also decided to study bolt on a very short section of only 0.05 m (Figure 6a,d), which resulted from the very thin-bed structure of the rock mass (Figure 2).

(**a**)

(**b**)

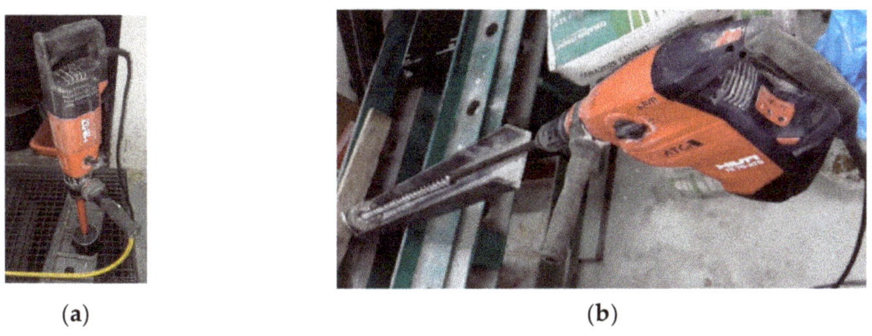

Figure 5. Drilling holes with application: (**a**) Crown drill; (**b**) Hammer drill bit with SDS MAX mounting.

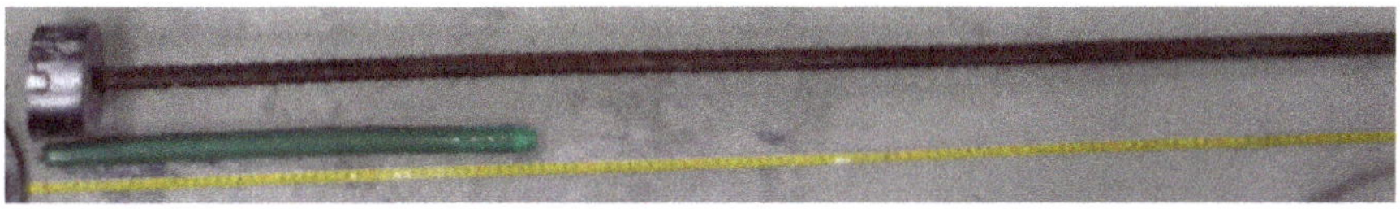

(**a**)

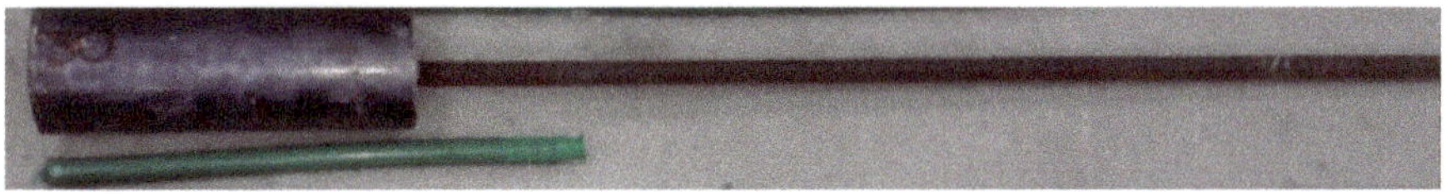

(**b**)

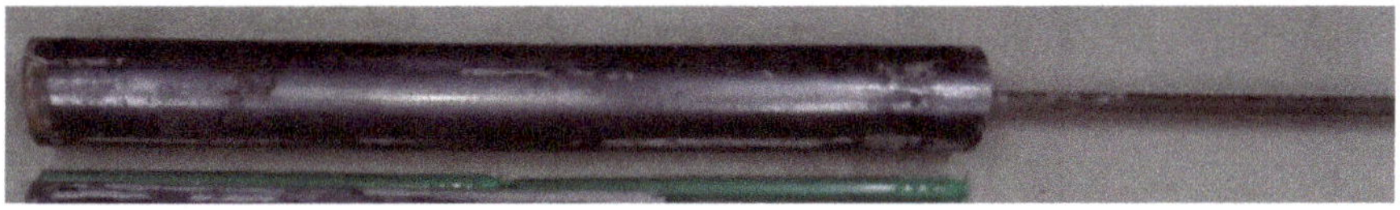

(**c**)

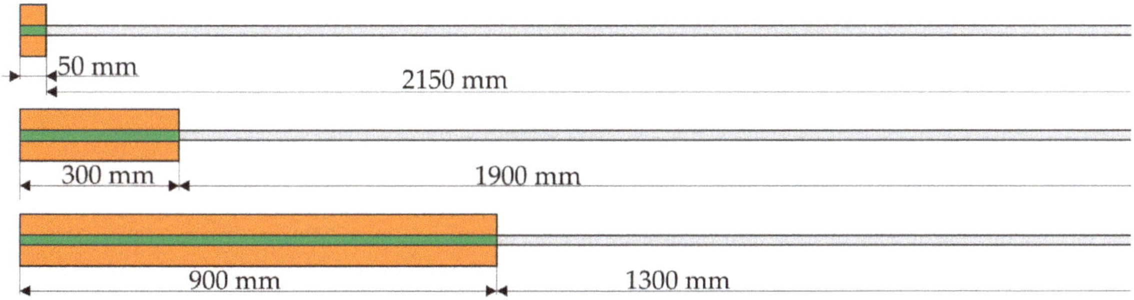

(**d**)

Figure 6. Bolt rod embedding lengths: (**a**) 0.05 m; (**b**) 0.3 m; (**c**) 0.9 m; (**d**) Block diagram.

3. Results

Static and quasi-static tensile tests of the partially embedded rock bolts were carried out in the bolting laboratory (Figure 7) at the Department of Mining Engineering and Occupational Safety at AGH University of Science and Technology in Kraków. The aim of the research was to determine the load-displacement characteristics, with the identification of the plastic range, maximum load and displacement (Figure 8), as well as the location of the material continuity interruption. For static and quasi-static studies, the load rate was 0.5 kN/s and 48 kN/s, respectively. This is the maximum value that was obtained with the use of the permissible power of the hydraulic system pump. The test results are presented in Figure 9a–d and in Table 2.

Figure 7. Tensile machine: 1—securing cylinder; 2—extendable discs with an internal blockade for a rod with a washer; 3—four force sensors spaced at 90°; 4—line encoder; 5—block of eight actuators; 6—support frame; 7—hydraulic hoses; 8—steel cylinder with a fixed bolt.

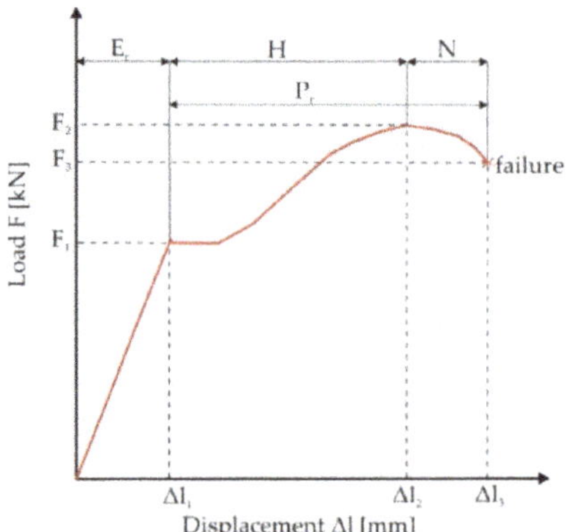

Figure 8. General diagram of the load-displacement characteristic: F_1—yield point; F_2—maximum load; F_3—failure at break; Δl_1—displacement in the elastic range; Δl_2—displacement for maximum load; Δl_3—displacement corresponding to rupture; E_r—elastic range; H—hardening; N—neckling; P_r—plastic range.

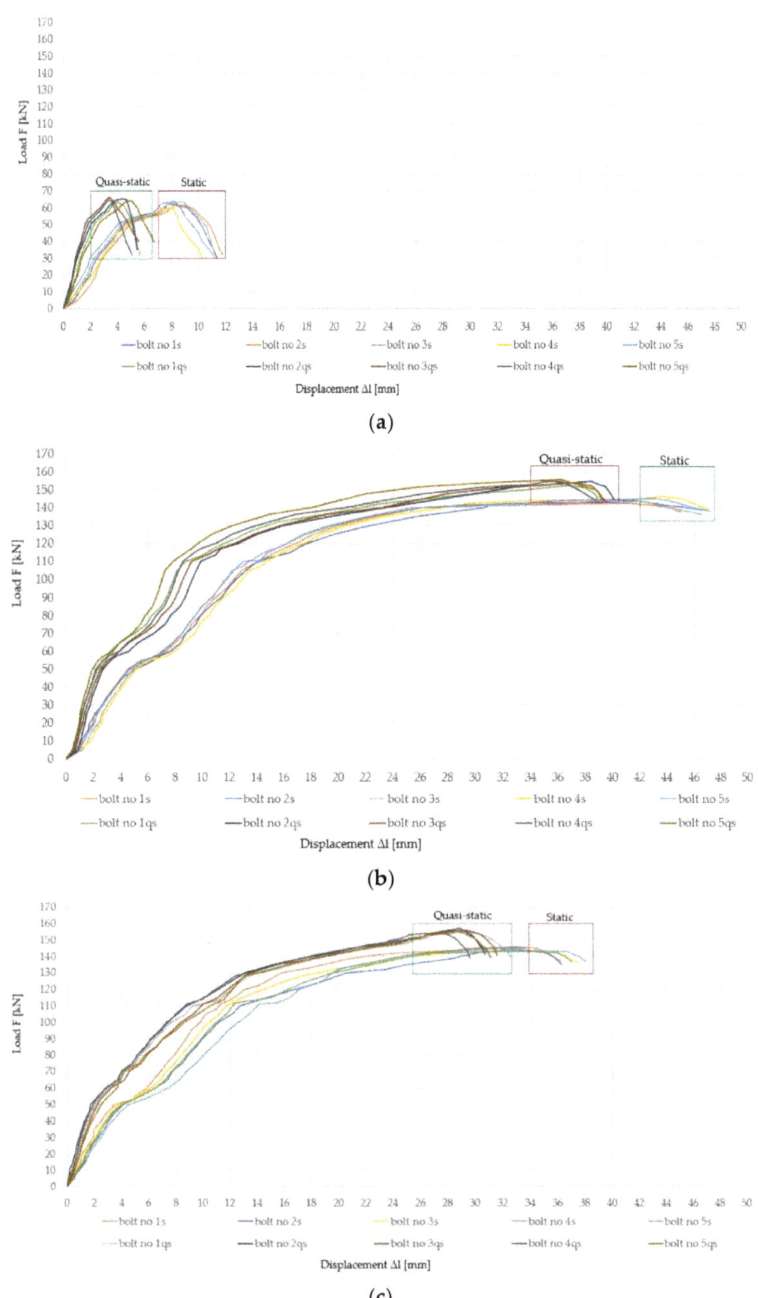

Figure 9. Load-displacement characteristics under static and quasi-static loading for the fixing length: (**a**) 0.05 m; (**b**) 0.3 m; (**c**) 0.9 m; s—static load; qs—quasi-static load.

Table 2. Summary of the average tensile values of the rock bolt partially embedded under static and quasi-static load.

Parameter	Unit	Embedded Length [m]					
		0.05 *		0.3		0.9	
		s	qs	s	qs	s	qs
load for the yield point, F_1	[kN]	-	-	108.5	113.8	110.6	114.4
maximum load, F_2	[kN]	62.4	64.6	143.8	154.3	145.2	155.5
load at failure, F_3	[kN]	-	-	137.5	144	136.4	140
displacement in the elastic range, Δl_1	[mm]	-	-	13.76	9.85	12.34	10.7
displacement for maximum load, Δl_2	[mm]	8.2	3.9	44.2	36.7	33.18	28.64
displacement corresponding to the failure, Δl_3	[mm]	-	-	46.4	39.2	37	31.1

* extension from the hole; s—static; qs—quasi-static.

The worst results were obtained for bolts embedded over a distance of 0.05 m. For 10 tests, the bolts slid out of the hole. The loss of adhesion occurred at the contact of the rod with the resin. For both static and quasi-static tests, the maximum load results were very close to each other. The main difference can be seen in the displacement, which for a quasi-static load was more than two times smaller compared to a static load. This is not a complete support characteristic because, as shown in Figure 9b,c, the yield strength is several dozen kilonewtons higher. In the case of the bolts installed at lengths of 0.3 m and 0.9 m, the rock bolt broke in the smallest cross-section of the thread core each time (Figure 10a,b). Despite the fact that for the embedded length of 0.05 m, the bolt support did not work in the full range of the load-displacement characteristics, it is still a clue for further tests. In particular, future research should concern the improvement of the components of the adhesive cartridge: resin, hardener, setting accelerators or retarders. In addition, special attention should be paid to improving the efficiency of mixing the components of the adhesive cartridge in the glove, and their cooperation with both the bolt rod and the surrounding rock. Equally important in future research may be the modification of the bolt rib geometry, in particular their height, inclination, width and spacing along the bolt rod.

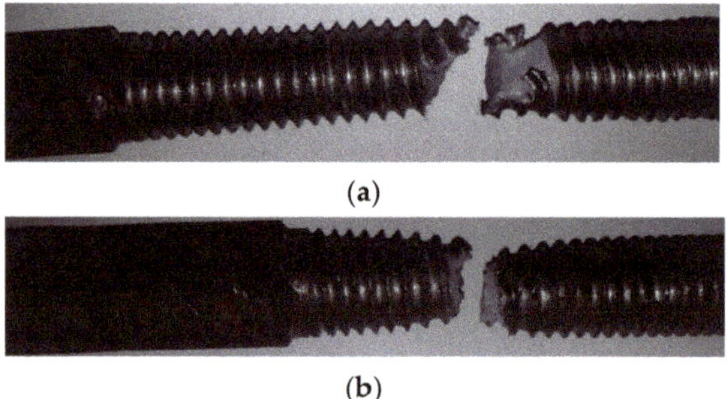

(a)

(b)

Figure 10. Failure on thread under load: (**a**) Static; (**b**) Quasi-static.

For embedded lengths of 0.3 m and 0.9 m, the yield point (F_1) under quasi-static load was higher by 4.88% and 3.43%, respectively, compared to the static load. At maximum load (F_2), the difference is much greater. Under quasi-static load, the bolt support takes up a greater load by 7.3% and 7.09%, respectively, compared to the static load. In the failure range (F_3), also at quasi-static loads, the load value is higher by 4.72% and 2.63%, respectively, in relation to static loads. However, for the length of 0.3 m, the displacements in the elastic range (Δl_1), the maximum load (Δl_2) the corresponding failure (Δl_3), they were about 39.69%, 20.43% and 18.36% higher compared to the quasi-static load, respectively. For the length of 0.9 m, the displacement values (Δl_1; Δl_2; Δl_3) were higher for a static load by 22.54%, 15.85% and 18.97%, respectively.

4. Discussion

Based on the load-displacement curves (Figure 9a–c), the stress-strain characteristics were determined (Figure 11a–f). The test results are summarized in Table 3. Additionally, in Figure 12a–f, mean and standard errors for stress and strain are presented. For the lengths of 0.3 m and 0.9 m, the failure occurred each time on the smallest cross-section of the thread core. In the calculation of the tensile stress for the yield stress (σ_1) of the maximum stress (σ_2) and the failure stress (σ_3), the diameter of the thread core was assumed to be 16.45 mm for the calculation of the surface area (A). The value of the tensile stress (σ_t) was calculated according to formula (1).

$$\sigma_t = \frac{F}{A} \tag{1}$$

where:

σ_t—tensile stress [MPa],
F—applied load [N],
A—cross-sectional area [mm^2].

Table 3. Summary of stress and strain results for partially embedded bolt support.

Parameter	Unit	Embedded Length [m]					
		0.05 *		0.3		0.9	
		s	qs	s	qs	s	qs
tensile stress for the yield point, σ_1	[MPa]	-	-	510.78	535.92	518.78	536.67
maximum stress, σ_2	[MPa]	293.8	304.3	678.84	726.48	679.31	732.04
stress at failure, σ_3	[MPa]	-	-	648.90	677.99	642.12	659.07
strain in the elastic range, ε_1	[%]	-	-	0.62	0.45	0.56	0.48
strain for maximum stress, ε_2	[%]	0.4	0.2	1.92	1.67	1.50	1.3
strain corresponding to the failure, ε_3	[%]	-	-	2.11	1.78	1.68	1.41

* extension from the hole; s—static; qs—quasi-static.

The value of specific strain (ε) was calculated according to the formula (2).

$$\varepsilon = \frac{\Delta l}{l_0} = \frac{l - l_0}{l_0} \cdot 100\% \tag{2}$$

where:

ε—strain [%],
l—length of the bolt rod after strain [mm],
l_0—initial length of the bolt rod [mm].

One of the most important factors in the performance of a rock bolt is the good adhesion of the ribs to the resin. It enables the correct transfer of the tensile stresses from the resin to the bolt rod, and in the case of higher stresses, it provides a certain slip. For the embedded length of 0.05 m, the rock bolt support worked only in the elastic range. Increasing the length to 0.3 m contributed to the achievement of full stress-strain characteristics. Hoien et al. [37] on the basis of pull-out tests of bolt rods made of steel grade (B500NC with a diameter of 20 mm) fixed in a concrete block on a cement binder at lengths of 0.1 m, 0.15 m, 0.2 m, 0.3 m, 0.45 m and 0.6 m, suggested critical embedment length of a grouted rebar bolt. Authors found two critical embedment related to elastic and plastic steel deformation. Furthermore, they stated that for minimum embedment length equal to 0.3 m, there is a strong relationship between load and water-cement index. Skrzypkowski et al. [36], by testing bolts made of the B500SP steel grade, proved that the minimum embedded length for which the bolt transfers full loads is 0.2 m. Yu et al. [38] modeled the rock mass with a concrete block, in which they installed 3 m long rock bolts fixed at the following lengths: 0.3 m, 0.9 m, 1.35 m, 1.5 m, 1.8 m, 2.1 m, 2.4 m and 2.7 m in order to detect mounting defects. The use of resin cartridges is closely related to the incomplete interaction between the bolt rod, resin and rock mass. The adhesive is surrounded by a glove, which very often causes poor mixing of the components and constitutes a kind of blockage. As a result of improper mixing, numerous sections along the bolt rod may appear that do not fulfill their function, and such embedment can then be considered as multi-point. The described phenomenon requires research in order to determine the minimum embedment length even with a length of at least 0.3 m. Xu et al. [39] tested 2 m long rod bolts which were fitted at lengths of 0.5 m, 1.0 m and 1.5 m with the use of resin cartridges and a setting time ranging from 90 to 180 s. Tests showed that the longer the anchorage length, the greater the load on the bolt rod.

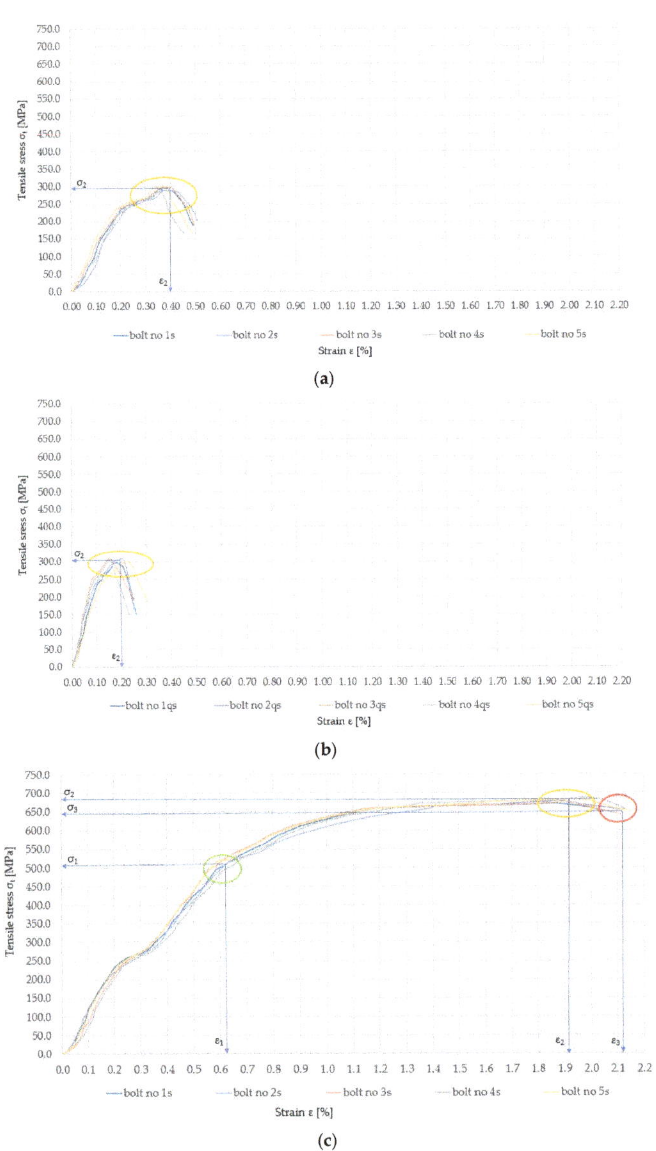

Figure 11. *Cont.*

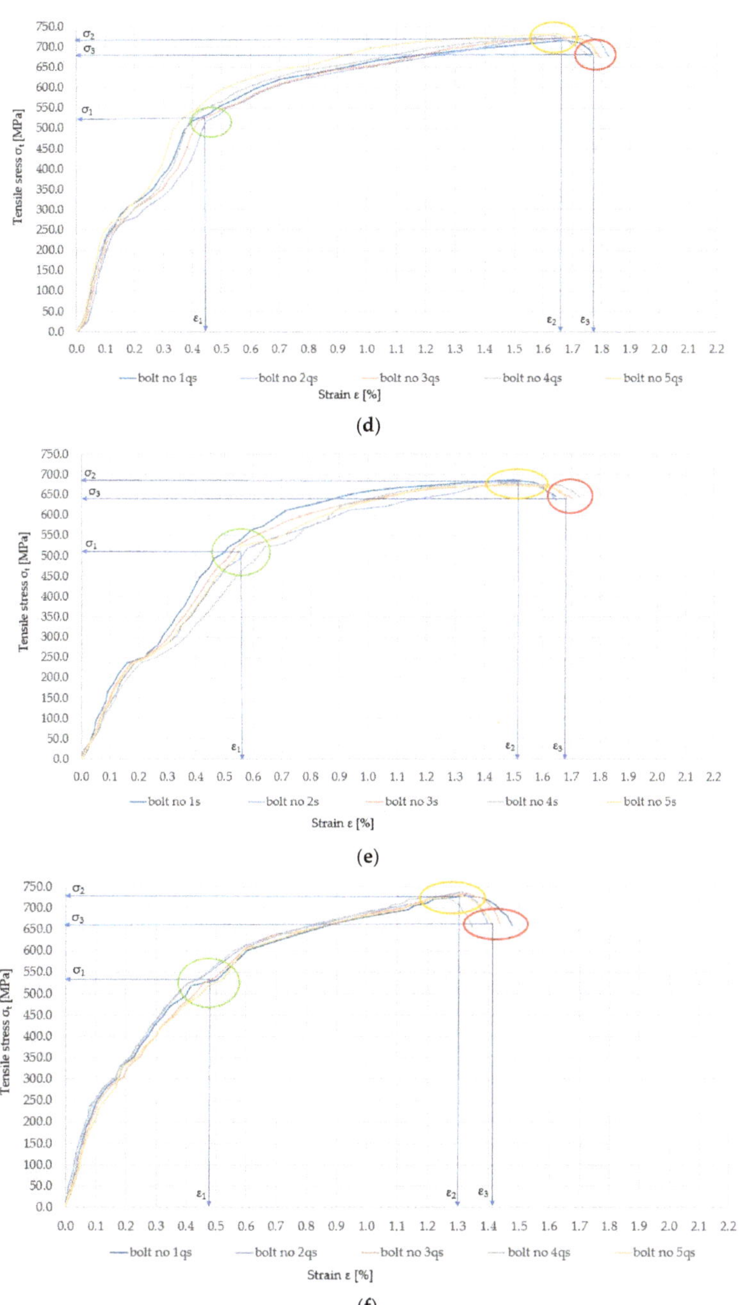

Figure 11. Stress-strain characteristics for partially embedded rock bolt under load: (**a**) Static over a distance of 0.05 m; (**b**) Quasi-static over a distance of 0.05 m; (**c**) Static over a distance of 0.3 m; (**d**) Quasi-static over a distance of 0.3 m; (**e**) Static over a distance of 0.9 m; (**f**) Quasi-static over a distance of 0.9 m.

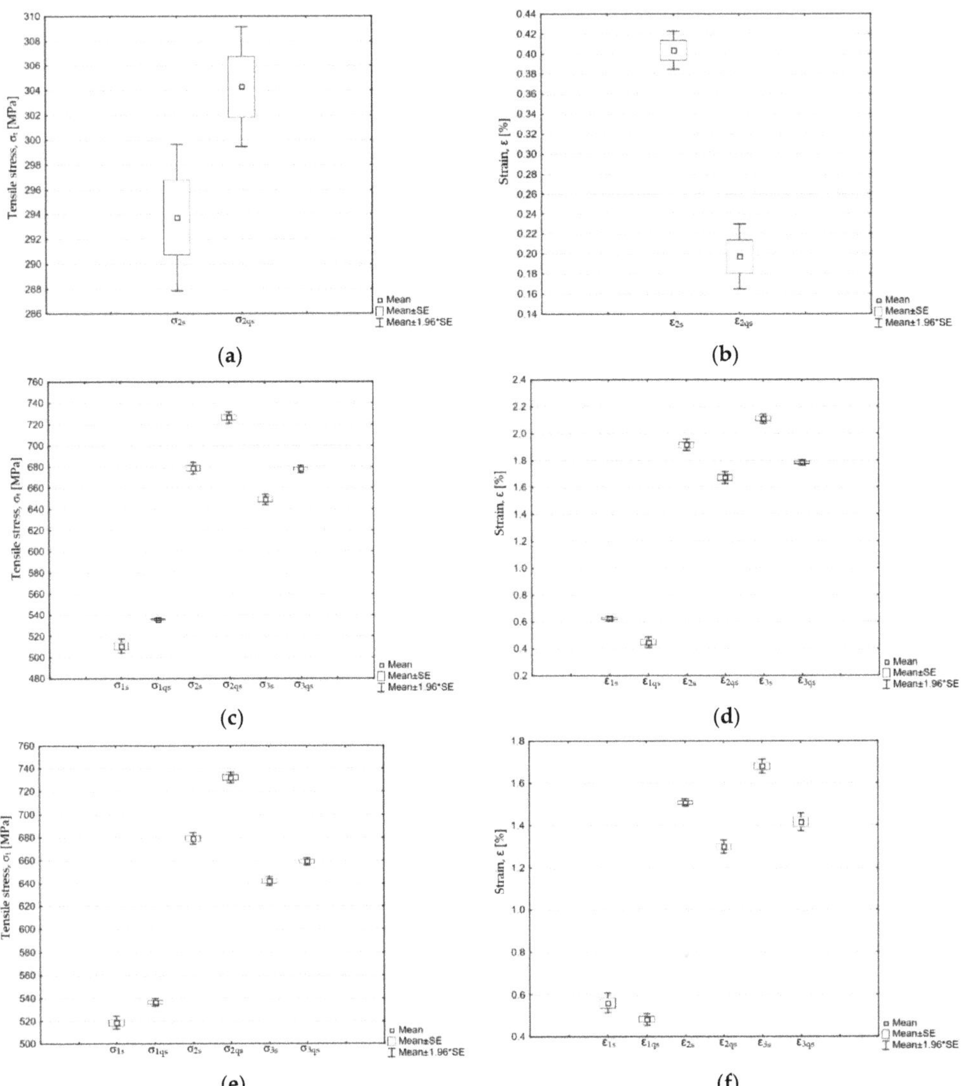

Figure 12. Mean and standard error for embedded lengths: (**a**) Stress for 0.05 m; (**b**); Strain for 0.05 m; (**c**); Stress for 0.3 m; (**d**) Strain for 0.3 m; (**e**) Stress for 0.9 m; (**f**) Strain for 0.9 m; s—static; qs—quasi-static.

Quasi-Static Coefficient for Stress and Strain

Quasi-static stresses (σ_{qs}) can be expressed by the product of the static stress (σ_s) and the quasi-static coefficient (C_{qs}), which is a multiplier not less than one, according to the formula (3):

$$\sigma_{qs} = C_{qs} \cdot \sigma_s \tag{3}$$

The values of the quasi-static coefficient for stress (σ_1) are: 1.04 and 1.03, respectively, for the lengths of 0.3 m and 0.9 m. For maximum stress (σ_2), the ratio of the coefficient was much larger compared to the stress for the yield point (σ_1). These are values of 1.07 for both lengths. In the case of failure stress, the difference in the quasi-static coefficient is only 0.018 in favor of the embedment length of 0.3 m. Much greater differences are found in the strain. In order to determine the quasi-static strain (ε_{qs}), just like for the stress, it can be expressed by the product of the static strain (ε_s) and the quasi-static coefficient (S_{qs}), which is a multiplier not greater than one according to formula (4):

$$\varepsilon_{qs} = S_{qs} \cdot \varepsilon_s \tag{4}$$

The values of the quasi-static coefficient for strain (ε_1) are 0.72 and 0.85, respectively, for the fixing lengths of 0.3 m and 0.9 m. The strain (ε_2) for the maximum stress is equal to 0.86. Also, in the case of the strain (ε_3) associated with the failure, the differences are very small, amounting to 0.84 and 0.83 for the fixing lengths of 0.3 m and 0.9 m, respectively. The thread turned out to be the weakest point of the rock bolt. Similar conclusions were obtained by Kang et al. [40] by testing bolts made of steel grades B335, B500 and B700. Within the limits of elasticity of the rock bolt material, a specific value of strain was obtained for a precisely defined tensile stress. These strains appeared immediately after the direct increase in tensile stresses. The research showed that the amount of strain is influenced by the duration of the load. In strictly dynamic studies [41,42] the value of the dynamic coefficient is much higher than the value of 1. Nevertheless, the presented test results under quasi-static loading indicate a slight increase in stress and a reduction in strain.

5. Conclusions

This laboratory study presents a comparison of load-displacement and stress-stress characteristics under static and quasi-static loading. Several conclusions can be drawn:

- The load capacity of a rock bolt with a thread is not determined by the breaking strength of the rod, but by the thread strength;
- For embedded length of 0.3 m, the tensile stress at the points of yield, maximum and failure for quasi-static stress are higher by 4.92%, 6.94% and 4.48%, respectively, compared to the static stress;
- For embedded length of 0.9 m, the tensile stress at the points of yield, maximum and failure for quasi-static stress are higher by 3.44%, 7.76% and 2.63%, respectively, compared to the static stress;
- For embedded length of 0.3 m, the value of strain in the elastic range, the strain at maximum stress and the strain corresponding to failure for quasi-static stress are lower by 27.42%, 13.03% and 15.64%, respectively, compared to the static strain;
- For embedded length of 0.9 m, the value of strain in the elastic range, the strain at maximum stress and the strain corresponding to failure for quasi-static stress are lower by 14.29%, 13.34% and 16.08%, respectively, compared to the static strain.

The presented research results refer to both static and quick-changing loads occurring in underground mining. As a result of the seismic events, the rock bolt support may be additionally loaded, resulting in a change in the stress and strain state. By modeling the cooperation of the bolt support with the rock mass by using the maximum power of the hydraulic pump, it is possible to reflect to some extent the operating conditions of the bolts at variable loads. Future research should focus on the new construction materials featuring new steel grades, variable rock bolt rod geometry and new adhesive cartridges. The subject of interest will be the determination of the contact between the rock bolt, the resin cartridge and the rock mass model.

Funding: This research was prepared as part of AGH University of Science and Technology in Poland, scientific subsidy under number: 16.16.100.215.

Institutional Review Board Statement: Not applicable.

Informed Consent Statement: Not applicable.

Data Availability Statement: The data presented in this study are new and have not been previously published.

Conflicts of Interest: The author declares no conflict of interest.

References

1. Waclawik, P.; Snuparek, R.; Kukutsch, R. Rock bolting at the room and pillar method at great depths. *Procedia Eng.* **2017**, *191*, 575–582. [CrossRef]
2. Małkowski, P.; Niedbalski, Z. A comprehensive geomechanical method for the assessment of rockburst hazards in underground mining. *Int. J. Min. Sci. Technol.* **2020**, *30*, 345–355. [CrossRef]
3. Masoudi, R.; Sharifzadeh, M. Reinforcement selection for deep and high-stress tunnels at preliminary design stages using ground demand and support capacity approach. *Int. J. Min. Sci. Technol.* **2018**, *28*, 573–582. [CrossRef]
4. Li, C.C.; Mikula, P.; Simser, B.; Hebblewhite, B.; Joughin, W.; Feng, X.; Xu, N. Discussions on rockburst and dynamic ground support in deep mines. *J. Rock Mech. Geotech. Eng.* **2019**, *11*, 1110–1118. [CrossRef]
5. Rahimi, B.; Sharifzadeh, M.; Feng, X.-T. Ground behaviour analysis, support system design and construction strategies in deep hard rock mining e Justified in Western Australian's mines. *J. Rock Mech. Geotech. Eng.* **2020**, *12*, 1–20. [CrossRef]
6. Li, C.C. Principles and methods of rock support for rock burst control. *J. Rock Mech. Geotech. Eng.* **2020**, *12*, 1–14. [CrossRef]
7. Kim, H.; Rehman, H.; Ali, W.; Muntaqim Naji, A.; Kim, J.-J.; Kim, J.; Yoo, H. Classification of Factors Affecting the Performance of Fully Grouted Rock Bolts with Empirical Classification Systems. *Appl. Sci.* **2019**, *9*, 4781. [CrossRef]
8. Ozturk, C.A.; Dogan, E.; Donmez, M.S.; Karaman, E. Ground Support Design of Underground Openings in Seismically Active Mine: A Case Study from a Lead and Zinc Underground Mine. *Geotech. Geol. Eng.* **2021**, *39*, 1–21. [CrossRef]
9. Wang, J.; Apel, D.B.; Pu, Y.; Hall, R.; Wei, C.; Sepehri, M. Numerical modeling for rock bursts: A state-of-the-art review. *J. Rock Mech. Geotech.* **2020**. [CrossRef]
10. Cao, J.; Zhang, N.; Wang, S.; Qiana, D.; Xiea, Z. Physical model test study on support of super pre-stressed anchor in the mining engineering. *Eng. Fail. Anal.* **2020**, *118*, 104833. [CrossRef]
11. Feng, X.; Zhang, N.; Lv, C. Effects of Interface Damage Resulting from the Separation of Layered Strata on Bolt Anchoring Systems. *Shock. Vib.* **2016**, *2016*, 2590816. [CrossRef]
12. Feng, X.; Zhang, N.; Li, G.; Guo, G. Pullout Test on Fully Grouted Bolt Sheathed by Different Length of Segmented Steel Tubes. *Shock. Vib.* **2017**, *2017*, 4304190. [CrossRef]
13. Skrzypkowski, K.; Korzeniowski, W.; Zagórski, K.; Zagórska, A. Flexibility and load-bearing capacity of roof bolting as functions of mounting depth and hole diameter. *Energies* **2019**, *12*, 3754. [CrossRef]
14. Bacić, M.; Kovacević, M.S.; Jurić Kaćunić, D. Non-Destructive Evaluation of Rock Bolt Grouting Quality by Analysis of Its Natural Frequencies. *Materials* **2020**, *13*, 282. [CrossRef] [PubMed]
15. Zhan, Y.; Li, N.; Wang, H.; Zheng, P.; Zhang, J.; Liu, Q. The Mechanism of Interface Dilatancy of Cement Mortar Rockbolts. *Adv. Civ. Eng.* **2020**, *2020*, 8838488. [CrossRef]
16. Zhao, J.-y.; Zhang, D.-s.; Lu, W.-c.; Yang, C.-z. Design and experiment of hydraulic impact loading system for mine cable bolt. *Procedia Earth Planet. Sci.* **2009**, *1*, 1337–1342. [CrossRef]
17. He, M.; Gong, W.; Wang, J.; Qi, P.; Tao, Z.; Du, S.; Peng, Y. Development of a novel energy-absorbing bolt with extraordinarily large elongation and constant resistance. *Int. J. Rock Mech. Min. Sci.* **2014**, *67*, 29–42. [CrossRef]
18. Zhao, X.; Zhang, S.; Zhu, Q.; Li, H.; Chen, G.; Zhang, P. Dynamic and static analysis of a kind of novel J energy-releasing bolts. *Geomat. Nat. Hazards Risk* **2020**, *11*, 2486–2508. [CrossRef]
19. Morozov, M.; Tian, G.Y.; Withers, P.J. The pulsed eddy current response to applied loading of various aluminum alloys. *NDT E Int.* **2010**, *43*, 493–500. [CrossRef]
20. Withers, P.J.; Turski, M.; Edwards, L.; Bouchard, P.J.; Buttle, D.J. Recent advances in residual stress measurement. *Int. J. Pres. Ves. Pip.* **2008**, *85*, 118–127. [CrossRef]
21. Wilson, J.W.; Tian, G.Y.; Barrans, S. Residual magnetic field sensing for stress measurement. *Sens. Actuators A* **2007**, *135*, 381–387. [CrossRef]
22. Skrzypkowski, K.; Korzeniowski, W.; Zagórski, K.; Dominik, I.; Lalik, K. Fast, non-destructive measurement of roof-bolt loads. *Stud. Geotech. Mech.* **2019**, *41*, 93–101. [CrossRef]
23. Crompton, B.R.; Sheppard, J. A practical design approach for an improved resin-anchored tendon. *J. S. Afr. Inst. Min. Metall.* **2020**, *120*, 7–13. [CrossRef]
24. Feng, X.; Xue, F.; Jiang, W.; Wang, M.; Song, W. Re-think of solving the gloving problem in bolting systems by adulterating steel particles. *Constr. Build. Mater.* **2021**, *268*, 121179. [CrossRef]
25. Xu, C.; Li, Z.; Wang, S.; Wang, S.; Fu, L.; Tang, C. Pullout Performances of Grouted Rockbolt Systems with Bond Defects. *Rock Mech. Rock Eng.* **2018**, *51*, 861–871. [CrossRef]
26. Campbell, R.; Mould, R.J. Impacts of gloving and un-mixed resin in fully encapsulated roof bolts on geotechnical design assumptions and strata control in coal mines. *Int. J. Coal Geol.* **2005**, *64*, 116–125. [CrossRef]
27. Burtan, Z.; Zorychta, A.; Cieślik, J.; Chlebowski, D. Influence of mining operating conditions on fault behavior. *Arch. Min. Sci.* **2014**, *59*, 691–704. [CrossRef]

28. Regulation of the Minister of the Environment Regarding Natural Hazards in Mining Plants. Available online: https://sip.lex.pl/akty-prawne/dzu-dziennik-ustaw/zagrozenia-naturalne-w-zakladach-gorniczych-17955795 (accessed on 15 January 2021).
29. Drzęźla, B.; Dubiński, J.; Fajklewicz, Z.; Goszcz, A.; Marcak, H.; Pilecki, Z.; Zuberek, W. *Mining Geophysics Handbook*; PAN/AGH Publishing: Kraków, Poland, 1995; Volume II, p. 61.
30. Skrzypkowski, K.; Korzeniowski, W.; Zagórski, K.; Zagórska, A. Adjustment of the Yielding System of Mechanical Rock Bolts for Room and Pillar Mining Method in Stratified Rock Mass. *Energies* **2020**, *13*, 2082. [CrossRef]
31. Skrzypkowski, K. Case Studies of Rock Bolt Support Loads and Rock Mass Monitoring for the Room and Pillar Method in the Legnica-Głogów Copper District in Poland. *Energies* **2020**, *13*, 2998. [CrossRef]
32. Butra, J.; Dańda, Z.; Katulski, A. The causes of roof rock falls and method control in ore mining. *Work Saf. Environ. Prot. Min.* **1997**, *6*, 151–157.
33. Polish Standard: PN-EN 10002-1:2004. In *Metallic Materials—Tensile Testing—Part 1: Method of Test at Ambient Temperature*; Polish Committee for Standardization: Warszawa, Poland, 2004.
34. Korzeniowski, W.; Skrzypkowski, K.; Herezy, Ł. Laboratory method for evaluating the characteristics of expansion rock bolts subjected to axial tension. *Arch. Min. Sci.* **2015**, *60*, 209–224. [CrossRef]
35. Epstal. Available online: http://epstal.pl/en/reinforcing-steel/properties (accessed on 10 January 2021).
36. Skrzypkowski, K.; Korzeniowski, W.; Zagórski, K.; Zagórska, A. Modified Rock Bolt Support for Mining Method with Controlled Roof Bending. *Energies* **2020**, *13*, 1868. [CrossRef]
37. Hoien, A.H.; Li, C.C.; Zhang, N. Pull-out and Critical Embedment Length of Grouted Rebar Rock Bolts-Mechanisms When Approaching and Reaching the Ultimate Load. *Rock Mech. Rock Eng.* **2021**, *54*, 1–18. [CrossRef]
38. Yu, J.-D.; Lee, J.-S. Smart Sensing Using Electromagnetic Waves for Inspection of Defects in Rock Bolts. *Sensors* **2020**, *20*, 2821. [CrossRef] [PubMed]
39. Xu, X.; Tian, S. Load transfer mechanism and critical length of anchorage zone for anchor bolt. *PLoS ONE* **2020**, *15*, e227539. [CrossRef] [PubMed]
40. Kang, H.; Meng, X. Tests and analysis of mechanical behaviours of rock bolt components for China's coal mine roadways. *J. Rock Mech. Geotech.* **2015**, *7*, 14–26. [CrossRef]
41. Player, J.; Thompson, A.; Villaescusa, E. Dynamic testing of reinforcing system. In Proceedings of the 6th International Symposium on Ground Support in Mining and Civil Engineering Construction, SAIMM, SANIRE and ISRM, Cape Town, South Africa, 30 March–3 April 2008; pp. 597–622.
42. Darlington, B.; Rataj, M.; Roach, W. A new method to evaluate dynamic bolts and the development of a new dynamic rock bolt. In Proceedings of the Ninth International Conference on Deep and High Stress Mining, The Southern African Institute of Mining and Metallurgy, Johannesburg, South Africa, 24–25 June 2019; pp. 205–216. [CrossRef]

Article

Enhancement of Machinery Activity Recognition in a Mining Environment with GPS Data

Paulina Gackowiec *, Edyta Brzychczy and Marek Kęsek

Faculty of Civil Engineering and Resource Management, AGH University of Science and Technology, 30-059 Cracow, Poland; brzych3@agh.edu.pl (E.B.); kesek@agh.edu.pl (M.K.)
* Correspondence: gackowiec@agh.edu.pl

Abstract: Fast-growing methods of automatic data acquisition allow for collecting various types of data from the production process. This entails developing methods that are able to process vast amounts of data, providing generalised knowledge about the analysed process. Appropriate use of this knowledge can be the basis for decision-making, leading to more effective use of the company's resources. This article presents the approach for data analysis aimed at determining the operating states of a wheel loader and the place where it operates based on the recorded data. For this purpose, we have used several methods, e.g., for clustering and classification, namely: DBSCAN, CART, C5.0. Our approach has allowed for the creation of decision rules that recognise the operating states of the machine. In this study, we have taken into account the GPS signal readings, and thanks to this, we have indicated the differences in machine operation within the designated states in the open pit and at the mine base area. In this paper, we present the characteristics of the selected clusters corresponding to the machine operation states and emphasise the differences in the context of the operation area. The knowledge obtained in this study allows for determining the states based on only a few selected most essential parameters, even without consideration of the coordinates of the machine's workplace. Our approach enables a significant acceleration of subsequent analyses, e.g., analysis of the machine states structure, which may be helpful in the optimisation of its use.

Keywords: sensor data; mining machinery; activity recognition; clustering; GPS data

Citation: Gackowiec, P.; Brzychczy, E.; Kęsek, M. Enhancement of Machinery Activity Recognition in a Mining Environment with GPS Data. *Energies* **2021**, *14*, 3422. https://doi.org/10.3390/en14123422

Academic Editor: Nikolaos Koukouzas

Received: 6 May 2021
Accepted: 3 June 2021
Published: 10 June 2021

Publisher's Note: MDPI stays neutral with regard to jurisdictional claims in published maps and institutional affiliations.

Copyright: © 2021 by the authors. Licensee MDPI, Basel, Switzerland. This article is an open access article distributed under the terms and conditions of the Creative Commons Attribution (CC BY) license (https://creativecommons.org/licenses/by/4.0/).

1. Introduction

Nowadays, companies are looking for innovative methods and techniques to maximise the efficiency of their operations and optimise the usage of their fixed assets. Because of progressive technological development and increasing hardware capabilities of data acquisition and storage, approaches based on the analysis of machinery data are gaining importance. The gathered data enables monitoring and ongoing insight into the operation of the utilised equipment, as well as its comprehensive and complex evaluation. Currently, very detailed and precise machine-specific data are available to companies, characterising the operation of all the main components of the machine, often recorded continuously. This constitutes a valuable opportunity to understand a machine's performance from a broader perspective, trying to discover new patterns and specific work behaviours based on the analysis of its various parameters. The discovered dependencies can be used to improve the effectiveness and increase safety by indicating the activities and states generating the highest load for the machine. As a result of the acquired knowledge, the company may undertake real changes of unfavourable work parameters and, therefore, obtain notable benefits such as reduction of fuel consumption, an extension of the machine's operating time, or minimisation of extreme and dangerous states of the machine's operation.

Increasing expectations of the business environment, and also in the mining industry, stimulate the implementation of innovative solutions in the scope of productivity, work safety, and rational use of assets. An additional factor determining the searching and development of innovative methods is the fact that heavy machinery is characterised by considerable complexity of operating parameters, often imposed by the highly demanding conditions under which it operates. These aspects result in the fact that often defining the cause of an undesirable event or explicitly indicating patterns in the specifics of the machine operation depends on many variables and conditions and can be a challenging task. The existing techniques and analytical software currently used in data analysis allow comprehensive analyses to be conducted, enabling the specificity and complexity of the analysed machine operation data to be taken into account. In view of these considerations, it has been concluded that complementing the existing methods with additional assumptions reflecting the specificity of machine operation may contribute to more accurate and promising analysis results.

One of the interesting directions of machinery data analysis is its activity recognition based on raw sensor data. Various techniques can be used in the activity recognition task, especially from a common data mining task, namely clustering. This task aims to identify groups of observations with characteristics that are as similar as possible within a particular cluster and dissimilar across different clusters. Clustering is often applied in machine learning, social network analysis, geosciences, decision making, document retrieval, and image segmentation [1,2]. In scientific literature, several clustering approaches have been developed: partitional, hierarchical or density-based methods, and other paradigms such as nearest neighbour-based clustering, fuzzy clustering, and neural network-based clustering [2,3].

The paper is devoted to enhancing activity recognition of the wheel loader operation with data from the Global Positioning System (GPS). We have assumed that broadening the analysis of the loader's operating states with the location data may improve the quality of the information on wheel loader operation and its efficiency. In this research, we aimed to fill the gap in the field of activity recognition of mining machinery based on sensor and GPS data by using clustering and classification techniques. Results of the analysis provide additional knowledge about the operating characteristics of the equipment and may contribute to a better understanding of the specific operation of the equipment and the effectiveness of the operations carried out. The main contribution of our work is a demonstration that localisation data should be taken into consideration in activities' analysis.

In our approach for activity recognition, we used density-based clustering with the DBSCAN algorithm. The variables included in the analysis are related mainly to engine performance, driving system, bucket statuses that were selected from a broader set based on an assessment of data completeness, and principal component analysis (PCA). Discovered clusters were named based on statistical analysis results and, subsequently, named activities were analysed regarding identified working areas (mine base and open pit). We also prepared a description of defined activities as a rule set to enable labelling the raw sensor data, e.g., for process monitoring needs, using tree-based classifiers.

Obtained results confirmed a statistically significant difference in distributions of variables characterising defined activities in the identified areas. Thus, during process monitoring and activities analysis, these findings can bring more in-depth knowledge about machinery operation that can be helpful in the decision-making process regarding equipment management.

The paper is organised as follows: Section 2 provides an overview of the most important scientific literature in the field of machine condition recognition and data analysis with the use of GPS data. In Section 3, we introduce the dataset used for the analysis and the methodology applied to assign data to operation areas and a brief description of data mining methods used in our research. The results of the conducted analyses are presented in Section 4, along with a discussion. Finally, in Section 5, we formulate concluding remarks.

2. Related Works

The application of analytical techniques based on data obtained from sensors in state, activities, and operating conditions of machinery identification is commonly discussed in the literature. Especially in recent years, numerous publications have been published that deal with the topic of evaluating and monitoring machines based on their performance data. The literature analysis indicated examples of practical implementations of analytical methods in the field of monitoring parameters of various machines [4,5], including machines working in the mining industry [6]. In addition, some examples of the application of data mining techniques for monitoring working conditions in mining areas to improve safety by identifying microseismic events (based on classification techniques) can be found in [7]. Other work proposes monitoring and evaluating mine climate conditions based on sensor data to predict potential hazards [8]. In the more detailed view, an example of applying an analytical approach to identify different types of activities during construction equipment operation is presented in [9]. The authors use a data fusion and machine learning algorithm approach applied to audio and kinematic data to monitor the operations of single pieces of equipment. In [10], the authors present the application of activity recognition to the analysis of the construction equipment operation illustrated by the example of a front-end loader based on supervised machine learning classifiers in [10]. An interesting application of an analytical approach to operational data of construction equipment is presented in the publication [11]. The authors use a recurrent neural network to recognise the activities of an excavator and a front-end loader based on synthetic data. Another example illustrating the use of operational data in the field of analysis and activity recognition for construction machinery is the reference publication [12]. Wu C. et al., carried out analyses of data extracted from a smartphone in terms of a behaviour model for operations and to identify patterns of agricultural machinery [13]. In the paper [14], the authors analysed the operation of an LHD (load, haul, dump) machine from an underground copper ore mine based on statistical analysis and temperature data in the context of maintenance activities. Langroodi et al., have proposed in their work a new Fractional Random Forest machine learning method that can be applied to machine activity recognition based on a limited dataset [15]. This method has been applied to data for excavators and rollers.

In the literature, there are also current examples of analyses based on sensor data acquired from the loader and complemented by the specifics of the device operator's work. The authors of [16] addressed the issue of analysing the characteristics of the machine operation in different working conditions for a wheel loader considering the driver's influence. The authors analysed the operation of the device based on the data of the boom head cylinder pressure and proposed a method for evaluation of the difficulty level of the operating conditions based on the radar chart and clustering analysis. The analysis of braking strategies by deep learning methods for an automatic wheel loader based on driving data and operator work specifics was undertaken in [17]. Other examples of using operational data for a wheel loader machine to optimise its performance are given in [18]. The problem of finding the optimum for the wheel loader work cycle in terms of fuel efficiency was discussed in the article [19]. The paper presented an algorithm for improving fuel efficiency and productivity of a loader, which can be applied in the operator work support or system optimisation and concept selection for loaders.

The subject of clustering sensor data that characterise the operational states of machines has been addressed in the literature in various papers. J. Amutha et al. conducted a comprehensive literature review of data clustering methods and algorithms, including classical optimisation and machine learning techniques [20]. One can find other references presenting literature research on clustering methods in the field of sensor data analysis in [2,3,21]. The area of machine operation data clustering is widely addressed in the literature, especially in the context of equipment condition diagnostics based on various techniques such as correlation-based clustering [22], clustering maintenance records of excavator buckets [23], improved K-means clustering for detecting power transformer abnormal state [24], mean shift clustering in anomaly detection for machine tools [25], and k-means clustering algorithms for mining shovel failure prognostics [26]. An approach based on methods such as time series segmentation, clustering, and classification for analysing the operating states of wheel loader machines to detect anomalies in the time series dataset was proposed in [27].

In this work, we aimed to enrich the standard analysis of the obtained clusters with location data from the GSP signal, which, combined with the map of the working area, allowed us to distinguish distinct sub-areas within the territory where the machine operates. Distinguished areas allowed for a more detailed analysis and an indication of more specific subclusters concerning the operating site. Numerous examples in the literature review use location-based data, mainly for road and pedestrian traffic or travel behaviour detection [1,28,29]. Cheng et al., highlight the benefits of using location data to understand worksite operations better and to analyse the productivity of machines working in the field effectively [30]. An example of using GPS data to analyse and infer the working of construction equipment is in [31]. This paper proposes a method to identify workstations for a group of heavy vehicles, including wheel loaders, excavators and dump trucks based on GPS data. The proposed method determines the locations of different types of workstations with a probability density function. The paper [32] presents case studies using GPS data to analyse construction equipment performance, the job site layout and to visualise the GPS data through a developed user interface.

The subject of sensor data analysis for industrial machines is an issue that has been frequently addressed in scientific publications in recent years concerning a wide range of applications, from issues related to improving efficiency, determining patterns of machine operation, or investigating anomalous states in order to extend machine lifetime. The research examples cited above are mainly concerned with the implementation of data mining techniques for detecting operating states and monitoring machine performance from sensor data, visual data, and audio data. Other examples of cited publications focus on the use of machine data to support the work of operators. In the context of machine operation analysis, the primary and common application of GPS location data is a route optimisation and ongoing fleet monitoring in management systems. Known methods of detecting machine conditions do not take into account information about the device's current location during operation in terms of more accurate detection of activities. On the other hand, data on current operating parameters are used mainly to assess the effectiveness of work or monitor abnormal operating conditions. The combination in the analysis of both the data on the defined area in which the device is located and selected parameters of the machine operation allows for developing specific rules to identify the machine's state during the working process.

The challenge in evaluating machine performance remains to define the value-added activities and separate the nonvalue-added activities. The proposed approach, using the definition of activities in the context of the current location, provides an advantage in the analysis of machine performance by allowing easy identification of workspace-specific activities in addition to the main activities. The ability to easily determine the current operating status can provide a basis for more detailed analyses of machine operation from a process analysis point of view.

3. Materials and Methods

In this section, we briefly present methods and materials used in our research, namely: the analysed data set, marking of location perspective in data, as well as selected data mining techniques applied in our work.

3.1. Data Set

The original raw data set includes 9,810,934 observations and 432 variables covering six months of wheel loader operation; however, due to quality issues (observations only with timestamps), we had to select the best samples for further analysis (Figure 1).

For the best candidate of the sample, we selected observations from one week (156,863 observations). In the sample, we identified main groups of variables related, among others, to engine characteristics (e.g., speed, fuel pressure, fuel temperature, crankcase pressure), driving system (e.g., speed of the vehicle, acceleration pedal position, parking brake switch), bucket statuses and other variables (e.g., GPS position). From the original variable set, 208 variables were excluded from further analysis (containing 100% missing values).

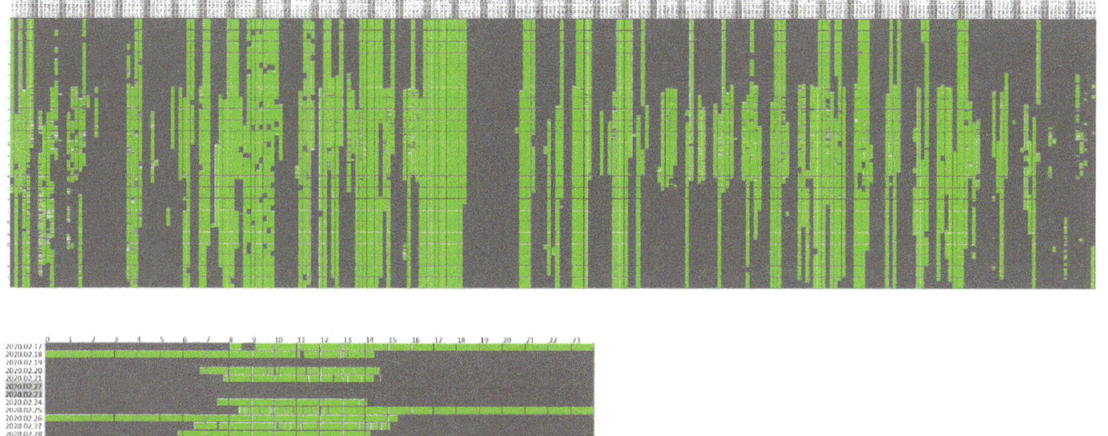

Figure 1. Data quality scan visualisation (grey colour denotes missing data).

The principal component analysis (PCA) method was used to select the appropriate target set of variables for analysis. PCA is a multivariate statistical technique for extracting information from a data set and representing it as new orthogonal variables, referred to as principal components. PCA is used to analyse a table of inter-correlated variables. The main premise of the PCA method is to reduce the dimension of the data used in the analysis, assuming that the reduction of multidimensionality is carried out maintaining most of the variability in the data set. This process is often carried out as a preliminary step before proceeding to further analyses [33,34].

After verification of 208 variables (some of them had a high rate of NAs), 68 variables were examined. Based on their dependency analysis and PCA analysis, we chose 19 variables with 47,093 observations as the final set used for activity recognition. Selected variables (due to IP requirements) are presented in Table 1 and in Figure 2.

Table 1. Fragment of variables list used in machine activity recognition.

Item	Variable Name	Unit
1	wheel-based vehicle speed	[km/h]
2	engine speed	[rpm]
3	accelerator pedal position	[%]
4	engine fuel rate	[l/h]
5	engine fuel temperature	[°C]
6	lifting bucket state	[-]
7	current weight	[kg]
8	parking brake switch	[-]
9	engine shutdown	[-]

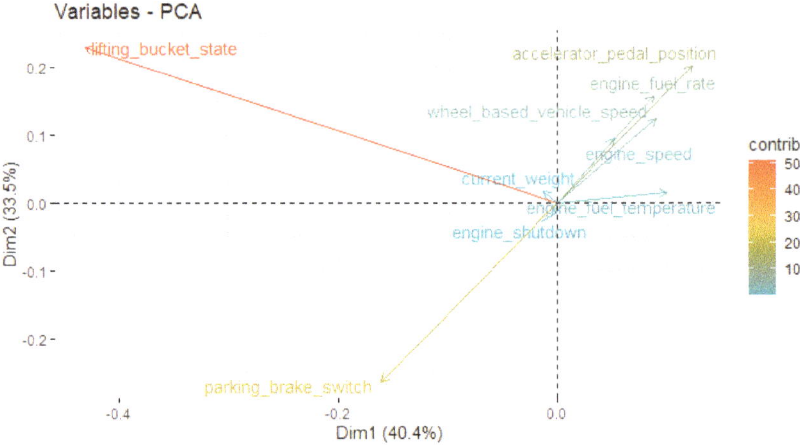

Figure 2. Visualisation of PCA analysis (selected variables).

Additionally, we created a variable denoting the area of wheel loader operation, based on GPS data, with the approach described in the next section.

3.2. Markings of Location Perspective in Data

The assignment of data to the working areas based on the records of GPS coordinates was performed using the PNPOLY (Point Inclusion in Polygon Test) method. The method is based on leading a ray horizontally from the tested point and then switching the in/out status at each polygon edge encountered. An odd number of intersections indicates the location of the test point inside the polygon.

The inpoly function for the R language [35], which implements the PNPOLY method, was developed based on the program code included in the work [36]. There is a point.in.polygon function in R (in the *sp* package), but it is not suitable for use in pipe mode. This was the main reason for creating the proprietary inpoly function. The function returns the true value if the point lies inside the polygon, which is set by an additional data frame with successive coordinates of forming points.

The data analysed includes the geographic coordinates of the current machine position (gps_position). The format of this record requires an appropriate transformation for the extraction of the desired coordinates.

The extraction of longitude and latitude comprised of the following operations were performed in R:

- extracting a string containing GPS coordinates,
- creating a list with three separated coordinates,
- conversion from a list to a vector,
- converting data type to numeric.

After the operations mentioned above, the GPS coordinates and the given area become input parameters to the function inpoly that returns information about belonging to the area.

Considering the geographical location of a mine and analysing the machine movement, we distinguished the mine base (1) and two mine exploitation areas (2,3). Other locations were marked as 0. The mutual position of the areas is shown in Figure 3, as well as the points describing the areas used in the inpoly function to assign the machine's current position to the working areas.

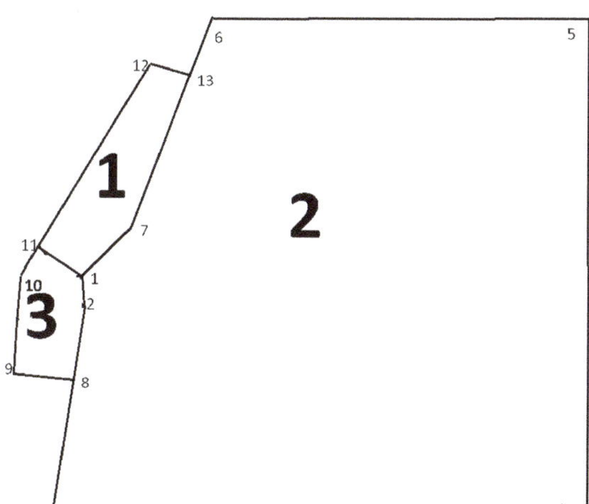

Figure 3. Wheel loader operation areas (1—base, 2—mine, 3—minor mine).

The distribution of identified working areas in the sample data set is presented in Table 2.

Table 2. Distribution of working areas in data set from one week.

Working Area	Frequency	Percentage (%)	Cumulative Frequency	Cumulative Percentage (%)
0	1463	0.9	1463	0.9
1	110,933	70.7	112,396	71.7
2	30,592	19.5	142,988	91.2
3	175	0.1	143,163	91.3
NAs	13,700	8.7	156,863	100.0

The main operation area of the analysed machine is the mine base (70% of observations). One can notice that area no 2 (the exact open pit mine) occurs almost in 20% of observations. Area no 3 is underrepresented; thus, we decided to join this data with area no 2. In the case of 1% of observations, the machine worked out of the mine area; for about 9% of observations, we could not identify the working area due to missing data.

Identified areas will be used for the enhancement of machinery activity recognition, presented in Section 4.

3.3. Selected Data Mining Techniques for Clustering and Results Explanation

Currently, for industrial enterprises, Internet of Things (IoT) systems are a primary source of vast data gathered during operations, allowing for insight into the process and its comprehensive analysis. The collected data can be used in the knowledge discovery process, which can be automated with appropriate data mining methods [37].

In general, data mining approaches can be divided into supervised, unsupervised, or reinforcement learning [38]. Among different unsupervised learning methods, clustering is one of the most popular tasks, which has the advantage of uncovering hidden, often unexpected groups in a data set without any prior knowledge or input about the partition. Cluster analysis at the stage of exploratory data analysis allows a better understanding or summary of the data [2].

Clustering is widely implemented in activity recognition tasks [21]. The primary purpose of a clustering task is to divide instances into different groups, determined by their similarity [37]. Researchers have proposed many clustering methods, such as partitioning clustering (e.g., k-mean, k-medoids, PAM, CLARANS, and CLARA algorithm), hierarchical clustering (e.g., CURE, BIRCH, and ROCK algorithm), density-based clustering (e.g., DBSCAN, OPTICS, and DBCLASD), grid-based clustering (e.g., STRING) or model-based (e.g., Self-Organized Map algorithm) [38,39]. Clustering techniques differ in several assumptions, such as the procedure for calculating similarity within and between clusters, setting the threshold for identifying cluster elements, or the methodology for grouping objects belonging to different degrees into one or more clusters [40]. Depending on the conditions and type of data, different clustering analysis algorithms result in different clusters. A comparison of the most popular clustering algorithms with the main assumptions and limitations of these methods can be found in [41,42].

Considering the characteristics of our data set, we selected the DBSCAN (Density-Based Spatial Clustering of Applications with Noise) algorithm because of its main advantages, such as applicability to multidimensional data, robustness to noisy data, ability to discover clusters of different shapes and sizes, and the lack of requirement to determine the number of clusters in advance [40].

DBSCAN is recommended as a leading algorithm for clustering high dimensional data and is the most commonly used density-based algorithm [40].

The main assumption of density clustering approaches is determining and separating high- and low-density regions [43]. The DBSCAN algorithm was proposed by Ester, M. et al., in response to observed challenges for clustering algorithms, such as the ability to define clusters of arbitrary shape, achieving high efficiency with large data sets, and requiring little domain knowledge to determine input parameters [44]. The authors defined a cluster as a set of connected density points that is maximal with respect to density reachability. The algorithm employs two main input parameters: the neighbourhood radius—Eps (neighbourhood of a point) and the minimum number of points in the neighbourhood—MinPts to determine a density threshold; based on that, the data are aggregated into clusters. The DBSCAN algorithm with correctly specified parameters can produce clusters of any shape [45]. For a point in a cluster, a neighbourhood of a specified radius must contain at least a minimum number of points, i.e., the density must exceed the mentioned threshold, and the chosen distance function indicates the shape of the neighbourhood. There are two types of points within a cluster: core points, which are located inside the cluster, and border points, which are points on the border of the cluster [44]. The DBSCAN algorithm is also used to efficiently discover noise in the data, which is defined as a set of points in the database that do not belong to any of the clusters [44,46], so-called outliers.

In an analysis of discovered clusters, as an explanation of results, we used decision trees which enabled the formulation of rules for assigning observations to proper clusters in the analysed area. A method for solving classification problems such as decision trees is characterised primarily by its intuitiveness, simplicity of interpretation of results, and reasonable accuracy [47], enabling prediction of categorical outputs with tree or rule structures. Trees are graphical representations of the decision-making process in which the structure consists of internal nodes representing attribute tests (decision nodes) and leaf nodes corresponding to predicted class labels [48,49]. There are many algorithms for classification using decision trees, the most popular of which are: CART [50], CHAID [51], QUEST [52], and C5.0 [53,54] as a successor of the C4.5 algorithm [55], which is based on ID3 [56]. The main improvement of the C5.0 is boosting technology, allowing the addition of each sample weight to determine its importance [53].

In explanation of clustering results, we tested CART and C5.0 algorithms (with various settings). The CART algorithm is the most commonly used decision-tree technique [50] which allows the detection of structures even in complex data and the construction of accurate and reliable models [57]. Based on labelled data, CART trees enable the discovery of rules that can be used to classify new data. This method is an example of binary recursive partitioning using the GINI index. Binary partitioning can be performed repeatedly, and instances in a node can only be divided into two groups [58]. The second algorithm used to describe the obtained clusters is the C5.0 algorithm (with the rule model option). The C5.0 algorithm employs an information gain rate to build a decision tree. Information gain is determined for each feature in the data set to identify the best split points [54,59].

In order to increase the accuracy of a classifier like a classification tree, pruning techniques aim to reduce the size of the tree by eliminating the less frequent sections of the tree that are considered non-critical and have low classification ability. Pruning reduces the complexity of the tree and consequently improves the classification performance by preventing overfitting [60]. In the case of used algorithms, we tested the following parameters for pruning: cp—complexity parameter (CART) and Min Cases parameter as the smallest number of samples that must be put in at least two of the splits (C5.0). The complexity parameter (cp) denotes the minimum improvement in the model needed at each node. It is based on the cost complexity of the model. The cp parameter helps speed up the search for splits because it can identify splits that do not meet this criterion and prune them before the tree becomes too wide [61].

4. Results and Discussion

Our data set with 18 variables (the time variable was omitted in clustering task) was analysed with R library *dbscan* [62]. Since the DBSCAN algorithm is sensitive to Eps and MinPts settings, we ran multiple calculations with various values of parameters. The Eps parameter setting started from the analysis of the KNN plot for k = 18 (number of dimensions) (Figure 4). At first, we tested a value that corresponded to the curve's inflexion point (that is 0.5). In the beginning, we changed the value of the Eps parameter with step 0.05. When the number of clusters decreased, we adaptively decreased the Eps value. In the case of MinPts value, firstly, we assumed a number of points equal to a number of dimensions +1 (19); however, we obtained many outliers. We repeated the calculations by doubling and tripling this number. Finally, we chose MinPts as 57 points and Eps value as 0.6.

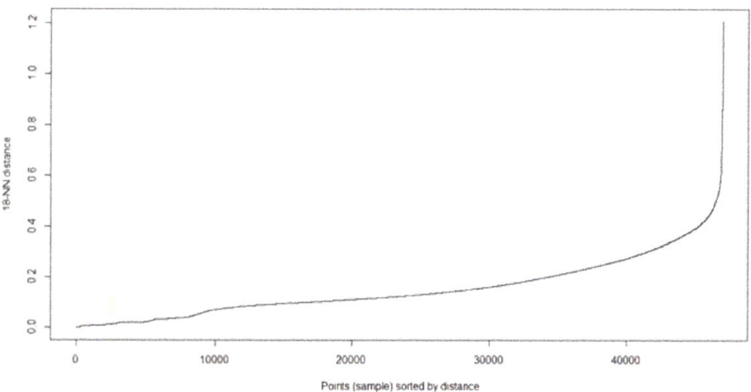

Figure 4. KNN plot for k = 18.

The DBSCAN algorithm, with defined parameters as above, found five clusters in our data set (Figure 5). Black dots in the figure denote outliers (not clustered observations).

The distribution of observations among discovered clusters is presented in Figure 6. Cluster 0 contains outliers—only 145 observations were collected in this cluster (0.3% of observations). The largest, cluster 1, contains almost 52% of observations (24,338), and cluster 3 contains 32% observations (15,303). Smaller clusters, no 2 and no 4, contain 8% (3763) and almost 7% of observations (3251), respectively. Finally, cluster 5 contains 0.6% of observations.

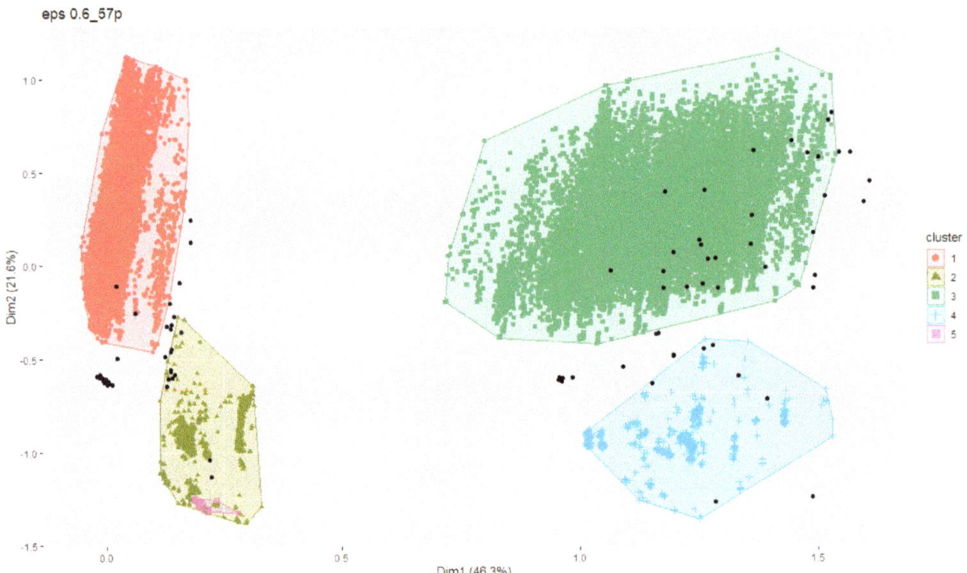

Figure 5. Clustering results (with outliers marked as black dots).

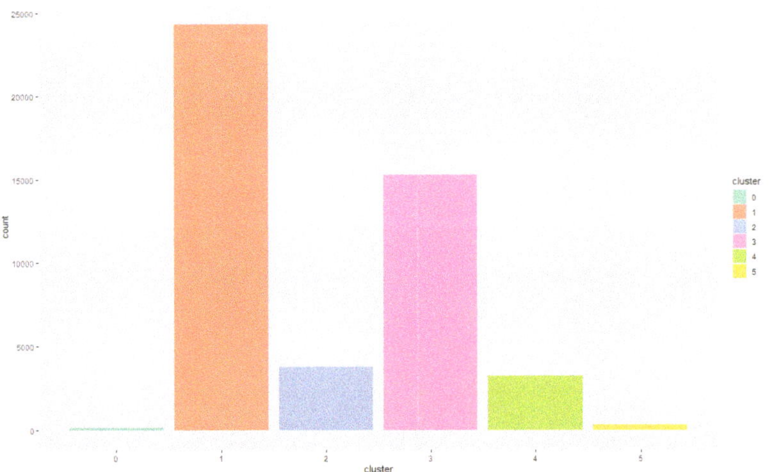

Figure 6. Distribution of observations in clusters.

Selected statistics of the discovered clusters are presented in Table 3.

Based on presented statistics, the following description of activities was prepared:

- Cluster 1—Moving/travelling;
- Cluster 2—Stoppage with engine ON;
- Cluster 3—Normal work with loading;
- Cluster 4—Stoppage with loading;
- Cluster 5—Engine OFF.

Distributions of discovered clusters in the identified working areas are presented in Figure 7.

Table 3. Basic statistics of the discovered clusters.

Variable Name	Cluster 1			Cluster 2			Cluster 3			Cluster 4			Cluster 5		
	Min	Median	Max	Min	Median	Max	Min	Median	Max	Min	Median	Max	Min	Median	Max
wheel-based vehicle speed	0	5	21	0	0	0	0	4	20	0	0	0	0	0	0
engine speed	621	1128	2015	0	791	1495	645	1117	1699	0	793	1606	0	0	52
accelerator pedal position	0	29	72	0	0	57	0	29	72	0	0	53	0	0	0
engine fuel rate	0	15	55	0	3	27	0	15	55	0	3	24	0	0	0
engine fuel temperature	0	21	26	0	21	28	0	19	23	0	6	25	0	16	17
lifting bucket state	0	0	0	0	0	0	1	1	1	1	1	1	0	0	0
current weight	0	0	0	0	0	0	−24,608	1072	29,353	−4210	−2233	6213	0	0	0
parking brake switch	0	0	0	1	1	1	0	0	0	1	1	1	1	1	1
shutdown engine	0	0	0	0	0	0	0	0	0	0	0	0	1	1	1

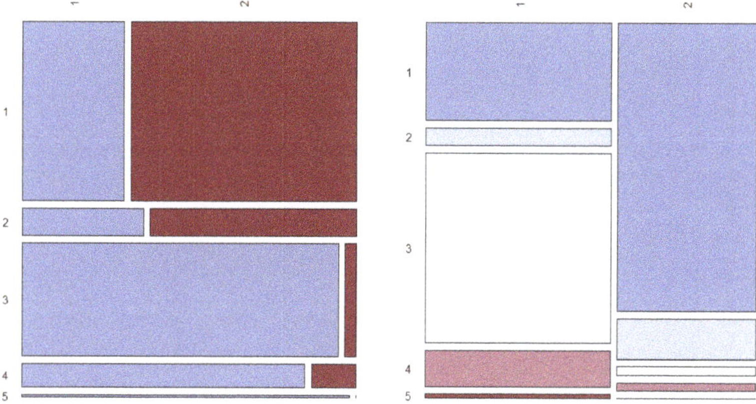

Figure 7. Distribution of clusters and observations in identified working areas (x axis = area, y axis = cluster).

As can be observed, the main activity in the mine base area (area no 1) is regarding Normal work with loading, while the dominant activity in the open pit area is Moving (Figure 7—right figure). Thus, we can conclude that the main area of efficient work is the mine base area, and usage of wheel loader in the open pit has a rather auxiliary character, e.g., raw material moving (distribution of activity no 2 and 4—Stoppage with/or without loading). Activity Engine OFF is mainly observed in the mine base area.

In further investigations, we analysed whether there was a difference between machine behaviour in areas of work. In other words, whether a place of work influences the characteristics of the discovered activity. We performed a statistical analysis of distributions of variables in each cluster versus working area.

In Table 4, we present the statistical analysis results regarding the comparison of selected numerical variable distributions in the working areas for each activity. Since none of the numerical variables in the data set is normally distributed, to test the differences within groups, we chose the non-parametric Wilcoxon test. Colours in the table indicate statistically significant test values.

Comparison of variable boxplots in each cluster and area are presented in Figure 8.

Statistical analysis of selected variables in each cluster confirmed that there is a statistically significant difference in their distributions in the defined areas for most variables. Thus, during the process monitoring and activities analysis, these differences should be taken into consideration.

Table 4. Results of the Wilcoxon test for group comparison.

Variable Name		Wheel-Based Vehicle Speed	Engine Speed	Accelerator Pedal Position	Engine Fuel Rate	Engine Fuel Temperature	Current Weight
Cluster 1	median 1	5.0	1120	29.6	16.6	22.0	0
	median 2	5.9	1131	28.8	14.6	21.0	0
	p value	<0.01	<0.01	<0.01	<0.01	<0.01	-
Cluster 2	median 1	0	800.5	0.8	5.4	5.0	0
	median 2	0	651.0	0	2.6	21.0	0
	p-value	-	<0.01	<0.01	<0.01	<0.01	-
Cluster 3	median 1	4.7	1122	29.6	15.6	19.0	1127.5
	median 2	1.6	915	10.0	5.9	21.0	665
	p value	<0.01	<0.01	<0.01	<0.01	<0.01	<0.05
Cluster 4	median 1	0	798.2	0	4.6	5	−2236
	median 2	0	649.9	0	2.6	22	5996
	p-value	-	<0.01	>0.05	<0.01	<0.01	<0.01
Cluster 5	median 1	0	0	0	0	16.0	0
	median 2	0	0	0	0	-	-
	p-value	-	-	-	-	-	-

For a better description of discovered clusters and considering revealed differences, we attempted to discover rules enabling raw data labelling. In this task, we examined selected classifiers with CART and C5.0 algorithms.

Firstly, we divided the data set randomly (without data of cluster 0, denoting outliers) into train and test subsets with the proportion of 80% (37,558 observations) and 20% (9390 observations), respectively. Subsequently, we trained classifiers and checked them on the test data set. In Tables 5 and 6, we present the main parameters and the obtained results (the most similar results obtained for the two classification algorithms are marked with bolded font).

Table 5. Parameters of CART classifiers.

No of Classifier	Cp	Tree Size (Nsplits)	Accuracy on Train Dataset	Accuracy on Test Dataset
1	0.000250	115	0.9098	0.9085
2	0.000375	87	0.9045	0.9043
3	0.000600	34	0.8890	0.8882
4	0.000750	30	0.8874	0.8858
5	0.001000	27	0.8854	0.8851
6	**0.002000**	**17**	**0.8758**	**0.8765**
7	0.002500	9	0.8602	0.8608

The simplest classification tree (no 7) obtained using the CART algorithm is presented in Figure 9.

The performed tests have shown that both algorithms have built the tree models enabling accurate prediction of the tested class: wheel loader activity in the working area. The most valuable rules, due to their confidence, obtained from comparable-in-size classifiers: CART with 18 rules (no 6) and C5.0 with 17 rules (no 5), are presented in Table 7.

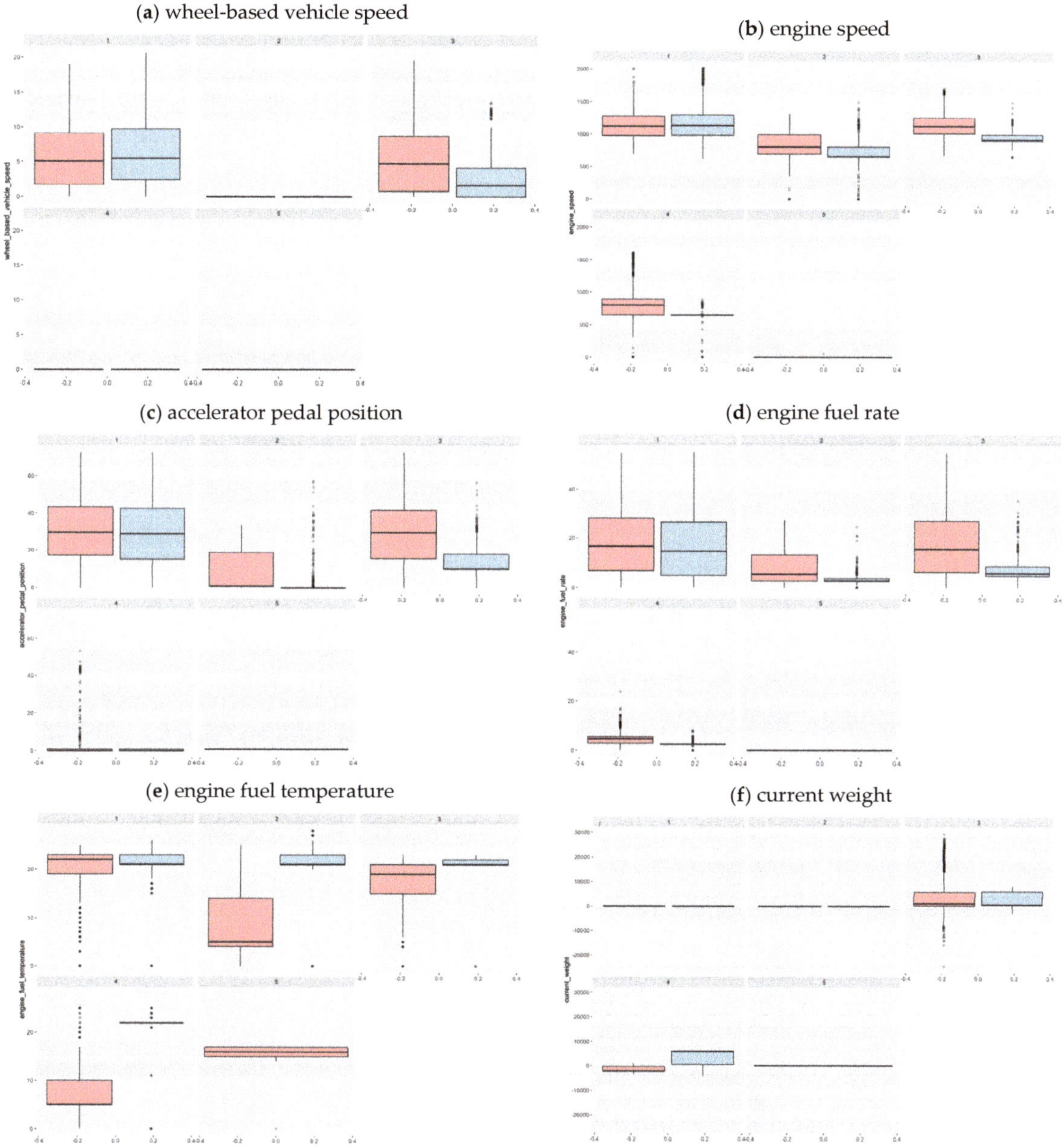

Figure 8. Boxplots of selected variables (**a–f**) in the perspective of clusters and working areas. (Red colour denotes area 1, blue colour denotes area 2).

According to Table 7, we can observe that obtained rules are characterised by a high or very high confidence level (0.73–1.0). Rules responsible for classifying machine characteristics as a particular activity are very similar for both applied classifiers; however, in some cases are different in the level of variable occurrence in the tree structure.

Table 6. Parameters of C5.0 classifiers.

No of Classifier	Min Cases	Numbers of Rules	Accuracy on Train Dataset	Accuracy on Test Dataset
1	50	41	0.9040	0.8954
2	100	22	0.8870	0.8825
3	150	17	0.8830	0.8797
4	200	16	0.8800	0.8778
5	**250**	**17**	**0.8770**	**0.8729**
6	300	13	0.8760	0.8732
7	600	9	0.8480	0.8436

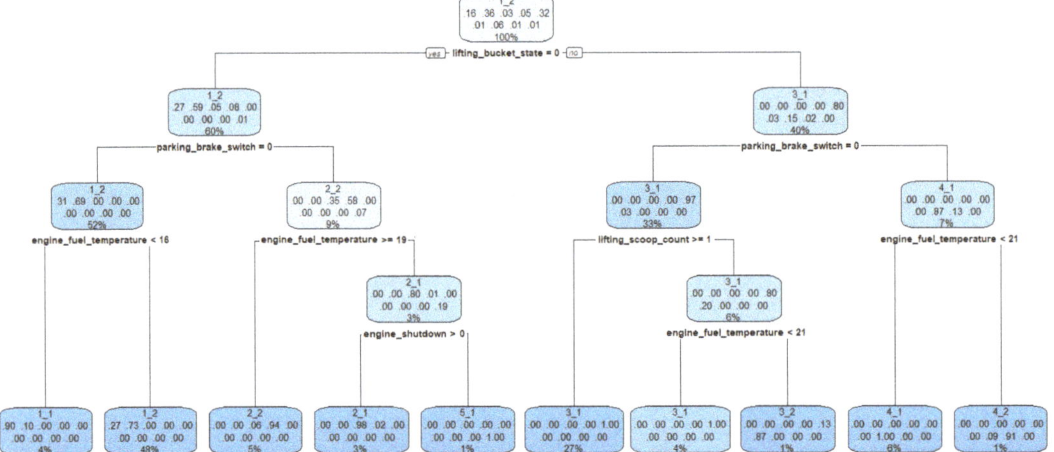

Figure 9. Classification tree CART.

The key machine operating parameters for assigning an observation to activity 1, identified similarly by both types of classifiers, are primarily lifting_bucket_state and parking_brake_switch (less than 0.5; in practice, only 0 or 1 are possible). Based on engine_fuel_temperature, classification to area 1 is made for temperatures less than 15/16 °C and area 2 for temperatures greater than 15/16 °C. However, in the C5.0 tree, additional rules have been formulated related to vehicle speed. The rules determining assignment to activity 2 are based on the parameters such as parking_brake_switch > 0 and lifting_bucket_state <= 0. Similar to activity 1, depending on engine_fuel_temperature, observations are classified into area 1 (temperature less than 19 °C) or area 2 otherwise. Classification to activity 3 is based, among other criteria, on the parking_brake_switch variable, whose value, in this case, equals 0. Depending on the values taken by the lifting_scoop_weight variable in the C5.0 tree, the observations are classified in area 1 (lifting_scoop_weight > 0) or area 2 when this variable takes a value less than 0. In the CART tree, another variable related to lifting is taken into consideration—lifting_scoop_weight. Assignment to activity 4 is based largely on 3 variables, parking_brake_switch, lifting_bucket_state, and engine_fuel_temperature, where the first two variables should be greater than 0, while temperature greater than 20 °C determines the classification of the observation as activity 4 in area 2, while less than 20 °C is in area 1. The rule for classifying into cluster 5 in the C5.0 tree is mainly based on variable engine_shutdown greater than 0. CART tree extends rules with additional variables. The general conclusion can be made, comparing similarities and dissimilarities of variables used in tree classifiers, that indication of the operation of wheel loader in the open pit area is related to the greater temperature of engine fuel, which can be explained by more inconvenient conditions of work as downhill rides and elevation differences that impact loading of the engine.

Table 7. Rules describing wheel loader activities in the working areas.

Activity	CART		C5.0	
	Rule	Confidence	Rule	Confidence
1_1	lifting_bucket_state = 0 parking_brake_switch = 0 engine_fuel_temperature < 16	0.897	engine_fuel_temperature <= 15 parking_brake_switch <= 0 lifting_bucket_state <= 0	0.904
1_2	lifting_bucket_state = 0 parking_brake_switch = 0 engine_fuel_temperature >= 16	0.735	wheel_based_vehicle_speed > 9.85 engine_fuel_temperature> 15 lifting_bucket_state <= 0	0.874
2_1	lifting_bucket_state = 0 parking_brake_switch > 0 engine_fuel_temperature < 19 engine_shutdown = 0	0.981	engine_fuel_temperature<= 19 parking_brake_switch > 0 engine_shutdown <= 0 lifting_bucket_state <= 0	0.983
2_2	lifting_bucket_state = 0 parking_brake_switch > 0 engine_fuel_temperature >= 19	0.938	engine_fuel_temperature > 19 parking_brake_switch > 0 lifting_bucket_state <= 0	0.941
3_1	lifting_bucket_state > 0 parking_brake_switch = 0 lifting_scoop_count >= 1	0.998	engine_fuel_temperature <=20 lifting_scoop_weight > 0 parking_brake_switch <= 0	1.0
3_2	lifting_bucket_state > 0 parking_brake_switch <= 0 lifting_scoop_weight < 1 engine_fuel_temperature >= 21	0.868	engine_fuel_temperature > 20 lifting_scoop_weight <= 0 parking_brake_switch <= 0 lifting_bucket_state > 0	0.873
4_1	lifting_bucket_state > 0 parking_brake_switch > 0 engine_fuel_temperature < 21	0.995	engine_fuel_temperature <= 20 parking_brake_switch > 0 lifting_bucket_state > 0	0.995
4_2	lifting_bucket_state > 0 parking_brake_switch > 0 engine_fuel_temperature >= 21	0.909	engine_fuel_temperature > 20 parking_brake_switch > 0 lifting_bucket_state > 0	0.911
5_1	lifting_bucket_state = 0 parking_brake_switch >= 0 engine_fuel_temperature < 19 engine_shutdown > 0	0.996	engine_shutdown > 0	0.966

Rules describing conditions (variable values) enable data labelling considering area of operation. The labelled data can be further used for visualisation and process analysis, e.g., in the machinery monitoring systems. Such labelled activities can be further analysed, among others, with process mining techniques and used for process modelling and analysis, e.g., with process maps (Figure 10).

Analysis of activities in different areas can support understanding specific conditions of operation needed to optimise equipment efficiency and its usage. For example:

- longer travelling time in the mine area can cause a loss in the effective time of working; one can consider leaving machines in the mine area and arranging for a faster means of transport for service personnel to the machine—it can result in saving working time, increasing the use of the loader, and reducing the cost of fuel used by the loader to get to the job site.

- automatic detection of the loading activity can allow for its correlation with the fuel consumption associated with this activity, and detailed data on the weight transferred in the bucket can be the basis for determining the optimal weight that should be loaded on the bucket to minimise the cost of loading.
- the more frequent activity of loading in the base area may be a reason for purchasing an additional machine.
- the analysis of changes in the structure of activities over time may show significant changes in the use of machines that will justify (demonstrate the need of) a decision on periodic leasing (renting, etc.) of machines from external entities, which will reduce operating costs.

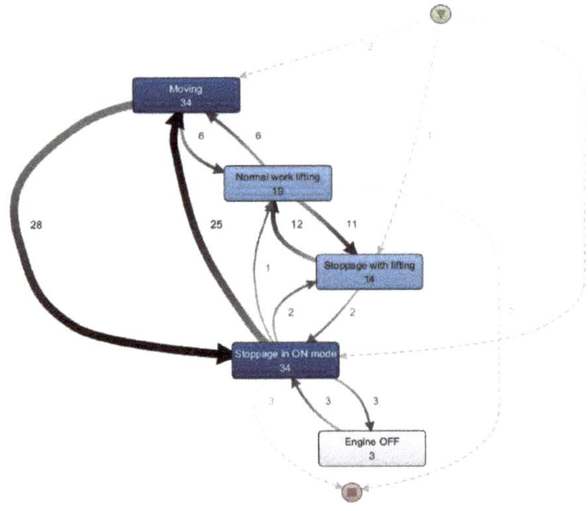

Figure 10. A process model of wheel loader operation.

In the further works, we plan to investigate the abovementioned issues related to a more in-depth analysis of dependencies between characterised activities and selected KPIs, as well as overall efficiency based on process-oriented analytics.

5. Conclusions

In this paper, an analysis of sensor data characterising the operation of a wheel loader in an open pit mine was presented. The purpose of the analysis was to effectively identify machine activities based on a real operational dataset. A subset of the entire dataset was considered in the study, including variables related to engine operation, driving system, bucket statuses and other variables, which have been clustered using the density method and the DBSCAN algorithm. We further extended the analysis with GPS data, which allowed us to divide the working area into subareas (mine base and open pit). Based on the statistical characteristics of the obtained clusters, we have named them, and together with the identified working areas, we have identified the statistical differences of variable distributions in clusters (activities) performed in various areas. The analysis results encouraged us to develop classifiers describing clusters in the form of rules, which can be helpful in raw data labelling in the future. In this part, we used selected classifiers based on the CART and C5.0 algorithms. As a result, we presented the most valuable rules for wheel loader activity recognition.

Obtained results showed that density clustering methods can provide an efficient multidimensional space search for compact and sensible clusters of observations in a real, noisy dataset. Moreover, the introduction to analysis of the location variable enabled us to identify the statistically significant differences in machinery operation in two defined working areas. These findings have also been positively validated by tree classifiers, with high accuracy rates on train and test datasets. From built classifiers, one can extract valuable rules (with high confidence factor), enabling definition/recognition of activity during process monitoring.

The identified rules require validation in a real-life environment and verification on-site during activity execution, which is planned in the next stage of our research. Then, after positive validation result, rules can be applied, e.g., in machinery or process monitoring systems.

The discovered dependencies can be used to improve the effectiveness and increase the safety of operation by indicating the activities and states generating the highest load for the machine. As a result of the acquired knowledge, the company may undertake real changes of unfavourable work parameters and, therefore, obtain notable benefits such as reduction of fuel consumption, an extension of the machine's operating time, or minimisation of extreme and dangerous states of the machine's operation.

Remarkably, distinguishing the states of a machine's work is a necessary condition for developing an algorithm that automatically determines the usage of a machine and calculates detailed indicators that consider the diversity of these states. The values of these parameters can be a premise in the decision-making process to change the utilisation of a machinery park in a mining company. In the presented analysis, consideration of the identified working areas allowed to prove the existing differences in the quantitative characteristics of activity concerning the place of machine operation. It should be noted that the analysed machine is one of many used in the mine, so automation of its activity recognition should be an integral part of the online analysis related to the machinery park.

In addition, the analysis conducted in the article is one of the steps in the development of a system that allows the calculation of the optimal operating conditions based on data collected from multiple machines. Processed data provide information about the different stages of production. They can be the basis for determining the real-time indicators of the effectiveness of used machinery and also can provide a basis for decisions about the timing of repairs or maintenance ahead of equipment failure. Awareness of such a need, strengthened by the results of data analysis, allows for planning the appropriate time and duration of activities related to the replacement of worn-out machine elements during scheduled downtime. This can help to avoid losses generated as a result of unplanned downtime.

The findings obtained from the analysis presented above can contribute to the knowledge base for MES (Manufacturing Execution System) systems in many areas specified by the international standardization organization MESA International (Manufacturing Enterprise Solutions Association International).

Author Contributions: Conceptualisation, E.B. and P.G.; methodology, E.B., M.K.; investigation, P.G., E.B. and M.K., data curation, M.K. and E.B.; writing—original draft preparation, P.G., E.B. and M.K.; writing—review and editing, P.G., E.B. and M.K.; visualisation, E.B. and M.K.; supervision, E.B. All authors have read and agreed to the published version of the manuscript.

Funding: This research was funded by EIT Raw Materials, grant number 18327—project smartHUB—smart courier field data IIoT radio network and big data analytics.

Data Availability Statement: Data sharing not applicable.

Acknowledgments: In the main research, we used the following R libraries: DescTools, dbscan, ggplot2, rpart, and C5.0.

Conflicts of Interest: The authors declare no conflict of interest.

References

1. Zhou, X.; Gu, J.; Shen, S.; Ma, H.; Miao, F.; Zhang, H.; Gong, H. An automatic K-Means clustering algorithm of GPS data combining a novel niche genetic algorithm with noise and density. *ISPRS Int. J. Geo. Inf.* **2017**, *6*, 392. [CrossRef]

2.   Bhattacharjee, P.; Mitra, P. A survey of density based clustering algorithms. *Front. Comput. Sci.* **2021**, 15. [CrossRef]

3.   Ahmad, A.; Khan, S.S. Survey of State-of-the-Art Mixed Data Clustering Algorithms. *IEEE Access* **2019**, *7*, 31883–31902. [CrossRef]

4.   Ingrao, C.; Evola, R.S.; Cantore, P.; De Bernardi, P.; Del Borghi, A.; Vesce, E.; Beltramo, R. The contribution of sensor-based equipment to life cycle assessment through improvement of data collection in the industry. *Environ. Impact Assess. Rev.* **2021**, *88*, 106569. [CrossRef]

5.   Frieß, U.; Kolouch, M.; Friedrich, A.; Zander, A. Fuzzy-clustering of machine states for condition monitoring. *CIRP J. Manuf. Sci. Technol.* **2018**, *23*, 64–77. [CrossRef]

6.   Polak, M.; Stefaniak, P.; Zimroz, R.; Wylomanska, A.; Sliwinski, P.; Andrzejewski, M. Identification of Loading Process Based on Hydraulic Pressure Signal. *Int. Multidiscip. Sci. GeoConf. SGEM Surv. Geol. Min. Ecol. Manag.* **2016**, *2*, 459–466.

7.   Dong, L.; Wesseloo, J.; Potvin, Y.; Li, X. Discrimination of Mine Seismic Events and Blasts Using the Fisher Classifier, Naive Bayesian Classifier and Logistic Regression. *Rock Mech. Rock Eng.* **2016**, *49*, 183–211. [CrossRef]

8.   Jha, A.; Tukkaraja, P. Monitoring and assessment of underground climatic conditions using sensors and GIS tools. *Int. J. Min. Sci. Technol.* **2020**, *30*, 495–499. [CrossRef]

9.   Sherafat, B.; Rashidi, A.; Lee, Y.C.; Ahn, C.R. Automated activity recognition of construction equipment using a data fusion approach. In *Computing in Civil Engineering 2019: Data, Sensing, and Analytics*; American Society of Civil Engineers: Reston, VA, USA, 2019; pp. 1–8.

10.   Akhavian, R.; Behzadan, A.H. Construction equipment activity recognition for simulation input modeling using mobile sensors and machine learning classifiers. *Adv. Eng. Inform.* **2015**, *29*, 867–877. [CrossRef]

11.   Redlich, D.; Molka, T.; Gilani, W.; Blair, G.; Rashid, A. Data-Driven Process Discovery and Analysis. *Lect. Notes Bus. Inf. Process.* **2015**, *237*, 79–106.

12.   Cheng, C.F.; Rashidi, A.; Davenport, M.A.; Anderson, D.V. Activity analysis of construction equipment using audio signals and support vector machines. *Autom. Constr.* **2017**, *81*, 240–253. [CrossRef]

13.   Wu, C.; Chen, Z.; Wang, D.; Kou, Z.; Cai, Y.; Yang, W. Behavior modelling and sensing for machinery operations using smartphone's sensor data: A case study of forage maize sowing. *Int. J. Agric. Biol. Eng.* **2019**, *12*, 66–74. [CrossRef]

14.   Stefaniak, P.; Śliwiński, P.; Poczynek, P.; Wyłomańska, A.; Zimroz, R. The automatic method of technical condition change detection for LHD machines—Engine coolant temperature analysis. *Appl. Cond. Monit.* **2019**, *15*, 54–63.

15.   Langroodi, A.K.; Vahdatikhaki, F.; Doree, A. Activity recognition of construction equipment using fractional random forest. *Autom. Constr.* **2021**, *122*, 103465. [CrossRef]

16.   Wang, S.; Hou, L.; Lee, J.; Bu, X. Evaluating wheel loader operating conditions based on radar chart. *Autom. Constr.* **2017**, *84*, 42–49.

17.   Shi, J.; Sun, D.; Hu, M.; Liu, S.; Kan, Y.; Chen, R.; Ma, K. Prediction of brake pedal aperture for automatic wheel loader based on deep learning. *Autom. Constr.* **2020**, *119*, 1–12. [CrossRef]

18.   Zhang, W.; Wang, S.; Hou, L.; Jiao, R.J. Operating data-driven inverse design optimization for product usage personalization with an application to wheel loaders. *J. Ind. Inf. Integr.* **2021**, *23*, 100212.

19.   Frank, B.; Kleinert, J.; Filla, R. Optimal control of wheel loader actuators in gravel applications. *Autom. Constr.* **2018**, *91*, 1–14. [CrossRef]

20.   Amutha, J.; Sharma, S.; Sharma, S.K. Strategies based on various aspects of clustering in wireless sensor networks using classical, optimization and machine learning techniques: Review, taxonomy, research findings, challenges and future directions. *Comput. Sci. Rev.* **2021**, *40*, 100376. [CrossRef]

21.   Hennig, M.; Grafinger, M.; Gerhard, D.; Dumss, S.; Rosenberger, P. Comparison of time series clustering algorithms for machine state detection. *Procedia CIRP* **2020**, *93*, 1352–1357. [CrossRef]

22.   Yoo, Y.J. Data-driven fault detection process using correlation based clustering. *Comput. Ind.* **2020**, *122*, 103279. [CrossRef]

23.   Yang, Z.; Baraldi, P.; Zio, E. A novel method for maintenance record clustering and its application to a case study of maintenance optimization. *Reliab. Eng. Syst. Saf.* **2020**, *203*, 107103. [CrossRef]

24.   Liang, X.; Wang, Y.; Li, H.; He, Y.; Zhao, Y. Power Transformer Abnormal State Recognition Model Based on Improved K-Means Clustering. In Proceedings of the 2018 IEEE Electrical Insulation Conference (EIC), San Antonio, TX, USA, 17–20 June 2018; pp. 327–330.

25.   Netzer, M.; Michelberger, J.; Fleischer, J. Intelligent anomaly detection of machine tools based on mean shift clustering. *Procedia CIRP* **2020**, *93*, 1448–1453. [CrossRef]

26.   Dindarloo, S.R.; Siami-Irdemoosa, E. Data mining in mining engineering: Results of classification and clustering of shovels failures data. *Int. J. Min. Reclam. Environ.* **2017**, *31*, 105–118. [CrossRef]

27.   Krogerus, T.; Hyvönen, M.; Huhtala, K. Recognition of Operating States of a Wheel Loader for Diagnostics Purposes. *SAE Int. J. Commer. Veh.* **2013**, *6*, 412–418. [CrossRef]

28.   Liao, L.; Fox, D.; Kautz, H. Extracting places and activities from GPS traces using hierarchical conditional random fields. *Int. J. Rob. Res.* **2007**, *26*, 119–134. [CrossRef]

29.   Wang, Y.; Qin, K.; Chen, Y.; Zhao, P. Detecting anomalous trajectories and behavior patterns using hierarchical clustering from Taxi GPS Data. *ISPRS Int. J. Geo-Inf.* **2018**, *7*, 25. [CrossRef]

30.   Cheng, T.; Venugopal, M.; Teizer, J.; Vela, P.A. Performance evaluation of ultra wideband technology for construction resource location tracking in harsh environments. *Autom. Constr.* **2011**, *20*, 1173–1184. [CrossRef]

31. Fu, J.; Jenelius, E.; Koutsopoulos, H.N. Identification of workstations in earthwork operations from vehicle GPS data. *Autom. Constr.* **2017**, *83*, 237–246. [CrossRef]
32. Pradhananga, N.; Teizer, J. Automatic spatio-temporal analysis of construction site equipment operations using GPS data. *Autom. Constr.* **2013**, *29*, 107–122. [CrossRef]
33. Abdi, H.; Williams, L.J. Principal component analysis. *Wiley Interdiscip. Rev. Comput. Stat.* **2010**, *2*, 433–459. [CrossRef]
34. Ringnér, M. What is principal components analysis? *Nat. Biotechnol.* **2008**, *26*, 303–304. [CrossRef] [PubMed]
35. Kęsek, M. Analysing data with the R programming language to control machine operation. *Inż. Miner.* **2019**. [CrossRef]
36. PNPOLY—Point Inclusion in Polygon Test, W. Randolph Franklin (WRF). Available online: https://wrf.ecse.rpi.edu/Research/Short_Notes/pnpoly.html (accessed on 10 April 2021).
37. Dogan, A.; Birant, D. Machine learning and data mining in manufacturing. *Expert Syst. Appl.* **2021**, *166*, 114060. [CrossRef]
38. Sunhare, P.; Chowdhary, R.R.; Chattopadhyay, M.K. Internet of things and data mining: An application oriented survey. *J. King Saud Univ. Comput. Inf. Sci.* **2020**. [CrossRef]
39. Anand, N.; Vikram, P. Comprehensive Analysis & Performance Comparison of Clustering Algorithms for Big Data. *Rev. Comput. Eng. Res.* **2017**, *4*, 54–80.
40. Valarmathy, N.; Krishnaveni, S. A novel method to enhance the performance evaluation of DBSCAN clustering algorithm using different distinguished metrics. *Mater. Today Proc.* **2020**. [CrossRef]
41. Garima; Gulati, H.; Singh, P.K. Clustering techniques in data mining: A comparison. In Proceedings of the 2015 2nd International Conference on Computing for Sustainable Global Development (INDIACom), New Delhi, India, 11–13 March 2015; pp. 410–415.
42. Archana Patel, K.M.; Thakral, P. The best clustering algorithms in data mining. In Proceedings of the 2016 International Conference on Communication and Signal Processing (ICCSP), Melmaruvathur, India, 6–8 April 2016; pp. 2042–2046.
43. Smiti, A.; Eloudi, Z. Soft DBSCAN: Improving DBSCAN clustering method using fuzzy set theory. In Proceedings of the 2013 6th International Conference on Human System Interactions (HSI), Sopot, Poland, 6–8 June 2013; pp. 380–385.
44. Ester, M.; Kriegel, H.-P.; Sander, J.; Xu, X. A density-based algorithm for discovering clusters in large spatial databases with noise. In Proceedings of the International Conference on Knowledge Discovery and Data Mining, Portland, Oregon, USA, 2–4 August 1996; pp. 226–231.
45. Hou, J.; Liu, W. Evaluating the density parameter in density peak based clustering. In Proceedings of the 2016 Seventh International Conference on Intelligent Control and Information Processing (ICICIP), Siem Reap, Cambodia, 1–4 December 2016; pp. 68–72.
46. Smiti, A. A critical overview of outlier detection methods. *Comput. Sci. Rev.* **2020**, *38*, 100306. [CrossRef]
47. Srivastava, A.; Han, E.H.S.; Kumar, V.; Singh, V. Parallel formulations of decision-tree classification algorithms. *Proc. Int. Conf. Parallel Process.* **1998**, 237–244. [CrossRef]
48. Otero, F.E.B.; Freitas, A.A.; Johnson, C.G. Inducing decision trees with an ant colony optimization algorithm. *Appl. Soft Comput. J.* **2012**, *12*, 3615–3626. [CrossRef]
49. Maleki, F.; Ovens, K.; Najafian, K.; Forghani, B.; Reinhold, C.; Forghani, R. Overview of Machine Learning Part 1: Fundamentals and Classic Approaches. *Neuroimaging Clin. N. Am.* **2020**, *30*, e17–e32. [CrossRef] [PubMed]
50. Rutkowski, L.; Jaworski, M.; Pietruczuk, L.; Duda, P. The CART decision tree for mining data streams. *Inf. Sci.* **2014**, *266*, 1–15. [CrossRef]
51. Kass, G.V. An Exploratory Technique for Investigating Large Quantities of Categorical Data. *Appl. Stat.* **1980**, *29*, 119. [CrossRef]
52. Loh, W.Y.; Shin, Y.S. Split selection methods for classification trees. *Stat. Sin.* **1997**, *7*, 815–840.
53. Pang, S.; Gong, J. C5.0 Classification Algorithm and Application on Individual Credit Evaluation of Banks. *Syst. Eng. Theory Pract.* **2009**, *29*, 94–104. [CrossRef]
54. Balamurugan, M.; Kannan, S. Performance analysis of cart and C5.0 using sampling techniques. In Proceedings of the 2016 IEEE International Conference on Advances in Computer Applications (ICACA), Coimbatore, India, 24 October 2016; pp. 72–75.
55. Quinlan, J.R. *C4.5: Programs for Machine Learning*; Morgan Kaufmann Publishers Inc.: Burlington, MA, USA, 1993.
56. Quinlan, J.R. Induction of decision trees. *Mach. Learn.* **1986**, *1*, 81–106. [CrossRef]
57. Bin, S.; Sun, G. Data Mining in census data with CART. In Proceedings of the 2010 3rd International Conference on Advanced Computer Theory and Engineering (ICACTE), Chengdu, China, 20–22 August 2010; pp. 260–264.
58. Lewis, R.J.; Street, W.C. An Introduction to Classification and Regression Tree (CART) Analysis. In Proceedings of the Annual Meeting of the Society for Academic Emergency Medicine in San Francisco, Francisco, CA, USA, 22–25 May 2000.
59. Guo, J.; Liu, H.; Luan, Y.; Wu, Y. Application of birth defect prediction model based on c5.0 decision tree algorithm. In Proceedings of the 2018 IEEE International Conference on Internet of Things (iThings) and IEEE Green Computing and Communications (GreenCom) and IEEE Cyber, Physical and Social Computing (CPSCom) and IEEE Smart Data (SmartData), Halifax, NS, Canada, 30 July–3 August 2018; pp. 1867–1871.
60. Hssina, B.; Merbouha, A.; Ezzikouri, H.; Erritali, M. A comparative study of decision tree ID3 and C4.5. *Int. J. Adv. Comput. Sci. Appl.* **2014**, *4*, 13–19. [CrossRef]
61. Decision Trees (rpart) in, R. Available online: http://www.learnbymarketing.com/tutorials/rpart-decision-trees-in-r/ (accessed on 15 April 2021).
62. Hahsler, M.; Piekenbrock, M.; Doran, D. dbscan: Fast Density-Based Clustering with R. *J. Stat. Softw.* **2019**, *91*. [CrossRef]

energies

Article

Determination of the Backfilling Time for the Zinc and Lead Ore Deposits with Application of the BackfillCAD Model

Krzysztof Skrzypkowski

Faculty of Mining and Geoengineering, AGH University of Science and Technology, Mickiewicza 30 av., 30-059 Kraków, Poland; skrzypko@agh.edu.pl; Tel.: +48-126-172-160

Abstract: This article introduces a BackfillCAD model that relates to the determination of the backfilling time. The relationship of the individual model modules using a flow chart are characterized and presented. The main aim of the research was to determine the time of backfilling for the prospective deposits of zinc and lead ores in the Olkusz region in Poland. In the first stage of the research, laboratory tests were carried out on the backfilling mixture consisting of sand and water in a 1:1 volume ratio. In the laboratory tests, the content of grains below 0.1 mm, the washability, water permeability, and compressibility of the backfilling mixture were determined. After the standard requirements were met by the backfilling mixture, the arrangement of one-way and bidirectional strip excavations was designed. In the next stages, by means of computer aided-design MineScape software, maximum thicknesses of the ore-bearing dolomite layer (T21_VI) for four geological cross-sections were determined. The height of the first backfilled layer with a thickness of 5 m was analyzed. Taking into account the geometrical parameters of the strip—the maximum length and its width and height, as well as the capacity of the backfilling installation—this study calculated the backfilling times for the future strip excavations.

Keywords: hydraulic sand backfilling; BackfillCAD model; MineScape; strip mining

check for **updates**

Citation: Skrzypkowski, K. Determination of the Backfilling Time for the Zinc and Lead Ore Deposits with Application of the BackfillCAD Model. *Energies* **2021**, *14*, 3186. https://doi.org/10.3390/en14113186

Academic Editor: Maxim Tyulenev

Received: 5 May 2021
Accepted: 27 May 2021
Published: 29 May 2021

Publisher's Note: MDPI stays neutral with regard to jurisdictional claims in published maps and institutional affiliations.

Copyright: © 2021 by the author. Licensee MDPI, Basel, Switzerland. This article is an open access article distributed under the terms and conditions of the Creative Commons Attribution (CC BY) license (https://creativecommons.org/licenses/by/4.0/).

1. Introduction

Underground exploitation of useful minerals poses a number of problems related to the necessity to fill various types of voids. Some of them, such as the backfilling of goafs, due to their common use in underground mines, can be classified as basic issues in the field of mining technology [1]. Feng et al. [2] stated that solid backfilling mining is a green mining technology. Other cases of the necessity to use a backfilling result from the occurrence of failures related to the existing natural hazards, in particular rock bursts, or from special conditions and technology of exploitation, e.g., the need to maintain and strengthen the roadways. Zhang et al. [3] proved that backfilling reduces the risk of roof-caving face bursts. Wang et al. [4] pointed out the correct cooperation between geogrid and backfill surfaces even for saturated sand in the case of dynamic loading. Mining backfills are also used to eliminate shallow voids that pose a threat to surface structures. Fly ashes and cement materials are added to the filling mixture in order to improve strength parameters and to counteract surface subsidence [5]. On the other hand, Wang et al. [6] stated that the use of hydraulic backfilling is associated with high costs. The variety of special applications of backfillings makes it necessary to use different types and technologies. Lingga et al. [7] discovered shear strength envelopes for cemented rockfills, which play an important role in blasthole stopping and cut and fill mining methods. Sivakugan et al. [8] highlighted cemented and uncemented backfilling strategies in underground mining methods. Li et al. [9] proposed filling the post-excavation space with waste rocks in the longwall panel. Nujaim et al. [10] studied interaction of barricade and the backfill in relation to the mining excavation. Zhou et al. [11] suggested adding glass fibers to the cemented paste backfill in order to improve its properties. Hefni et al. [12] revealed that the new solution in the form of foam mine backfilling may be used in mines. Chen et al. [13] distinguished materials for hydraulic backfilling into river sand, hill sand, gravel, and gangue and stated that river sand is the best option in order to reduce subsidence. In turn, Huang et al. [14] divided solid backfilling mining materials into gangue, fly ash, aeolian sand, loess, and mineral waste residue. In some cases, technology is used to eliminate the post-mining space in the case of a highly gaseous rock mass [15]; in others, it is required to adapt, for example, pneumatic backfilling of shallow voids, or a completely separate technology is developed, for example, to minimize the impact of underground mining on the built-up area of the site [16]. Huang et al. [17] stated that backfilling material selection plays a special role in shallow depth. Filling the post-exploitation voids is primarily aimed at equalizing the stress distribution in the whole bed loaded by roof rocks. The use of a backfilling strengthens the inter-room pillars. Mo et al. [18] strongly indicated correlation between cohesive and non-covesive backfilling in reference to coal pillar strength. Skrzypkowski [19], by means of numerical modelling, proved that sand backfilling contributes to the increase of the stability of excavations in the room and pillar methods. Backfilling is often used to extract thick deposits. Deng et al. [20] applied cemented backfilling during coal seam deposit with a thickness 21 m, which was divided into six slices. Wu [21] stated that for medium-thick ore body, losses and dilution can be reduced by backfilling, especially for sublevel open stopping methods. Furthermore, cemented paste backfilling with aeolian sand in particular is used to recovery of the residual coal pillar [22]. Wang et al. [23] projected backfilling strip-mining technique under thick unconsolidated layers. Lu et al. [24] presented backfilling mining mode for strip in two stages of exploitation and liquidation. Under certain geological and mining conditions, mining methods with backfilling allow for obtaining greater output without leaving operational losses in comparison with strip methods [25]. Zhao et al. [26] stated that backfilling is particularly important in the case of strongly inclined deposits. The role of backfilling is to improve the stability of the roof conditions. Gonet et al. [27] pointed out that the backfilling process is sometimes carried out in a mixed manner, alone and with the addition of a cement binder. Raffaldi et al. [28] marked that in the underhand cut and fill mining method, the role of backfilling plays a significant role in maintaining the stability of excavations. Dzimunya et al. [29] presented several backfilling technologies to optimize pillar recovery.

Hydraulic backfilling is one of the technologies of filling post-exploitation voids, consisting in the transport of solids (backfilling material) by means of a stream of water in pipelines to the backfilled space [30]. Sivakugan [31] stated that hydraulic backfilling is characterized by good drainage characteristics. A mixture of backfilling materials with water is called a backfill mixture. After the mixture flows into the filling space, the materials sediment and the water flows into the drainage and treatment system. The backfill process continues until the material to be filled is completely filled. During the backfill process, a failure may occur due to clogging of the pipeline. Reasons for a backfilling failure can include non-mixed items entering the backfilling, excessive mixture density, pipeline rupture or gasket failure, insufficient flushing of the pipeline prior to the backfilling process or its late completion, damage to the pipeline by a rock fall, and the formation of air bags. The above-mentioned causes significantly contribute to the increase of the backfilling time; therefore, the formation of blockage is observed by the rapid growth of the mixture in the filling funnel. For the risk of blockage to be eliminated, the backfilling material must meet the requirements of standards and regulations. According to the Polish standard [32], there are three classes of backfilling material (Table 1), for which permissible values are defined. Non-combustible and non-toxic material can be used as filling materials: sand, gravel, slag, waste rock, industrial waste, and mixtures of various backfilling materials.

Table 1. Physical properties of backfill materials based on the work of [32].

Material Class	The Content of Particles of Size Less than 0.1 mm (at Most) (%)	Maximal Dimension of Grains (mm)	Compressibility at Pressure 15 MPa (at Most) (%)	Water-Permeability (Least) (cm/s)	Washability (at Most) (%)	Bulk Density (kg/dm^3)	The Content of Visible Plant Parts
I	10		5	0.007			
II	15	60	10	0.002	20	1.3	It should not content
III	20		15	0.0004			

The literature review showed a very small number of studies on the time of backfilling underground excavations. Regardless of the type of backfilling, the tests always concern determination of geotechnical parameters in laboratory conditions, numerical modelling, or industrial research on the behavior of excavations filled with backfilling material. In underground mining, computer-aided design programs are increasingly being used, which are equipped with modules for deposit modeling, mining methods, and scheduling. However, there is still some space to be developed related to the timing of backfilling. Therefore, the article presents for the first time a research model combining analytical calculations and laboratory tests with computer-aided design using the MineScape software. For the prospective zinc and lead ore deposit, this study modeled the deposition of rock mass layers. The thickness of the ore-bearing layer was determined with special distinction, for which mining blocks from 15 to 65 m long and from 15 to 55 m high were designed. Due to the shallow deposition of the ore deposit, about 71÷135 m below the ground surface, a series of laboratory tests was carried out with filling mixture, the purpose of which was to determine the material class. For the first time, sand with a grain size of 0.5 mm was used in the tests, which was also included in the analytical calculations. The use of the MineScape program made it possible to determine the places of arrangement of preparatory and operational excavations, for which the times of backfilling of strip excavations both in one-way and bidirectionally were calculated.

2. BackfillCAD Model

A very important factor in the success of a mining project related to the time of backfilling excavations is building an interdisciplinary model and adapting it to the intended goals. An important component of a correct model is to determine the relationship between individual studies. The model named BackfillCAD consists of four main modules (Figure 1). The first module deals with building a stratigraphic model using computer-aided design. On the basis of drilling cores taken from the prospective deposit region, this study determined the types of layers that make up the rock mass and the depth of their deposition. The drill holes database was then built and imported into MineScape. By using the stratigraphic module, this study was able to arrange the individual test holes for which the cross-sections were made. The main goal for these cross-sections was to determine the thickness of the ore-bearing layer between individual exploratory drill holes. The second module concerned the determination of the geometry of the strip or room excavation that will be backfilled. After determining the maximum allowable dimensions of the designed excavation, this study started its distribution on the cross-sections made in the MineScape program. The third module is related to laboratory testing of the backfilling mixture. The purpose of the tests was to determine the class of the filling material in accordance with the standard requirements. In particular, tests defined the content of grains below 0.1 mm, in terms of the washability, water-permeability, and compressibility of the backfilling mixture. The fourth module of the BackfiillCAD model is related to the analytical calculations of the backfilling time. In particular, the difference in height between the inlet (surface coordinate) and the outlet (backfill coordinate) of the backfilling mixture must be determined. The shortest and longest flow paths of the backfilling are indicated by means of a stratigraphic model. The parameter, which is also closely related to laboratory tests, is the diameter of the sand grains, which is taken into account in the calculation of the critical velocity and the motion confidence index of the backfilling mixture. After the possibility of the motion of the backfilling mixture is determined, the next step is to determine the backfilling time.

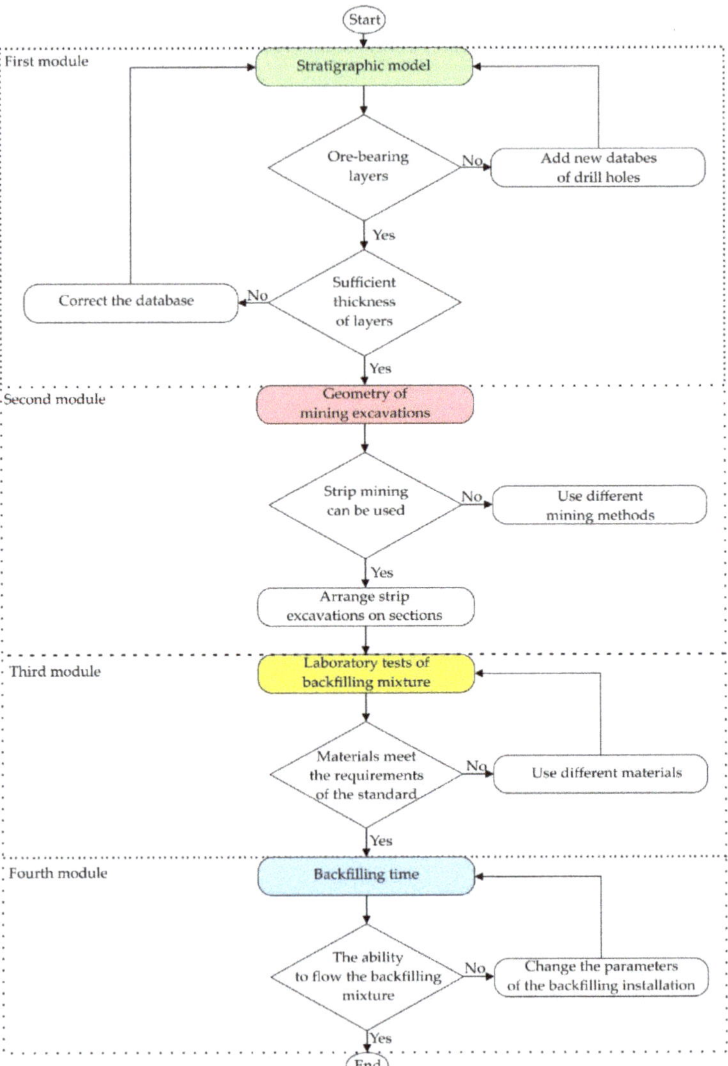

Figure 1. Flowchart for BackfillCAD model.

In designing a hydraulic backfilling for filling the post-mining space, this study's main task was to determine the efficiency of the backfilling installation, as well as the quality of the backfilling material and the flow conditions of the mixture. During the flow of the backfilling mixture, the solids are in a state of continuous or intermittent suspension, dispersed more or less evenly across the cross-section of the flows. Such a state can be achieved only during a turbulent flow of a mixture of solids and water, and the term kinetic specific gravity and volume concentration of the backfilling mixture can be used in relation to this. While the water itself can flow at a very low velocity, a flow velocity greater than the so-called critical velocity below which the solids begin to settle is necessary to keep the solids suspended in the water. The backfilling time for the post-exploitation space can be calculated according to Equation (1) [33]:

$$B_t = \frac{V_s}{C_b},\tag{1}$$

where

B_t—backfill time (h);
V_s—volume of the strip excavation (m^3);
C_b—capacity of the backfilling installation (m^3/h), according to Equation (2);

$$C_b = S \cdot w_v \cdot 3600 \cdot q,\tag{2}$$

where

S—cross-sectional area of the backfilling pipeline (m^2);
w_v—working velocity (m/s);
q—coefficient of backfilling efficiency, ($q = 1.3424$).

The value of q is the product of the coefficient 0.839, which is used in Polish underground mines, and the average specific weight of the backfilling material.

In order for it to be concluded that the movement of the mixture in the backfilling installation will be possible, the motion confidence index (M) should be calculated according to Equation (3) With the value of $M = 1$, a slight increase in local resistance causes a reduction of the operating velocity below the critical value, contributing to the disturbance of operation or even blockage of the pipeline. Therefore, it is assumed that the minimum value of the M index for sand filling should be greater than the value of 1.1.

$$M = \frac{w_v}{c_v},\tag{3}$$

where

w_v—working velocity (m/s), according to Equation (4);
c_v—critical velocity (m/s).

$$w_v = \sqrt{\frac{10^4 \cdot U_{el} + 2122.9 - 0.20578 \cdot \gamma_k}{58.1 - 0.0013 \cdot \gamma_k}},\tag{4}$$

where:

U_{el}—unit energy losses (Pa/s), according to Equation (5);

$$U_{el} = \frac{h \cdot \gamma_k \cdot \alpha}{L_0},\tag{5}$$

where

h—height difference of the inlet and outlet of the filling mixture (m);
α—hydraulic efficiency factor of backfilling installation, ($\alpha = 0.8$), (/);
L_0—equivalent length of the backfilling installation (m);
γ_k—kinetic specific gravity (N/m^3).

For the quotient of L_0 and h greater than 6.8, the kinetic specific gravity is 1.8 kN/m^3. However, for a value lower than 6.8, kinetic specific gravity is expressed in the following Equation (6):

$$\gamma_k = 22715 - 695 \cdot \omega,\tag{6}$$

where

ω—parameter characterizing the spatial arrangement of the backfilling installation (/), according to Equation (7);

$$\omega = \frac{L_0}{h},\tag{7}$$

Critical velocity (c_v) of the flow filling mixture, taking into account the kinetic specific gravity of the feed and the maximum grain dimensions (N/m^3), is expressed by the Equation (8):

$$c_v = \frac{-8.491}{d + 1.284} + 5.04,\tag{8}$$

where

d—maximum grain size (mm).

3. Laboratory Tests of the Backfilling Mixture

Laboratory tests of the backfilling mixture were performed in the laboratories of the Faculty of Mining and Geoengineering, AGH University of Science and Technology in Kraków, Poland. The main goal of the research was to define the class of the backfilling material. Quartz sand was used in the tests. In particular, the research focused on determining sand grains below 0.1 mm and calculating the washability, water-permeability, and compressibility of the backfilling mixture.

3.1. Determination of the Grains below 0.1 mm

Depending on the class of the backfilling material, the content of grains smaller than 0.1 mm should not exceed 20%. Small fractions below 0.1 mm make it difficult for water to drain from the material deposited in the dammed post-exploitation space, extending the time of material settling and increasing the compressibility of the material. In extreme cases, the material behind the dam may exist in the form of a liquid, for a long period of time, exerting a large pressure on backfilling dams and creating a risk of damaging the dams and flooding active mine workings. In addition, fine material fractions are washed away by the draining backfilling water and the field settling tanks are quickly filled with sludge. The average content of grains with dimensions below 0.1 mm was determined using an Analyzer 22 MicroTec Plus laser gauge (Idar-Oberstein, Germany), which enables the examination of particle sizes from 0.08 to 2000 μm [34]. The measurement results are presented in Table 2 and in Figure 2a–c.

Tests 1, 2, and 3 were performed on samples of the same sand. As the results of the tests were very similar, the tests were repeated only three times. From Table 2, it can be seen that the average percentage contribution of particles for 0.1 mm grain class was 1.2%, and the average cumulative percentage contribution of particles was 9.7%. The average cumulative percentage contribution of particles for a specific grain class was calculated by adding the average percentage contribution of particles to the value of average cumulative percentage contribution of particles from the previous row. On the basis of the obtained results, this study concluded that the tested sand met the requirements presented in Table 1 with regard to the grain content below 0.1 mm. For the first class of backfill material, the tested sand was characterized by an almost 10 times lower value.

3.2. Determination of the Washability of the Backfilling Material

The washability of the material was determined in a tubular container with a diameter of 150 mm and a length of 500 mm (Figure 3). The container was blind on one side and had a sealing lid on the other. After the material sample was dried at 105 °C, 2000 g of sand was weighed and mixed with water in a 1:1.5 volume ratio. This mixture was poured into a container. The closed container was then placed on a shaker at 200 strokes/min and shaken for a period of 30 min.

Table 2. Grain distribution.

Particle Class, x, (μm)	Average Cumulative Percentage Contribution of Particles in Respective Grain Class, Q3(x), (%)	Standard Deviation from Three Tests, CV, (%)	Test No. 1	Test No. 2	Test No. 3	Average Percentage Contribution of Particles in Respective Grain Class, dQ3(x), (%)
0.5	0.4	11	0.4	0.4	0.5	0.4
1	1.1	9.6	1.1	1	1.2	0.7
2	2.3	9.8	2.3	2	2.6	1.2
4	3.7	10.7	3.6	3.2	4.3	1.4
6	4.4	10.6	4.2	3.9	5	0.7
8	4.9	10.5	4.7	4.4	5.6	0.5
10	5.4	10.6	5.1	4.8	6.1	0.5
11	5.6	10.6	5.4	5	6.4	0.2
12	5.8	10.6	5.6	5.2	6.7	0.2
14	6.3	10.4	6.1	5.6	7.2	0.5
16	6.7	10.1	6.6	6	7.6	0.4
18	7.1	9.7	6.9	6.3	8	0.4
20	7.3	9.4	7.2	6.5	8.2	0.2
25	7.5	8.9	7.4	6.8	8.4	0.2
30	7.5	8.8	7.4	6.8	8.4	0.0
35	7.5	8.8	7.4	6.8	8.4	0.0
40	7.5	8.8	7.4	6.8	8.4	0.0
45	7.5	8.8	7.4	6.8	8.4	0.0
63	7.5	8.7	7.4	6.8	8.4	0.0
71	7.6	8.5	7.5	6.9	8.5	0.1
80	7.9	7.9	7.7	7.2	8.7	0.3
90	8.5	7.0	8.3	7.9	9.3	0.6
100	9.7	6.0	9.4	9.1	10.5	1.2
120	13.4	4.2	13.1	12.9	14.2	3.7
140	18.1	3.0	17.9	17.6	18.9	4.7
160	22.8	2.3	22.6	22.2	23.5	4.7
180	26.9	1.9	26.8	26.3	27.6	4.1
200	30.7	1.5	30.7	30.2	31.3	3.8
250	43.4	0.8	43.3	43.1	43.9	12.7
300	61.3	0.4	61	61.1	61.6	17.9
350	78.4	0.3	78.2	78.4	78.7	17.1
400	89.5	0.2	89.3	89.4	89.7	11.1
500	100	0	100	100	100	10.5

After the end of shaking (washing), the contents of the container were placed on a sieve with a capacity of 4 dm^3 and mesh size of 0.1 mm and washed with water. The sieve residue was dried at 105 °C and weighed to the nearest 0.01 g. The content of grains (G_c) with dimensions below 0.1 mm was calculated according to Equation (9). The results of five measurements are shown in Figure 4.

$$G_c = \frac{m - m_s}{m} \cdot 100 \ [\%] \,, \tag{9}$$

where

m—weight of the sample to be screened, (g);

m_s—weight of the residue on the sieve, (g).

The mean content of grains (G_c) smaller than 0.1 mm for a sample of 2000 g was 0.25%. The washability of the backfilling material was much less than 20%; therefore, the requirements of Table 1 were met.

3.3. Determination of the Compressibility of the Backfilling Mixture

Compressibility refers to the change in the height of the backfilling layer under load. The compressibility of the material was determined in an oedometer. Holes with a diameter of 2 mm were made in the bottom of the oedometer to allow drainage of water from the backfilling mixture. The sand was mixed with water. The mixture prepared in this way was poured into the cylinder of the oedometer (Figure 5a) to the height (h) (Figure 5b). Then, the cylinder with the piston was placed in the testing machine (Figure 5c). The sample was loaded with a load increase of about 0.04 MPa/s. The tests were performed for five samples, each time recording the change in height at a given load level. The sample was a mixture of sand and water in a 1:1 volume ratio. Sand with a grain size of up to 0.5 mm was mixed with water with a temperature of 12 °C. One day before the tests, sand and water were placed in the oedometer. After this time, the water was filtered through the sand and then the compressibility tests were started. In each case, the distance between the top plane of the backfilling mixture and the cylinder did not exceed 10 mm. The measurement results are shown in Figure 6.

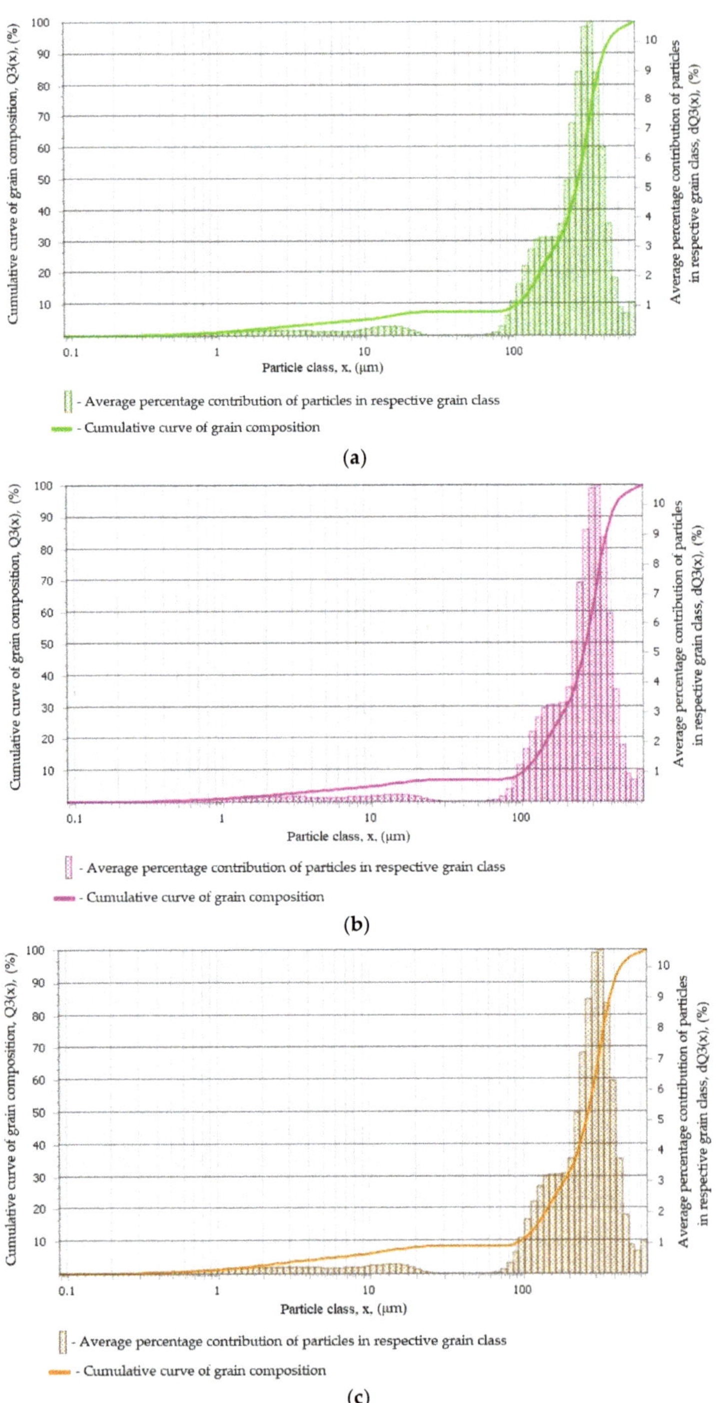

(**a**)

(**b**)

(**c**)

Figure 2. Particles distribution curve for (**a**) test no. 1; (**b**) test no. 2; (**c**) test no. 3.

Figure 3. Laboratory shaker.

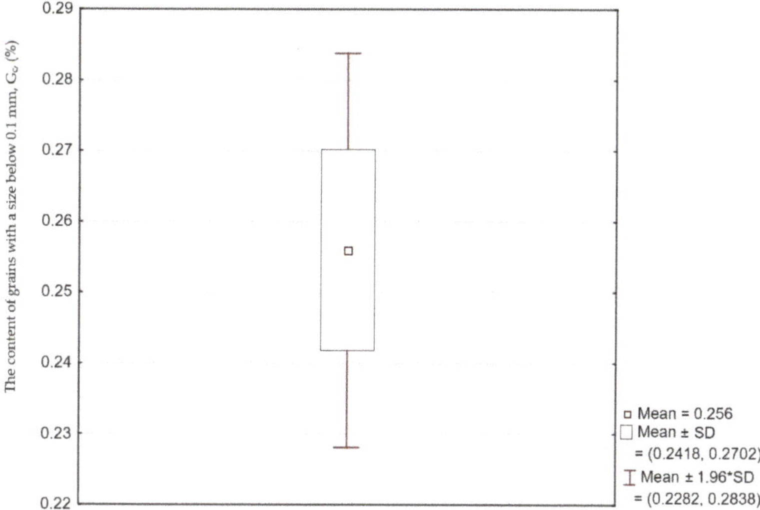

Figure 4. Content of grains less than 0.1 mm.

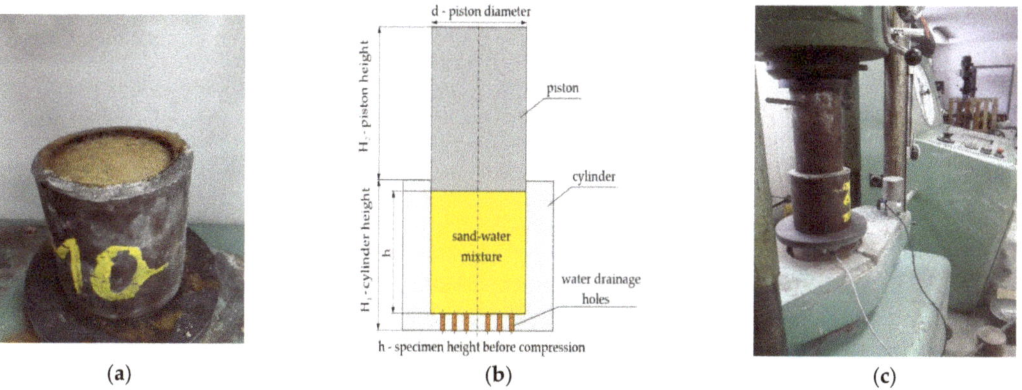

Figure 5. Compressibility test of the backfilling mixture: (**a**) general view of the filled cylinder; (**b**) elements of the oedometer; (**c**) compression test of the mixture by means of oedometer.

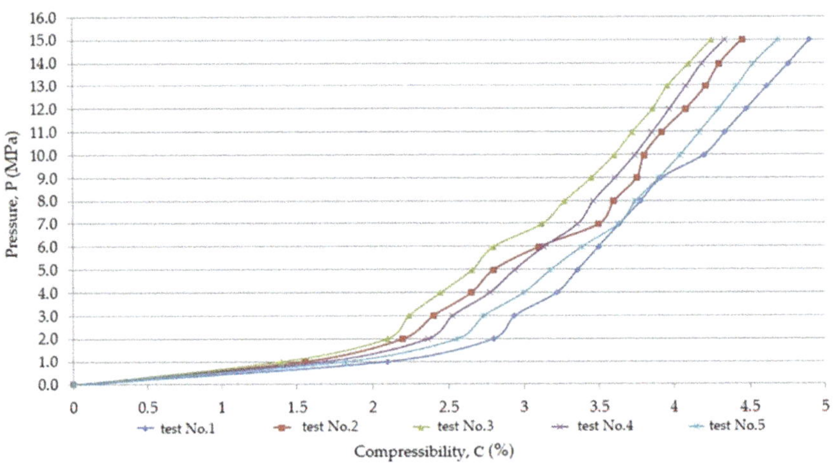

Figure 6. Pressure/compressibility characteristics of the backfilling mixture.

The average value of the compressibility of the backfilling mixture at a pressure of 15 MPa was 4.5%. Such a result means that the material was included in the first class of backfilling material, according to Table 1.

3.4. Determination of the Water Permeability of the Backfilling Material

Water permeability is the ability of water to flow through a cross-section of a backfilling mixture per unit time. It is expressed in centimeters per second. The higher the water permeability value, the faster the water drains from the sand. At high values above 0.04 cm/s, the mixture flowing out of the pipeline has a limited range of spreading in the post-exploitation space. Characteristic mounds of backfilling material form at the outlet. During the flow of the backfilling mixture, the solids are in a state of continuous or intermittent suspension, dispersing more or less evenly across the cross section of the flowing streams. This state can only be achieved during a turbulent flow of a mixture of solids (sand) and water, which is related to the volume concentration of the backfilling mixture. While the water itself can flow very slowly, a flow velocity greater than the critical velocity below which the solids (sand) begin to settle is necessary to keep the solids (sand) suspended in the water. Hence, the critical flow velocity of the backfilling mixture is essential for the reliability of the backfilling installation. The lower the water permeability values, the more favorable is the spread of the mixture and the tighter the filling of the post-mining space. With water permeability below 0.0004 cm/s, the material is difficult to filter and cannot be classified as a backfilling material. Water permeability tests were performed in a special measuring cylinder (Figure 7a,b). First, a sample of the backfilling mixture was poured into the inner cylinder. Then, after the surface of the deposited material was levelled, it was covered with a sieve with 0.5 mm square meshes and loaded with a weight of 5000 g. One hour after sample preparation, water was supplied through the inlet tube, keeping the water level 200 mm higher than the surface of the filling mixture in the cylinder. The excess water was drained off through the overflow pipe. On the other hand, the water from the outer cylinder was filtered through the tested material sample and through the drainpipe, and then it was collected in the measuring cylinder. One test lasted 60 min, during which the volume of filtered water was recorded every 10 min. The measurement results are shown in Figure 8. The water permeability was calculated according to Equation (10):

$$\mathrm{WP} = \frac{W_v}{I_c \cdot t \cdot h_p \cdot (0.7 + \alpha \cdot t_w)},\tag{10}$$

where

WP—water permeability, (cm/s);
W_v—water volume, (cm^3);
I_c—inner cylinder cross-sectional area, (cm^2);
t—test time, (s);
h_p—hydraulic drop, (/);
α—change in water viscosity with a temperature change by 1 °C, (1 / °C), according to Equation (11);

$$\alpha = 0.03 \cdot \frac{1}{t_w - 10}, \tag{11}$$

where

t_w—water temperature, (°C).

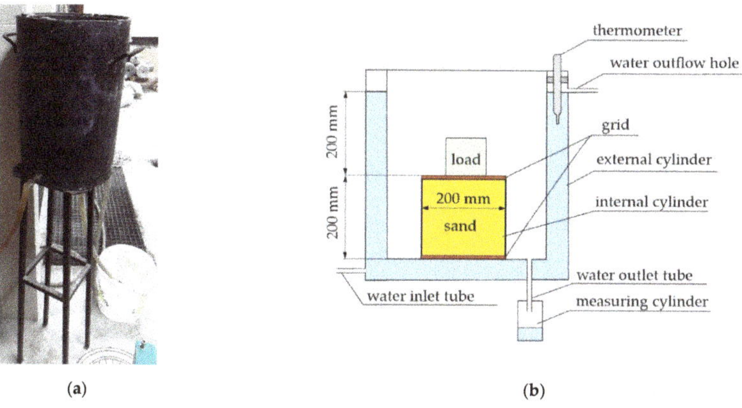

Figure 7. Measuring cylinder for testing water permeability: (**a**) general view; (**b**) cylinder components.

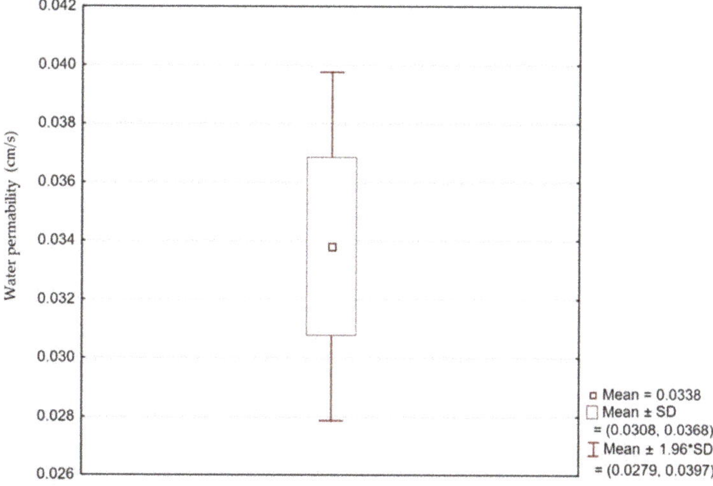

Figure 8. Mean and standard deviation for the water-permeability tests.

The mean value of the water permeability was 0.033 cm/s. The obtained value qualified the backfilling material to the first class, according to Table 1.

4. Backfilling Time for the Future Exploitation Area and Discussion

The deposits of zinc and lead ores in the Olkusz region in Poland have the form of a lens and a pocket. Only the deposits that can be exploited for economic benefit under the existing geological and mining conditions are of economic importance and industrial value. In the case of zinc and lead ore deposits, it is very important to identify the deposit on the basis of which further stages related to preparation and operation can be started. It should be considered that a deposit is not suitable for favorable exploitation if the amount of ore in the deposit is small or if the ore contains too little metal. In the case of the exploitation of zinc and lead ore deposits in Poland, the metal content often changes in individual parts of the deposit (from 1.5% to 3%). On the basis of the exploratory holes made in the Olkusz region, the distribution of individual layers of the rock mass was modeled. MineScape software (Denver, CO, United States of America) in a version 7 was used for modeling, which is more and more often used in underground mining [35]. A characteristic feature of the MineScape program is, inter alia, its modular structure, which includes creating the geometry of excavations, importing data from various graphic programs and building stratigraphic and block models. Individual modules can operate independently, contributing to the division of duties and assigning tasks for specialists responsible for planning and scheduling mining works. One of the greatest advantages of the program is the possibility of visualizing the spatial arrangement of exploratory drill holes or the entire structure of the mine, including access, preparatory excavations, and mining excavations. Moreover, the program includes functions that make it possible to model discontinuous deformations in the form of faults [36]. First, a future research area with 21 exploratory drill holes was selected. The area was 1000 m × 1000 m. For this area, drill hole coordinates were entered into MineScape (Figure 9). The ore deposit named Laski 1 is located in the Olkusz poviat in close proximity to the Pomorzany deposit. The geological resources of the deposit documented in the C1 and C2 categories total 10.76 million Mg of zinc and lead ore, including 424.86×10^3 Mg of zinc and 67.81×10^3 Mg of lead [37].

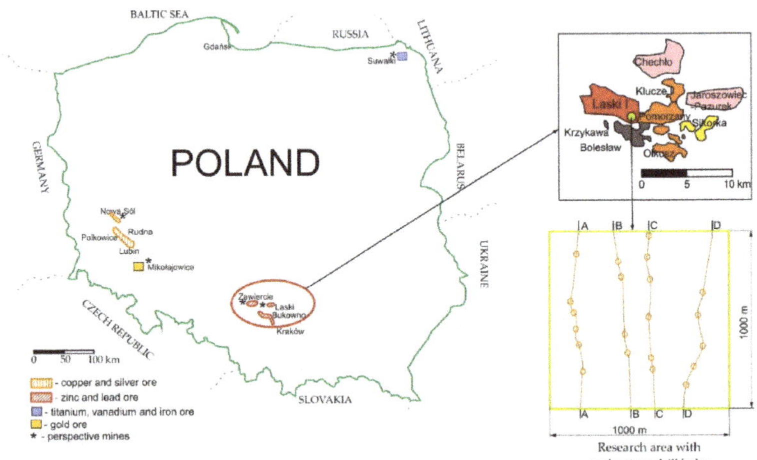

Figure 9. Research area.

In the selected research area, cross-sections through drill holes A-A, B-B, C-C, and D-D were made, on which the thicknesses of the ore-bearing dolomite layers (layer T21_VI) were marked (Figure 10a–d). Ore-bearing dolomites occur in the form of gray, fine-crystalline dolomites with an uneven fracture. In the oxidized zones, dolomites are strongly broke, and the voids and fissures are partially or completely filled with zinc, lead, and iron sulfides, as well as waste minerals such as vein calcite; dolomite; and, less frequently, barite and grenocite. In addition, the sections show mining blocks, the width of which included one strip in a one-way direction and bidirectionally, and the width of a haulage room equal to 5 m. Moreover, it was assumed that the thick orebody would be exploited in horizontal layers with a thickness of 5 m. The length of the strip (L_s), which was equal to the maximum step of backfill, was calculated according to Equation (12) [38]:

$$L_s = \sqrt{\frac{h_1 \cdot \gamma_1 + h_2 \cdot \gamma_2}{h_1 \cdot \gamma_1} + \left[\frac{b_s \cdot h_1^2}{6 \cdot (h_1 \cdot \gamma_1 + h_2 \cdot \gamma_2)} - 0.7 \right]}, \tag{12}$$

where

L_s—length of the strip excavation (backfilling length) (m);
h_1—thickness of the direct roof rocks (m);
h_2—thickness of the overburden rocks (m);
γ_1—unit weight of the direct roof (MN/m^3);
γ_2—unit weight of the overburden rocks (MN/m^3);
b_s—bending strength of the direct roof rocks (b_s = 4.62), (MPa).

For the strip with length of 30 m, this study assumed that the block width for a one-way exploitation was 35 m (Figure 11a), while for bidirectional mining it was equal to 65 m (Figure 11b). However, the width of the strip (W_s) was calculated according to Equation (13):

$$W_s = \left(2.4 \cdot ctg\delta + 0.2 \cdot \varepsilon \cdot \sqrt{\frac{C_s \cdot a_1}{\gamma}} \right) \cdot \frac{1}{cos\alpha}, \tag{13}$$

where

W_s—width of the strip excavation, (m);
δ—rock refraction angle for mines in the Olkusz region (δ = 60), (°);
ε—creep coefficient for deposit rocks (ε = 0.8), (−);
C_s—compressive strength of the direct roof (C_s = 35) (MPa);
a_1—distance between the main fissure systems (a_1 = 0.4), (m);
γ—unit weight of the direct roof (γ = 0.026) (MN/m^3);
cos—cosine trigonometric function;
α—the angle of inclination of the stratification (α = 5), (°).

The strip width was 5.11 m. For further calculations, the result was rounded down to 5 m.

Depending on the extent of dolomitization, the average thickness of ore-bearing dolomites determined in the MineScape program was very variable and ranged from 15 to 55 m. The bottom of the T21_VI layer was also located at different depths. The maximum depth of deposition was 136.8 m, while the lowest was 69.1 m. Due to the shallow deposition of the deposit, this study required the use of mining systems with roof protection. For this purpose, the mixture of sand and water, which is a hydraulic backfill, was tested in laboratory conditions. In the backfilling time calculation for the strip-mining method, this study assumed that the backfilling mixture would be transported in a 150 mm diameter backfilling pipeline. From the 130 m long backfilling shaft, the mixture would be transported by gravity to the 250 m long drift. Then, the mixture would flow to the transportation roadway, which would be connected to four inclined drifts spaced from each other every 250 m. From four inclined drifts, each 500 m long, haulage room would be made (Figure 12), from which one-way or bidirectional strip excavations would be made.

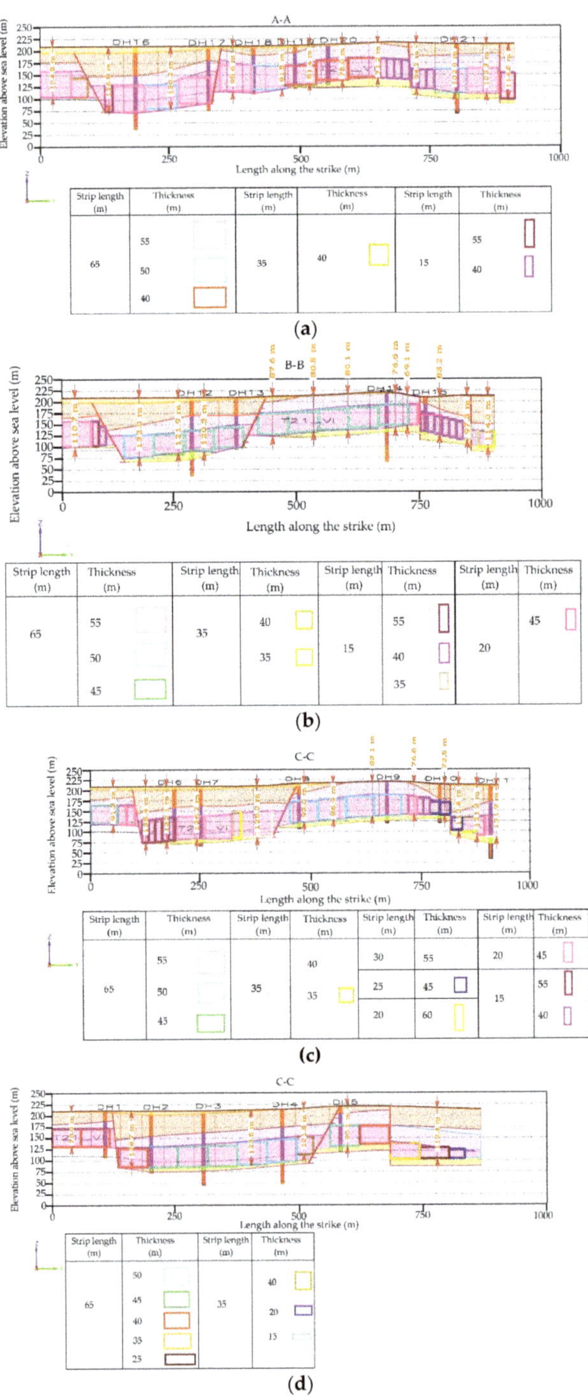

Figure 10. Cross-section through the exploratory drill holes: (**a**) A-A section; (**b**) B-B section; (**c**) C-C section; (**d**) D-D section; T21_V1—ore-bearing dolomites; DH1-21—numbers of drill holes.

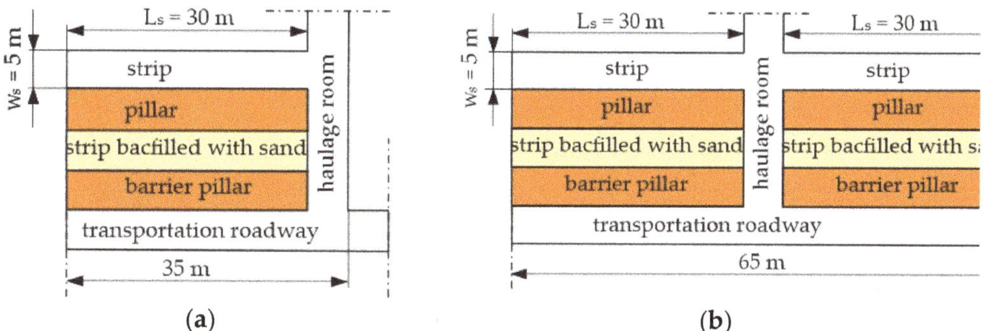

Figure 11. Block diagram of the strip-mining excavation: (**a**) one-way; (**b**) bidirectional.

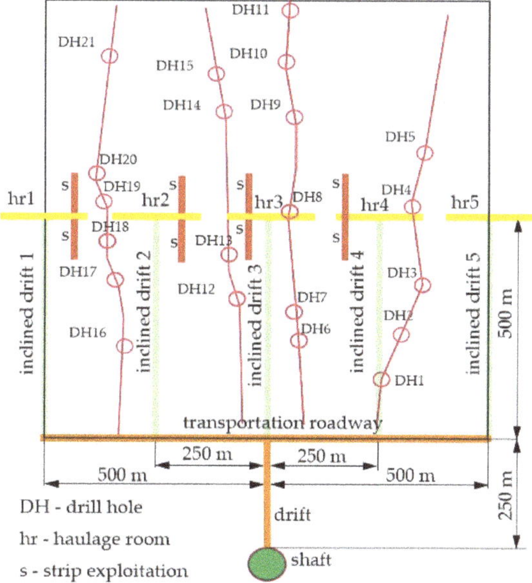

Figure 12. Structure scheme with mining excavations.

The calculations assumed that the operation would start from half of the future production block. For sections A-A, B-B, C-C, and D-D (Figure 10a–d), the mean depths of deposition were 91.5, 120.3, 93, and 123.5 m, respectively. For these depths, the backfilling times were calculated, taking into account the length of the strips 15, 20, 25, 30, 35, and 65 m. The data for the calculations were placed in Equation (1). The results of the calculations are shown in Figure 13.

Comparing the average depths of the ore-bearing dolomite layer presented in Figure 10a–d, this study found that for the sections A-A and C-C, the depths were practically the same. The difference was only 1.5 m. A similar situation can be observed for sections B-B and D-D, where the difference was only 3.2 m. It would seem that the filling times should be similar. Meanwhile, the backfilling time depended directly on the number of preparatory excavations. On the basis of Figure 11, one can conclude that the backfilling time for the driving excavation for the A-A cross-section was 24.56% greater in relation to the C-C cross-section. It was directly related to the longest distance that the backfilling mixture must travel from the shaft to the placement site. In the case of the B-B and D-D sections, the difference was significantly smaller. The backfilling time for the B-B section was

13.1% shorter than for the D-D section. Very similar times for backfilling were obtained for the B-B and C-C sections. The difference was only 2.33%. Despite the significant difference in the height of the deposit, over 27 m, the decisive parameter determining the backfilling time was the length of the installed pipelines.

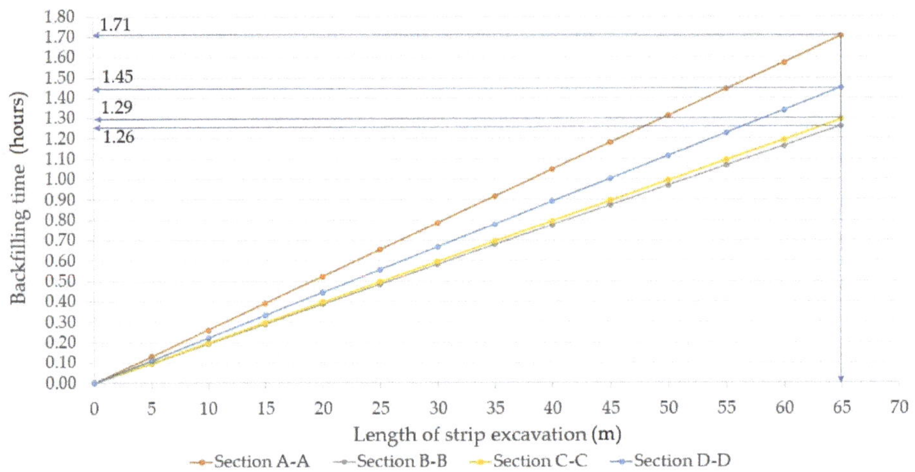

Figure 13. Time-length characteristics for average depth in relation to sections: A-A, B-B, C-C, and D-D.

Exploitation of a single layer of the deposit causes deformations of the terrain surface, the size of which is a function of the surface of the exploited deposit, depth, and thickness of the layer. For a specific method of liquidation of the post-exploitation space, the coefficients characterizing the influence of exploitation on the ground surface should be determined. In engineering calculations, this most often includes values of the maximum subsidence, slope, and horizontal deformation of the terrain. In order to calculate these values, one must know the exploitation coefficient and the angle of the main influence range. In the conditions of underground exploitation of zinc and lead ore deposits with a hydraulic backfilling in the Olkusz region, these parameters were equal to 0.15 and 60°, respectively [38]. Calculating the terrain inclination by dividing the maximum subsidence (the product of exploitation coefficient and the thickness of selected layer) and the radius range of principal influence (the quotient of the depth of exploitation and the tangent of angle of the main influence range), one finds that for the shallowest and the deepest mining of 69.1÷136.8 m (Figure 10) for selecting the first layer with a thickness of 5 m, the inclination values will be 18.79 and 9.49 mm/m for the shallowest and deepest exploitations, respectively. The permissible inclination values for the third category of surface protection ranged from 5 to 10 mm/m, while for the fifth category, they were values above 15 mm/m [39]. The fifth category of surface protection concerns undeveloped areas, i.e., forests and meadows, while the first category had the greatest restrictions (historic buildings). Mining thick deposits of zinc and lead ores lying at shallow depths requires the division of the deposit into layers so as to accurately fill the post-mining space. It is worth noting that the selection of subsequent layers is significantly delayed in time, which contributes to the gradual delamination of the roof, which is often the reason why it is difficult to select higher layers. The use of a hydraulic backfilling to use in the post-mining space undoubtedly stiffens the spatial structure and reduces the deformation of the main roof and contributes to the reduction of surface deformation. Nevertheless, the surface protection categories should be taken into account and the exploitation system should be adjusted to the limit values.

5. Conclusions

The article presents the BackfillCAD model, which is a combination of analytical laboratory tests and computer-aided design, the purpose of which is to determine the backfilling time. By combining the various stages of the research, this study specified four characteristic modules: stratigraphic, geometry of strip excavations, laboratory tests of the backfilling mixture, and the backfilling time. The original contribution to the research was the creation of a model for the backfilling process adequately to the level of deposit recognition and, above all, the performance of tests minimizing the risk associated with the incorrect selection of the backfilling material. The model provides the basis for planning and organizing the preparation of the deposit for exploitation, and therefore it has a direct impact on the method of managing the process of backfilling in post-mining areas.

On the basis of the laboratory tests for the backfilling mixture consisting of sand with a grain size of 0.5 mm and water in a 1:1 volume ratio, this study concluded that

- The average percentage contribution of particles for 0.1 mm grain class was 1.2% and the average cumulative percentage contribution of particles was 9.7%;
- The average value of the compressibility of the backfilling mixture at a pressure of 15 MPa was 4.5%;
- The mean value of the water-permeability was 0.033 cm/s.

On the basis of the design research using the MineScape program for the prospective zinc and lead ore deposit in the Olkusz region in Poland, this study concluded that

- The minimum and maximum depths of ore-bearing dolomites modeled in the MineScape program ranged from 69.1 to 136.8 m;
- Deposit thickness was very variable and ranged from 15 m to 55 m;
- The maximum length of the strip excavation was 35 m.

The time to fill in was found to be most influenced by height difference of the mixture inlet and outlet of the backfill; equivalent length of the backfilling installation; kinetic-specific gravity; and unit energy losses, working velocity, critical velocity, and maximum grain size. Calculated backfilling times did not take into account the time associated with shortening the pipeline as the backfilling was filled. In addition, when calculating the total time needed to eliminate the selected post-mining space, one should also take into account the technological processes related to the flushing time of the backfilling installation, which takes into account the actual length and diameter of the pipeline and the construction of backfilling dams. The presented BackfillCAD model can be useful at the planning and scheduling stage for both operating mining plants and, above all, for new prospective deposits of mineral resources.

Funding: This research was prepared as part of AGH University of Science and Technology in Poland, scientific subsidy under number: 16.16.100.215.

Data Availability Statement: The data presented in this study are new and have not been previously published.

Conflicts of Interest: The author declares no conflict of interest.

References

1. Huang, Y.; Zhang, J.; Yin, W.; Sun, Q. Analysis of Overlying Strata Movement and Behaviors in Caving and Solid Backfilling Mixed Coal Mining. *Energies* **2017**, *10*, 1057. [CrossRef]
2. Feng, J.; Peng, H.; Shuai, G.; Meng, X.; Lixin, L. A roof model and its application in solid backfilling mining. *Int. J. Min. Sci. Technol.* **2017**, *27*, 139–143. [CrossRef]
3. Zhang, J.; Li, B.; Zhou, N.; Zhang, Q. Application of solid backfilling to reduce hard-roof caving and longwall coal face burst potential. *Int. J. Rock Mech. Min. Sci.* **2016**, *88*, 197–205. [CrossRef]
4. Wang, L.; Chen, G.; Chen, S. Experimental study on seismic response of geogrid reinforced rigid retaining walls with saturated backfill sand. *Geotext. Geomembr.* **2015**, *43*, 35–45. [CrossRef]
5. Bai, E.; Guo, W.; Tan, Y.; Yang, D. The analysis and application of granular backfill material to reduce surface subsidence in China's northwest coal mining area. *PLoS ONE* **2018**, *13*, e0201112. [CrossRef]

6. Wang, F.; Ma, Q.; Li, G.; Wu, C.; Guo, G. Overlying Strata Movement Laws Induced by Longwall Mining of Deep Buried Coal Seam with Superhigh-Water Material Backfilling Technology. *Adv. Civ. Eng.* **2018**, *2018*, 1–10. [CrossRef]
7. Lingga, B.A.; Apel, D.B. Shear properties of cemented rockfills. *J. Rock Mech. Geotech. Eng.* **2018**, *10*, 635–644. [CrossRef]
8. Sivakugan, N.; Rankine, R.M.; Rankine, K.J.; Rankine, K.S. Geotechnical considerations in mine backfilling in Australia. *J. Clean. Prod.* **2006**, *14*, 1168–1175. [CrossRef]
9. Li, J.; Yin, Z.Q.; Li, C.M. Waste rock filling in fully mechanized coal mining for goaf-side entry retaining in thin coal seam. *Arab. J. Geosci.* **2019**, *12*, 509. [CrossRef]
10. Nujaim, M.; Belem, T.; Giraud, A. Experimental Tests on a Small-Scale Model of a Mine Stope to Study the Behavior of Waste Rock Barricades during Backfilling. *Minerals* **2020**, *10*, 941. [CrossRef]
11. Zhou, N.; Du, E.; Zhang, J.; Zhu, C.; Zhou, H. Mechanical properties improvement of Sand-Based cemented backfill body by adding glass fibers of different lengths and ratios. *Constr Build Mater.* **2021**, *280*, 122408. [CrossRef]
12. Hefni, M.; Hassani, F. Experimantal development of a novel mine backfill material: Foam mine fill. *Minerals* **2020**, *10*, 564. [CrossRef]
13. Chen, S.; Yin, D.; Cao, F.; Liu, Y.; Ren, K. An overview of integrated surface subsidence-reducing technology in mining areas of China. *Nat Hazards* **2016**, *81*, 1129–1145. [CrossRef]
14. Huang, P.; Zhang, J.; Yan, X.; Spearing, A.J.S.; Li, M.; Liu, S. Deformation response of roof in solid backfilling coal mining based on viscoelastic properties of waste gangue. *Int. J. Min. Sci. Technol.* **2021**, *31*, 279–289. [CrossRef]
15. Zhang, J.; Zhang, Q.; Spearing, A.J.S.; Miao, X.; Guo, S.; Sun, Q. Green coal mining technique integrating mining-dressing-gas draining-backfilling-mining. *Int. J. Min. Sci. Technol.* **2017**, *27*, 17–27. [CrossRef]
16. Pu, H.; Zhang, J. Research on protecting the safety of buildings by using backfill mining with solid. *Procedia Environ. Sci.* **2012**, *12*, 191–198. [CrossRef]
17. Huang, P.; Spearing, S.; Ju, F.; Jessu, K.V.; Wang, Z.; Ning, P. Control Effects of Five Common Solid Waste Backfilling Materials on In Situ Strata of Gob. *Energies* **2019**, *12*, 154. [CrossRef]
18. Mo, S.; Canbulat, I.; Zhang, C.; Oh, J.; Shen, B.; Hagan, P. Numerical investigation into the effect of backfilling on coal pillar strength in highwall mining. *Int. J. Min. Sci. Technol.* **2018**, *28*, 281–286. [CrossRef]
19. Skrzypkowski, K. Decreasing Mining Losses for the Room and Pillar Method by Replacing the Inter-Room Pillars by the Construction of Wooden Cribs Filled with Waste Rocks. *Energies* **2020**, *13*, 3564. [CrossRef]
20. Deng, X.J.; Zhang, J.X.; Zhou, N.; Wit, B.; Wang, C.T. Upward slicing longwall-roadway cemented backfilling technology for mining an extra-thick coal seam located under aquifers: A case study. *Environ. Earth Sci.* **2017**, *76*, 789. [CrossRef]
21. Wu, J. Research on sublevel open stoping recovery processes of inclined medium thick orebody on the basis of physical simulation experiments. *PLoS ONE* **2020**, *15*, e0232640. [CrossRef]
22. Zhou, N.; Yan, H.; Jiang, S.; Sun, Q.; Ouyang, S. Stability Analysis of Surrounding Rock in Paste Backfill Recovery of Residual Room Pillars. *Sustainability* **2019**, *11*, 478. [CrossRef]
23. Wang, F.; Jiang, B.; Chen, S.; Ren, M. Surface collapse control under thick unconsolidated layers by backfilling strip mining in coal mines. *Int. J. Rock Mech. Min. Sci.* **2019**, *113*, 268–277. [CrossRef]
24. Lu, B.; Li, Y.; Fang, S.; Lin, H.; Zhu, Y. Cemented Backfilling Mining Technology for Gently Inclined Coal Seams Using a Continuous Mining and Continuous Backfilling Method. *Shock. Vib.* **2021**, *2021*, 1–12. [CrossRef]
25. Qiang, Z.; Jixiong, Z.; Shuai, G.; Rui, G.; Weikang, L. Design and application of solid, dense backfill advanced mining technology with two pre-driving entries. *Int. J. Min. Sci. Technol.* **2015**, *25*, 127–132. [CrossRef]
26. Zhao, T.; Zhang, Z.; Yin, Y.; Tan, Y.; Liu, X. Ground control in mining steeply dipping coal seams by backfilling with waste rock. *J. S. Afr. Inst. Min. Metall.* **2018**, *118*, 15–26. [CrossRef]
27. Gonen, A.; Kose, H. Stability analysis of open stopes and backfill in longhole stoping method for Asikoy underground copper mine. *Arch. Min. Sci.* **2011**, *56*, 375–387.
28. Raffaldi, M.J.; Seymour, J.B.; Richardson, J.; Zahl, E.; Board, E. Cemented Paste Backfill Geomechanics at a Narrow-Vein Underhand Cut-and-Fill Mine. *Rock. Mech. Rock. Eng.* **2019**, *52*, 4925–4940. [CrossRef]
29. Dzimunya, N.; Radhe, K.; Chanda, M.; William, C.M. Design and dimensioning of sublevel stoping for extraction of thin ore (<12 m) at very deep level: A case study of konkola copper mines (kcm), Zambia. *Math. Model. Eng. Probl.* **2018**, *5*, 27–32. [CrossRef]
30. Zhang, Q.L.; Hu, G.Y.; Wang, X.M. Hydraulic calculation of gravity transportation pipeline system for backfill slurry. *J. Cent. South Univ. Technol.* **2008**, *15*, 645–649. [CrossRef]
31. Sivakugan, N.; Veenstra, R.; Naguleswaran, N. Underground Mine Backfilling in Australia Using Paste Fills and Hydraulic Fills. *Int. J. Geosynth. Ground Eng.* **2015**, *1*, 1–18. [CrossRef]
32. Polish Committee for Standardization. *Polish Standard: PN-G-11010*; Mining—Materials for hydraulic backfill—Requirements and tests; Polish Committee for Standardization: Warszawa, Poland, 1993. (In Polish)
33. Piechota, S. *Technika Podziemnej Eksploatacji złóż i Likwidacji Kopalń (Technique of Underground Mining Deposits and Liquidation of Mines)*; AGH Publishing House: Kraków, Poland, 2008; pp. 252–256. (In Polish)
34. Laser Particle Sizer ANALYSETTE 22 MicroTec Plus. Available online: https://www.ulprospector.com/en/asia/Coatings/Detail/6647/204545/Laser-Particle-Sizer-ANALYSETTE-22-MicroTec-Plus (accessed on 17 May 2021).
35. MineScape. Available online: https://www.dataminesoftware.com/solutions/minescape-geological-modelling-mine-planning (accessed on 18 May 2021).

36. Skrzypkowski, K. The Influence of Room and Pillar Method Geometry on the Deposit Utilization Rate and Rock Bolt Load. *Energies* **2019**, *12*, 4770. [CrossRef]
37. Polish Geological Institute, National Research Institute. Available online: https://www.pgi.gov.pl/oferta-inst/wydawnictwa/serie-wydawnicze/bilans-zasobow-kopalin.html (accessed on 18 May 2021).
38. *Geological Documentation of the Olkusz-Pomorzany Mine*; Mining and Metallurgy Plant: Bukowno, Poland, 2020. (In Polish)
39. Kowalski, A. *Surface Deformation in the Upper Silesian Coal Basin*; Central Mining Institute Publisher: Katowice, Poland, 2015; p. 105. (In Polish)

Article

Failure Analysis of a Pre-Excavation Double Equipment Withdrawal Channel and Its Control Techniques

Chen Li, Xiaofei Guo *, Xiaoyong Lian and Nianjie Ma

School of Energy & Mining Engineering, China University of Mining and Technology (Beijing), Beijing 100083, China; bqt1800101008@student.cumtb.edu.cn (C.L.); bqt2000101013@student.cumtb.edu.cn (X.L.); 108031@cumtb.edu.cn (N.M.)

* Correspondence: b201901@student.cumtb.edu.cn; Tel.: +86-152-0129-0185

Received: 16 November 2020; Accepted: 30 November 2020; Published: 2 December 2020

Abstract: The use of pre-excavation equipment withdrawal channels (EWCs) at the stop-production line is important for the rapid withdrawal of coal mining equipment. However, during the final mining period, the dynamic pressure of a pre-excavated double EWC is severe, which leads to instability of the surrounding rock around the EWCs. Therefore, in this paper, the methods of field monitoring, theoretical analysis, and numerical simulation were used to systematically study the stress and plastic zone evolution of a double EWC during the final mining period. Firstly, the distribution characteristics of mining abutment pressure and roadway failure modes under the action of mining abutment pressure were analyzed theoretically. Afterward, a FLAC3D mining numerical model was established according to the distribution of rock strata obtained from roof detection. Finally, the evolution laws of the stress fields and plastic zones of the EWCs during final mining were obtained by numerical simulation. The present study suggests that asymmetric stress distribution dominates asymmetric failure of the surrounding rock around the EWCs during the final mining period, and deformation failure within 10 m from the working face to the main EWC (MEWC) accounted for most of the roadway deformation. Based on the research results combined with actual production experience, the stability control technique of the surrounding rock with reinforcement of anchor cables and double-row buttress hydraulic support for the MEWC was put forward. After the field application, the ideal result was obtained.

Keywords: equipment withdrawal channel; stress distribution; plastic zone; surrounding rock control

1. Introduction

Underground mining of coal accounts for more than 75 percent of China's current coal production, and the inevitable part of underground mining is the withdrawal of mining equipment when the work is finished [1,2]. A pre-excavation double equipment withdrawal channel (EWC) has two EWCs that are arranged near the stop-production line while preparing the working face: There is a main equipment withdrawal channel (MEWC) and an auxiliary withdrawal channel (AEWC), as shown in Figure 1. The EWC is excavated and formed by using an EBZ135A integrated mechanized roadheader, which made in XuZhou Construction Machinery Group, Xuzhou China (as shown in Figure 2). This type of roadheader has the advantages of fast excavation speed, high efficiency, and small disturbance to surrounding rock. The speed of the equipment's safe withdrawal is directly related to the profit of a coal mine. To shorten the withdrawal time of the mining equipment and relieve the tension of replacing a working face, more and more coal mines use the layout of pre-excavation EWCs [3].

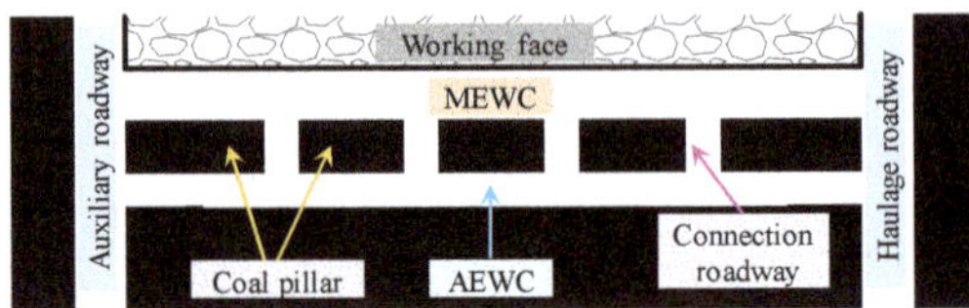

Figure 1. Schematic diagram of the pre-excavation equipment withdrawal channels (EWCs). MEWC: main equipment withdrawal channel; AEWC: auxiliary equipment withdrawal channel.

Figure 2. EBZ135A integrated mechanized roadheader.

In the process of underground mining of coal, the front of the working face is always affected by moving abutment pressure [4–6]. Advanced microseismic monitoring equipment can obtain the distribution characteristics of the stress field around goaf intuitively [7,8]. Some scholars have studied and determined the variables affecting stress redistribution in the supercritical longwall formation, which is of great significance to the study of the stress field [9,10]. Kang et al. studied abnormal stress under the coal pillar with residual support and obtained its distribution law, which played a major role in the optimization of roadway layout [10]. From another point of view, there is a certain correlation between the coal pillar size and the final mining period of the EWC. By means of numerical calculation, Yang et al. obtained the stress distribution state and plastic zone failure characteristics of roadway-surrounding rocks under different pillar widths. The results show that the stress is in the decreasing zone when the coal pillar is less than 8 m, which is consistent with the distribution of the stress field in the EWC zone during the final mining [11,12]. The research on the influence of rock bolts on roof stability under the action of abutment pressure makes it more clear that bolt support is important for roadway support [13–15]. The surrounding rock is affected by the mining and will inevitably undergo deformation and failure, and the related tests also proved the failure rules of rock by abutment pressure [16,17]. Some scholars have obtained the deformation and failure mechanism of the surrounding rock of mining roadways by combining stress distribution under the influence of mining with the failure mode of the roadway [18–20]. The study on the mechanical properties of the rock and the boundary equation of the roadway plastic zone under non-constant pressure conditions further reveals the evolution mechanism of the plastic zone under mining [21,22]. Many scholars have studied stability control measures of the surrounding rock [23–26], such as the surrounding rock control of super-large section tunnels, roadway support in special structural zones, reinforcement support of dynamic pressure roadways, and so on. In addition, numerical simulation analysis plays an increasingly important role in the field of geotechnical engineering [27]. Many scholars have obtained rock failure mechanics theories based on numerical simulation analysis and applied the results to guide engineering practice [28–32].

Previous studies are important for investigating the mechanism of dynamic disasters in the roadway. However, the Lijiahao Coal Mine has not studied the continuous dynamic pressure disturbance of the pre-excavated double EWC. The layout of the pre-excavation double EWC plays a crucial role in production improvement. In the final mining period, the speed of advancement is slowed down, which leads to an increase of mining pressure and a large deformation of the surrounding rock. Faced with such a situation, the distribution characteristics of the moving stress field and the plastic failure mode of the surrounding rock around EWCs is not well understood. Therefore, this paper systematically studied the dynamic evolution law of the regional stress field during the final mining period and the distribution characteristics of the plastic zone during the service period of the EWCs.

2. Engineering Background

2.1. Geological Survey

The Lijiahao Coal Mine is located in the south-central region of the Dongsheng coalfield of Ordos city, south of the Inner Mongolia Autonomous Region in China, with an annual output of 6 Mt. The coal seams in the mining area has the characteristics of shallow burial, complex overlying bedrock structure, thin and soft bedrock. The research object was panel 22116 of the 2-2 middle coal seam in the Lijiahao Coal Mine. The length of the working face is 300 m. The thickness of the coal seam is between 2.88 and 3.21 m, averaging 3.0 m. The average depth of the coal seam is 185 m. The partial roof contains gangue. Generally speaking, the geological conditions are relatively complex. The roof and floor of the coal seam are dominated by mudstone and sandy mudstone. The roof has the characteristics of weak rock, low rock strength, unstable occurrence, and poor stability [33].

2.2. Project Profile

The distance between the MEWC and AEWC of panel 22116 was 25 m. The combined supports of anchor bolts and anchor cables was adopted in the EWC. According to previous engineering experience, the double-row buttress hydraulic support should be installed when the working face is about 300 m away from the MEWC. The MEWC and AEWC sections (width × height) of panel 22216 were 5.2 × 3.0 m and 5.5 × 3.0 m, respectively. The initial support system of the MEWC used anchor cables as "3-2-3" with 2000 mm spacing and 2000 mm row spacing, and the diameter and length of the anchor cables were 17.8 mm and 8000 mm, respectively. The rebar bolts were 20 mm in diameter and 2500 mm in length and were used in the roof with a spacing of 960 mm between bolts and spacing of 1000 mm between rows. The initial support system of the AEWC used anchor cables as "2-2" with 2000 mm spacing and 2000 mm row spacing, and the diameter and length of anchor cables were 17.8 mm and 8000 mm, respectively. The rebar bolts were 20 mm in diameter and 2500 mm in length and were used in the roof in a grid pattern of 1000 mm × 1000 mm. The ribs bolts of both the MEWC and AEWC were used with a spacing of 800 mm between bolts and spacing of 1000 mm between rows. The rib bolts of the MEWC near the working face were fiberglass reinforced plastic (FRP) bolts with 22 mm in diameter and 2000 mm in length and the rib bolts were rod bolts with 16 mm diameter and 2200 mm length. The supporting section is shown in Figure 3.

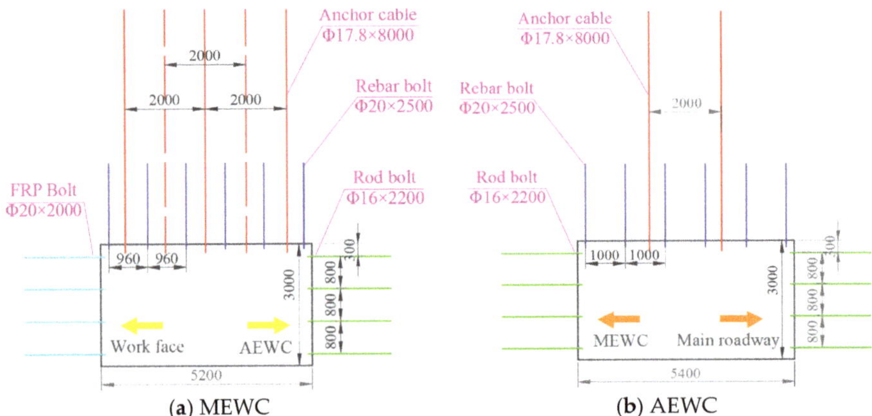

Figure 3. Support profile of the EWCs (mm).

2.3. Roof Structure Detection

The roof structure and fracture development of the MEWC and AEWC were monitored by using a TYGD10 rock drilling detector when the working face was 200 m away from the MEWC. The borehole was designed to have a depth of 8 m and a diameter of 28 mm. The roof of MEWC and AEWC were evenly arranged with 6 and 4 monitoring boreholes, respectively. The roof structure obtained by borehole detection of the MEWC is shown in Figure 4.

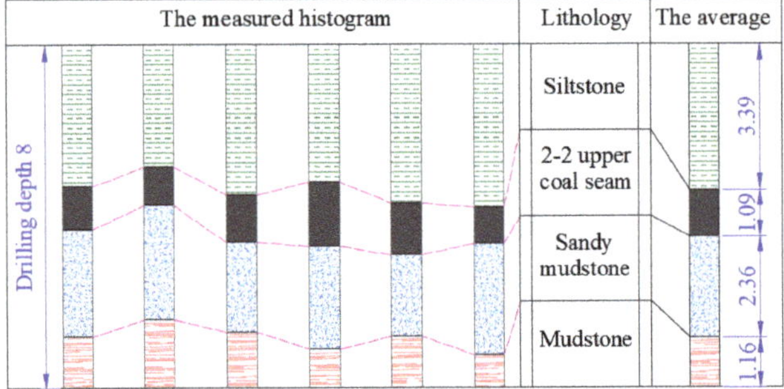

Figure 4. Detection results of the MEWC roof structure (m).

The following information was obtained from the borehole probe. The range of 0–1.16 m above the roof of the MEWC and AEWC is mudstone with little lithologic change. Most roof fractures developed in this section, and it is the main failure zone of the roof. The range of 1.16–3.52 m above the roof is sandy mudstone. The surrounding rock of this section is complete and partially contains coal lines. The range of 3.52–4.61 m above the roof is the 2-2 upper coal seam. The roof above 4.61 m is siltstone with good rock integrity. The representative screenshot of borehole detection is shown in Figure 5. The mechanical parameters of the rock strata of the EWC are shown in Table 1.

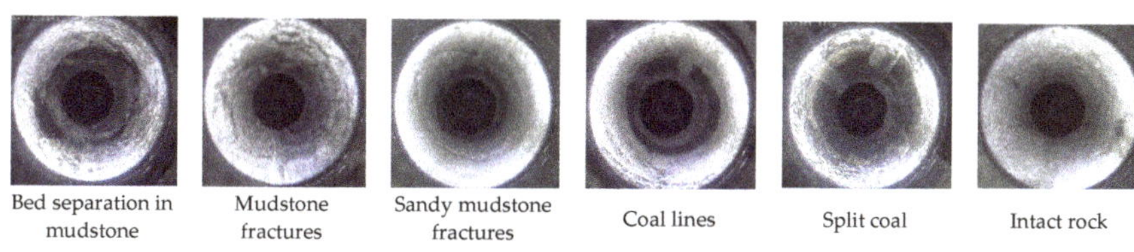

| Bed separation in mudstone | Mudstone fractures | Sandy mudstone fractures | Coal lines | Split coal | Intact rock |

Figure 5. Borehole imaging screenshot.

Table 1. Lithology and rock mechanics parameters.

Lithology	Density (kg/m^3)	Bulk Modulus /10^3 MPa	Shear Modulus /10^3 MPa	Friction Angle/(°)	Cohesion /MPa	Tensile Strength/MPa
Fuyan	2500	3.18	1.6	32	2.21	1.22
Sandy-mudstone 5	2400	3.68	1.8	35	2.63	1.13
Siltstone 2	2600	2.92	1.9	31	1.52	1.08
Fine-sandstone 3	2500	2.52	1.7	34	2.53	1.17
Siltstone 1	2600	3.91	1.9	30	1.94	1.13
2-2 upper-coal	1400	1.89	0.6	25	1.56	0.93
Sandy-mudstone 4	2200	2.76	1.6	32	1.72	1.26
Mudstone	2300	1.76	0.8	27	1.44	0.86
2-2 middle-coal	1500	1.89	0.6	25	1.52	0.93
Sandy-mudstone 3	2400	3.81	2.2	30	1.83	1.02
Fine-sandstone 2	2400	3.66	1.8	28	1.76	1.15
Sandy-mudstone 2	2500	2.53	2.7	32	2.11	1.21
Fine-sandstone 1	2400	3.68	1.8	31	1.92	1.12
Sandy-mudstone 1	2500	2.53	1.7	33	2.52	1.06

3. Theoretical Analysis

3.1. Analysis of the Stress Field in the Mining Area

The stress field around the goaf is redistributed after the coal is extracted [34,35], as shown in Figure 6. The tangential stress (σ_t) increases sharply in front of the working face, then decreases gradually after reaching a peak, and finally tends to the in situ stress. This is consistent with the distribution law of vertical stress [1]. The radial stress (σ_r) is zero at the free surface of the goaf coal wall and increases gradually along the front of the coal wall. Finally, it also tends to in situ stress, and the overall variation range is small. This is similar to the horizontal stress distribution in mining. According to the existing research results, the coefficient abutment pressure K in front of the working face can reach 3–5 [1,6]. Therefore, the high deviational stress area is formed in a certain range in front of the working face [16].

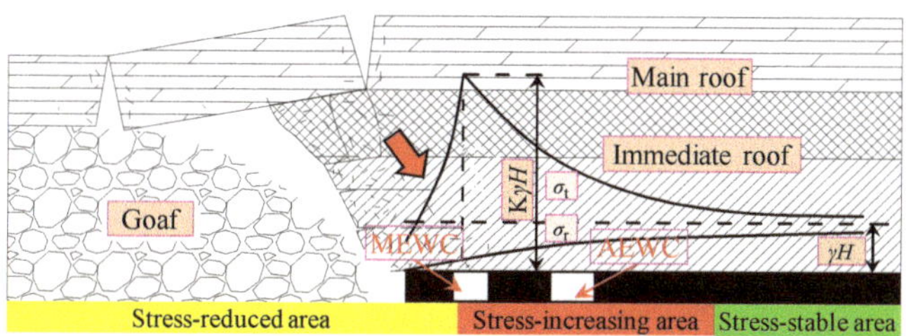

Figure 6. Abutment pressure model in front of the working face.

With the advance of the working face, the immediate roof of the goaf caves in directly and the main roof breaks. It leads to stress release in the upper part of goaf and stress transfer around goaf. Because the collapse and fracture of the overlying strata of the goaf are not synchronous, in other words, the failure of the immediate roof and main roof are asynchronous, the rock blocks are hinged and twisted after the roof is broken, and a huge thrust is formed in the oblique direction of goaf side, as shown in Figure 6. This causes the regional stress field of the surrounding rock to deflect. However, the range of the stress field affected by the above situation is limited. Therefore, the spatial position relationship between the EWC and the goaf dominates the distribution law of the regional stress field and the evolution of the surrounding rock failure mode during final mining.

3.2. Failure Pattern of the Roadway under Asymmetric Pressure

The modified Fenner formula or Kastner formula applies to bidirectional isobaric stress field conditions, i.e., lateral coefficient (λ) is 1. Therefore, the obtained plastic boundary of a circular hole is no longer applicable to the mining roadway [36]. Based on the non-uniform stress environment of mining, the recessive equation of the surrounding rock plastic boundary is obtained by using the theory of elastic mechanics and the Mohr–Coulomb criterion [18,36,37], as shown in Equation (1). The mechanical model of a non-isobaric circular roadway and its plastic zone is shown in Figure 7.

$$\begin{aligned}
9(1-\lambda)^2(\tfrac{a}{R_0})^8 &- [12(1-\lambda)^2 + 6(1-\lambda^2)\cos 2(\theta-\alpha)](\tfrac{a}{R_0})^6 + \Big\{2(1-\lambda)^2\big[\cos^2 2(\theta-\alpha)(5-2\sin^2\varphi) - \sin^2 2(\theta-\alpha)\big] \\
&+ (1+\lambda)^2 + 4(1-\lambda^2)\cos 2(\theta-\alpha)\Big\}(\tfrac{a}{R_0})^4 - [4(1-\lambda)^2\cos 4(\theta-\alpha) + 2(1-\lambda^2)\cos 2(\theta-\alpha)(1-2\sin^2\varphi) \\
&- \tfrac{4}{P}(1-\lambda)C\cos 2(\theta-\alpha)\sin 2\varphi](\tfrac{a}{R_0})^2(1-\lambda)^2 - \sin^2\varphi\Big(1+\lambda+\tfrac{2C}{P}\tfrac{\cos\varphi}{\sin\varphi}\Big)^2 = 0
\end{aligned} \tag{1}$$

where a is the radius of the roadway; R_0 is the boundary radius of the plastic zone; α is the vertical deflection angle of the maximum principal stress; φ and C are the internal friction angle and cohesion of rock media, respectively; and P is the vertical load of the roadway.

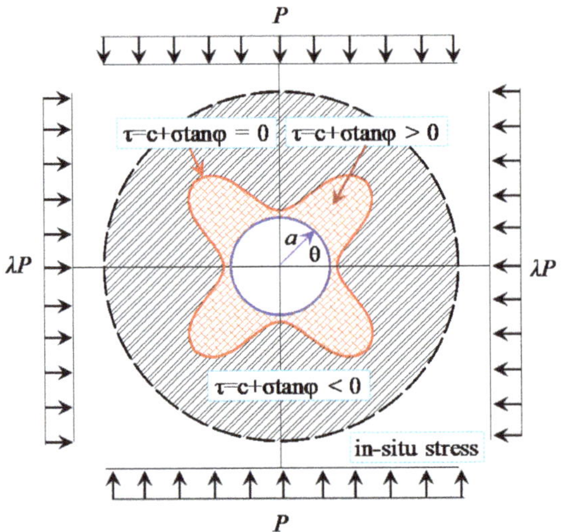

$$P = 20 \text{ MPa}, \ \lambda = 0.4, \ a = 2 \text{ m}, \ C = 2.7 \text{ MPa}, \ \varphi = 24°, \ \alpha = 0$$

Figure 7. Mechanical model of the circular roadway.

Matlab was used to calculate the plastic zone of the circular roadway using Equation (1), and FLAC^{3D} was used to simulate the shapes of different roadway cross-sections, as shown in Figure 8. All the initial parameters in Figure 8 remain unchanged except stress state and roadway cross-section shape.

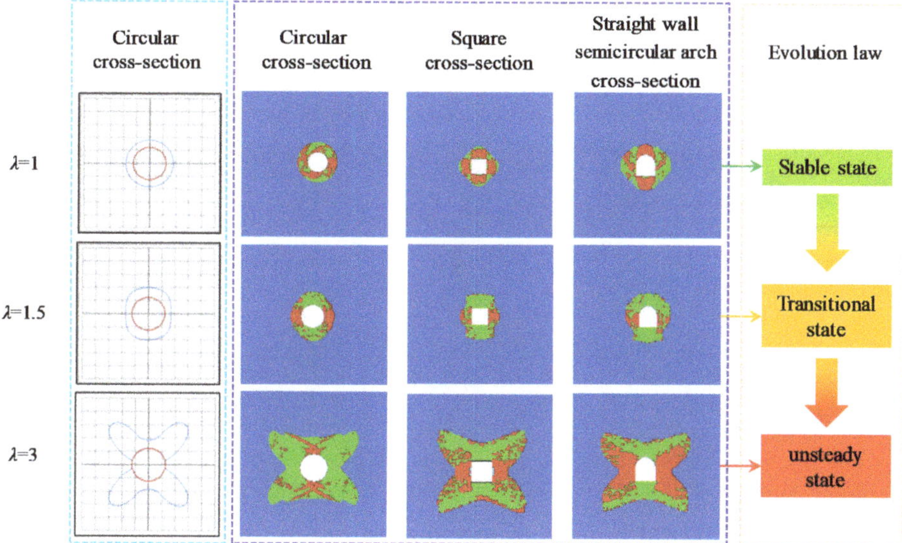

Figure 8. Evolution of the rock failure patterns surrounding the roadway.

The evolution law of the plastic zone of the roadway under different stress states was obtained by theoretical calculation and numerical simulation. It can be seen from Figure 8 that the theoretical calculation is consistent with the numerical simulation results [38]. Under the condition of bidirectional isostatic stress, the plastic zone of the surrounding rock is circular, and the roadway is in a stable state. When there is little difference in pressure values between the two directions, the surrounding rock plastic zone of the roadway evolves into an approximate elliptic shape, and the roadway is in a transitional state. Subsequently, as the pressure ratio of the two directions continues to increase, the boundary of the plastic zone presents a butterfly-shaped distribution, and the roadway is in a state of instability. Different cross-section shapes of roadway have the same response to the same stress state, and there is little difference in the morphology and evolution law of the plastic zone under different cross-sections.

4. Numerical Simulation

FLAC^{3D} is a simulation software based on three-dimensional fast Lagrange difference analysis. It can simulate stress characteristics of the three-dimensional structure of soil and rock, which play an increasingly important role in the field of geotechnical engineering [21,39]. To obtain the stress distribution law in front of the working face during mining, a large FLAC^{3D} numerical model was established. The length, width, and height of the model were 500, 400, and 150 m, respectively. The grid before and after the MEWC was divided into 0.5 m, and the grid in other areas was divided into 1–5 m. The model fixed the lower boundary and the surrounding boundary, and the upper boundary compensated for the 2 MPa vertical stress of the overlying rock. The coefficient of horizontal pressure was 1.2 based on relevant geological data [33,40]. The model calculation was based on the Mohr–Coulomb criterion. The numerical calculation model is shown in Figure 9, and the rock mechanics parameters are shown in Table 1.

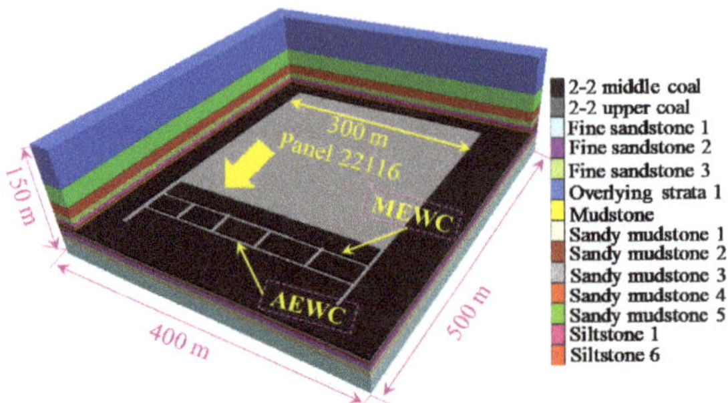

Figure 9. 3D view of the numerical model.

4.1. Evolution Law of Mining-Induced Stress during Final Mining

After the initial stress balance of the model, panel 22216 was excavated for 300 m. After the stress was re-balanced, the vertical and horizontal stresses in front of the working face were extracted. The results are shown in Figure 10.

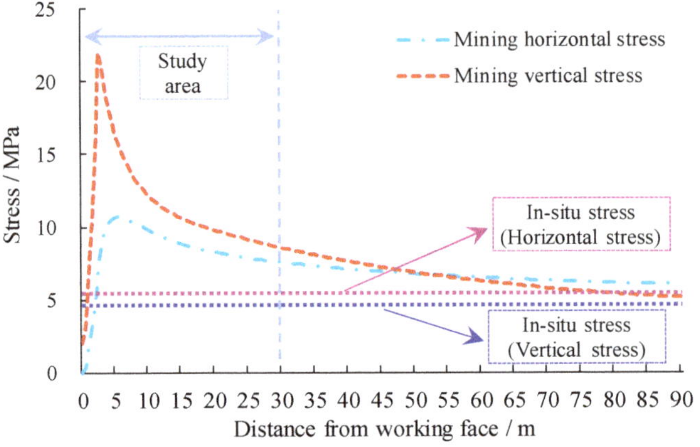

Figure 10. Mining stress distribution characteristics.

It can be seen from Figure 10 that the vertical and horizontal stresses within 3 m of the working face are lower than the in situ stress. Then, the vertical stress increases sharply and reaches its peak at 4 m in front of the working face. The maximum vertical stress is 22 MPa, which is 4.73 times the in situ stress. Subsequently, the stress decreases, but the decline rate becomes smaller and smaller. The vertical stress at 30 m in front of the working face is 1.82 times the in situ stress, which is 8.45 MPa. The vertical stress at 50 m in front of the working face is 6.71 MPa, which is 1.44 times the in situ stress. The variation trend of horizontal stress is the same as that of the vertical stress, but the stress peak is much smaller than that of the vertical stress. The influence range of the mining stress on panel 22216 is greater than 100 m, but the violent influence range is 30 m.

According to the above analysis, the stress distribution in the final mining period was studied. When the working face is within 30 m of MEWC, the stress distribution is shown in Figure 11. When the working face is 30–20 m away from MEWC, a high-stress concentration occurs in the coal pillar between the working face and MEWC, and the stress is asymmetrical. The stress on the side close to the working face is larger (as shown in Figure 12). There are two stress peaks of vertical stress in the coal pillar between the working face and MEWC, while there was only one peak of the horizontal stress. When the working face is 10 m away from MEWC, the vertical and horizontal stresses in the coal pillar between the working face and MEWC further increase. The stress changes from asymmetric distribution to uniform and the vertical stress is still bimodal. When the working face is 5 m away from the MEWC, the vertical and horizontal stresses in the coal pillar between the working face and the MEWC decrease and present symmetrically distribution with a single peak. In the process of the working face gradually approaching the MEWC from 30 m away, the stress in 25 m coal pillar between the MEWC and AEWC gradually increases, but the distribution law is unchanged. Vertical stress is an asymmetrical bimodal distribution in the coal pillar from beginning to end, but the horizontal stress is asymmetrical and unimodal. The stress on the non-pillar side of the AEWC increases with the approach of the working face.

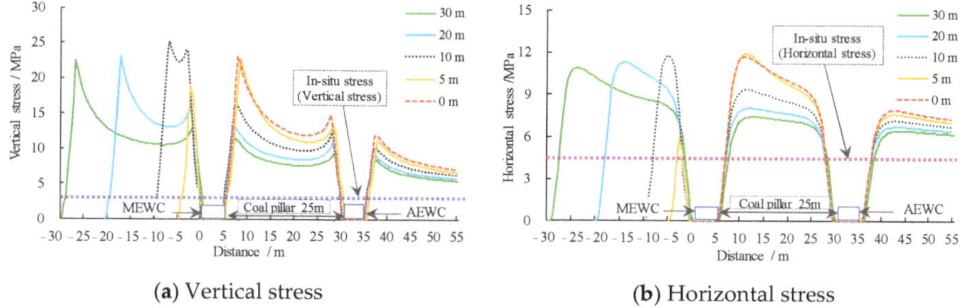

(**a**) Vertical stress   (**b**) Horizontal stress

Figure 11. Law of stress evolution during final mining.

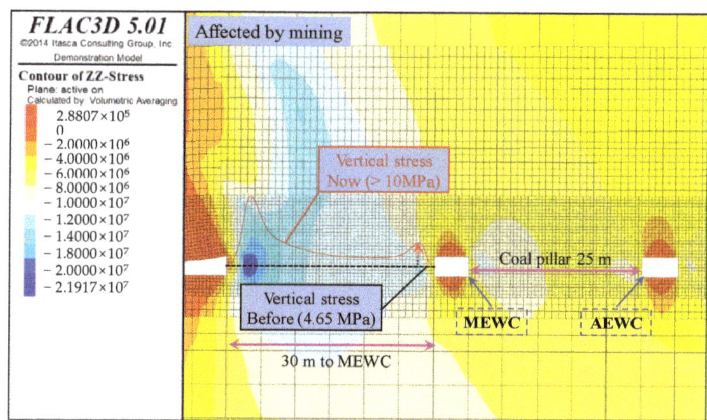

Figure 12. The vertical stress nephogram.

4.2. Evolution Law of the Plastic Zone during Final Mining

The above numerical model was used to calculate the plastic failure law of the EWCs in different periods, as shown in Figure 13.

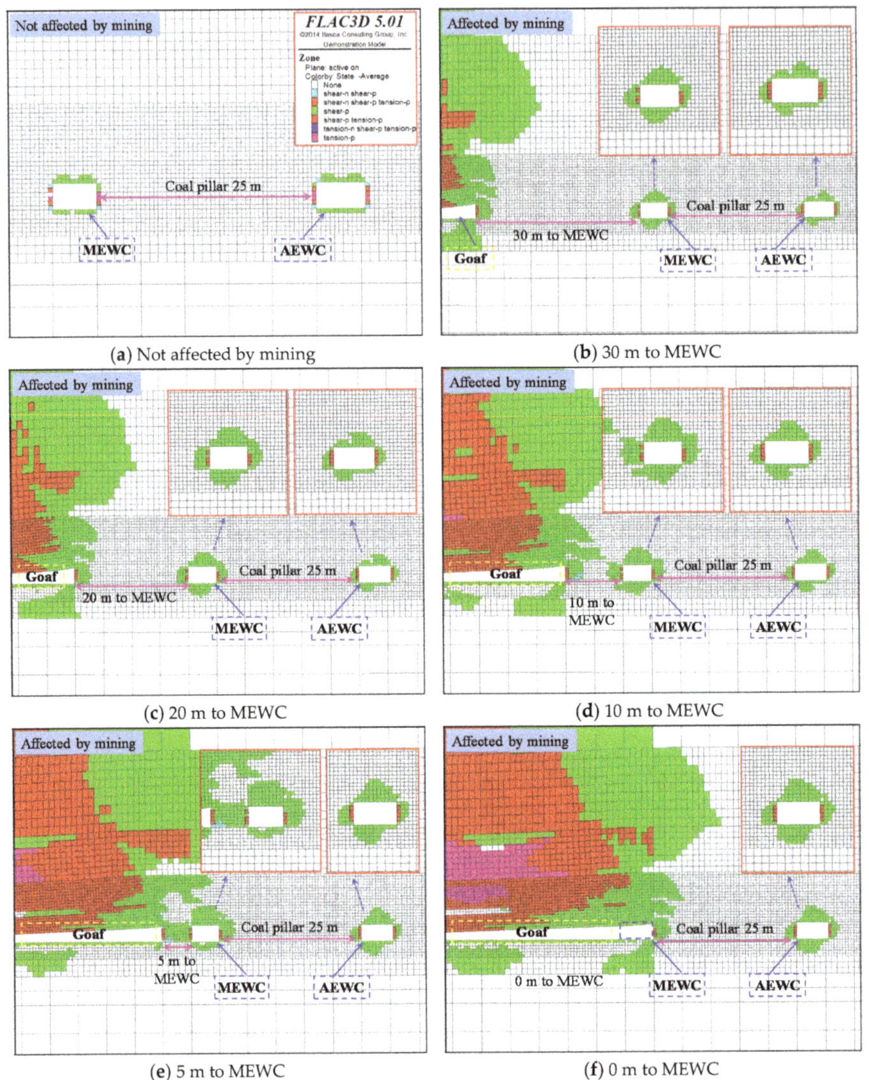

Figure 13. Evolution of the plastic zone.

Before the EWCs are affected by mining, their plastic zone is uniformly and symmetrically distributed, and the size of the plastic zone is only 1 m, as shown in Figure 13a. When the working face is 30 m away from MEWC, the MEWC is within the scope of working face's violent influence, and the plastic zone of surrounding rock around the MEWC is expanded. The failure depth of the roof and rib of the MEWC is 2.0 m and 1.5–2.0 m, respectively. The distribution of the plastic zone is no longer in a symmetrical shape. This is mainly affected by the asymmetrically distributed stress. The AEWC is far from the working face and less affected by mining, so the plastic zone of surrounding rock is generally smaller than that of the MEWC, as shown in Figure 13b,c. With the advance of the working face, the size of the plastic zone of the EWCs increases gradually. When the working face is 10 m away from the MEWC, the stress in front of the working face is symmetrically distributed, so the plastic zone distribution of the MEWC evolves into an asymmetrical shape. In addition, due to the influence of huge abutment pressure, the coal body is damaged, and the plastic zone of goaf is connected to the plastic zone of the MEWC, as shown in Figure 13d. When the working face is only 5 m away from the MEWC, the coal pillar between them completely enters the plastic failure state, so the bearing capacity of the coal pillar decreases (Figures 11 and 13e). At this time, the plastic zone of roof and ribs of the MEWC is greater than 3 m (the bolts lay in the plastic zone of surrounding rock), and the ribs and roof plastic zone of the AEWC are 1.5–2 m and 2.5 m, respectively. When the working face is pushed to the MEWC, the roof of the MEWC completely enters a plastic state and the rib plastic zone is greater than 3 m, the plastic zone of roof and ribs in the AEWC are 3 m and 2 m, respectively. The plastic zone of the AEWC is still within the support range of the bolts and cables.

5. Surrounding Rock Control Scheme and Application

5.1. EWC Support Technology during the Final Mining Period

According to numerical simulation, we know that the instability of the surrounding rock during the final mining period starts about 30 m before the connection. In the process of the working face approaching the MEWC, the peaks in the regional vertical stress fields on both sides of the AEWC increase by 4.73 MPa and 3.76 MPa (10.31 MPa→15.04 MPa, 8.27 MPa→12.03 MPa), respectively, while the peaks in the regional vertical stress fields on both sides of the MEWC increase by 14.35 MPa and 12.26 MPa (10.68 MPa→25.03 MPa, 10.92 MPa→23.18 MPa), respectively. The plastic zone of the MEWC changes more during the final mining period, while the plastic zone of the AEWC is always within the function range of the support body. After the working face is pushed to connect with the MEWC, the whole service cycle of the AEWC is about 2–3 weeks.

Based on the above analysis, we allow the AEWC to produce large deformations as long as the roof does not fall during the final mining period. Therefore, during the final mining period, the key point of stability control of the surrounding rock should focus on the MEWC to prevent partial roof collapse and wall caving during the process of withdrawal of the mining equipment.

(1)   Reinforcement support of anchor cable

The short length of the bolts limits the supporting scope during the final mining period. When the plastic zone of the surrounding rock around the MEWC is larger than the length of the bolt, the bolt cannot effectively control the displacement of the shallow surrounding rock, but also produces overall displacement within the shallow surrounding rock [15]. Therefore, the solid coal rib of the MEWC should be strengthened. In this way, the shallow failure of the surrounding rock can be anchored in the deep stable area to avoid wall collapsing.

With the working face gradually approaching the MEWC, the plastic zone of the roof gradually expands and finally completely fails, which leads to instability of the surrounding rock. Therefore, high-extension anchor cables should be strengthened before they are affected by mining, the roof rock strata should be combined as a common carrier, and the residual bearing capacity of the rock strata within the anchorage range should be improved while the roof deformation is allowed. This effectively prevents the failure of hydraulic supports caused by partial roof collapse [41,42]. The reinforcement support scheme of the MEWC is shown in Figure 14.

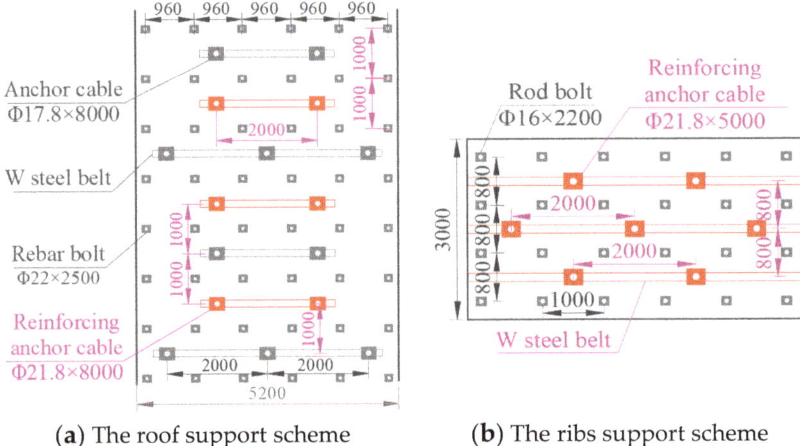

(**a**) The roof support scheme   (**b**) The ribs support scheme

Figure 14. Reinforcement support scheme.

(2)   Support of buttress hydraulics

To avoid severe roof sag of the MEWC caused by severe mining pressure during the final mining period, buttress hydraulic supports are installed in the MEWC before it is affected by mining. The buttress hydraulic supports adopt an installation mode of double-row parallel, as shown in Figure 15. A pair of supports were also installed in the connection roadway near the MEWC. It is required that the distance between the working face and the MEWC should be greater than 100 m when the buttress hydraulic support installation is completed. In this way, support resistance can be provided in time to avoid large-scale separation failure of the surrounding rock that is affected by mining.

Figure 15. Photos of buttress hydraulic supports on site.

5.2. Field Monitoring of the MEWC

During the final mining period, anchor cables reinforcement and buttress hydraulic supports were adopted. The engineering practice proves that the EWCs are stable and the withdrawal of the working face is successful. During the withdrawal the roof did not fall and the walls did not cave in. Pressure monitoring during the final mining period is shown in Figures 16 and 17 below.

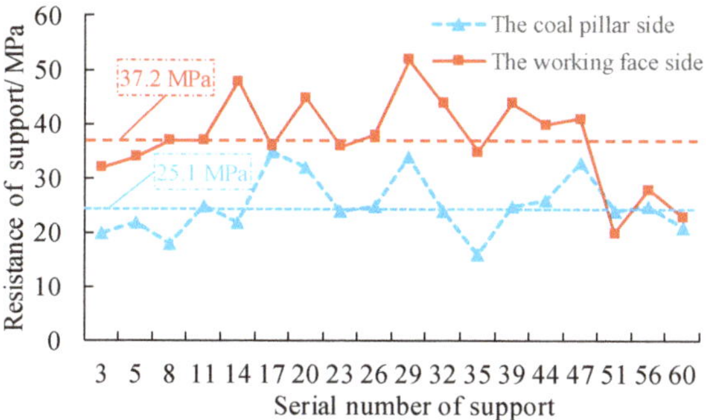

Figure 16. Working resistance of buttress hydraulic supports.

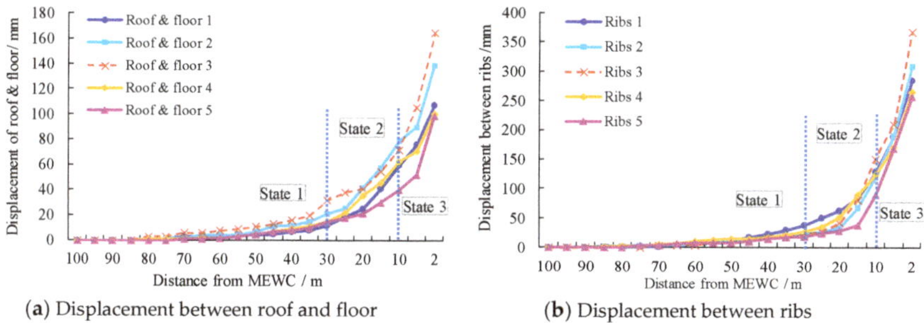

(**a**) Displacement between roof and floor   (**b**) Displacement between ribs

Figure 17. The curve of displacement monitoring.

(1)   Working resistance monitoring of buttress hydraulic supports

When the working face was 1 m away from the MEWC of 22116, the working resistance monitoring of the two-row buttress hydraulic supports are shown in Figure 16. From this, it can be concluded that the working resistance of buttress hydraulic supports near the working face are generally greater than that near the coal pillar. The average working resistance of buttress hydraulic supports close to the working face was 37.2 MPa, while close to the coal pillar it was 25.1 MPa. During the final mining period, the buttress hydraulic supports provided stable working resistance, no failure occurred in the MEWC, and the stability of EWCs was good.

(2)   Surface displacement of the roadway

Five observation points were evenly arranged in the MEWC, and observation started when the working face was 100 m away from the MEWC. The monitoring data are shown in Figure 17. According to the figure, the deformation of the MEWC during the final mining period was divided into three stages. They were a stable stage (state 1), slow deformation stage (state 2), and severe deformation stage (state 3. In other words, if the distance between the working face and the MEWC is greater than 30 m, it is state 1; if the distance between the working face and the MEWC is 10–30 m, it is state 2; and if the distance between the working face and the MEWC is less than 10 m, it is state 3. According to the monitoring data, the average deformation ratio of the roof and floor is 15.5%, 35.8%, and 48.7%, in states 1, 2, and 3, respectively, and the average deformation ratio of ribs is 8.9%, 32.7%, and 58.4%, respectively. Although the MEWC had a large deformation during withdrawal, there was no roof fall or wall caving, and the roadway had good stability.

6. Discussion

With the continuous excavation of the working face, the abutment pressure in front of the working face is constantly moving forward, which is the moving abutment pressure. Compared with the reserved roadway, which is only affected by the fixed abutment pressure from the side of the goaf, the EWCs in front of the working face are affected by the moving abutment pressure during the final mining period. As a result, the stress distribution and the shape of plastic zone of the EWCs are constantly changing during the final mining period.

Based on the voussoir beam theory, this paper briefly analyzed the cause of the abutment pressure around goaf. The stress field of the surrounding rock under abutment pressure is non-uniformly distributed. Therefore, based on elastic mechanics, the formula of the roadway's plastic zone under the influence of a non-uniform stress field was deduced by using the Mohr–Coulomb failure criterion. The theoretical formula calculation results are in good agreement with the FLAC^{3D} calculation results. The numerical simulation results show that the formula has guiding significance for different cross-section shapes of roadways. The plastic zone evolution diagram of the roadway is shown in Figure 18. The failure of the surrounding rock under different stress states occurs within a mile.

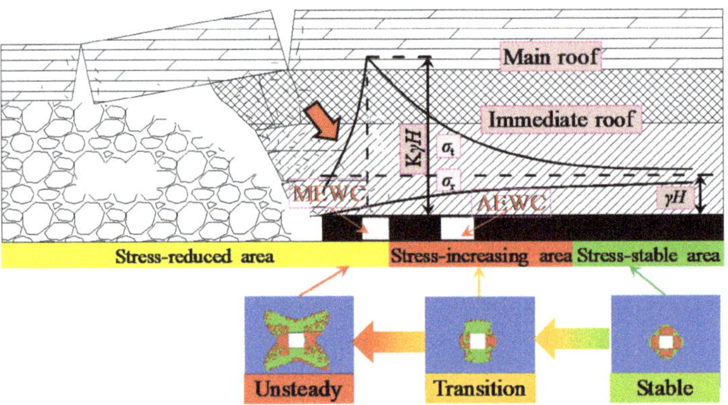

Figure 18. Relationship between mining stress and the plastic zone.

We used FLAC3D to numerically simulate the evolution of the regional stress field and the distribution characteristics of the plastic zone around the EWCs during the final mining period. We found that the stress disturbance on the MEWC is more complicated. The stress on the goaf side of the MEWC gradually evolved from bimodal symmetrical distribution to bimodal asymmetrical distribution, then to bimodal symmetrical distribution, and finally to a unimodal symmetrical distribution. However, the stress on the other side that is close to the AEWC always presented a bimodal asymmetric distribution, only the stress value increased. The shape and size of the plastic zone of the EWCs also developed, and finally presented butterfly-shaped distribution. The results of the numerical simulation were highly consistent with the butterfly shaped plastic zone theory proposed by us. The surrounding rock stability control scheme based on this theory also effectively ensured the smooth progress of the project.

Butterfly-shaped plastic zone theory can better explain the large deformation of the surrounding rock, coal, and gas outburst, and rockburst in mining engineering. However, the limited accuracy of the detection equipment meant it was not possible to detect the actual shape and size of the plastic zone of the mine's surrounding rock. In terms of roadway support, the author conducted some theoretical studies on the mechanisms of the anchor cable support. In roadway engineering, the anchor cable support will not play a significant role in the improvement of the surrounding rock plastic zone and the excavation stress field [16,43], as shown in Figure 19. However, the laboratory mechanism research of anchor cable support still needs further study.

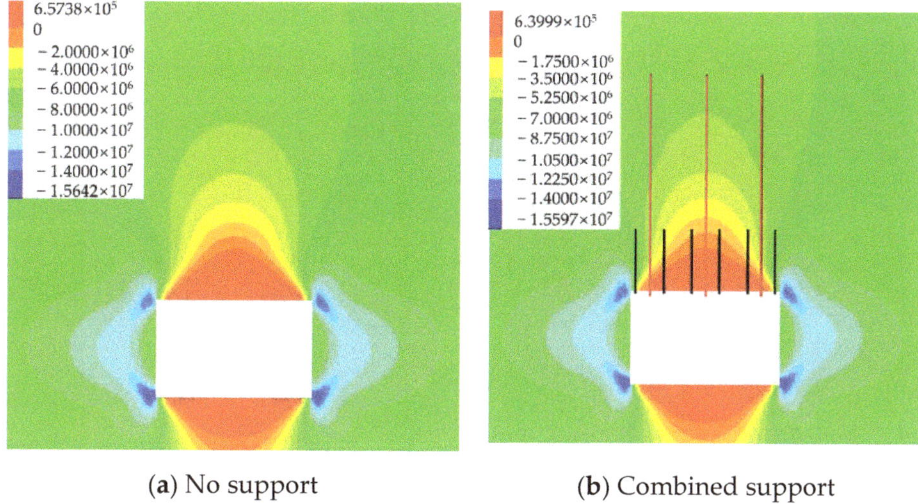

(a) No support
(b) Combined support

Figure 19. Vertical stress distribution of the roadway with and without support.

7. Conclusions

1. The extraction of coal leads to a stress superposition of the surrounding rock around the goaf, and the maximum stress concentration coefficient can reach 3~5. Under such a non-uniform stress environment, the plastic zone of the surrounding rock is no longer a circular distribution but gradually evolves into a butterfly shape with the increase of the lateral pressure coefficient. The different cross-section shapes of the roadway do not affect the final plastic zone shape.

2. The stress environment of the MEWC is more complex than that of the AEWC. The stress peak value on the side close to the working face of the MEWC first increases and then decreases, and its distribution features are first asymmetric and then symmetric. Meanwhile, the stress on the side close to the coal pillar is always asymmetrical and the stress value keeps increasing. The stress distribution around the AEWC remains almost unchanged; the value increases only slightly.

3. When the MEWC is not affected by mining, the plastic zone around it is only 1 M and symmetrically distributed. With the advance of the working face, the EWC plastic zone gradually expands and presents an asymmetric distribution. Finally, the plastic zone depth of the MEWC is greater than the length of the supporting body, and the surrounding rock tends to be unstable.

4. After analyzing the damage characteristics of the EWCs during the final mining, the use characteristics of EWCs were further analyzed. The service cycle of the EWCs is short, only about one month. Therefore, we proposed a stability control scheme of the MEWC's surrounding rock, i.e., the reinforcement technologies of anchor cables and buttress hydraulic supports. After the field application, there was no roof fall or wall caving, and the equipment withdrawal process was successful.

Author Contributions: Conceptualization, C.L. and X.L.; Methodology, C.L. and X.G.; Investigation, C.L. and X.L.; Writing, C.L. and X.G.; review and editing, N.M. All authors have read and agreed to the published version of the manuscript.

Funding: This work was partially supported by the National Natural Science Foundation of China (Grant No. 52004289) and the National Key Research and Development Program (Grant No. 2016YFC0600708).

Conflicts of Interest: The authors declare no conflict of interest.

References

1. Qian, M.G.; Shi, P.W.; Xu, J.L. *Mining Pressure and Strata Control*; China University of Mining and Technology Press: Xuzhou, China, 2010.
2. Wang, B.N. *Mechanism of Surrounding Rock Deformation and Failure and Control. Method Research in Retracement Roadway*; Xi'an University of Science and Technology: Xi'an, China, 2017.
3. Ti, Z.Y.; Zhang, F.; Pan, J. Accident analysis of bracket and roof collapse before retracement under high face mining. *Chin. J. Geol. Hazard. Control.* **2019**, *30*, 78–83. [CrossRef]
4. Diederichs, M.S.; Kaiser, P.K. Stability of large excavations in laminated hard rock masses: The voussoir analogue revisited. *Int. J. Rock Mech. Min.* **1999**, *36*, 97–117. [CrossRef]
5. Shabanimashcool, M.; Li, C.C. Vertical stress changes in multi-seam mining under supercritical longwall panels. *Int. J. Rock Mech. Min.* **2013**, *106*, 39–47. [CrossRef]
6. Ren, Y.F.; Ning, Y. Changing Feature of Advancing Abutment Pressure in Shallow Long Wall Working Face. *J. China Coal Soc.* **2014**, *39*, 38–42. [CrossRef]
7. Zhang, W.L.; Qu, X.C.; Li, C.; Wu, Z. Fracture analysis of multi-hard roofs based on microseismic monitoring and control techniques for induced rock burst: A case study. *Arab. J. Geosci.* **2019**, *12*, 784. [CrossRef]
8. Zhang, W.L.; Ma, N.; Ma, J.; Li, C.; Ren, J. Mechanism of Rock Burst Revealed by Numerical Simulation and Energy Calculation. *Shock Vib.* **2020**, *2020*, 8862849. [CrossRef]
9. Cohen, T.; Masri, R.; Durban, D. Analysis of circular hole expansion with generalized yield criteria. *Int. J. Solids Struct.* **2009**, *46*, 3643–3650. [CrossRef]
10. Kang, J.Z.; Shen, W.L.; Bai, J.B.; Yan, S.; Wang, X.Y.; Li, W.F.; Wang, R.F. Influence of abnormal stress under a residual bearing coal pillar on the stability of a mine entry. *Int. J. Min. Sci. Technol.* **2017**, *27*, 945–954. [CrossRef]
11. Yang, R.S.; Zhu, Y.; Li, Y.L.; Li, W.Y.; Lin, H. Coal pillar size design and surrounding rock control techniques in deep longwall entry. *Arab. J. Geosci.* **2020**, *13*, 1–14. [CrossRef]
12. Yang, K.; Gou, P.F. Research on Reasonable Width of Coal Pillars in High Strength Mining Roadway in Wantugou Mine. *Geotech. Geol. Eng.* **2020**, *2020*, 1–9. [CrossRef]
13. Osgoui, R.R.; Unal, E. An empirical method for design of grouted bolts in rock tunnels based on the Geological Strength Index (GSI). *Geotech. Geol. Eng.* **2009**, *107*, 154–166. [CrossRef]
14. Ghabraie, B.; Ren, G.; Zhang, X.Y.; Smith, J. Physical modelling of subsidence from sequential extraction of partially overlapping longwall panels and study of substrata movement characteristics. *Int. J. Coal Geol.* **2015**, *140*, 71–83. [CrossRef]
15. Kang, H.P.; Li, J.Z.; Yang, J.H.; Gao, F.Q. Investigation on the Influence of abutment pressure on the stability of rock bolt reinforced roof strata through physical and numerical modeling. *Rock Mech. Rock Eng.* **2017**, *50*, 387–401. [CrossRef]
16. Li, C.; Zhang, W.L.; Wang, N.; Hao, C. Roof stability control based on plastic zone evolution during mining. *J. Min. Saf. Eng.* **2019**, *36*, 753–761. [CrossRef]
17. Ren, F.Q.; Zhu, C.; He, M.C. Moment tensor analysis of acoustic emissions for cracking mechanisms during schist strain burst. *Rock Mech. Rock Eng.* **2020**, *53*, 153–170. [CrossRef]
18. Ma, N.J.; Li, J.; Zhao, Z.Q. Distribution of the deviatoric stress field and plastic zone in circular roadway surrounding rock. *J. China U. Min. Techno.* **2015**, *44*, 206–213. [CrossRef]
19. Li, J.; Ma, N.J.; Ding, Z.W. Heterogeneous large deformation mechanism based on change of principal stress direction in deep gob side entry and control. *J. Min. Saf. Eng.* **2018**, *35*, 670–676. [CrossRef]
20. Yan, H.; He, F.L.; Yang, T.; Li, L.Y.; Zhang, S.B.; Zhang, J.X. The mechanism of bedding separation in roof strata overlying a roadway within a thick coal seam: A case study from the Pingshuo Coalfield, China. *Eng. Fail. Anal.* **2015**, *62*, 75–92. [CrossRef]
21. Molladavoodi, H.; Rahmati, M. Dilation angle variations in plastic zone around tunnels in rocks-constant or variable dilation parameter. *J. Cent. South. Univ.* **2018**, *25*, 2550–2566. [CrossRef]
22. Wang, W.J.; Dong, E.Y.; Yuan, C. Boundary equation of plastic zone of circular roadway in non-axisymmetric stress and its application. *J. China Coal Soc.* **2019**, *44*, 105–114. [CrossRef]

23. Chen, H.; Jiang, Y.D.; Deng, D.X.; Shao, M.M. Study on deformation characteristics and support optimization of weak surrounding rock in fault zone. *J. Min. Sci. Technol.* **2018**, *3*, 543–552. [CrossRef]
24. Yang, J.; Gao, Y.B.; Liu, S.Q.; Peng, Y.H. Study on failure mechanism and control techniques of the preparation roadway induced by dynamic mining disturbance. *J. Min. Sci. Technol.* **2018**, *3*, 451–460. [CrossRef]
25. Gao, X.C.; Luan, Y.C.; Hu, C.; Lu, W.; Sun, H. Study on bearing mechanism and coupling mechanism of steel arch-concrete composite structure of initial support system of large section tunnel. *Geotech. Geol. Eng.* **2019**, *37*, 4877–4887. [CrossRef]
26. Pan, R.; Wang, Q.; Jiang, B.; Li, S.C. Model test on failure and control mechanism of surrounding rocks in tunnels with super large sections. *Arab. J. Geosci.* **2019**, *12*, 1–17. [CrossRef]
27. Yang, H.Z.; Guo, Z.; Chen, D.; Wang, C.; Zhang, F.; Du, Z. Study on Reasonable Roadway Position of Working Face under Strip Coal Pillar in Rock Burst Mine. *Shock Vib.* **2020**, *2020*, 8832791. [CrossRef]
28. Pellet, F.; Roosefid, M.; Deleruyelle, F. On the 3D numerical modelling of the time-dependent development of the damage zone around underground galleries during and after excavation. *Tunn. Undergr. Sp. Tech.* **2009**, *24*, 665–674. [CrossRef]
29. Bai, Q.X.; Tu, S.H.; Wang, F.T. Field and numerical investigations of gateroad system failure induced by hard roofs in a longwall top coal caving face. *Int. J. Coal Geol.* **2017**, *173*, 176–199. [CrossRef]
30. Jiang, L.S.; Wang, P.; Zhang, P.P. Numerical analysis of the effects induced by normal faults and dip angles on rock bursts. *C. R. Mec.* **2019**, *345*, 690–705. [CrossRef]
31. Jia, H.S.; Wang, L.Y.; Fan, K.; Peng, B.; Pan, K. Control technology of soft rock floor in mining roadway with coal pillar protection: A case study. *Energies* **2019**, *12*, 3009. [CrossRef]
32. Li, A.; Liu, Y.; Dai, F.; Liu, K.; Wei, M.D. Continuum analysis of the structurally controlled displacements for large-scale underground caverns in bedded rock masses. *Tunn. Undergr. Sp. Tech.* **2020**, *97*, 103288. [CrossRef]
33. Yang, G.R. *Study on the Roof Caving Prediction Principle and Method of the Hazard. Area at Starting Cut Lijiahao Mine*; China University of Mining and Technology (Beijing): Beijing, China, 2013.
34. Kaiser, P.K.; Yazici, S.; Maloney, S. Mining-induced stress change and consequences of stress path on excavation stability—A case study. *Int. J. Rock Mech. Min.* **2001**, *38*, 167–180. [CrossRef]
35. Rezaei, M.; Hossaini, M.F.; Majdi, A. Development of a time-dependent energy model to calculate the mining-induced stress over gates and pillars. *J. Rock. Mech. Geotech.* **2015**, *7*, 306–317. [CrossRef]
36. Zhao, Z.Q. *Mechanism of Surrounding Rock Deformation and Failure and Control. Method Research in Large Deformation Mining Roadway*; China University of Mining and Technology (Beijing): Beijing, China, 2014.
37. Zhao, Z.Q.; Ma, N.J.; Liu, H.; Guo, X.F. A butterfly failure theory of rock mass around roadway and its application prospect. *J. China Univ. Min. Technol.* **2018**, *47*. [CrossRef]
38. Guo, X.F.; Zhao, Z.Q.; Gao, X.; Wu, X.Y.; Ma, N.J. Analytical solutions for characteristic radii of circular roadway surrounding rock plastic zone and their application. *Int. J. Min. Sci. Technol.* **2019**, *29*, 263–272. [CrossRef]
39. Liang, P.; Gao, Y.T. Numerical investigation on cracking behavior of granite with intersecting two-flaws: A flat-joint modeling method. *Lat. Am. J. Solids Struct.* **2020**, *17*, 1–14. [CrossRef]
40. Liu, J. In-situ stress measurements and study on distribution characteristics of stress fields in underground coal mines in yitai mining area. *J. China Coal Soc.* **2011**, *36*, 562–566. [CrossRef]
41. Jia, H.S.; Ma, N.J.; Zhu, K. Mechanism and control method of roof fall resulted from butterfly plastic zone penetration. *J. China Coal Soc.* **2016**, *41*, 1384–1392. [CrossRef]
42. Lin, C.D.; Lu, S.L.; Shi, Y.W. Study on supporting effect of soft roof bolt in coal roadway. *J. China Coal Soc.* **2000**, *25*, 482–485. [CrossRef]
43. Li, C.; Wu, Z.; Zhang, W.L.; Sun, Y.H.; Zhu, C.; Zhang, X.H. A case study on asymmetric deformation mechanism of the reserved roadway under mining influences and its control techniques. *Geomech. Eng.* **2020**, *22*, 449–460. [CrossRef]

Publisher's Note: MDPI stays neutral with regard to jurisdictional claims in published maps and institutional affiliations.

© 2020 by the authors. Licensee MDPI, Basel, Switzerland. This article is an open access article distributed under the terms and conditions of the Creative Commons Attribution (CC BY) license (http://creativecommons.org/licenses/by/4.0/).

Article

Innovative Method for Calculating the Break-Even for Multi-Assortment Production

Dariusz Fuksa

Faculty of Mining and Geoengineering, AGH University of Science and Technology, Mickiewicza 30 Av., 30-059 Kraków, Poland; fuksa@agh.edu.pl

Abstract: The subject of the article is a new method that I have developed for calculating a multi-asset break-even for multi-assortment production, extended by a percentage threshold and a current sales ratio (which was missing in previously published methods). The percentage threshold provides unambiguous information about the economic health of a company. As a result, it became possible to use it in practice to evaluate the activities of economic entities (mines) and to perform modelling and optimisation of production plans based on different variants of customer demand scenarios. The publication addresses the complexity of the problem of determining the break-even in multi-assortment production. Moreover, it discusses the practical limitations of previous methods and demonstrates the usefulness of the proposed method on the example of hard coal mines.

Keywords: multi-assortment break-even; coal mines; percentage threshold

Citation: Fuksa, D. Innovative Method for Calculating the Break-Even for Multi-Assortment Production. *Energies* **2021**, *14*, 4213. https://doi.org/10.3390/en14144213

Academic Editor: Wen-Hsien Tsai

Received: 29 May 2021
Accepted: 9 July 2021
Published: 12 July 2021

Publisher's Note: MDPI stays neutral with regard to jurisdictional claims in published maps and institutional affiliations.

Copyright: © 2021 by the author. Licensee MDPI, Basel, Switzerland. This article is an open access article distributed under the terms and conditions of the Creative Commons Attribution (CC BY) license (https://creativecommons.org/licenses/by/4.0/).

1. Introduction

The break-even analysis in the production of two products is already very complicated. Most companies produce between a few and a dozen products, which further complicates the interpretation of the results. The break-even calculation methods proposed in the literature only allow calculating threshold quantities of individual assortments and total revenue values for which a zero financial result is achieved. Each method provides different solutions, as each method differently approaches the calculation of the quantitative and valuable threshold. In other words, there are as many different solutions as there are methods. In fact, the set of admissible solutions is infinitely large. The lack of an unambiguous solution makes these methods unsuitable for practical use in analysis or production planning.

There are generally three different methods of analysis. The choice of a particular method is determined by the possibilities of estimating fixed costs, which in turn is influenced by the cost accounting in the company and the accuracy of the methods of separating fixed and variable costs. Hence, in different methods, the fixed costs are:

- fully accounted for between individual grades of products [1–3];
- charged in full to the company [3], including the graphical determination of the break-even [4] and the method based on the weighted average contribution margin [1,5]; and
- in part accounted for between the individual grades of products and in part related to the company—segmental analysis [1,2,6,7].

There are also proposals to calculate the break-even point for companies based on single-assortment threshold formulas [8,9]; this approach is unfortunately a major simplification. Due to the fact that practically any enterprise does not produce a single assortment, the analysis of the single-assortment threshold currently should remain only in the pedagogical aspect. My research on the methods of improving the operational efficiency of, among other things, coal mining companies [10] contributed to the development of my own method for calculating the break-even [11]. My intention was to find a way of recognising the threshold that would give an unambiguous value. The result of this research is the break-even expressed as percentage, which I have developed. This recognition of the threshold has been missing in previous methods. With it, it is possible to assess the economic health of a company, compare companies with one another, and quickly assess whether the current sales volume is profit- or loss-making.

2. The Essence of the Break-Even

Break-even (BE) analysis involves examining the so-called break-even at which revenue from sales exactly agrees with the incurred costs. The company's financial result is then zero, thus no profit or loss is made.

In single-assortment production, the break-even is a single point. According to the definition above, the break-even is at the point where the value of sales (S) equals the level of total costs (Kc), which can be represented by Equation (1) [4,6,8,9,11]:

$$S = Kc \tag{1}$$

whereby:

$$S = P \cdot c \tag{2}$$

And

$$Kc = Ks + P \cdot kjz \tag{3}$$

where:

P—the amount of production (sales) (Mg),
c—unit selling price (PLN*/Mg),
Ks—total fixed costs (PLN*),
kjz—unit variable costs (PLN*/Mg),
*—national currency.

By substituting Equations (2) and (3) into Equation (1) we obtain the relation:

$$P \cdot c = Ks + P \cdot kjz \tag{4}$$

Based on Equation (4), the break-even point can be calculated in terms of:

- quantitative:

$$BEP = \frac{Ks}{c - kjz} \cdot (\text{Mg}) \tag{5}$$

Based on Formula (5), an obtained answer is clear to what quantity of production guarantees the mine (enterprise) a zero profit. The enterprise producing and selling a smaller amount will make a loss, while selling more will be profitable. This unambiguity of the result is possible only with the production of one assortment. With the production of at least two products, there will be infinitely many similar solutions (quantities) guaranteeing the achievement of the break-even point. In current times, hardly any enterprise produces a single product.

- value:

$$BEP\prime = \frac{Ks}{c - kjz} \cdot c = BEP \cdot c \cdot (\text{PLN}) \tag{6}$$

Formula (6) provides information on the (critical) incomes required to cover total costs at break-even; this information has limited practical use.

- as a degree of use of production capacity:

$$BEP'' = \frac{Ks}{P_m \cdot (c - kjz)} \cdot 100 = \frac{BEP}{P_m} \cdot 100\% \tag{7}$$

where: P_m—maximum production (sales) (Mg).

On the basis of Equation (7), the most important information is obtained which is what percentage of production capacity is needed to cover the incurred costs, and what is remaining to generate profit. This allows different enterprises to be compared with each other and a preliminary estimate of the economic situation in the company.

Whereas, in production of many different products BE is a set of finitely many points. The alignment of total costs with sales revenue can be achieved with many different combinations of the quantitative product structure, as can be seen in Equation (8):

$$\sum_{i=1}^{n} P_i \cdot c_i = \sum_{i=1}^{n} P_i \cdot kjz_i + Ks \tag{8}$$

where:

i—product type (assortment), i = 1, 2, ... , n.

Let us to consider hypothetically the sale of two sortiments of coal by the mine "X" (presented in Chapter 4). Table 1 contains the modified, for the purpose of this example, structure of monthly production of the mine "X", as well as information of the prices, variable unit costs of particular coal sort, and fixed costs (changed for the purposes of this case). To the values given in PLN their equivalent in EUR was added according to the average exchange rate of 4 July 2021 (1 EUR = 4.52 PLN).

Table 1. Assumed value of production and economic indicators of mine "X" in a monthly take.

Coal Size Grade	Quantity	Unit Sale Price	Variable Cost	Total Cost
	(Mg)	(PLN/Mg)	(PLN/Mg)	(PLN)
Cobble	25,500	550 (€121.68)	185 (€40.93)	11,563,515.55
Nut Coal	38,250	485 (€107.30)	185 (€40.93)	(€2,558,299.90)
Total	63,750	-	-	-

In this case, the break-even can be reached with a finite number of combinations of the production structure. It will be a set of combinations of quantities of particular sortiments (set of points) lying on a segment, the beginning and end of which we determine from Equation (8).

We assume hypothetically that we will produce only the cobble sort, then on the basis of Equation (8) it is possible to determine its quantity (Pc), the sale of which at a given price and cost will result in the mine reaching break-even:

$$550 \cdot P_C + 485 \cdot 0 = 11,563,515.55 + 185 \cdot P_C + 185 \cdot 0$$

$$P_C = \frac{11,563,515.55}{365} = 31,680.86 (Mg)$$

The threshold quantity of sales of the cobble sortiment for the analysed mine will be 31,680.86 Mg. On the other hand, producing and selling only nut coal, the break-even will be reached at 38,545.05 Mg, according to the calculations:

$$550 \cdot 0 + 485 \cdot P_N = 11,563,515.55 + 185 \cdot 0 + 185 \cdot P_N$$

$$P_M = \frac{11,563,515.55}{300} = 38,545,05 (Mg)$$

Figure 1 presents a graphical solution for the analysed example. The determined boundary quantities of coal sortiments are marked in the diagram with letters A (for the cobble sortiment) and B (for the nut coal sortiment). These are points of intersection with the axes representing quantities of the analysed sortiments. Connecting points A and B leads to a segment AB which is a finite set of points (combinations of quantities of cobble and nut coal sortiments) fulfilling Equation (8). Each point of this segment guarantees finding it in the break-even.

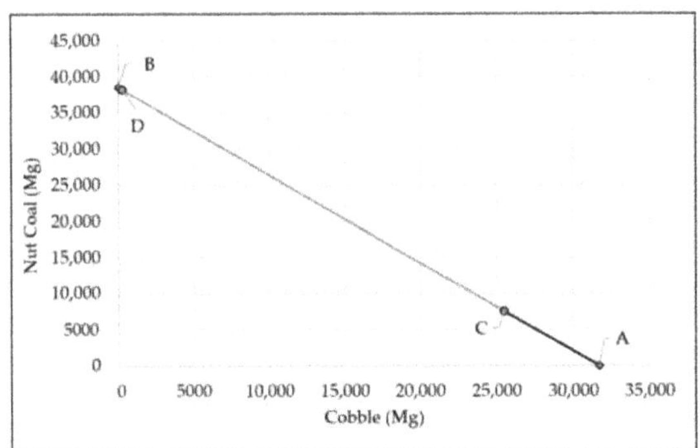

Figure 1. Graphic presentation of the BE on the analysed case.

However, this is not an acceptable solution due to the established production structure: 25,500 Mg of the cobble sortiment and 38,250 Mg of the nut coal sortiment. In this situation, an acceptable solution for the threshold quantities of the analysed sortiments will be a fragment of section AB, namely section CD, which is the result of the adopted production structure (Figure 1). In Table 2, I have listed exemplary quantities of the cobble and nut coal sortiments, the sale of which also guarantees reaching the break-even (these are points lying on the CD segment). If we sum up the quantity of sales of sortiments (Table 2, column 4) for each variant, we would obtain the summarised break-even. As can be seen (Table 2) in each variant this total quantity is different. We do not obtain one specific value. Therefore, it is difficult to determine, having the information on actual sales at a given moment, whether the company is above the threshold (making profits), below the threshold (incurring losses), or perhaps at the threshold (zero profit).

Table 2. Example threshold quantities for cobble and nut coal assortments, sum of threshold quantities, and revenues.

Variants	Cobble	Nut Coal	Summarised Break-Even	Revenue	Revenue
	(Mg)	(Mg)	(Mg)	(PLN)	(€)
Variant 1	4000	33,678.39	37,678.39	18,534,016.8	(€4,100,446.19)
Variant 2	1000	37,328.39	38,328.39	18,654,266.8	(€4,127,050.80)
Variant 3	12,400	23,458.39	35,858.39	18,197,316.8	(€4,025,955.04)
Variant 4	11,132.92	25,000.00	36,132.92	18,248,105.6	(€4,037,191.50)
Variant 5	25,105.52	8000	33,105.52	17,688,037.1	(€3,913,282.54)
Variant 6	242.51	38,250	38,492.51	18,684,629.6	(€4,133,767.61)

In conclusion, the determination of threshold production (sales) quantities for individual products is a very complex issue, as there are an infinite number of combinations of their quantities that guarantee the company zero profit. Each method of determining a multi-assortment threshold proposed so far in the literature gives a single, unique solution. This is related to the way the authors choose to calculate the individual quantities of products from Equation (8). Each solution of the different methods is contained in a comprehensive set of solutions to Equation (8). For this reason, information about the specific threshold quantities that can be calculated by these methods is of little practical use and cannot be used as a basis for making any important production decisions. Their uselessness is due to the fact that actual sales of the product are very unlikely to approach the threshold quantities determined by any of these methods.

The methods presented in the literature also allow the calculation of the value of critical revenues (valuable threshold). Unfortunately, this is not one specific value, but again a finite, huge set. Columns 5 and 6 in Table 2 show the critical revenues for each variant of the threshold quantities of the cobble and nut coal sort (calculated by multiplying these quantities by the sales prices for the coal sortiment). Since there are an infinite number of variants of sales of particular sortiments and for each of them we obtain an equally numerous set of revenues, this information is not useful in practical application. Consequently, the methods on the basis of which we can calculate it are not useful either. It is worth noticing that the methods used so far make it possible to calculate only threshold quantities of sales of particular assortments and the value of revenue. So far, other than myself, no researcher has attempted to determine the percentage threshold. The percentage threshold provides one specific value characterising the condition of a given enterprise, as is discussed in Section 3.

As the number of produced assortments increases, the analysis becomes more complicated because of the dependencies between the individual assortments. It becomes even impossible to analyze the break-even.

3. Author's Concept of Multi-Asset Break-Even Analysis

The method I have developed makes it possible to calculate the threshold percentage, which distinguishes it from the methods presented in the literature to date. The most important is that the percentage threshold is a single value, specific to a particular company. It provides information on the economic health of the company and, above all, enables quick determination of the financial situation of the company for the actual volume of sales at a given time.

I propose to determine its value according to the Equation (9):

$$PR(P) = \frac{Ks}{\sum\limits_{i=1}^{n} P_i \cdot (c_i - kjz_i)} \cdot 100\% \tag{9}$$

PR(P) provides the following data: how much of the global gross margin goes to cover fixed costs. Topping up to 100% determines the achievable profit for the company. Its value, e.g., 60%, means that 60% of the global margin covers costs, and 40% brings the company a profit. The lower its value the better the financial health of the company.

I propose to use the following Equation (10) in order to determine the recognition of the quantitative threshold:

$$PR(I) = PR(P) \cdot P_m \; (\text{Mg}) \tag{10}$$

where:

P_m—maximum production (sale) (Mg).

Whereas the threshold quantity of any assortment according to Equation (11) is:

$$Pp_i = PR(P) \cdot P_{msi} \; (\text{Mg}) \tag{11}$$

where:

P_{ms}—maximum output of the product (assortment) (Mg).

I propose that the valuable threshold be determined as the product of the percentage threshold and the maximum revenue in Equation (12):

$$PR(W) = PR(P) \cdot \sum\limits_{i=1}^{n} P_{msi} \cdot c_i \; (\text{PLN}) \tag{12}$$

4. Results

The possibilities of practical use of the developed method will be demonstrated on the example of coal mines.

Table 3 shows the initial structure of the monthly production of mine "X", as well as information on prices, variable unit costs of individual coal grades, and fixed costs. Whereas Table 4 consists of analogous data for the mine "Y".

Table 3. Assumed value of production and economic indicators of mine "X" in a monthly take.

Coal Size Grade	Quantity	Unit Sale Price	Variable Cost	Total Cost
	(Mg)	(PLN/Mg)	(PLN/Mg)	(PLN)
Cobble	25,500	550 (€121.68)	185 (€40.93)	22,927,031.10
Nut Coal	38,250	485 (€107.30)	185 (€40.93)	(€5,072,352.01)
Fine Coal I	106,250	395 (€87.39)	185 (€40.93)	-
Fine Coal IIA	85,000	330 (€73.01)	185 (€40.93)	-
Fine Coal II	170,000	310 (€68.58)	185 (€40.93)	-
Total	425,000	-	-	-

Table 4. Assumed value of production and economic indicators of mine "Y" in a monthly take.

Coal Size Grade	Quantity	Unit Sale Price	Variable Cost	Total Cost
	(Mg)	(PLN/Mg)	(PLN/Mg)	(PLN)
Cobble	11,920	550 (€121.68)	197 (€43.58)	27,843,956.43
Nut Coal	23,833	485 (€107.30)	197 (€43.58)	(€6,160,167.35)
Fine Coal I	13,867	395 (€87.39)	197 (€43.58)	-
Fine Coal IIA	167,050	330 (€73.01)	197 (€43.58)	-

The percentage threshold (Equation (9)) for mine "X" is:

$$PR(P) = \frac{22,927,031.10}{25,500 \cdot (550 - 185) + 38,250 \cdot (485 - 185) + \ldots + 170,000 \cdot (310 - 185)} \cdot 100 = 29.90(\%)$$

Whereas for mine "Y":

$$PR(P) = \frac{27,843,956.43}{11,920 \cdot (550 - 197) + 23,833 \cdot (485 - 197) + \ldots + 167,050 \cdot (330 - 197)} \cdot 100 = 77.30(\%)$$

Mine "X" has a low break-even. Only about 30% of the gross margin covers fixed costs and 70% perhaps represents the maximum profit (achievable). It can be determined from Equation (13):

$$\text{Profit} = (100\% - PR(P)) \cdot \sum_{i=1}^{n} P_{msi} \cdot (c_i - kjz_i) \ (\text{PLN}) \tag{13}$$

Hence,

$$\text{Profit} = (100\% - 22.90\%) \cdot [25,500 \cdot (550 - 185) + \ldots + 170,000 \cdot (310 - 185)] = 53,742,968.90(\text{PLN})(€11,890,037.37)$$

Hence, it can be considered that we are dealing with a rather high threshold in the case of Mine "Y" since as much as 77.30% of the gross margin is covered by fixed costs. The maximum achievable profit represents just under 23% from revenue less variable costs, as shown in Equation (13):

$$\text{Profit} = (100\% - 77.30\%) \cdot [11,920 \cdot (550 - 197) + \ldots + 167,050 \cdot (330 - 197)] = 8,191,023.57(\text{PLN})(€1,812,173.36)$$

The proposed percentage threshold clearly defines the financial health of the company. In the case under review, the mine with the lower percentage threshold is in better financial condition. To confirm this, the profitability ratio (*GPM*) for both mines was calculated from Equation (14):

$$GPM = \frac{grossprofit}{netsales} \cdot 100(\%) \tag{14}$$

For mine "X" it is:

$$GPM = \frac{53,742,968.90}{25,500 \cdot 550 + \ldots 170,000 \cdot 310} \cdot 100 = 34.61(\%)$$

And for mine "Y":

$$GPM = \frac{8,191,023.57}{11,920 \cdot 550 + \ldots 167,050 \cdot 330} \cdot 100 = 10.41(\%)$$

Further analysis will be carried out for mine "X". The threshold quantities of individual coal grades for mine "X" (calculated on the basis of Equation (11)) are as follows:

$$Pp_1 = 29.90\% \cdot 25,500 = 7625.40 (Mg) for cobble,$$

$$Pp_2 = 29.90\% \cdot 38,250 = 11,438.10 (Mg) for nutcoal,$$

$$Pp_1 = 29.90\% \cdot 25,500 = 7625.40 (Mg) for cobble,$$

$$Pp_4 = 29.90\% \cdot 85,000 = 25,417.99 (Mg) for fine coal IIA,$$

$$Pp_5 = 29.90\% \cdot 170,000 = 50,835.99 (Mg) for fine coal II.$$

Selling exactly these quantities will result in zero profit (reaching break-even). The results were intentionally left with decimal places, as rounding to whole tonnes will result in placing the mine in the profit zone (as shown below).

Do the calculated threshold quantities Pp_1–Pp_5 provide relevant information, which could be useful in practice? The answer is simple—unfortunately not. The probability of actual sales in threshold quantities Pp_1–Pp_5 is virtually zero. Only if actual sales of each coal grade were less than the threshold quantities could a mine be said to be below the threshold and making a loss. There is an infinitely large set of such solutions (threshold quantities) as follows from Equation (8). Table 5 summarises six example combinations of sales of individual coal grades, the sale of which in such quantities also guarantees that the mine reaches break-even.

Table 5. Summary of the threshold quantities of the individual coal grades, the total quantity threshold and the ad valorem threshold.

Cobble	Nut Coal	Fine Coal I	Fine Coal IIA	Fine Coal II	PR(I)	PR(W)	PR(W)
(Mg)	(Mg)	(Mg)	(Mg)	(Mg)	(Mg)	(PLN)	(€)
7625.40	11,438.10	31,772.49	25,417.99	50,835.99	127,089.97	46,438,675.65	10,274,043.29
0.00	0.00	48,462.05	0.00	102,000	150,462.05	50,762,510.88	11,230,644.00
9500	5360	61,578.72	21,000	15,000	112,438.72	43,728,194.22	9,674,379.25
12,200	14,000	33,221.58	43,000	8500	110,921.58	43,447,522.79	9,612,283.80
2000	0.00	105,700.15	0.00	0.00	107,700.15	42,851,558.50	9,480,433.30
15,000	34,256.77	0.00	15,000	40,000	104,256.77	42,214,533.62	9,339,498.59

Total volume break-even (calculated from Equation (10)):

$$PR(I) = \sum_{i=1}^{n} Pp_i = 127,089.97 (Mg)$$

also makes up the set of finitely many admissible solutions (column 6 of Table 5, (PR(I)). As can be seen, each solution gives a different value for the aggregate quantitative threshold. It is not one specific value, so we are not able to say whether actual sales are profitable or loss-making.

The situation is analogous with the valuable threshold (column 7 of Table 5, PR(W))—calculated from Equation (12):

$$PR(W) = 29.90\% \cdot [7625.40 \cdot 550 + 11,438.10 \cdot 485 + 31,772.49 \cdot 395 + 25,417.99 \cdot 330 + 50,835.99 \cdot 310]$$
$$PR(W) = 46,438,675.65(PLN)(\text{€}10,274,043.29)$$

There are as many solutions as there are values for the valuable threshold. Unfortunately, this is not one particular value, but again, a finitely numerous set. Therefore, information on the value of income is also not useful in practice.

Let us consider different hypothetical variants for the sale of coal grades of mine "X" (Table 6).

Table 6. Variants of sales of size grades of coal of mine "X" on a monthly basis (Mg).

Sales Scenario	Cobble	Nut Coal	Fine Coal I	Fine Coal IIA	Fine Coal II
	(Mg)	(Mg)	(Mg)	(Mg)	(Mg)
Threshold quantity	7625.40	11,438.10	31,772.49	25,417.99	50,835.99
I	7626	11,439.00	31,773.00	25,418.00	50,836.00
II	7600	11,440.00	31,780.00	25,418.00	50,870.00
III	9500	5360	32,000.00	21,000.00	64,000.00
IV	5000	5000	30,000.00	20,000.00	90,000.00
V	0.00	0.00	48,462.05	0.00	102,000.00

The presented variants (combinations) of sales I–V do not explicitly state whether mine "X" is in the profit, loss, or zero zone. To determine this, the proposed percentage threshold, i.e., knowledge of one characteristic quantity, will be useful. First, the proposed Equation (15) is used by calculating the current sales ratio:

$$WAS = \frac{\sum\limits_{i=1}^{n} P_i \cdot (c_i - kjz_i)}{\sum\limits_{i=1}^{n} P_{mi} \cdot (c_i - kjz_i)} \cdot 100(\%) \tag{15}$$

Based on Equation (15), the percentage share of the currently realised gross margin in relation to the global margin (achievable with total sales of all products) is determined. The calculated value of current sales ratio (WAS) should be compared with the proposed percentage threshold. We immediately receive an unambiguous answer:

- *WAS* < *PR(P)*—the company is incurring losses,
- *WAS* = *PR(P)*—the company has met the break-even (profit is 0),
- *WAS* > *PR(P)*—the company is making profit.

For variant I (Table 6), the *WAS* value (according to Equation (15)) is:

$$WAS = \frac{7626 \cdot (550 - 185) + 11,439 \cdot (485 - 185) + \ldots + 50,836 \cdot (310 - 185)}{25,500 \cdot (550 - 185) + 38,250 \cdot (485 - 185) + \ldots + 170,000 \cdot (310 - 185)} \cdot 100 = 29.904(\%)$$

WAS > *PR(P)* so selling in such quantities makes a profit for the mine.

To assess what financial result (profit/loss) a mine achieves for a given volume of sales, it should be calculated according to Equation (16):

$$WF = (WAS - PRP) \cdot \sum\limits_{i=1}^{n} P_{msi} \cdot (c_i - kjz_i) \ (PLN) \tag{16}$$

For variant I, the following financial result is obtained after substituting the data (according to Equation (16)):

$$WF = (29.904 - 29.903) \cdot [25,500 \cdot (550 - 185) + 38,250 \cdot (485 - 185) + \ldots + 170,000 \cdot (310 - 185)]$$
$$WF = 598.90 \text{ (PLN)} (\text{€}132.50)$$

When analyzing variant II, it should be stated that the *WAS* value is (according to Equation (15)):

$$WAS = \frac{7600 \cdot (550 - 185) + 11,440 \cdot (485 - 185) + \ldots + 50,870 \cdot (310 - 185)}{25,500 \cdot (550 - 185) + 38,250 \cdot (485 - 185) + \ldots + 170,000 \cdot (310 - 185)} \cdot 100 = 29.899(\%)$$

WAS < *PR(P)* so selling in such quantities makes the mine a loss (Equation (16)):

$$WF = (29.899 - 29.903) \cdot [25,500 \cdot (550 - 185) + 38,250 \cdot (485 - 185) + \ldots + 170,000 \cdot (310 - 185)]$$
$$WF = -2,871.10(\text{PLN})(-\text{€}635.20)$$

In variant III, the *WAS* value is (Equation (15)):

$$WAS = \frac{9500 \cdot (550 - 185) + 5360 \cdot (485 - 185) + \ldots + 64,000 \cdot (310 - 185)}{25,500 \cdot (550 - 185) + 38,250 \cdot (485 - 185) + \ldots + 170,000 \cdot (310 - 185)} \cdot 100 = 29.791(\%)$$

WAS < *PR(P)* so selling in such quantities also makes the mine a loss (Equation (16)):

$$WF = (29.791 - 29.903) \cdot [25,500 \cdot (550 - 185) + 38,250 \cdot (485 - 185) + \ldots + 170,000 \cdot (310 - 185)]$$
$$WF = -86,531.10 \text{ (PLN)} (-\text{€}19,144.05)$$

For variant IV, the *WAS* value is:

$$WAS = \frac{5000 \cdot (550 - 185) + 5000 \cdot (485 - 185) + \ldots + 90,000 \cdot (310 - 185)}{25,500 \cdot (550 - 185) + 38,250 \cdot (485 - 185) + \ldots + 170,000 \cdot (310 - 185)} \cdot 100 = 31.009(\%)$$

WAS > *PR(P)* so selling in such quantities makes a profit for the mine (according to Equation (16)):

$$WF = (31.009 - 29.903) \cdot [25,500 \cdot (550 - 185) + 38,250 \cdot (485 - 185) + \ldots + 170,000 \cdot (310 - 185)]$$
$$WF = 847,968.89 \text{ (PLN)} (\text{€}187,603.74)$$

In variant V, the *WAS* value is:

$$WAS = \frac{0 \cdot (550-185) + 0 \cdot (485-185) + 48,462.05 \cdot (395-185) + 0 \cdot (330-185) + 120,000 \cdot (310-185)}{25,500 \cdot (550-185) + 38,250 \cdot (485-185) + \ldots + 170,000 \cdot (310-185)} \cdot 100$$
$$WAS = 29.903(\%)$$

WAS = *PR(P)* so selling in such quantities makes the mine zero profit. The mine is at the break-even.

5. Discussion and Conclusions

As can be observed, the proposed *WAS* indicator is an important element of the multi-asset threshold analysis, which allows a quick assessment of the financial result on the currently realised sales volume by comparing *WAS* with *PR(P)*. The practical usefulness increases when analysing numerous variants of mine production scenarios (the author analysed scenarios of 1000 and 10,000). The methods of multi-asset break-even analysis available in the literature, based solely on the quantitative and valuable recognition, do not meet the practical usefulness.

The author hopes that the examples cited have helped the reader to understand the issue of the multi-assortment threshold and prove the thesis that the proposed method is of practical use, especially that the percentage threshold is relevant to solving this issue.

The method I propose for calculating and analysing the break-even in multi-assortment production is an effective tool for supporting rational production decisions. Its usefulness lies in its simplicity and the possibility of obtaining an unambiguous result. It has been positively verified during my research for its practical applicability in the mining industry.

The sole knowledge of the quantitative threshold and the aggregate ad valorem threshold does not make it possible to state unequivocally in what position (in relation to the threshold level) the mine is. Only the proposed percentage threshold creates such clarity.

The percentage threshold I developed makes it possible to assess the company's economic health, analyse its dynamics, and allow comparisons with other mines in the industry.

Conditions, both internal and external, force a verification of the set objectives and significantly affect the sales level of individual assortments, their share in the production structure, and the sales price. All these considerations necessitate the importance of determining the percentage of break-even.

Funding: This research was prepared as part of AGH University of Science and Technology in Poland, scientific subsidy under number: 16.16.100.215.

Institutional Review Board Statement: Not applicable.

Informed Consent Statement: Not applicable.

Data Availability Statement: The data presented in this study are new and have not been previously published.

Conflicts of Interest: The author declares no conflict of interest.

References

1. Nowak, E. *Rachunkowość Zarządcza*; CeDeWu Sp. z o.o.: Warszawa, Poland, 2018; pp. 88–95. (In Polish)
2. Nowak, E. *Zaawansowana Rachunkowość Zarządcza*; PWE: Warszawa, Poland, 2003; pp. 84–86. (In Polish)
3. Orlov, O.; Ryasnykh, R.; Dumanska, K.; Savchenko, O. Scientific approach to quantitative measurement and economic processes research in corporate management. *SHS Web Conf.* **2021**, *107*, 1–7. [CrossRef]
4. Wermut, J. *Rachunkowość Zarządcza*; ODDK: Gdańsk, Poland, 2000; pp. 171–191. (In Polish)
5. Correa, H. Break-even analysis for multiple product firms. *Socio Econ. Plan. Sci.* **1984**, *18*, 211–217. [CrossRef]
6. Sobańska, I.; Czarnecki, J.; Wnuk-Pel, T. *Rachunek Kosztów i Rachunkowość Zarządcza*; C.H.BECK: Warszawa, Poland, 2003; pp. 164–169. (In Polish)
7. Zwirbla, A. *Nowe Podejście do Analizy Progu Rentowności. Produkcja Wieloasortymentowa*; PWN: Warszawa, Poland, 2015; pp. 123–142. (In Polish)
8. Alnasser, N.; Shaban, O.S.; Al-Zubi, Z. The Effect of Using Break-Even-Point in Planning, Controlling, and Decision Making in the Industrial Jordanian Companies. *Int. J. Acad. Res. Bus. Soc. Sci.* **2014**, *4*, 626–636. [CrossRef]
9. Oppusunggu, L.S. Importance of Break-Even Analysis for the Micro, Small and Medium Enterprises. *Int. J. Res. Granthaalayah.* **2020**, *8*, 212–218. [CrossRef]
10. Fuksa, D. *Metoda Oceny Wpływu Zmiennego Zapotrzebowania Odbiorców Węgla Kamiennego na Efektywność Funkcjonowania Wielozakładowego Przedsiębiorstwa Górniczego*; Wydawnictwa AGH: Kraków, Poland, 2012. (In Polish)
11. Fuksa, D. Concept of determination and analysis of the break-even point for a mining enterprise. *Arc. Min. Sci.* **2013**, *58*, 395–410. [CrossRef]

Article

Estimation of Damage Induced by Single-Hole Rock Blasting: A Review on Analytical, Numerical, and Experimental Solutions

Mahdi Shadabfar [1,*], Cagri Gokdemir [2,*], Mingliang Zhou [2], Hadi Kordestani [3] and Edmond V. Muho [4]

[1] Department of Civil Engineering, Sharif University of Technology, Tehran 145888-9694, Iran
[2] Department of Geotechnical Engineering, College of Civil Engineering, Tongji University, Shanghai 200092, China; zhoum@tongji.edu.cn
[3] Structural Vibration Group, Qingdao University of Technology, Qingdao 266033, China; hadi@qut.edu.cn
[4] Department of Disaster Mitigation for Structures, College of Civil Engineering, Tongji University, Shanghai 200092, China; edmondmuho@tongji.edu.cn
* Correspondence: mahdi.shadabfar@sharif.edu (M.S.); cagri.gokdemir@gmail.com (C.G.)

Abstract: This paper presents a review of the existing models for the estimation of explosion-induced crushed and cracked zones. The control of these zones is of utmost importance in the rock explosion design, since it aims at optimizing the fragmentation and, as a result, minimizing the fine grain production and recovery cycle. Moreover, this optimization can reduce the damage beyond the set border and align the excavation plan with the geometric design. The models are categorized into three groups based on the approach, i.e., analytical, numerical, and experimental approaches, and for each group, the relevant studies are classified and presented in a comprehensive manner. More specifically, in the analytical methods, the assumptions and results are described and discussed in order to provide a useful reference to judge the applicability of each model. Considering the numerical models, all commonly-used algorithms along with the simulation details and the influential parameters are reported and discussed. Finally, considering the experimental models, the emphasis is given here on presenting the most practical and widely employed laboratory models. The empirical equations derived from the models and their applications are examined in detail. In the Discussion section, the most common methods are selected and used to estimate the damage size of 13 case study problems. The results are then utilized to compare the accuracy and applicability of each selected method. Furthermore, the probabilistic analysis of the explosion-induced failure is reviewed using several structural reliability models. The selection, classification, and discussion of the models presented in this paper can be used as a reference in real engineering projects.

Keywords: rock explosion; explosion-induced damage; crushed and cracked zones; failure probability

Citation: Shadabfar, M.; Gokdemir, C.; Zhou, M.; Kordestani, H.; Muho, E.V. Estimation of Damage Induced by Single-Hole Rock Blasting: A Review on Analytical, Numerical, and Experimental Solutions. *Energies* **2021**, *14*, 29. https://dx.doi.org/10.3390/en14010029

Received: 17 November 2020
Accepted: 14 December 2020
Published: 23 December 2020

Publisher's Note: MDPI stays neutral with regard to jurisdictional claims in published maps and institutional affiliations.

Copyright: © 2020 by the authors. Licensee MDPI, Basel, Switzerland. This article is an open access article distributed under the terms and conditions of the Creative Commons Attribution (CC BY) license (https://creativecommons.org/licenses/by/4.0/).

1. Introduction

To fracture in-situ rock mass and prepare it for subsequent drilling and transport, explosion is widely used in the mining industry. In such conditions, run of mine fragmentation is assumed to be good when it is fine and loose enough to provide an efficient digging and loading operation [1]. Thus, significant attention has been drawn to estimating explosion-induced damage size in rock mass. The primary objective of research in this area has been to tailor blast fragmentation as well as to optimize mineral extraction and recovery cycle [2].

It should be noted that large amounts of fine materials are also produced by the crushed zone induced around the blast hole [3]. Thus, increasing the amount of fines multiplies handling and processing costs and, in many cases, reduces product value. Additionally, in some cases, such as quarry production, generated fines are recognized as waste. The volume of such wasted fines in Europe alone has been estimated to be about 500 million tons per year [4]. Thus, determining the size of the crushed zone and produced fines appears to be necessary [5]. Moreover, damage size at the perimeter of an excavation is of importance once the so-called drill and blast techniques are used for rock excavation [6]. In this area, minimizing explosion-induced damage is the main objective. This principle also needs to be considered, for example, in walls of drifts and other underground openings as well as slopes of surface mines. The damage penetrated through the walls and slopes is taken into account as unwanted damage or overbreak. This type of damage caused by explosion can thus have a direct impact on the stability and performance of the main structure [7]. Accordingly, diminishing such damages can:

- prevent possible damage to adjacent structures [8],
- enhance the stability of roof and side walls [9],
- improve excavation rate,
- reduce manufacturing expenses, and
- cut down operating costs [10].

In summary, by controlling the size of the crushed zone, one can optimize blasting fragmentation and minimize produced fine materials and recovery cycle. At the same time, optimizing the cracked zone can lead to a reduction in damage beyond the expected excavation boundary, control undesirable damage, and fit the explosion scheme to the geometric design [11]. That is why one main objective in rock blasting operations has been to keep unwanted damage under control [12]. To meet this objective, it is essential to understand and predict destruction caused by explosion [13].

In this paper, various methods associated with measuring and estimating explosion-induced rock damage are carefully studied and classified into different groups. To this end, this review is organized as follows: (1) the whole process of single-hole blasting is described from detonation initiation to wave propagation and rock vibration, and details are separately provided for each step (Section 2); (2) rock explosion damage is classified into different groups and illustrated schematically (Section 3); (3) different models are classified for assessing the size and severity of rock damage and presented in three different groups of analytical, numerical, and laboratory methods (Section 4); (4) the most commonly used methods are adopted in 13 case studies, and their results are compared (Section 5.1); (5) probabilistic methods are examined for calculating failure probability caused by rock explosion, and their potential differences are described compared to deterministic methods (Section 5.2).

2. A Review on Explosion Mechanism

In this section, the single-hole blasting process is examined and then the impact of induced detonation wave is described on surrounding rock medium. For this purpose, consider a single blast hole containing an explosive charge, as shown in Figure 1. Assume that a gauge is placed at point B to record explosion history. The results would be similar to those plotted in Figure 2a,b, wherein pressure–time (p–t) and pressure–distance (p–y) graphs induced by the explosion are schematically presented [4].

Detonation begins from the bottom of the borehole, i.e., point A, which corresponds to point O in the p–t graph. At this moment, the detonation pressure is still zero because a portion of the explosion is yet to be recorded at point B. Then, the detonation wave travels from the bottom of the hole to point B. This part corresponds to OE on the p–t graph, wherein the pressure is still zero. Once the wave front arrives at point B, the p–t graph encounters a sudden peak and the induced pressure reaches its maximum value. This peak point is called the Von-Neumann spike. Next, the detonation wave passes point B toward point C. Consequently, the p–t graph drops sharply from E to F. Once the detonation wave touches point C, some part of the wave goes through stemming, with the remaining part reflecting back into the blast hole. Afterwards, the detonation wave moves through stemming, reaches the collar, and subsequently moves back from point D to point C. In the meantime, the pressure at point B decreases from F to G. After that, the induced wave travels toward point B and consequently the pressure drops from G to H. Following this step, the detonation wave moves toward the bottom of the blast hole and then gradually dissipates through the surrounding cracks and damages and leaks away from the system. This process continues until the borehole pressure reaches the atmospheric pressure. Correspondingly, the p–t graph is slowly reduced from point H to zero.

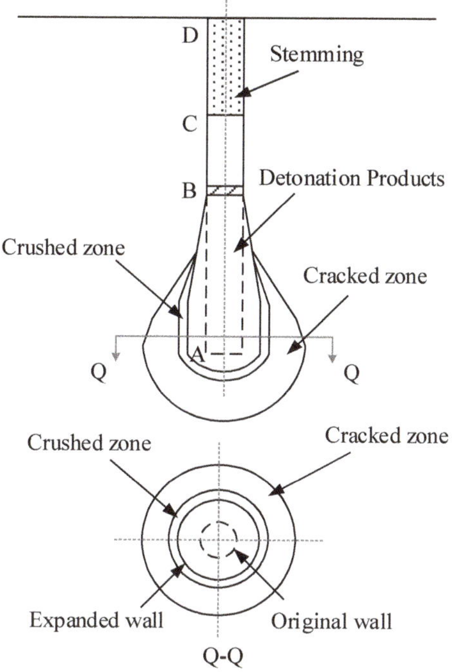

Figure 1. The blast hole and surrounding damages.

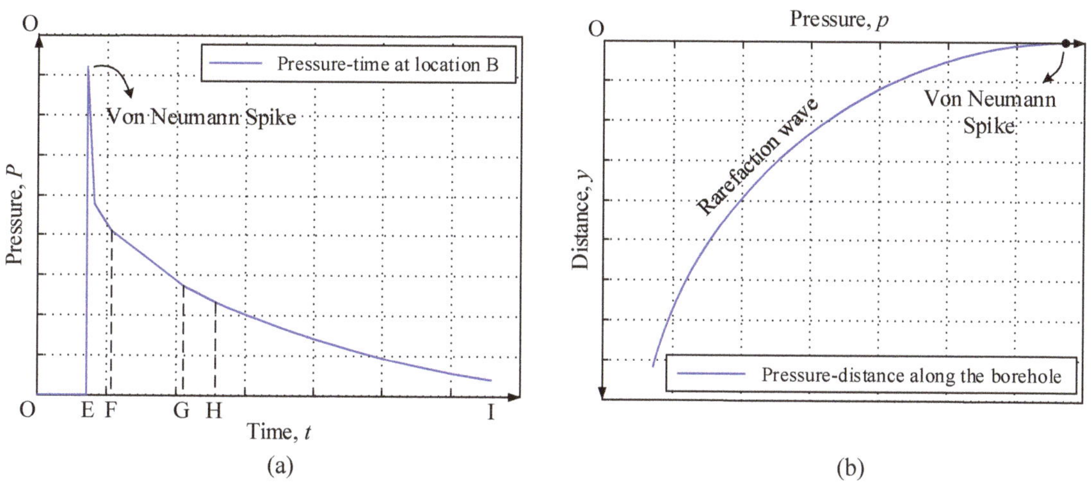

(a)

(b)

Figure 2. The rock mass response to explosion wave in the form of (**a**) p–t and (**b**) p–y graphs.

Pressure changes for different positions along the borehole from point B (wherein the p–t graph is at the maximum point) downward are illustrated in Figure 2b. As demonstrated, the pressure decreases sharply as it goes farther from point B due to borehole expansion, crushing and cracking of surrounding rock medium, explosion-induced gas leakage, etc.

It should be also noted that the process described above is more accurate for cases, in which the blast hole is long enough and the reflected wave in the borehole is neglected. In practice, multiple reflections of the wave from the bottom and the collar as well as the interaction between the wave and the lateral boundaries of the borehole can produce some fluctuations in pressure history. Thus, the actual p–t and p–y graphs are not perfectly smooth and exhibit some fluctuations. The next point to highlight is the role of stemming in the extension of the p–t graph. In fact, stemming causes the detonation wave to trap into the blast hole, making the detonation energy focused on fragmentation and breakage; this issue extends pressure history and consequently enhances explosion efficiency. More precisely, denoting stemming wave velocity and stemming length by C_s and L_{cd}, respectively, the time duration of pressure history increases as $t = 2L_{cd}/C_s$, provided that stemming is correctly placed. Without stemming, the explosion-induced gas tends to escape from the collar and the described t cannot be saved anymore. It will then waste energy and decrease explosion efficiency [4,14,15]. However, it should be noted that calculating the exact optimal stemming length is very difficult and challenging as the flow of energy during the explosion cannot be modeled accurately and usually accompanies with a margin of uncertainty. What can be mentioned with certainty is that the more the so-called energy leakage gaps are reduced, the better the energy released by the explosion can be used for fragmentation purposes [16].

3. Damage Pattern

Following the explosion, the pressure waves are rapidly released and strongly vibrate the rock environment [11,17]. The resulting vibration, occurring in a fraction of a second, stimulates mechanical and dynamic characteristics of rock mass [18]. This stimulation correspondingly generates a series of tensions and stresses on existing discontinuities, and also contributes to the opening and expansion of joints, depending on rock toughness [19,20].

First, the blast hole is relatively expanded [21,22]. Then, discontinuities increase and lead to formation of an unstable crushed zone due to the growth of fine cracks [23]. On the other hand, cracks affected more by the blast shock go beyond the crushed zone and penetrate radially into the surrounding environment [24,25]. Beyond the crushed zone and radial cracks, the effects of the explosion are observed in the form of ground vibration [26,27]. These three sections are shown in Figure 3.

Thus, the effects of single-hole explosion can be summarized in four steps:

- The blast hole is expanded.
- A crushed zone is formed surrounding the blast hole.
- Radial cracks penetrate through the rock, causing a cracked zone.
- Explosion-induced waves affect the surrounding environment, producing some ground vibrations.

The pattern of damage generated around the explosion point is initially found in practical projects but later proved in experimental models. For instance, Olsson and Ouchterlony [28] showed the pattern generated in experimental models. However, it should be noted that, in practice, the zones mentioned above are interconnected without any sharp boundaries that help in distinguishing them from one another. However, the definition provided for damage pattern can greatly help in establishing models and calculating results.

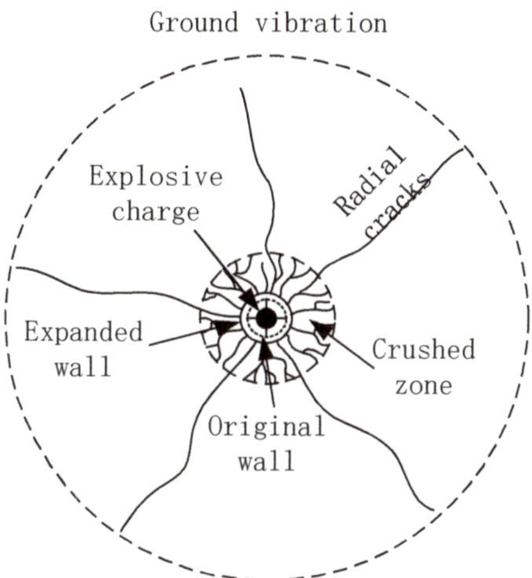

Figure 3. Crack formation around a blast hole.

4. Estimation of Induced Damage

A shock wave is initially generated once an explosion charge is fired. Subsequently, a stress wave affects the surrounding environment, creating two damage zones near the blast hole, namely crushed and cracked zones. As discussed, sizes of these two damage zones are of importance for an optimal blast design. As far as the project client can estimate the size of damage zones (i.e., crushed and cracked zones) as a function of input parameters such as rock properties and explosive characteristics, the optimal values of input parameters can be obtained using a blast design optimization. This optimization can be done through a try-and-error process to obtain the optimal values of target parameters or can be mathematically implemented in an optimization algorithm [29–31].

Different researchers have proposed various methods to approximate the induced damage. In a general case, the size of a damage zone can be assumed as a function of input parameters such as:

$$r = f(\theta_1, \theta_2, \ldots, \theta_n), \tag{1}$$

where r is the damage zone radius and θ_1 to θ_n represent the input parameters (either rock or explosion characteristics). The most important input parameters are outlined in Table 1.

Table 1. Main parameters involved in explosion-induced rock damages.

No.	Sources	Parameters	Description
1		E_d	Young's modulus of rock
2		v_d	Poisson's ratio of rock
3	Rock characteristics	σ_c	Uniaxial compressive strength of rock
4		F_c	Confined compressive strength of rock
5		T	Tensile strength of rock
6		ρ_0	Unexploded explosive density
7	Explosive characteristics	D_{CJ}	Ideal detonation velocity
8		r_0	Blast hole radius
9		Q_{ef}	Effective energy of explosive

This form of damage estimation can be simply applied to a particular case study with virtually no complexities. However, it is not always easy to provide a close-form function such as $f(\bullet)$ since several sources of uncertainties are included in explosion-associated problems [32–34]. Numerous research studies are also available in the related literature to estimate damage in rock and soil media [35–37]. In an overall view, these methods can be categorized into three main groups: analytical, numerical [38], and experimental [39].

Approaches toward problems in each of these three methods are not the same. In analytical techniques, a parameter such as peak particle velocity (PPV) [40–42], borehole pressure [43,44], or explosion pressure [45] is generally presented as a critical factor to estimate the size of a damage zone. Next, two different approaches are used to provide a solution; either an analytical calculation is employed to directly determine both the critical parameter and the damage zone, or the relationship between the parameter and the damage zone is estimated and the rest of the problem remains unsolved for the reader. In numerical methods, however, an algorithm such as finite element method (FEM), finite difference method (FDM), discrete element method (DEM), etc. is used to evaluate changes in the stress field surrounding an explosion point and examine consequent issues including induced cracks and damages [46–48]. In experimental approaches, some laboratory or in-situ tests are utilized to develop an empirical relationship to estimate the size and dimension of damages [49].

In the following, each of these three methods is separately addressed, and then related previously developed research works are listed and explained in more detail.

4.1. Analytical Approach

As previously described, in analytical approaches, a feature of a model is selected as a main parameter, and it is determined how this parameter is distributed around the blast hole. The relationship between the parameter and rock damage is then examined so that damage size can be measured for each parameter value. Peak ground velocity and borehole pressure are assumed as two parameters widely used for this purpose [50]. Analytical approaches based on PPV and borehole pressure are discussed in Sections 4.1.1 and 4.1.2, respectively.

4.1.1. Damage Prediction Using PPV

PPV is known as one of the critical parameters, used by several researchers, to estimate damage zones [51]. Accordingly, damage rate can be predicted if PPV is estimated in different areas in rock environments. Some of the PPV-based criteria for blast-induced damage in rocks are presented in Tables 2 and 3.

Table 2. PPV-based criteria for blast-induced damage in rock (adopted from Bauer and Calder [52]).

PPV (mm/s)	Effects of Damage
<250	No fracture of intact rock
250–635	Occurrence of minor tensile slabbing
635–2540	Strong tensile and some radial cracking
>2540	Complete break-up of rock mass

Table 3. PPV-based criteria for blast-induced damage in rock (adopted from Mojtabai and Beattie [53]).

Rock Type	Uniaxial Strength (MPa)	RQD (%)	PPV (mm/s)		
			Minor Damage	Medium Damage	Heavy Damage
Soft schist	14–30	20	130–155	155–355	>355
Hard schist	49	50	230–350	305–600	>600
Shultze granite	30–55	40	310–470	470–1700	>1700
Granite porphyry	30–80	40	440–775	775–1240	>1240

Two requirements need to be met when using PPV in damage estimation and structural control: (1) the PPV at the desired location should be predicted; (2) the relationship between the PPV and the damage state (i.e., fragility curve) should be provided. In practice, PPV vectors surrounding a blast hole are difficult to be developed, and there are not many sources available in this area. To further explain this issue, the proposed model by Holmberg and Persson can be noted [40,54], since they offered the following equation to estimate PPV:

$$V = K\frac{W^\alpha}{R^\beta},$$ (2)

where V shows PPV, K, α, and β are the empirical constants, W is the charge weight unit, and R denotes the distance unit from the charge. However, this equation is developed for areas far from the explosion point. In fact, the given equation is applicable for areas, in which R is so large that it makes blast hole dimension negligible, and cannot be used for areas close to the explosion point, where a blast hole dimension should be considered.

To solve this problem, Holmberg and Persson [54] assumed that the explosive charge has a cylindrical shape. Accordingly, they divided the explosive charge into small pieces so that each of the pieces has a length of dx and unit charge concentration of q (kg/m). Then, they stated that PPV in an arbitrary point such as (r_0, x_0) can be calculated as follows:

$$V = K\left(q \int_T^{H+J} \frac{dx}{(r_0^2 + (x - x_0)^2)^{\frac{\beta}{2\alpha}}}\right)^\alpha,$$ (3)

where T is the stemming depth (m) [55], H is the charge length (m), and J is the subdrill (m) [56], as demonstrated in Figure 4.

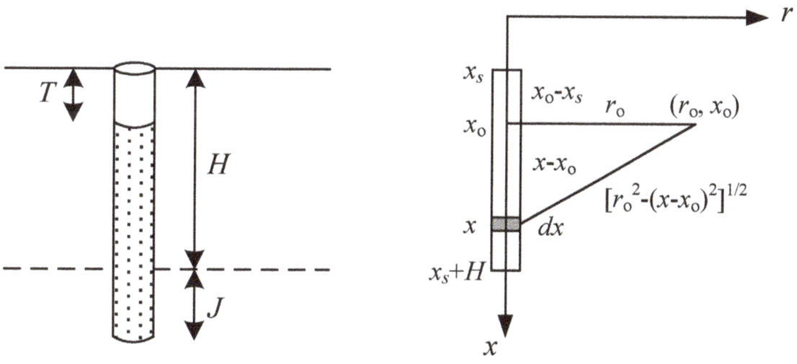

Figure 4. The explosive charge and partitioning.

The parameter α was assumed as follows:

$$\alpha = \frac{\beta}{2}.$$ (4)

After integrating from Equation (3), PPV was explicitly presented as follows:

$$V = K\left(\frac{q}{r}(\tan^{-1}(\frac{H+J-x_0}{r_0}) - \tan^{-1}(\frac{T-x_0}{r_0}))\right)^\alpha.$$ (5)

The values of K, α, and β for hard rock mass are 700, 0.7, and 1.4, respectively. Thus, by having q, the PPV amount can be calculated for any desired distance (r). In this respect, Changshou Sun [6] presented Table 4 for Scandinavian bedrock to approximate rock damage based on the induced PPV.

Table 4. Damage type based on induced PPV in a Scandinavian bedrock.

PPV (m/s)	Tensile Stress (MPa)	Strain Energy (J/kg)	Typical Effect in Hard Scandinavian Bedrock
0.7	8.7	0.25	Incipient swelling
0.1	12.5	0.5	Incipient damage
2.5	31.2	3.1	Fragmentation
5	62.4	12.5	Good fragmentation
15	187	112.5	Crushing

Although this appears to be a very simple and practical method, later, Hustrulid and Lu [57] reiterated that the main integral suffered from a basic mistake and the accurate form of the integral was:

$$V = Kq^{\alpha} \int_{T}^{H+J} \left(\frac{dx}{(r_0^2 + (x - x_0)^2)^{\frac{\beta}{2\alpha}}} \right)^{\alpha}, \tag{6}$$

which could not be solved obviously in an analytic manner. Moreover, the proposed method has other problems; for instance:

- Only the magnitude of the PPV is considered and the direction of the PPV is neglected.
- Only the explosive weight is taken into account, and other characteristics are ignored.
- To determine the parameters K, α, and β, further laboratory or in-situ tests are required, which are difficult to conduct.

For such reasons, these types of approaches have not been highly welcomed by research communities.

4.1.2. Damage Prediction Using Borehole Pressure

Borehole pressure is known as one of the common parameters in the estimation of rock damage. Accordingly, it is believed that the initiation and propagation of cracks in rocks are due to severe stress caused by explosion waves. Thus, borehole pressure is being used by many researchers to directly estimate the crushed zone size in rock environments. The next sections present a summary of such studies.

Mosinets' Model

Mosinets et al. [58] expressed the radius of damage zones (either cracked or crushed ones) surrounding the blast hole by the following equation:

$$r = k\sqrt[3]{q}, \tag{7}$$

where r is the damage zone radius, k represents the proportionality coefficient, and q refers to the charge weight in the TNT equivalent. Each damage zone also has its own coefficient k. For the crushed zone, the coefficient k is as follows:

$$k = \sqrt{\frac{V_s}{V_p}}, \tag{8}$$

where V_p shows the longitudinal wave velocity and V_s denotes the transverse wave velocity.

Drukovanyi' Model

Assuming an isotropic and incompressible granular medium with cohesion (derived from Il'yushin's model [59]), Drukovanyi et al. [60] examined behaviors of rocks in the zone of fine crushing. Considering a plane strain for detonation of a column of explosive material in rock mass, they theoretically developed the following relation to determine the crushed zone radius (r_c) close to the blast hole:

$$r_c = r_0 \left(\frac{P_b}{-\frac{C}{f} + (\sigma_c + \frac{C}{f})L^{\frac{f}{1+f}}} \right)^{\frac{1}{2\gamma}} \sqrt{L},$$ (9)

where r_0 stands for the borehole radius (mm), P_b is the borehole pressure (Pa), C shows cohesion (Pa), f refers to the internal friction coefficient ($f = tan(\varphi)$), σ_c is the uniaxial (unconfined) compressive strength (Pa), γ denotes the adiabatic expansion constant of explosive, and L represents a constant defined by:

$$E = \frac{\frac{E}{1+\nu}}{\sigma_c \left(1 + \ln(\frac{\sigma_c}{T})\right)},$$ (10)

where E is the Young's modulus (Pa), ν refers to the Poisson's ratio, and T shows the tensile strength (Pa) of the rock. Drukovanyi's model assumes that the borehole pressure is calculated as follows:

$$P_b = \frac{1}{8}\rho D^2,$$ (11)

where ρ shows the explosive density and D is the detonation velocity. They also referred to Il'yushin's model and considered $\gamma = 3$, $\rho = 0.9$ gr/cm^2, and $D = 4000$–6000 m/s.

It was noted by Drukovanyi et al. [60] that damage predicted using this model can give values higher than reality to rocks by the compressive strength less than 100 MPa. Moreover, it has been reported in the related literature that this approach is limited to cases, where the main mode of failure is compression [6].

Senuk's Model

Senuk [61] developed the following relation to estimate the cracked zone radius in the vicinity of a cylindrical explosive charge:

$$r_c = kr_0 \sqrt{\frac{P_b}{T}},$$ (12)

where k is the stress concentration factor in sharp cracks and joints, which is assumed to be approximately equal to 1.12, P_b shows the blast hole pressure approximated by $P_b = \rho_0 D_{CJ}^2/8$, and T represents the rock tensile strength. Thus, Equation (12) can be rewritten as follows:

$$r_c = 1.12r_0 \sqrt{\frac{\rho_0 D_{CJ}^2}{8T}}.$$ (13)

Szuladzinski's Model

According to the hydrodynamic theory for rock explosion, Szuladzinski [62] proposed a model to predict the crushed zone radius around a blast hole. In this model, the rock environment in the vicinity of the blast hole was assumed as an elastic medium with cracking and crushing capabilities. The effective explosion energy in the model was also roughly assumed as two-thirds of the total explosive energy, and decoupling effects were ignored. The relation presented in this study for the crushed zone radius is as follows:

$$r_c = \sqrt{\frac{2r_0^2 \rho_0 Q_{ef}}{F_c'}},$$ (14)

where r_0 (mm) is the borehole radius, ρ_0 (g/mm^3) shows the explosive density, Q_{ef} (Nmm/g) represents the effective explosive energy (assumed to be two-thirds of the complete reaction heat), and F_c' (MPa) refers to the confined dynamic compressive strength of rock mass (assumed to be approximately eight times of the unconfined static compressive strength, σ_c [63–66]).

SveBeFo Model

The Swedish Engineering Research Organization (SveBeFo) conducted several studies on the initiation and propagation of cracks in rock environments under explosion load. Based on the results of these studies, Ouchterlony [67] presented the following relation to calculate the length of induced radial cracks:

$$2\frac{r_{co}}{d_h} = \left(\frac{P_b}{P_{b,cr}}\right)^{2/(3(\frac{D}{c})^{0.25}-1)}, \tag{15}$$

where r_{co} denotes the unanchored radius of the cracked zone, d_h shows the borehole diameter, P_b is the borehole pressure, D represents the velocity of detonation (VOD), c is the sound speed in a rock environment, and $P_{b,cr}$ denotes an experimental parameter showing critical borehole pressure which can be estimated by the following equation:

$$P_{b,cr} = 3.3\frac{K_{IC}}{\sqrt{d_h}}, \tag{16}$$

where K_{IC} is the fracture toughness of rock mass and d_h shows the borehole diameter. Accordingly, Ouchterlony [67] introduced the following equation to estimate the borehole pressure:

$$P_b = \frac{\gamma^\gamma \rho_0}{(\gamma+1)^{\gamma+1}} D^2 \left(\frac{d_e}{d_h}\right)^{2.2}, \tag{17}$$

where γ is an isotropic exponent for a specific explosive (1.254–2.154), ρ_0 denotes the explosive density, D stands for VOD, and d_e/d_h represents the ratio of explosive diameter to borehole diameter called the decoupling ratio. Later, Ouchterlony et al. [68] provided the following correction coefficients to improve the given method:

$$r_c = R_{co}F_h F_t F_r F_b, \tag{18}$$

where R_{co} stands for the corrected damage zone radius, F_h is correction for hole spacing, F_t shows correction for time spread in initiation, F_r stands for correction for wet holes, and F_b is correction for fracturing. These coefficients could be determined based on the concept of fracture mechanics, which is complicated for conventional engineering design. Moreover, it is not easy to determine rock fracture toughness, i.e., K_{IC}, especially for weak rocks. These issues led to limited applications of the given method in research works and consequently no extensive use of the method in practical designs.

Quasi-Static Model

Assuming a balance between borehole pressure and stress distributed in the surrounding rock medium, Sher and Aleksandrova [69,70] provided a model to predict crack size around a cylindrical borehole. In this model, a dynamic process was approximated by a quasi-static method, and the following equations were then proposed to estimate the radius of the cracked zone:

$$\left(\frac{Y}{E\alpha} - \frac{q}{E}\right)\left(\frac{r_d}{r}\right)^{\alpha/(1+\alpha)} - \frac{Y}{E\alpha} - \frac{P_h}{E} = 0, \tag{19}$$

$$\frac{q}{E} = -\frac{\dfrac{\sigma_c}{E} + 2\left(\dfrac{\sigma}{E} \times \dfrac{\sigma_c}{T}\right)}{1 + \dfrac{\sigma_c}{T}}, \tag{20}$$

$$\frac{u_b}{r_h} = -(1+v)\frac{r_d}{r_h}\frac{q + 2\sigma(1-v)}{E}, \tag{21}$$

$$\left(\frac{r}{r_h}\right)^2 - 1 = \left(\frac{r_d}{r_h}\right)^2 - \left(\frac{r_d}{r_h} - \frac{u_b}{r_h}\right)^2, \tag{22}$$

$$\frac{P_h}{E} = \frac{P_{CJ}}{E}\left(\frac{r}{r_h}\right)^{-2\gamma_1}, \quad r \leq r^*, \tag{23}$$

$$\frac{P_h}{E} = \frac{P_{CJ}}{E}\left(\frac{r}{r_h}\right)^{-2\gamma_1}\left(\frac{r^*}{r_h}\right)^{-2\gamma_2}, \quad r > r^*, \tag{24}$$

where r_h denotes the initial hole radius, r shows the final hole radius, u_b is the elastic deformation of rock, r/r_h represents the ratio of final radius to initial radius of hole, r^* refers to the radius at which the adiabatic constant changes, γ_1 is the initial adiabatic expansion constant ($\gamma_1 = 3$), and γ_2 signifies the final adiabatic expansion constant ($\gamma_2 = 1.27$). Both α and β parameters can be calculated as follows:

$$\alpha = \frac{2\sin\phi}{1 - \sin\phi}, \tag{25}$$

$$\beta = \frac{2c \times \cos\phi}{1 - \sin\phi}, \tag{26}$$

where c is the cohesion and ϕ refers to the internal friction angle. Based on the equations presented above, Hustrulid [71] provided the following process to determine the cracked zone radius:

Step 1 Calculate q/E from Equation (20)

Step 2 Approximate a value for r_d/r_h (this value is approximated in this step and later modified in a cyclic process)

Step 3 Substitute q/E and r_d/r_h in Equation (21) and calculate u_b/r_h

Step 4 Substitute u_b/r_h in Equation (22) and determine r/r_h

Step 5 Select one of Equations (23) or (24) based on a comparison between r/r_h and $r^*/r_h = 1.89$ and then calculate P_h/E

Step 6 Substitute P_h/E in Equation (19) to assess if equality is achieved (if so, r_d/r_h is the final answer. Otherwise, the steps 2–6 should be repeated until the final answer is reached).

Djordjevic's Model

Based on the Griffith's failure criterion, Djordjevic [72] developed a model mostly applicable for brittle rocks [2]. The crushed zone radius proposed in this study is:

$$r_c = \frac{r_0}{\sqrt{24T/P_b}}, \tag{27}$$

where r_0 (mm) is the blast hole radius, T (Pa) shows the tensile strength of rock material, and P_b (Pa) represents the borehole pressure.

Kanchibotla Model

Kanchibotla et al. [73] considered damage around a blast hole as a function of blast hole radius, explosion pressure, and uniaxial compressive strength of rock mass. Using these parameters, they proposed a relation to estimate the crushed zone radius as follows:

$$r_c = r_0 \sqrt{\frac{P_d}{\sigma_c}}, \tag{28}$$

where r_0 (mm) is the borehole radius, P_d (Pa) refers to the detonation pressure, and σ_c (Pa) denotes the unconfined compressive strength of rock.

Johnson's Model

Johnson [74] considered four different zones near a blast point including borehole, crushed zone, cracked zone (also named as transition zone), and no-damage zone (also labeled as seismic zone). To calculate the crushed zone radius in this study, the following formula was proposed:

$$Q_{cd} = P_b \sqrt{\frac{r_0}{r_c}} e^{-(r_c - r_0)\lambda}, \tag{29}$$

where σ_{cd} is the dynamic compressive strength of rock, P_b represents the borehole pressure, r_c shows the crushed zone radius, r_0 stands for the borehole radius, and λ is the crush damage decay constant determined by laboratory experiments.

Modified Ash's Model

Hustrulid [71] improved the Ash's model [63–66] and provided the following equation to approximate the size of the cracked zone around a blast hole:

$$r_c = 25 r_0 \left(\frac{d_e}{d_h}\right) \sqrt{RBS} \sqrt{\frac{2.65}{\rho_r}}, \tag{30}$$

where r_0 is the borehole radius, d_e/d_h shows the decoupling ratio, RBS stands for the relative bulk strength (compared with ammonium nitrate/fuel oil (ANFO)), and ρ_r refers to the rock density. RBS can be calculated as follows:

$$RBS = \frac{\rho_0 S_{ANFO}}{\rho_{ANFO}}, \tag{31}$$

where S_{ANFO} is the weight strength of explosive relative to ANFO, ρ_{ANFO} stands for the density of ANFO, and ρ_0 shows the explosive density.

4.2. Numerical Approach

When a rock is idealized as a continuous medium, continuum-based approaches such as FEM [75–81] or FDM [82–87] are usually employed to simulate the model. However, the rock is modeled as a set of structural units (such as springs, beams, etc.) or as separate particles bonded at contact point if modeling of discontinuities is required. In such cases, DEM [88–93], the bonded particle model [94] or hybrid methods [95–99] are used for numerical analysis. Typically, due to the high volume of calculations required in such methods, calculations are performed using software packages to simulate the model. Various packages have thus been developed for geotechnical modeling. In this regard, two famous DEM codes, UDEC and 3DEC, are widely utilized for modeling two- and three-dimensional cases of jointed rock, respectively.

Wang and Konietzky [46] used UDEC to study the initiation and propagation of damage around the blast hole in rock. They first used general-purpose multiphysics simulation software package (LS-DYANA) to model an explosion in intact rock and calculate the explosion load imposed on the borehole wall. Then, they returned to UDEC and modeled a jointed rock environment. They assigned the obtained load as a radial velocity history to the borehole wall and studied the damage evolution. Their coupled model is schematically shown in Figure 5.

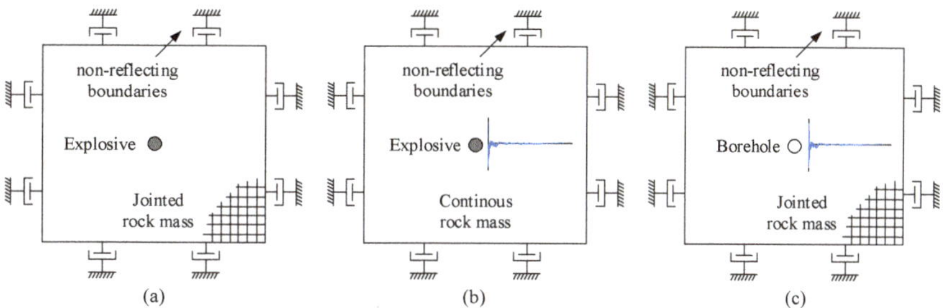

Figure 5. Coupled method illustrating the (**a**) physical model of explosion in jointed rock mass, (**b**) explosion history obtained via LS-DYNA, and (**c**) input of converted explosion history for UDEC simulation.

To model the jointed rock environment in UDEC, Wang and Konietzky [46] used two sets of joints: a randomly-generated series of polygonal joints (e.g., Voronoi joints) and two series of orthogonally aligned joints dipping at $-15°$ and $75°$. Having the load exerted on the blast hole, they began to study the growth of damage in rock. It was observed that some small damage formed around the blast hole and then penetrated through outer layers. By passing the time, the damage size continuously increased until $t = 5$ ms that it stopped growing further and reached at its maximum size. Later, it was noted that, at the time $t = 5$ ms, the maximum length of cracks reached to 2.5 m while the crushed zone radius varied between 7 to 9 times of the blast hole radius. This estimation is moderately higher than that in similar studies approximating the ratio below 5 between the crushed zone radius and the borehole radius (r_c/r_o).

Using the analytical code of the Realistic Failure Process Analysis 2D (RFPA2D), Liu et al. [100] analyzed rock mass behavior under explosion assuming different geometrical properties for joints including average distance from blast site to joints, length of joints, number of joints, and relative angles of joints. These properties are schematically shown in Figure 6.

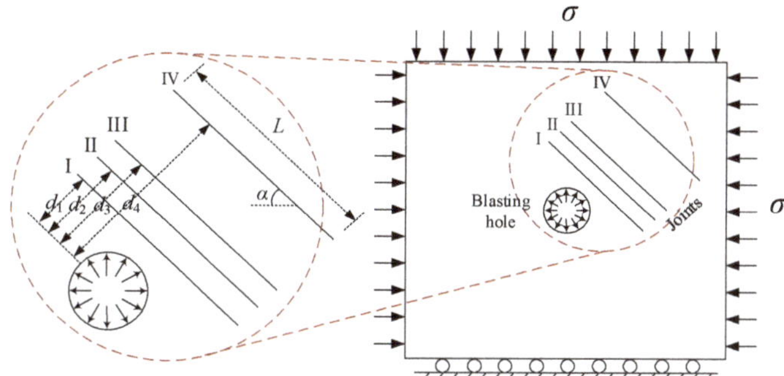

Figure 6. Single hole blasting model illustrating the model parameters including joints, the length of joints (L), distance from joints to blasting hole (d_i), and joint angle (α).

The results of this study showed that the effect of joints on blast-induced damage is slowly reduced as the average distance of joints from the blast hole increases. It was also observed that rock masses with long joints could experience more damage than those having short joints. Their results also revealed that the damage caused by the blast was significantly greater in rock masses with joints than in cases without joints. However, little change might be observed in damage intensity caused by an explosion when the number of joints exceeded 3. It was also concluded that joints could highly facilitate the propagation of cracks and play a significant role in determining their shapes, dimensions, and directions of propagation.

Moreover, Lak et al. [101] used a hybrid finite difference-boundary element method to investigate the propagation of blast-induced cracks around a wellbore. To this end, they planned two separate steps. First, they investigated the formation of radial cracks due to the propagation of shock waves caused by explosion using the dynamic finite difference method. Importing outputs into the second step, they then modeled the propagation of cracks caused by gas expansion using the quasi-static boundary element method.

Two types of cracks were considered in this study: type I and type II. The cracks along the direction of maximum horizontal stress (σ_{Hmax}) were considered as type I, and those along the direction of minimum horizontal stress (σ_{Hmin}) are marked as type II. The results of this study showed that the ratio of $\sigma_{Hmin}/\sigma_{Hmax}$ has an obvious effect on the radial cracks propagation. When $\sigma_{Hmin}/\sigma_{Hmax}$ has a small value near 0, the length of type I cracks is larger than that of type II cracks (roughly about twice, $a_2/a_1 = 0.5$). With the increase of the ratio of $\sigma_{Hmin}/\sigma_{Hmax}$, the length of type I cracks decreases and the length of type II cracks increases. Finally, in hydrostatic stress conditions, i.e., cases where $\sigma_{Hmin}/\sigma_{Hmax} = 1$, the length of type I and type II cracks turn out to be equal ($a_2/a_1 = 1$).

4.3. Experimental Approach

Explosion-induced damage in rocks occurs through a complex process. Therefore, laboratory studies have been exploited as one of the main methods to investigate this problem in the past few decades given the complexity of research in this area [102]. It should be noted that experimental studies of explosion phenomena as well as subsequent failures mainly deal with two main aspects [103]:

- Primary cracks due to the high amplitude of stress waves
- Further development of cracks due to gas penetration

In a study on stress-wave, Lownds [104] conducted a set of reduced-scale experiments in granite. A sensor hole was also placed at a distance of 150 mm away from a borehole with a diameter of 32 mm. A pressure transducer was subsequently placed in a water-coupled sensor hole with the same diameter as the borehole to measure stress waves. They recorded the effect of different coupling media on stress waves using the installed pressure transducer. In a similar study, Talhi et al. [105] measured the relative magnitude of stress pulses by placing a pressure gauge in a water-coupled sensor hole with a diameter of 8 mm to check the variation of stress waves caused by changes in decoupling ratio and borehole length. Through these methods, peak pressure at a relatively close distance to the borehole was measured providing a database of relative measurement for different borehole conditions. In addition, Teowee and Papilon [106,107] utilized piezoelectric sensors to measure the sympathetic pressure of adjacent deck charges and boreholes. For this purpose, they compared peak pressure between these piezoelectric gauges and concluded that the given sensors could measure peak pressure up to 138 MPa. Due to the relative size of the manometer and the applicability of measuring high pressure, piezoresistive gauges, such as carbon composite resistors, have been employed by different researchers to investigate pressure pulse caused by stress wave propagation [44,108].

Measuring damage size and pattern using induced gas pressure and penetration has also been investigated by different researchers. For example, the direct measurement of crack length was performed by Olsson and Bergqvist [109,110], Deghan Banadaki [111], and Nariseti [112] to reflect on blasting-induced rock fractures. Paventi and Mohanty [113] also shed light on the effect of coupling medium on crack pattern through laboratory-scale tests of samples with 10-cm diameter. Boreholes with diameters of 6 mm and 10 mm were thus equipped with pentaerythritol tetranitrate explosive in central position while containing different coupling media such as air, water, and clay. The results showed that fracture intensity was smallest in air-coupled boreholes followed by clay-coupled boreholes. Additionally, they reported that water-coupled boreholes led to intense fracturing near boreholes. Furthermore, Yamin [114] conducted reduced-scale experiments to examine damage extent around boreholes. To this end, pressure sensors were placed in sensor holes on concentric circles at different distances surrounding a borehole to measure gas pressure. It was consequently reported that gas penetration was observed from a 75-mm blast hole charged with a 40-mm emulsion cartridge up to a distance of about 15 borehole diameters. Additionally, Yamin [114], McHugh [115], and Brinkmann [116] investigated rock damage following gas penetration.

It should be noted that laboratory and experimental models to directly determine damage zones in rock under explosion load are difficult and costly [117]. The test site should also be safe and equipped with required facilities to measure rock behavior. Providing such conditions, however, is not simple. Therefore, there are few studies available in this area.

One of the existing approaches in this field is Esen's model [2]. To measure the crushed zone size around a blast hole, Esen et al. conducted a detailed laboratory test on 92 samples mostly made up of concrete with dimensions of 1.5 m $\times$ 1 m $\times$ 1.1 m. They consequently defined a crushing zone index (CZI) based on the crushing process around the explosion point as follows:

$$CZI = \frac{P_b^3}{K \times \sigma_c^2},\tag{32}$$

where P_b is the blast hole pressure (Pa), K stands for the stiffness of rock mass (Pa), and σ_c shows the uniaxial compressive strength of rock (Pa). The values of P_b and K are calculated as follows:

$$P_{CJ} = \frac{\rho_0 \times D_{CJ}^2}{4},\tag{33}$$

$$P_b = \frac{P_{CJ}}{2},\tag{34}$$

$$K = \frac{E_d}{1 + \nu_d},\tag{35}$$

where P_{CJ} is the ideal blast pressure (Pa), ρ_0 shows the unexploded explosive density (kg/m^3), D_{CJ} represents the ideal detonation velocity (m/s), E_d denotes the dynamic Young's modulus of rock (Pa), and ν_d refers to the dynamic Poisson's ratio of rock. These relationships are also used to approximate blast hole pressure and rock stiffness. Where more accurate values of these parameters are available through direct measurement or numerical modeling, they can be used instead of the presented equations [118]. After calculating CZI, Essen et al. [2] found a power relationship between this factor and the crushed zone radius as follows:

$$\frac{r_0}{r_c} = 1.23 \times CZI^{-0.219},\tag{36}$$

where CZI is the crushing zone index, r_c represents the crushed zone radius, and r_0 shows the blast hole radius. The crushed zone radius is then calculated as follows:

$$r_c = 0.812 \times r_0 \times CZI^{0.219},\tag{37}$$

5. Discussion

5.1. Comparison of Different Models

Various methods have been described to estimate sizes of damage zones around blast holes. In this section, the most commonly used methods are listed and their results are compared through several case studies. For this purpose, 13 rock explosion samples were selected from various studies, in which explosions were carried out in two different types of rocks, including clayey-limestone and basalt, with two types of explosives including

ANFO and water-resistant ANFO. The details of these samples including the characteristics of the rocks, explosives, and blast holes are outlined in Table 5.

Table 5. The characteristics of the rocks, explosives, and blast holes in the 13 studied samples.

Case No.	Rock	Explosive	P (g/cm^3)	q (MJ/kg)	D_{CJ} (km/s)	r_0 (mm)	r_c (mm)	P_b (GPa)
1	CL	ANFO	0.803	3.812	5.016	165	82.5	3.045
2	CL	ANFO	0.803	3.812	5.016	229	114.5	3.477
3	B	ANFO	0.803	3.812	5.016	102	51	2.061
4	B	ANFO	0.803	3.812	5.016	165	82.5	3.148
5	B	ANFO	0.803	3.812	5.016	229	114.5	3.595
6	CL	WR ANFO	0.994	3.918	5.829	51	25.5	2.016
7	CL	WR ANFO	0.994	3.918	5.829	102	51	4.033
8	CL	WR ANFO	0.994	3.918	5.829	165	82.5	4.974
9	CL	WR ANFO	0.994	3.918	5.829	229	114.5	5.44
10	B	WR ANFO	0.994	3.918	5.829	51	25.5	2.085
11	B	WR ANFO	0.994	3.918	5.829	102	51	4.169
12	B	WR ANFO	0.994	3.918	5.829	165	82.5	5.141
13	B	WR ANFO	0.994	3.918	5.829	229	114.5	5.623

In the following, five different methods were selected, including Esen's [2], Il'yushin's [59], Szuladzinski's [62], Djordjevic's [72], and Kanchibotla's [73] models, to calculate damage size corresponding to each of the 13 samples presented. The results of these models for the introduced case studies are displayed in the columns 2–6 of Table 6. The results are also plotted in Figure 7.

Table 6. Results of the selected methods in estimating damage radius for the 13 studied samples.

Case No.	Esen et al. [2]	Il'yushin [59]	Szuladzinski [62]	Djordjevic [72]	Kanchibotla [73]
1	372	1269	379	466	1192
2	564	1761	526	647	1654
3	67	402	108	139	339
4	143	651	175	225	549
5	217	903	242	312	762
6	88	441	132	186	476
7	277	881	264	372	953
8	513	1426	427	602	1541
9	756	1979	593	836	2139
10	34	239	61	90	219
11	107	478	122	179	439
12	198	774	197	290	710
13	291	1074	273	403	985

As can be observed from Figure 7, in almost all the 13 cases, Il'yushin's and Kanchibotla's models yielded relatively larger results than the other models. One possible reason for this issue is that both models used an ideal explosion assumption in their relationships, while the real cases are closer to a non-ideal one. This issue has been pointed out by other researchers as well. It has been argued in the related literature that the purpose of Kanchibotla's model is to study mine fragmentation and optimize mine excavation [2]. Therefore, it may not be able to provide a precise crack propagation estimation and damage zone determination. The other three models, i.e., Esen's, Szuladzinski's, and Djordjevic's models, correspondingly provide relatively close estimations. However, Esen's model is preferred to the other models in the literature. The advantage of using Esen's model compared with other ones is its development based on experimental and real-scale samples, whose results have also been validated by rock blasting real projects. However, the other three models have been developed theoretically through several simplifications and simple assumptions. Moreover, there are several comparisons between these models available in the literature, and almost all of them have addressed Esen's model as the most accurate approach to predict the crushed zone size. For instance, Amnieh and Bahadori [119,120] conducted several single-hole explosion tests at the Gotvand Olya Dam and compared the results with those of Ash's, Djordjevic's, Szuladzinski's, Kanchibotla's, and Esen's models. Eventually, they concluded that the results obtained from Esen's model were closer to real values compared with the other models in all the cases observed. Additionally, studying the crushed zone radius around a blast hole using laboratory tests, Changshou Sun [6] introduced Esen's model as the most complete set of data for the crushed zone extent and then applied it to validate their laboratory test results.

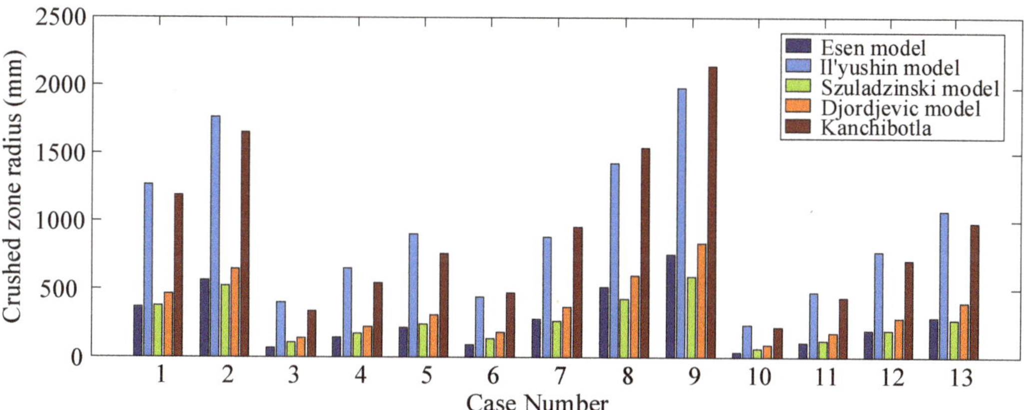

Figure 7. Comparison of different models in estimating damage radius obtained from the 13 studied samples.

5.2. Probabilistic Approaches

The models investigated in this review up to now have been addressing single-hole rock blasting in terms of a deterministic problem. Although the deterministic estimation of damage has been incorporated in a simple algorithm without any certain complexity, some uncertainties have remained unnoticed in the problem. Neither rock medium nor blast load characteristic values are deterministic and consequently cannot be determined with certainty. The deterministic estimation of damage zone is similar to assuming a 100% probability for a specific failure size, which does not appear to be a rational workaround. A better solution in this regard is to match each damage zone radius with a failure probability. In other words, instead of the deterministic estimation of damage zone, an exceedance probability is assumed as the goal of analysis. In fact, determinacy should be left aside and uncertainty needs to be modeled using random variables. As a result, one could estimate the probability required for a crack to exceed a certain length value. This issue has been formulated as a reliability problem in the related literature and investigated in some research works.

Defining the involved parameters as random variables with certain mean and standard deviation and establishing a limit state function for crush and crack zone radii, Shadabfar et al. [121–123] incorporated the Monte Carlo method to calculate failure probability. Based on their results, they concluded that exceedance probability was severely reduced following an increase in the crushed zone radius. Accordingly, the probability of the crushed zone radius longer than 0.5 m was reported to be less than 1%. Moreover, the comparison of different probabilistic models developed based on Esen's, Szuladzinski's, Djordjevic's, and Kanchinotla's models revealed that the results of Esen's model could exhibit a lower failure probability than those of the other models. In other words, Esen's model has a more optimistic estimation of the bearing capacity of rock mass compared with the other models.

Additionally, it was observed that Kanchibotla's model was much more different than the other models, yielding a much higher failure probability. This issue has also been noticed by other researchers [124]. It is reasoned that the objective of Kanchibotla's model is to study mine fragmentation and optimize mine drilling. Hence, it may not yield a precise crack propagation estimation and damage zone determination.

Furthermore, defining both the crushed and cracked zones in terms of a reliability problem in another study, Shadabfar et al. [125] took advantage of the first-order reliability method and calculated explosion-induced failure probability. The obtained results of this study were then represented in terms of exceedance probability versus damage zone radius (Figure 8).

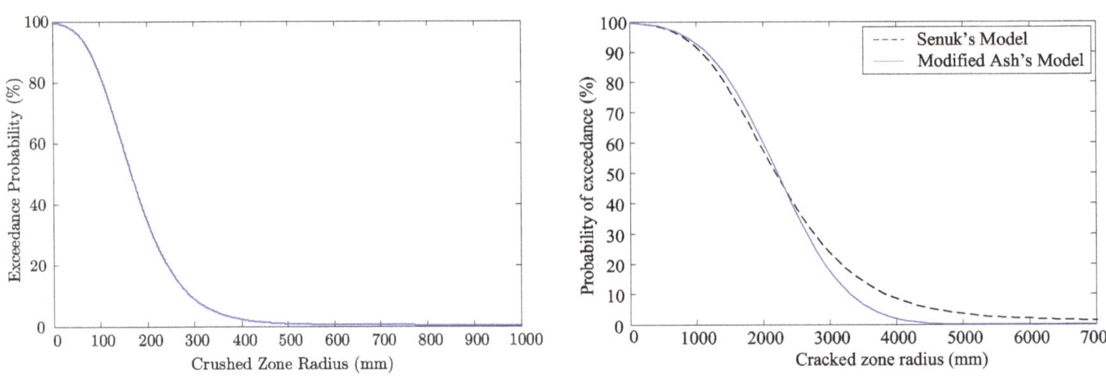

(a) Exceedance probability curve for crushed zone radius

(b) Exceedance probability curve for cracked zone radius

Figure 8. The exceedance probability curve for (**a**) crushed and (**b**) cracked zones.

The diagram was then drawn for all the points in the vicinity of the blast location and presented as exceedance probability contours (Figure 9).

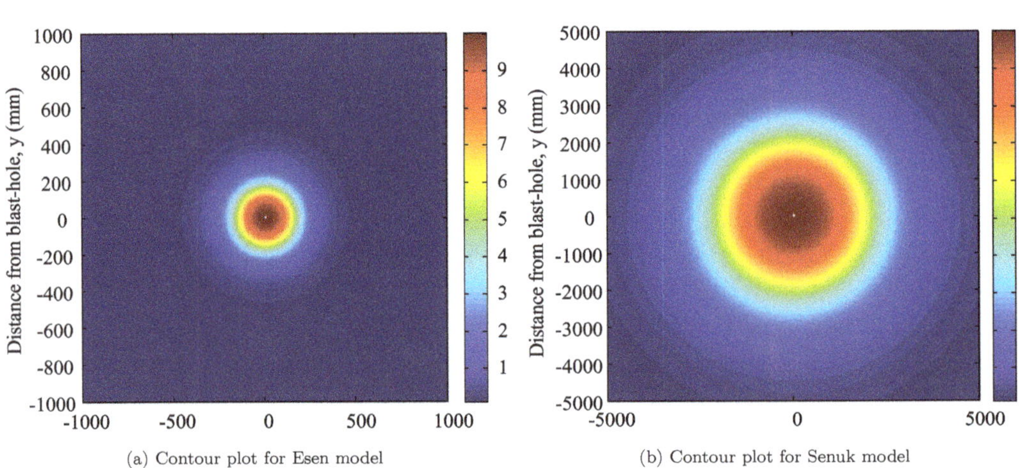

(a) Contour plot for Esen model

(b) Contour plot for Senuk model

Figure 9. The contour plot of exceedance probability for (**a**) crushed and (**b**) cracked zones.

As shown in Figure 9, failure probability is reduced via distancing from the blast location. Thus, failure probability for the crushed zone approaches 0 by exceeding 0.5 m.

In addition, using a parametric study, ShadabFar et al. [125] investigated the effects of decoupling on resulting failures. For this purpose, a number of reliability analyses were conducted and changes in failure probability were recorded accurately through incorporating different ratios of explosive charge to blast hole radii in the governing equation. According to their results, failure probability declined and the exceedance probability diagram showed a lower level of probability as the decoupling ratio was reduced. The reason is that the distance between explosive charge and blast hole wall increased through a reduction in the decoupling ratio, leading to the sharp damping of blast wave before propagating in the rock medium. Additionally, the results revealed that the highest impact of the decoupling ratio occurred in the range of small damage zone radii. More precisely, the decoupling ratio mostly affected failure radii in the range of 300–2350 mm, while its impact was severely reduced for larger failure radii.

Performing a set of reliability sensitivity analyses, ShadabFar et al. [125] also compared the influence of parameters involved in different models. The results of their study showed that the main parameters affecting the crushed zone radius in Esen's model were the uniaxial compressive strength of rock and the blast hole radius. Furthermore, according to Senuk's model, two main parameters influencing the cracked zone radius were the blast hole radius and the tensile stress of rock mass. The results of this analysis were then presented in a diagram in terms of sensitivity vector for each parameter involved in different models (Figure 10). This diagram presents the relative importance of each parameter in comparison with other parameters. Load variables (i.e., components whose growth would increase failure probability) are marked with a dark color, whereas resistance variables (i.e., components whose rising trend would decrease failure probability) are marked with a bright color.

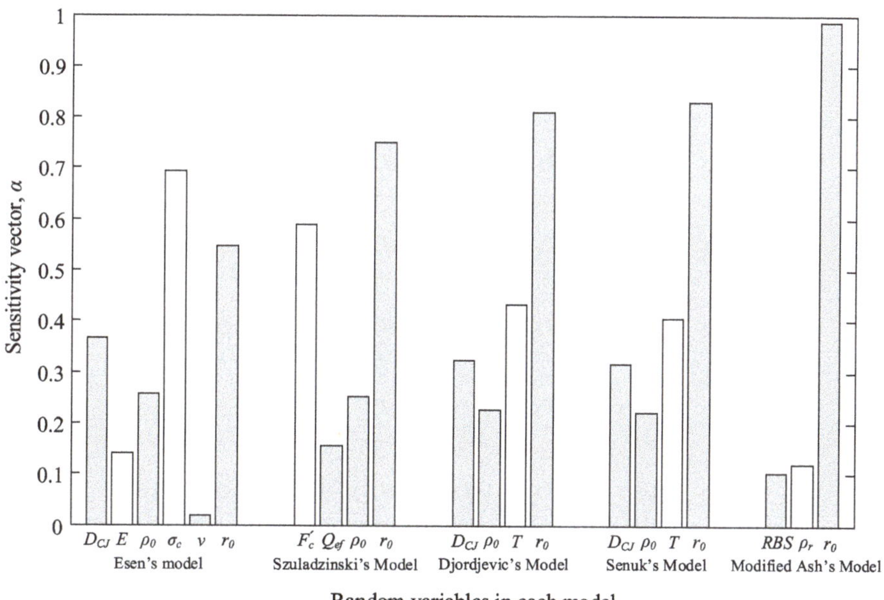

Figure 10. The relative importance of parameters involved in various models.

6. Conclusions

In this review, the most important existing models for the estimation of explosion-induced crushed and cracked zones were investigated. The models were categorized into three groups, i.e., analytical, numerical, and experimental approaches. First, the rock explosion mechanism was described from the detonation initiation and stress wave propagation to the rock failure. Subsequently, the induced damage was grouped into two forms of crushed and cracked zones and the most important parameters affecting these zones were reported. Then, the most important methods for estimating the dimensions of each damage zone were examined.

More specifically, the analytical models were presented based on the two main parameters of PPV and blast hole pressure. The numerical methods were addressed via the commonly used numerical codes, including the FEM, DEM, and FDM methods. The experimental models were divided into two general categories of primary cracks due to high-stress waves and deep cracks caused by gas penetration and discussed in detail. Finally, a number of empirical models drawn from laboratory results were used in a step-by-step approach.

The most commonly utilized models described in this review were selected to separately calculate the damage dimensions in 13 case studies, which were collected from the related literature and presented in an integrated form. All the results were compared, and their differences or similarities were discussed.

Thereafter, the probabilistic models available for analyzing the failure probability induced by the rock explosion were deliberated, and their advantages over the deterministic models were described. The comparisons were made between different models, and the relative importance of the involved parameters was investigated via the reliability sensitivity analysis.

This review categorized and reported the most important assumptions and key points of the related literature along with their prominent results to allow for more coherent and accurate use of their content. Hence, the results of the present study can be used as a comprehensive and categorized source for estimating the blast-induced damage in rocks. However, for the practical use of the methods presented here, their main sources should be utilized for more details. Finally, this paper only covered the single-hole blasting. In cases with multiple blasting, due to the interaction between the explosion waves released from each blast hole, a system of damage zones is generated, which can potentially overlap and cause more complex failure in the environment. This topic remains as the authors' concern for future research works.

Author Contributions: Conceptualization, data curation, formal analysis, investigation, methodology, project administration, resources, software, supervision, validation, visualization, writing—original draft, and writing—review & editing are done by M.S.; Funding acquisition, methodology, resources, validation, visualization, writing—original draft, and writing—review & editing are done by C.G.; Funding acquisition, validation, visualization, and writing—review & editing are done by M.Z.; Data curation, investigation, methodology, resources, validation, visualization, and writing—original draft are done by H.K.; Investigation, methodology, resources, validation, visualization, writing—original draft, and writing—review & editing are done by E.V.M. All authors have read and agreed to the published version of the manuscript.

Funding: The research presented in this paper was conducted with support from the Natural Science Foundation of China (NSFC) under the project number 02302340347.

Conflicts of Interest: The authors declare no conflict of interest.

References

1. Kanchibotla, S.; Morrell, S.; Valery, W.; Loughlin, P.O. Exploring the effect of blast design on SAG mill throughput at KCGM. In Proceedings of the Mine to Mill Conference, Brisbane, Australia, 11–14 October 1998.
2. Esen, S.; Onederra, I.; Bilgin, H. Modelling the size of the crushed zone around a blasthole. *Int. J. Rock Mech. Min. Sci.* **2003**, *40*, 485–495. [CrossRef]

3. Ouchterlony, F.; Moser, P. Lessons from single-hole blasting in mortar, concrete and rocks. In *Rock Fragmentation by Blasting*; Singh, P.K., Sinha, A., Eds.; CRC Press: London, UK, 2012; pp. 3–14.
4. Zhang, Z.X. *Rock Fracture and Blasting, Theory and Applications*; Butterworth-Heinemann: Amsterdam, The Netherlands, 2016.
5. Chi, L.Y.; Zhang, Z.X.; Aalberg, A.; Yang, J.; Li, C.C. Fracture processes in granite blocks under blast loading. *Rock Mech. Rock Eng.* **2019**, *52*, 853–868. [CrossRef]
6. Sun, C. Damage Zone Prediction for Rock Blasting. Ph.D. Thesis, University of Utah, Salt Lake City, UT, USA, 2013.
7. Zou, D. *Theory and Technology of Rock Excavation for Civil Engineering*, 1st ed.; Springer: Singapore, 2017.
8. Singh, S. Blast damage control in jointed rock mass. *Fragblast Int. J. Blasting Fragm.* **2005**, *9*, 175–187. [CrossRef]
9. Chen, M.; Lu, W.B.; Yan, P.; Hu, Y.G. Blasting excavation induced damage of surrounding rock masses in deep-buried tunnels. *KSCE J. Civ. Eng.* **2015**, *20*, 933–942. [CrossRef]
10. Kanchibotla, S. Optimum Blasting? Is it Minimum Cost Per Broken Rock or Maximum Value Per Broken Rock? *Fragblast Int. J. Blasting Fragm.* **2003**, *7*, 35–48. [CrossRef]
11. Mohanty, B.; Singh, V.K. (Eds.) *Performance of Explosives and New Developments*, 1st ed.; CRC Press: New Delhi, India, 2012.
12. Leng, Z.; Lu, W.; Chen, M.; Yan, P.; Hu, Y. A new theory of rock-explosive matching based on the reasonable control of crushed zone. *Engineering* **2014**, *12*, 32–38.
13. Trivino, L.; Mohanty, B. Assessment of crack initiation and propagation in rock from explosion-induced stress waves and gas expansion by cross-hole seismometry and FEM-DEM method. *Int. J. Rock Mech. Min. Sci.* **2015**, *77*, 287–299. [CrossRef]
14. Bhandari, S. *Engineer Rock Blasting Operations*, 1st ed.; CRC Press: Boca Raton, FL, USA, 1997.
15. Fairhurst, C. *Comprehensive Rock Comprehensive Rock Engineering: Principles, Practice and Projects, Analysis and Design Methods*, 2nd ed.; Pergamon Pr: Oxford, UK, 1995.
16. Qiu, X.; Shi, X.; Zhang, S.; Liu, B.; Zhou, J. Experimental Study on the Blasting Performance of Water-Soil Composite Stemming in Underground Mines. *Adv. Mater. Sci. Eng.* **2018**, *2018*. [CrossRef]
17. Xia, Y.; Jiang, N.; Zhou, C.; Luo, X. Safety assessment of upper water pipeline under the blasting vibration induced by Subway tunnel excavation. *Eng. Fail. Anal.* **2019**, *104*, 626–642. [CrossRef]
18. Ouchterlony, F.; Nyberg, U.; Olsson, M.; Bergqvist, I.; Granlund, L.; Grind, H. Where does the explosive energy in rock blasting rounds go? *Sci. Technol. Energetic Mater.* **2004**, *65*, 54–63.
19. Souissi, S.; Hamdi, E.; Sellami, H. Microstructure effect on hard rock damage and fracture during indentation process. *Geotech. Geol. Eng.* **2015**, *33*, 1539–1550. [CrossRef]
20. Fourney, W.L. The role of stress waves and fracture mechanics in fragmentation. In Proceedings of the 11th International Symposium on Rock Fragmentation by Blasting, Sydney, Australia, 24–26 August 2015; The Australasian Institute of Mining and Metallurgy: Carlton, Australia, 2015; pp. 27–40.
21. Persson, P.; Ladegaard-Pedersen, A.; Kihlstrom, B. The influence of borehole diameter on the rock blasting capacity of an extended explosive charge. *Int. J. Rock Mech. Min. Sci. Geomech. Abstr.* **1969**, *6*, 277–284. [CrossRef]
22. Brinkmann, J. An experimental study of the effects of shock and gas penetration in blasting. In Proceedings of the 3rd International Symposium on Rock Fragmentation by Blasting, Brisbane Australia, 26–31 August 1990; pp. 55–66.
23. Wang, C.; Zhu, Z.M.; Zheng, T. The fracturing behavior of detected rock under blasting loads. In *Applied Mechanics and Materials*; Trans Tech Publications: Stafa-Zurich, Switzerland, 2012; Volume 142, pp. 193–196. [CrossRef]
24. Saharan, M.; Mitri, H.; Jethwa, J. Rock fracturing by explosive energy: Review of state-of-the-art. *Fragblast Int. J. Blasting Fragm.* **2006**, *10*, 61–81. [CrossRef]
25. Banadaki, M.D.; Mohanty, B. Numerical simulation of stress wave induced fractures in rock. *Int. J. Impact Eng.* **2012**, *40–41*, 16–25. [CrossRef]
26. Adhikari, G.; Theresraj, A.; Venkatesh, H.; Balachander, R.; Gupta, R. Ground vibration due to blasting in limestone quarries. *Fragblast Int. J. Blasting Fragm.* **2010**, *8*, 85–94. [CrossRef]
27. Chi, L.Y.; Zhang, Z.X.; Aalberg, A.; Yang, J.; Li, C.C. Measurement of shock pressure and shock-wave attenuation near a blast hole in rock. *Int. J. Impact Eng.* **2019**, *125*, 27–38. [CrossRef]
28. Olsson, M.; Ouchterlony, F. *New Formula for Blast Induced Damage in the Remaining Rock (Title in Swedish: Ny skadezonsformel för skonsam sprangning)*; Technical Report SveBeFo Rapport 65; Swedish Rock Engineering Research: Stockholm, Sweden, 2003.
29. Asl, P.F.; Monjezi, M.; Hamidi, J.K.; Armaghani, D.J. Optimization of flyrock and rock fragmentation in the Tajareh limestone mine using metaheuristics method of firefly algorithm. *Eng. Comput.* **2018**, *34*, 241–251. [CrossRef]
30. Leng, Z.; Fan, Y.; Gao, Q.; Hu, Y. Evaluation and optimization of blasting approaches to reducing oversize boulders and toes in open-pit mine. *Int. J. Min. Sci. Technol.* **2020**, *30*, 373–380. [CrossRef]
31. Ma, L.; Lai, X.; Zhang, J.; Xiao, S.; Zhang, L.; Tu, Y. Blast-Casting Mechanism and Parameter Optimization of a Benched Deep-Hole in an Opencast Coal Mine. *Shock Vib.* **2020**, *2020*. [CrossRef]
32. Verma, H.; Samadhiya, N.; Singh, M.; Goel, R.; Singh, P. Blast induced rock mass damage around tunnels. *Tunn. Undergr. Space Technol.* **2018**, *71*, 149–158. [CrossRef]
33. Jang, H.; Kawamura, Y.; Shinji, U. An empirical approach of overbreak resistance factor for tunnel blasting. *Tunn. Undergr. Space Technol.* **2019**, *92*. [CrossRef]
34. Silva, J.; Worsey, T.; Lusk, B. Practical assessment of rock damage due to blasting. *Int. J. Min. Sci. Technol.* **2019**, *29*, 379–385. [CrossRef]

35. Srirajaraghavaraju, R.R. Transmitted Pressure and Resulting Crack Network in Selected Rocks from Single-Hole Blasts in Laboratory-Scale Experiments. Master's Thesis, University of Toronto, Toronto, ON, Canada, 2014.

36. Johnson, C.E. Fragmentation Analysis in the Dynamic Stress Wave Collision Regions in Bench Blasting. Ph.D. Thesis, University of Kentucky, Lexington, KY, USA, 2014.

37. Guo, Y.; Han, Z.; Guo, H.; Wang, T.; Liu, B.; Wang, D. Numerical simulation damage analysis of pipe-cement-rock combination due to the underwater explosion. *Eng. Fail. Anal.* **2019**, *105*, 584–596. [CrossRef]

38. Wei, X.; Zhao, Z.; Gu, J. Numerical simulations of rock mass damage induced by underground explosion. *Int. J. Rock Mech. Min. Sci.* **2009**, *46*, 1206–1213. [CrossRef]

39. Yang, R.; Wang, Y.; Ding, C. Laboratory study of wave propagation due to explosion in a jointed medium. *Int. J. Rock Mech. Min. Sci.* **2016**, *81*, 70–78. [CrossRef]

40. Persson, P.A.; Holmberg, R.; Lee, J. *Rock Blasting and Explosives Engineering*, 1st ed.; CRC Press: Boca Raton, FL, USA, 1993.

41. Jommi, C.; Pandolf, A. Vibrations induced by blasting in rock: A numerical approach. *Riv. Ital. Geotec.* **2008**, *2*, 77–94.

42. Fleetwood, K.G.; Villaescusa, E.; Li, J. Limitations of using ppv damage models to predict rock mass damage. In Proceedings of the Thirty-Fifth Annual Conference on Explosives and Blasting Technique, Denver, CO, USA, 8–11 February 2009; International Society of Explosives Engineers: Denver, CO, USA, 2009; Volume 1, pp. 1–15.

43. Duan, K.; Kwok, C.Y. Evolution of stress-induced borehole breakout in inherently anisotropic rock: Insights from discrete element modeling. *J. Geophys. Res. Solid Earth* **2016**, *121*, 2361–2381. [CrossRef]

44. Parra, L.F.T. Study of Blast-Induced Damage in Rock with Potential Application to Open Pit and Underground Mines. Ph.D. Thesis, University of Toronto, Toronto, ON, Canada, 2012.

45. Ma, G.; Hao, H.; Wang, F. Simulations of explosion-induced damage to underground rock chambers. *J. Rock Mech. Geotech. Eng.* **2011**, *3*, 19–29. [CrossRef]

46. Wang, Z.; Konietzky, H. Modelling of blast-induced fractures in jointed rock masses. *Eng. Fract. Mech.* **2009**, *76*, 1945–1955. [CrossRef]

47. Mitelman, A.; Elmo, D. Modelling of blast-induced damage in tunnels using a hybrid finite-discrete numerical approach. *J. Rock Mech. Geotech. Eng.* **2014**, *6*, 565–573. [CrossRef]

48. Bendezu, M.; Romanel, C.; Roehl, D. Finite element analysis of blast-induced fracture propagation in hard rocks. *Comput. Struct.* **2017**, *182*, 1–13. [CrossRef]

49. Onederra, I.A.; Furtney, J.K.; Sellers, E.; Iverson, S. Modelling blast induced damage from a fully coupled explosive charge. *Int. J. Rock Mech. Min. Sci.* **2013**, *58*, 73–84. [CrossRef] [PubMed]

50. Hu, Y.g.; Liu, M.s.; Wu, X.x.; Zhao, G.; Li, P. Damage-vibration couple control of rock mass blasting for high rock slopes. *Int. J. Rock Mech. Min. Sci.* **2018**, *103*, 137–144. [CrossRef]

51. Sun, X.J.; Sun, J.S. Research on the damage fracture of rock blasting based on velocity response spectrum. In *Applied Mechanics and Materials*; Trans Tech Publications Ltd.: Stafa-Zurich, Switzerland, 2012; Volume 105, pp. 1521–1527. [CrossRef]

52. Bauer, A.; Calder, P. *Open Pit and Blast Seminar*; Course No. 63321; Mining Engineering Department, Queens University: Kingston, ON, Canada, 1978.

53. Mojtabai, N.; Beatty, S. An empirical approach to assessment of and prediction of damage in bench blasting. *Trans. Inst. Min. Metall.* **1995**, *105*, A75–A80.

54. Holmberg, R.; Persson, P. The Swedish approach to contour blasting. In Proceedings of the Annual Conference on Explosives and Blasting Technique, New Orleans, LA, USA, 1–3 February 1978; International Society of Explosives Engineers: New Orleans, LA, USA, 1978; pp. 113–127.

55. Aleksandrova, N.I.; Sher, Y.N. Effect of stemming on rock breaking with explosion of a cylindrical charge. *J. Min. Sci.* **1999**, *35*, 483–493. [CrossRef]

56. Wang, M.; Yue, S.; Zhang, N.; Gao, K.; Wang, D. A method of calculating critical depth of burial of explosive charges to generate bulging and cratering in rock. *Shock Vib.* **2016**, 1–11. [CrossRef]

57. Hustrulid, W.; Lu, W. Some general design concepts regarding the control of blast-induced damage during rock slope excavation. In Proceedings of the 7th International Symposium on Rock Fragmentation by Blasting, Beijing, China, 11–15 August 2002; pp. 595–604.

58. Mosinets, V.; Gorbacheva, N. A seismological method of determining the parameters of the zones of deformation of rock by blasting. *Sov. Min.* **1972**, *8*, 640–647. [CrossRef]

59. Ilyushin, A. *The Mechanics of a Continuous Medium*; Izd-vo MGU: Moscow, Russia, 1971. (In Russian)

60. Drukovanyi, M.F.; Kravtsov, V.; Chernyavskii, Y.; Shelenok, V.; Reva, N.; Zverkov, S. Calculation of fracture zones created by exploding cylindrical charges in ledge rocks. *Sov. Min.* **1976**, *12*, 292–295. [CrossRef]

61. Senuk, V. The impulse from an explosion, and conditions for its greater utilization in crushing hard rock masses in blasting. *Sov. Min.* **1979**, *15*, 22–27. [CrossRef]

62. Szuladzinski, G. Response of rock medium to explosive borehole pressure. In *Rock Fragmentation by Blasting, Proceedings of The Fourth International Symposium on Rock Fragmentation by Blasting, FRAGBLAST-4, Vienna, Austria, 5–8 July 1993*; Balkema: Rotterdam, The Netherland, 1993; pp. 17–23.

63. Ash, R. The mechanics of rock breakage (part 1). *Pit Quarry* **1963**, *56*, 98–100.

64. Ash, R. The mechanics of rock breakage (part 2)-standards blasting design. *Pit Quarry* **1963**, *56*, 119–122.

65. Ash, R. The mechanics of rock breakage (part 3)-characteristics of explosives. *Pit Quarry* **1963**, *56*, 126–131.

66.   Ash, R. The mechanics of rock breakage (part 4)-material properties, powder factor, blasting cost. *Pit Quarry* **1963**, *56*, 109–118.
67.   Ouchterlony, F. Prediction of crack lengths in rock after cautious blasting with zero inter-hole delay. *Int. J. Blasting Fragm. Fragblast* **1997**, *1*, 417–444. [CrossRef]
68.   Ouchterlony, F.; Olsson, M.; Bergqvist, I. Towards new Swedish recommendations for cautious perimeter blasting. *Int. J. Blasting Fragm. Fragblast* **2002**, *6*, 235–261. [CrossRef]
69.   Sher, E.; Aleksandrova, N. Dynamics of development of crushing zone in elastoplastic medium in camouflet explosion of string charge. *J. Min. Sci.* **1997**, *33*, 529–535. [CrossRef]
70.   Sher, E.; Aleksandrova, N. Effect of borehole charge structure on the parameters of a failure zone in rocks under blasting. *J. Min. Sci.* **2007**, *43*, 409–417. [CrossRef]
71.   Hustrulid, W. Some comments regarding development drifting practices with special emphasis on caving applications. In *Second International Symposium on Block and Sublevel Caving*; Potvin, Y., Ed.; Australian Centre for Geomechanics: Perth, Australia, 2010; pp. 3–43.
72.   Djordjevic, N. A two-component model of blast fragmentation. In Proceedings of the 6th International Symposium for Rock Fragmentation by Blasting-Fragblast, Johannesburg, South Africa, 8–12 August 1999; pp. 213–219.
73.   Kanchibotla, S.; Valery, W.; Morrell, S. Modelling fines in blast fragmentation and its impact on crushing and grinding; Australasian Institute of Mining and Metallurgy Publication Series; In Proceedings of the Explo 99: A Conference on Rock Breaking, Kalgoorlie, WA, USA, 7–11 November 1999.
74.   Johnson, J. The Hustrulid Bar—A Dynamic Strength Test and Its Application to the Cautious Blasting of Rock. Ph.D. Thesis, The Unversity of Utah, Salt Lake, UT, USA, 2010.
75.   Saharan, M.R.; Mitri, H.S. Numerical Procedure for Dynamic Simulation of Discrete Fractures Due to Blasting. *Rock Mech. Rock Eng.* **2008**, *41*, 641–670. [CrossRef]
76.   Zhu, W.; Wei, J.; Zhao, J.; Niu, L. 2D numerical simulation on excavation damaged zone induced by dynamic stress redistribution. *Tunn. Undergr. Space Technol.* **2014**, *43*, 315–326. [CrossRef]
77.   Goodarzi, M.; Mohammadi, S.; Jafari, A. Numerical analysis of rock fracturing by gas pressure using the extended finite element method. *Pet. Sci.* **2015**, *12*, 304–315. [CrossRef]
78.   Hu, Y.; Lu, W.; Chen, M.; Yan, P.; Zhang, Y. Numerical simulation of the complete rock blasting response by SPH–DAM–FEM approach. *Simul. Model. Pract. Theory* **2015**, *56*, 55–68. [CrossRef]
79.   Yang, J.; Yao, C.; Jiang, Q.; Lu, W.; Jiang, S. 2D numerical analysis of rock damage induced by dynamic in-situ stress redistribution and blast loading in underground blasting excavation. *Tunn. Undergr. Space Technol.* **2017**, *70*, 221–232. [CrossRef]
80.   Hu, Y.; Lu, W.; Wu, X.; Liu, M.; Li, P. Numerical and experimental investigation of blasting damage control of a high rock slope in a deep valley. *Eng. Geol.* **2018**, *237*, 12–20. [CrossRef]
81.   Jayasinghe, L.; Shang, J.; Zhao, Z.; Goh, A. Numerical investigation into the blasting-induced damage characteristics of rocks considering the role of in-situ stresses and discontinuity persistence. *Comput. Geotech.* **2019**, *116*, 1–13. [CrossRef]
82.   Yilmaz, O.; Unlu, T. Three dimensional numerical rock damage analysis under blasting load. *Tunn. Undergr. Space Technol.* **2013**, *38*, 266–278. [CrossRef]
83.   Hu, R.; Zhu, Z.; Xie, J.; Xiao, D. Numerical study on crack propagation by using softening model under blasting. *Adv. Mater. Sci. Eng.* **2015**, *2015*, 1–9. [CrossRef]
84.   Zhao, Y.; Yang, T.; Zhang, P.; Zhou, J.; Yu, Q.; Deng, W. The analysis of rock damage process based on the microseismic monitoring and numerical simulations. *Tunn. Undergr. Space Technol.* **2017**, *69*, 1–17. [CrossRef]
85.   Jessu, K.; Spearing, A.; Sharifzadeh, M. A Parametric Study of Blast Damage on Hard Rock Pillar Strength. *Energies* **2018**, *11*, 1–18. [CrossRef]
86.   Lak, M.; Fatehi Marji, M.; Yarahmadi Bafghi, A.; Abdollahipour, A. Analytical and numerical modeling of rock blasting operations using a two-dimensional elasto-dynamic Green's function. *Int. J. Rock Mech. Min. Sci.* **2019**, *114*, 208–217. [CrossRef]
87.   Zhou, M.; Shadabfar, M.; Huang, H.; Leung, Y.F.; Uchida, S. Meta-modelling of coupled thermo-hydro-mechanical behaviour of hydrate reservoir. *Comput. Geotech.* **2020**, *128*, 103848. [CrossRef]
88.   Aliabadian, Z.; Sharafisafa, M.; Mortazavi, A.; Maarefvand, P. Wave and fracture propagation in continuum and faulted rock masses: distinct element modeling. *Arab. J. Geosci.* **2014**, *7*, 5021–5035. [CrossRef]
89.   Deng, X.; Zhu, J.; Chen, S.; Zhao, Z.; Zhou, Y.; Zhao, J. Numerical study on tunnel damage subject to blast-induced shock wave in jointed rock masses. *Tunn. Undergr. Space Technol.* **2014**, *43*, 88–100. [CrossRef]
90.   Sharafisafa, M.; Aliabadian, Z.; Alizadeh, R.; Mortazavi, A. Distinct element modelling of fracture plan control in continuum and jointed rock mass in presplitting method of surface mining. *Int. J. Min. Sci. Technol.* **2014**, *24*, 871–881. [CrossRef]
91.   Bai, Q.S.; Tu, S.H.; Zhang, C. DEM investigation of the fracture mechanism of rock disc containing hole(s) and its influence on tensile strength. *Theor. Appl. Fract. Mech.* **2016**, *86*, 197–216. [CrossRef]
92.   Yin, T.; Zhang, S.; Li, X.; Bai, L. A numerical estimate method of dynamic fracture initiation toughness of rock under high temperature. *Eng. Fract. Mech.* **2018**, *204*, 87–102. [CrossRef]
93.   Yuan, W.; Su, X.; Wang, W.; Wen, L.; Chang, J. Numerical study of the contributions of shock wave and detonation gas to crack generation in deep rock without free surfaces. *J. Pet. Sci. Eng.* **2019**, *177*, 699–710. [CrossRef]
94.   Potyondy, D.O.; Cundall, P.A. A bonded-particle model for rock. *Int. J. Rock Mech. Min. Sci.* **2004**, *41*, 1329–1364. [CrossRef]
95.   Fakhimi, A.; Lanari, M. DEM-SPH simulation of rock blasting. *Comput. Geotech.* **2014**, *55*, 158–164. [CrossRef]

96. An, H.; Liu, H.; Han, H.; Zheng, X.; Wang, X. Hybrid finite-discrete element modelling of dynamic fracture and resultant fragment casting and muck-piling by rock blast. *Comput. Geotech.* **2017**, *81*, 322–345. [CrossRef]

97. Zhu, J.B.; Li, Y.S.; Wu, S.Y.; Zhang, R.; Ren, L. Decoupled explosion in an underground opening and dynamic responses of surrounding rock masses and structures and induced ground motions: A FEM-DEM numerical study. *Tunn. Undergr. Space Technol.* **2018**, *82*, 442–454. [CrossRef]

98. Wang, Z.l.; Li, Y.c.; Wang, J. Numerical analysis of blast-induced wave propagation and spalling damage in a rock plate. *Int. J. Rock Mech. Min. Sci.* **2008**, *45*, 600–608. [CrossRef]

99. Wang, Z.; Konietzky, H.; Shen, R. Coupled finite element and discrete element method for underground blast in faulted rock masses. *Soil Dyn. Earthq. Eng.* **2009**, *29*, 939–945. [CrossRef]

100. Liu, C.; Yang, M.; Han, H.; Yue, W. Numerical simulation of fracture characteristics of jointed rock masses under blasting load. *Eng. Comput.* **2019**, *36*, 835–1851. [CrossRef]

101. Lak, M.; Marji, M.F.; Bafghi, A.Y.; Abdollahipour, A. A coupled finite difference-boundary element method for modeling the propagation of explosion-induced radial cracks around a wellbore. *J. Nat. Gas Sci. Eng.* **2019**, *64*, 41–51. [CrossRef]

102. Chi, L.Y.; Zhang, Z.X.; Aalberg, A.; Li, C.C. Experimental investigation of blast-induced fractures in rock cylinders. *Rock Mech. Rock Eng.* **2019**, *52*, 2569–2584. [CrossRef]

103. Ge, J.; Li, G.Q.; Chen, S.W. Theoretical and experimental investigation on fragment behavior of architectural glass panel under blast loading. *Eng. Fail. Anal.* **2012**, *26*, 293–303. [CrossRef]

104. Lownds, M. Measurement shock pressures in splitting of dimensional stone. In Proceedings of the 1st World Conference on Explosives and Blasting Technique, Munich, Germany, 6–8 September 2000; pp. 241–246.

105. Talhi, K.; Hadjaj-Aoul, E.; Hannachi, E.B. Design of a model blasting system to measure peak P-wave stress. *Acta Geod. Geophys. Hung.* **2004**, *39*, 427–438. [CrossRef]

106. Teowee, G.; Papillon, B. Measurement of borehole pressure during blasting. In Proceedings of the Rock Fragmentation by Blasting: Fragblast 10, New Delhi, India, 26–29 November 2012; pp. 599–605.

107. Teowee, G.; Papillon, B. Monitoring of dynamic borehole pressures. In Proceedings of the 39th Annual Conference on Explosives and Blasting Techniques, Fort Worth, TX, USA, 10–13 February 2013; pp. 1–10.

108. Austing, J.L.; Tulis, A.J.; Hrdina, D.J.; Baker, D.E.; Martinez, R. Carbon resistor gauges for measuring shock and detonation pressures. I. Principles of functioning and calibration. *Propellants Explos. Pyrotech.* **1991**, *16*, 205–215. [CrossRef]

109. Olsson, M.; Bergqvist, I. Crack lengths from explosives in small diameter boreholes. In Proceedings of the 4th International symposium on Rock Fragmentation by Blasting, Vienna, Austria, 5–8 July 1993; pp. 193–196.

110. Olsson, M.; Bergqvist, I. Crack lengths from explosives in multiple hole blasting. In Proceedings of the 5th International symposium on Rock Fragmentation by Blasting, Montreal, QC, Canada, 25–29 August 1996; pp. 187–193.

111. Dehghan-Banadaki, M. Stress-Wave Induced Fracture in Rock due to Explosive Action. Ph.D. Thesis, University of Toronto, Toronto, ON, Canada, 2010.

112. Nariseti, C. Quantification of Damage in Selected Rocks due to Impact with Tungsten Carbide Bits. Master's Thesis, University of Toronto, Toronto, ON, Canada, 2013.

113. Paventi, M.; Mohanty, B. Mapping of blast-induced fractures in rock. In Proceedings of the 7th International Symposium of Rock Fragmentation by Blasting, Beijing, China, 11–15 August 2002; Metallurgical Industry Press: Beijing, China, 2002; pp. 166–172.

114. Yamin, G. Field Measurements of Blast Induced Damage in Rock. Master's Thesis, University of Toronto, Toronto, ON, Canada, 2005.

115. McHugh, S. Crack extension caused by internal gas pressure compared with extension caused by tensile stress. *Int. J. Fract.* **1983**, *21*, 163–176. [CrossRef]

116. Brinkmann, J. Separating shock wave and gas expansion breakage mechanism. In Proceedings of the 2nd International Symposium on Rock Fragmentation by Blasting, Keystone, CO, USA, 23–26 August 1987; pp. 6–15.

117. Fullelove, I.; Onederra, I.; Villaescusa, E. Empirical approach to estimate rock mass damage from long-hole winze (LHW) blasting. *Min. Technol.* **2017**, *126*, 34–43. [CrossRef]

118. Hu, Y.; Lu, W.; Chen, M.; Yan, P.; Yang, J. Comparison of blast-induced damage between presplit and smooth blasting of high rock slope. *Rock Mech. Rock Eng.* **2014**, *47*, 1307–1320. [CrossRef]

119. Amnieh, H.B.; Bahadori, M. Numerical analysis for effects of single blast hole in mudstone rock-mass at Gotvand Olya dam. *Iran. J. Geophys. (IJG)* **2012**, *6*, 56–72.

120. Amnieh, H.B.; Bahadori, M. Numerical and field analysis of single-hole blasting mechanism in conglomerate rock mass of Gotvand Olya Dam. *Energy Eng. Manag.* **2012**, *2*, 22–31.

121. Shadab Far, M.; Wang, Y. Probabilistic analysis of crushed zone for rock blasting. *Comput. Geotech.* **2016**, *80*, 290–300. [CrossRef]

122. Shadab Far, M.; Wang, Y. Approximation of the Monte Carlo Sampling Method for Reliability Analysis of Structures. *Math. Probl. Eng.* **2016**, *2016*, 1–9. [CrossRef]

123. Shadabfar, M.; Huang, H.; Wang, Y.; Wu, C. Monte Carlo analysis of the induced cracked zone by single-hole rock explosion. *Geomech. Eng.* **2020**, *21*, 289–300. [CrossRef]

124. Lu, W.; Leng, Z.; Chen, M.; Yan, P.; Hu, Y. A modified model to calculate the size of the crushed zone around a blast-hole. *J. S. Afr. Inst. Min. Metall.* **2016**, *116*, 413–422. [CrossRef]

125. Shadab Far, M.; Wang, Y.; Dallo, Y.A.H. Reliability analysis of the induced damage for single-hole rock blasting. *Georisk Assess. Manag. Risk Eng. Syst. Geohazards* **2019**, *13*, 82–98. [CrossRef]

Article

Method for Determining the Utilization Rate of Thin-Deck Shearers Based on Recorded Electromotor Loads

Marek Kęsek and Romuald Ogrodnik *

Faculty of Mining and Geoengineering, AGH University of Science and Technology, 30-059 Kraków, Poland; kesek@agh.edu.pl
* Correspondence: rogrod@agh.edu.pl

Abstract: Mining machinery and equipment used in modern mining are equipped with sensors and measurement systems at the stage of their production. Measuring devices are most often components of a control system or a machine performance monitoring system. In the case of headers, the primary task of these systems is to ensure safe operation and to monitor its correctness. It is customary to collect information in very large databases and analyze it when a failure occurs. Data mining methods allow for analysis to be made during the operation of machinery and mining equipment, thanks to which it is possible to determine not only their technical condition but also the causes of any changes that have occurred. The purpose of this work is to present a method for discovering missing information based on other available parameters, which facilitates the subsequent analysis of machine performance. The primary data used in this paper are the currents flowing through the windings of four header motors. In the method, the original reconstruction of the data layout was performed using the R language function, and then the analysis of the operating states of the header was performed based on these data. Based on the rules used and determined in the analysis, the percentage structure of machine operation states was obtained, which allows for additional reporting and verification of parts of the process.

Keywords: mining shearer; underground mining; mining process; data mining

check for
updates

Citation: Kęsek, M.; Ogrodnik, R. Method for Determining the Utilization Rate of Thin-Deck Shearers Based on Recorded Electromotor Loads. *Energies* **2021**, *14*, 4059. https://doi.org/10.3390/ en14134059

Academic Editor: Krzysztof Skrzypkowski

Received: 28 May 2021
Accepted: 2 July 2021
Published: 5 July 2021

Publisher's Note: MDPI stays neutral with regard to jurisdictional claims in published maps and institutional affiliations.

Copyright: © 2021 by the authors. Licensee MDPI, Basel, Switzerland. This article is an open access article distributed under the terms and conditions of the Creative Commons Attribution (CC BY) license (https:// creativecommons.org/licenses/by/ 4.0/).

1. Introduction

The continuously growing demand for electricity in the world encourages the rational use of energy resources. The extraction of hard coal from thin seams is one of the possibilities for the rational management of natural resources, particularly in those countries where the deposits that are the most attractive in terms of extraction profitability, i.e., those located in medium and thick seams, have already been partially or completely mined. In recent years, the coal reserve of medium-thick and thick coal seams, which are thicker than 2 m, has decreased significantly [1].

Since no available energy source can be ignored, the possibility of the economically viable exploitation of thin deposits has received increasing attention in recent years [2–4]. Satisfactory technical and economic results are possible to obtain with the use of fully mechanized equipment [5].

In some countries, it is assumed that the lower limit of deposit thickness for thin seams is 0.4–0.5 m. In Ukrainian coal mining, the assumed thickness is 0.7–1.2 m [6–8], with a lower technical limit of 0.65 m [9]. In Chinese conditions, thin seams are those with a thickness of 0.8–1.3 m, while seams thinner than 0.8 m are classified as extremely thin [10]. In the case of Poland, thin seams are those with a thickness of 1–1.5 m.

The exploitation of thin seams is a point of focus not only for users but also for mining machinery manufacturers. Decades of efforts resulted in significant achievements, and the automation level of thin coal seam mining has been gradually improving. Currently, longwall shearers or static coal plows are most commonly used to mine thin seams. Their advantages, disadvantages, and adaptability for different geological conditions can be

111

found in the literature [11,12]. A longwall shearer is a component of a mechanized longwall system that is also equipped with a face conveyor and powered roof support. The mechanized longwall system enables the execution of the mining process and the loading and haulage of the excavated material from the longwall. Currently, regardless of the thickness of the seam, the most popular of the produced mining machines are two-arm, two-unit shearers moving on the conveyor by means of a chainless haulage system. A unique solution used for two-way, noncavity mining and loading of coal in thin seams is the Mikrus system, which is equipped with a GUŁ-500 mining and loading head. The data used for this study were acquired from the Mikrus system, for the period in which the shearer was tested under actual working conditions.

The use of the right mining machines for mining low seams is crucial to an efficient coal mining process. The possibility of obtaining relevant data from them is also important. The intensive digitization of the mining process is creating tremendous opportunities for mining companies. These opportunities are related to the collection, storage, and processing of massive amounts of data that can be used to derive new and useful knowledge about processes taking place in a mining company. Digitization of the mining industry translates into significant improvements in productivity [13].

The proper use of low-level data obtained from existing monitoring systems is very important. Properly used data analysis should result in the optimization of business management processes and procedures. Heterogeneous data are most often generated by a series of low-level sensors that monitor the operation of machines and equipment that are part of a larger process. The data must be preprocessed and supplemented with specific knowledge from the domain of processes. Additionally, sensor data need to be translated into a higher-level representation, e.g., event log [14–16]. Event logs are comprised of activities that have occurred during the execution of a process. They enable a process analyst to explore the process which generated a particular event log. In other words, the event log is the evidence of the process that produced it [17]. When making their decisions, managers of modern mining companies use data from different areas of the company as well as from other sources. Process data can come from multiple sources: for example, enterprise resource planning (ERP) systems, machinery and equipment monitoring systems, the work environment, or employee location. Different origins of data result in varying degrees of detail: from the most general (e.g., geometric dimensions of the excavation), through more detailed operating states of machines (work, alarm, standstill), to simple measurements (e.g., methane, currents in motors, transformer switching) [18]. It is essential that all data are assigned to a specific process and analyzed in its context to provide an in-depth analysis of performance and security.

There are several different approaches to analyzing data related to work environment processes and conditions in the scientific literature. Many authors favor an approach that includes a broad spectrum of data mining algorithms that are used in the classification of phenomena [19–21], prediction [22–25], and description tasks [26–28]. For process improvement, process-oriented methods should be used, such as process mining (PM), which is derived from workflow analysis. Process mining is most commonly used in business process management. Tools used in PM include such techniques as process model discovery, compliance verification, process model repair, role discovery, bottleneck analysis, and prediction of the remaining flow time [18,29].

In real-world conditions, we do not always obtain all the data we need to determine the process flow, and thus we do not have the complete information needed to optimize the process. This can be caused by sensor misalignment, sensor failure, disrupted data transfer, or, finally, incorrect readings. This paper presents an algorithm for reconstructing the original data layout using the R programming language and discovering parts of the process based on other parameters, i.e., presentation of a set of results showing the structure of a machine's operating states based on readings of changes in current intensities. This approach allows new information to be acquired, existing information to be verified, and further process analysis to be conducted.

The article is structured as follows: the second section presents the description and characteristics of the "Mikrus" longwall system from which the analyzed data were derived. The section also presents the R language functions used in the calculations. The next section presents the data used for the analysis and how they were prepared. The section also contains the author's analysis of shearer states and the structure of the machine operating states. The final section presents a summary of the work and prospects for further research in this area.

2. Materials and Methods

This section consists of two subsections. The first subsection presents a description and basic data on the Mikrus longwall complex. The second subsection describes the basic functions of the R language used in further calculations.

2.1. The "Mikrus" System

The subject of the analysis is data collected during the operation of the "Mikrus" longwall system from the period when the system was tested under real working conditions. The "Mikrus" longwall system is designed for working thin seams with a deposit thickness of 1.1–1.5 m. It is equipped with a GUŁ-500 cutting and loading head which is moved on a face conveyor along the coal wall. The head is moved using a linkage system of mining organs underneath a powered roof support shield—Figure 1.

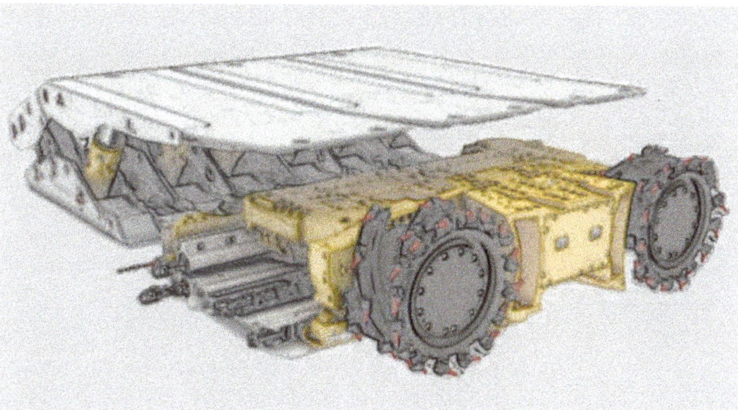

Figure 1. "Mikrus" longwall system [30].

The system is controlled by an operator using a central console located in the temporary storage gallery. The basic technical parameters of the system are shown in Table 1.

Table 1. Basic technical parameters of the system [30,31].

Parameter	Value
Cutting height	1100–1700 mm
Longwall length	260 m
Longitudinal longwall inclination	±35°
Transverse longwall inclination	±20°
Cutting head diameter	1200 ÷ 1600 mm
Working depth	600 mm
Supply voltage	3300 V
Overcut and undercut	50 mm
Minimum height of the cutting-loading head above the conveyor	850 mm

Table 1. *Cont.*

Parameter	Value
Minimum height in the working field	1000 mm
Maximum total installed power	633 kW
Maximum power of cutting head motor	500 kW
Maximum feed motor power	2 × 60 kW
Maximum winch motor power	13 kW
Maximum motor power of the conveyor drive	2 × 200/400 kW
Feed force (0–50 Hz)	2 × 320 kN
Feed rate	0–27 m/min
Longwall conveyor S-850N with the height of trough profile	220 mm
Conveyor scraper width	800 mm
Total weight of cutting and loading head	approx. 19.2 tons
Powered roof support Tagor-08,/16-POz	0.85 m–1.6 m
Assumed hourly capacity for 40 MPa coal strength	560 t/h
Assumed hourly capacity for 10 MPa coal strength	800 t/h

For more information on the "Mikrus" complex, see [30,31].

2.2. R Functions Used in Calculations

The R language was used to perform the calculations. It allows easy manipulation of large data sets by placing them in structures called data frames. The data prepared in this way are then processed using the functions of the R language. From the beginning of the development of this language, many libraries with functions covering a wide spectrum of calculations have been created. The significant fact is that it is an open-source language which allows researchers to add new features to its libraries as data science and analysis develop. Access to the libraries of the R system is provided by the CRAN archive. CRAN is a network of FTP and web servers around the world that store identical, up to date versions of code and documentation for R [32]. This language can run on the most popular operating systems, such as macOS, Windows, and most Linux distributions.

In addition to the standard functions built into the R language, the following libraries were used in the calculations: *ggpolt2, dplyr, runner*, and the *replace_na_with_last* function.

The *ggplot2* library is a tool for creating advanced graphs. It allows for overlapping successive layers of graphs, assigning shapes and colors of objects depending on the value of attributes, automatic calculation and presentation of statistics, and creating panels and histograms [33]. The following functions were used in the presented research: *ggplot, geom_line, geom_point*, and *geom_histogram*.

Dplyr is a grammar of data manipulation, providing a consistent set of verbs that help one to solve the most common data manipulation challenges. [34]. From this package, the functions that were used in the calculations are: *mutate, select, filter, arrange*, and *group_by*.

Runner is a lightweight library for rolling windows operations. The package enables full control over the window length, window lag, and time indices [35]

In addition, the *replace_na_with_last* [36] function was used when restoring compressed variable values. In this function, a missing value of a variable can be filled with its last known value until the next known value appears.

3. Results and Discussion

The data used for the calculations were obtained when the shearer was tested under actual working conditions. They cover a period of two months and describe the values of the currents flowing through the windings of four shearer motors. One of the motors drives the cutting organs, two other motors are used to drive the shearer's feed, and the fourth one drives the mechanism responsible for properly laying the power cables.

3.1. Data Preparation

The original data format contained 554,352 records with information about changes in current intensities, the time of their occurrence, and the code of the motor to which the change pertained, which appropriately reduced the amount of transmitted and collected data, but for its analysis, the original data layout had to be reconstructed.

R language functions were used to recreate the original data layout.

The processed dataset includes 5,254,993 records containing a timestamp, and four records containing the corresponding information about current. A fragment of the reconstructed set is shown in Table 2.

Table 2. Excerpt from the dataset.

tsu	ouf	npf	ngf	nuf
1364880939	31	32.3	32.4	16
1364880940	31	32.3	33.0	16
1364880941	31	33.6	33.6	16
1364880942	30	37.3	36.3	16
1364880943	32	41.6	39.5	16
1364880944	39	41.6	40.1	16
1364880945	48	41.6	39.1	16
1364880946	39	40.3	39.6	16

Source: Own study.

In Table 2, the column headings indicate, respectively:

Tsu—time (unix timestamp);
ouf—current of the shearing organ motor A;
npf—current of the auxiliary drive motor A;
ngf—current of the main drive motor A;
nuf—current of the stacker motor A.

The data thus prepared were the basis for further calculations.

3.2. Preliminary Data Analysis—A Study of Current Intensity Distributions

Calculations were performed to determine histograms of current intensities for each motor. This allowed for revealing the nature of data variability. Calculation results are illustrated in Figure 2.

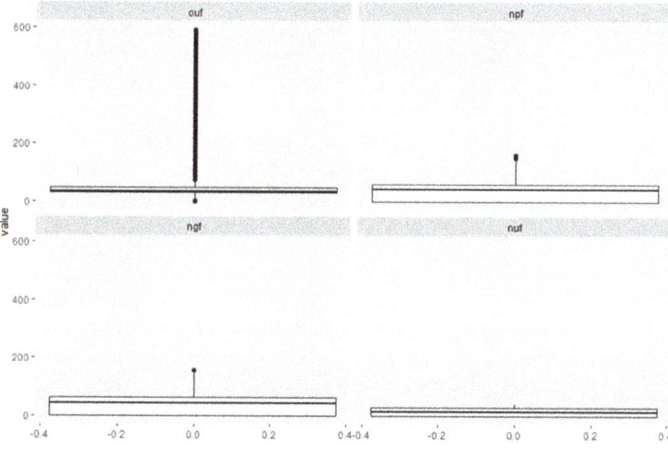

Figure 2. A box plot of the variability of the recorded parameters. Source: Own study.

As can be seen in Figure 2, a significant portion of the measured values of the cutting organ current (ouf) is at the lower end of the range of observed intensities, with outliers as high as 600 [A]. These are isolated situations associated with the starting of a particular motor. In order to better illustrate the variation of the remaining parameters, the values of current exceeding 200 [A] and equal to zero were eliminated from the graph, and another graph was then created (Figure 3).

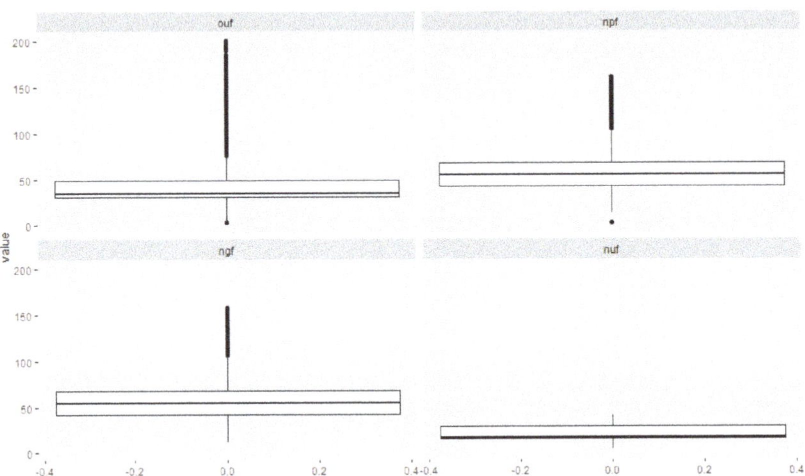

Figure 3. A box plot of the variability of the recorded parameters after rejecting extreme values. Source: Own study.

An analysis of the graph in Figure 3 reveals that the ranges of current intensities of the motors (ngf, npf) are very close to each other. The average value of these intensities is approximatly 42 A, and the interquartile range is ca. 24 A. The range of currents flowing through the motor of the cutting drive (ngf) is much larger and reaches up to 600 A (Figure 1), while the interquartile range is smaller than that of the drive motors and amounts to 17 A, and the mean of the observed values is ca. 35 A. As can be seen from the analysis of distributions, it should be noted that the terms 'main engine' and 'auxiliary engine' are conventional because the observed currents flowing through these engines are similar, and, moreover (as can be seen from other analyses), they operated simultaneously throughout the analyzed period.

3.3. Data Illustration

Using the ggplot function available in the R language, it is possible to generate graphs that can be customized in any possible way. A graph mapping the changes in currents as a function of time, recorded in the database, is presented below (Figure 4).

In Figure 4, one can observe the periods of both work and standstill of the shearer. Purple is used for the values of current of the cutting unit motor (ou). In the presented period, there is a peak reaching the value exceeding 400 [A] connected with the starting of this motor. This is confirmed by the different increases in the currents flowing through the other motors. This figure illustrates the mutual similarity of the current values flowing through the two feed drive motors and the relatively constant current values flowing through the stacker drive motor.

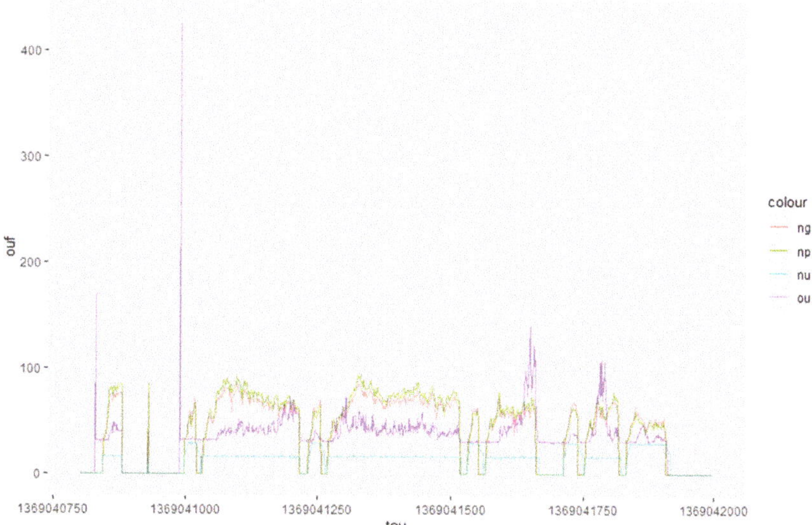

Figure 4. A plot of the variation of recorded current intensities as a function of time. Source: Own study.

3.4. Shearer Status Analysis

To investigate the nature of the load distribution of shearer motors in more detail, histograms of the values of the currents flowing through the main drive motor (Figure 5) and through the main feed drive motor (Figure 6) were generated.

Figure 5. A histogram illustrating the current distribution of the cutting organ motor. Source: Own study.

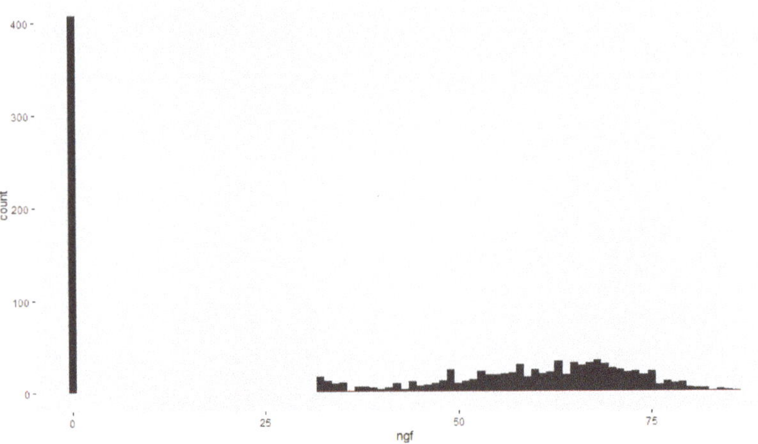

Figure 6. A histogram illustrating the current distribution of the main feed motor. Source: Own study.

Figure 5 presents a large number of observed zero values associated with device stoppage. In addition, there is a large number of observations of 30 [A] values which correspond to the movement of the cutting organ without the mining load. Higher values are observed during cutting.

In interpreting the distribution of current intensities in Figure 6, we can observe, similarly as before, a large number of zero value occurrences. However, in this case, it is difficult to clearly distinguish the values of current intensity corresponding to driving the shearer's feed without the use of a cutting unit.

A better understanding of the operating state of the shearer can be obtained when the above histograms are overlayed. Figure 7 presents a graph resulting from superimposing the observations in the space formed by the currents of the cutter motor and the currents of the feed motor.

Figure 7. An illustration of the number of observations in the space of current intensities for two shearer motors. Source: Own study.

The number of observations corresponding to the values of these currents is mapped by the degree of blackness of a point in this coordinate system. This was accomplished by overlapping points with a high degree of transparency. In Figure 6, four operating states of the machine can be distinguished. The first one, in which the currents of both motors are zero, is the machine shutdown state. The second state is the idle state, which is a situation where the current of the feed motor is zero, while the current of the cutting unit motor is approximately 30 [A]. The third state is the maneuvering state. In this state, a current equal to approximately 30 [A] passes through both motors. It stands out in Figure 6 as a dark circle at the bottom of the point cloud. The last operating state of the shearer is the cutting state. In this state, the current of both the motor driving the cutting unit and the feed mechanism exceeds 30 [A].

Based on the boundaries assumed above, it is possible to assign the recorded readings of current intensity to the distinguished shearer working states and then to calculate the percentage structure of these states in the analyzed period of time. Taking economic calculations into account, the state of cutting is expected to occupy the largest part in this structure. The rules for assigning cutter operating states are summarized in Table 3.

Table 3. Assignment of operating states to observed current intensities.

ngf\ouf	0	0–33	>33
0	Off	Idle	X
>32	Maneuvering	Maneuvering	Extraction

Source: Own study.

The state marked with an "X" in Table 3 deserves an additional comment. It is a prohibited state. In this state, the shearer would mine coal without feed. Observations of this state are possible but only to analyze the incorrect operation of the shearer. An example would be to start the drive of the cutting organ that is hogged into a coal bed. This action can very likely result in damage to the machine.

The operation of the feed drive motors is analyzed in the next part of the study. Observations of the distributions of currents flowing through these motors and their waveforms observed on the graphs suggest a high similarity of these values. To confirm this, a graph was constructed (Figure 8) whose axes mark the currents of both motors.

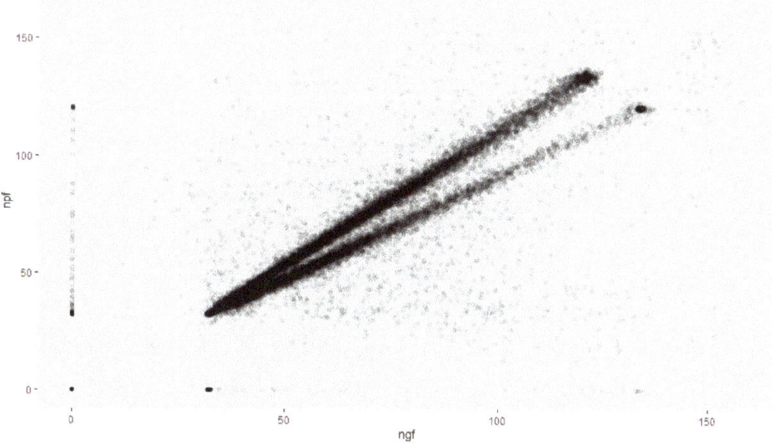

Figure 8. Mutual dependence of the loads on feed motors. Source: Own study.

In Figure 8, it can be seen that the points align along two distinct straight lines. The top line represents the cases where the power consumed by the main motor is greater than the power consumed by the auxiliary motor, while the bottom line represents the opposite case. It can also be seen that the difference between the currents flowing through the motors increases proportionally to their load. To better understand the structure of these cases, their percentages were calculated. The results are shown in Table 4. This list, as well as the rest of the analysis, does not include cases in which the feed drives were not working, because it concerns the states of a device in motion.

Table 4. Percentages of cases.

Case	Share (%)
main > auxiliary	52
main = auxiliary	1
main < auxiliary	47

Source: Own study.

Since the structure of these cases turned out to be uniform, a histogram was determined in the next step (Figure 9) which shows the distribution of differences between the currents of the two motors.

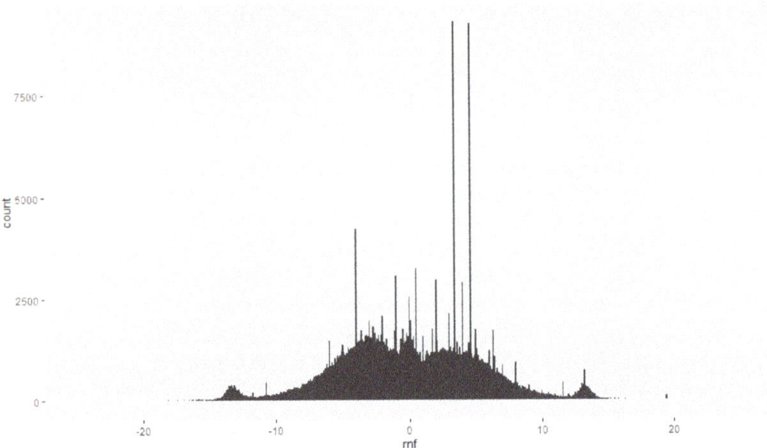

Figure 9. A histogram of differences in motor loads. Source: Own study.

The values to the right of zero are cases where the main motor draws more power than the auxiliary motor. This histogram is characterized by high symmetry, which, together with the percentage distribution calculated earlier, suggests that, to a significant extent, these motors swap roles in driving the shearer's feed.

This observation is the basis for the hypothesis that shearer working directions can be distinguished. Since a significant longwall slope is likely to occur, there may be differences in loads between the motors driving the shearer feed. These differences may be increased by the activity of loading the excavated material onto a scraper conveyor, which is additionally performed by the moving shearer, and which may depend on the direction of the feed. Similar differences resulting from the direction of the shearer's movement, but observed in the engines of the cutting units, were described in [37].

Partial confirmation of this hypothesis can be provided by the graph shown in Figure 10.

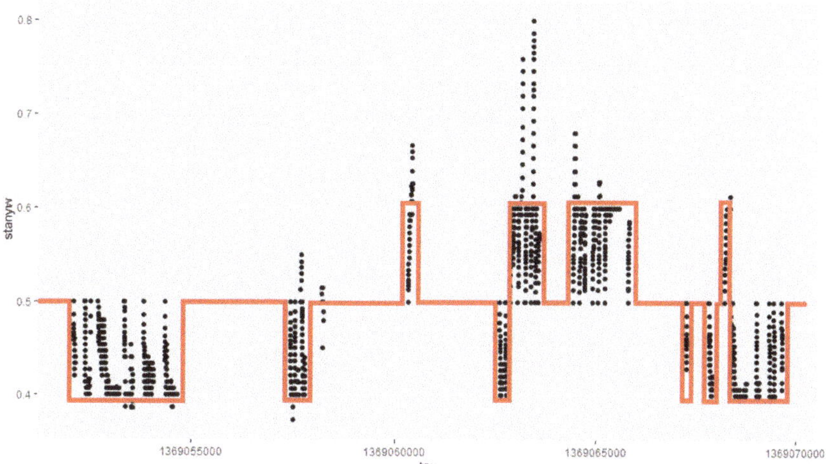

Figure 10. Changes in direction of shearer movement. Source: Own study.

Figure 10 was developed by introducing an additional variable (state) into the data set, based on the intervals determined in Figure 9. This variable takes values according to Table 5.

Table 5. Values of the 'state' variable in individual ranges of different currents of feed drive motors.

Interval	Value
$(-\infty, -12.5>$	0.2
$(-12.5, 2>$	0.4
$(-2, 2)$	0.5
$<2, 12.5)$	0.6
$<12.5, \infty)$	0.8

Source: Own study.

To increase the clarity of the graph, the values of this variable were averaged within a fifteen-second window. The graph in Figure 10 covers a period of 5 h 30 min. The red line indicates the probable directions of the shearer's movement (values of 0.4 and 0.6), and a value of 0.5 indicates its standstill.

3.5. Structure of Machine Operating States

Based on the rules determined in the analysis and collected in Table 3, a set of results (Table 6) was generated, showing the percentage structure of the machine operating states observed on consecutive days.

Table 6. Extract from the results of the machine structure.

Observation Day No.	Off	M	I	E
36	96.77	0.08	0.00	3.15
37	84.90	4.31	1.12	9.67
38	83.55	2.74	1.35	12.36
39	78.52	5.80	2.79	12.89
40	89.93	0.85	0.75	8.47
41	100	0	0	0
42	100	0	0	0

Source: Own study.

Table 6 presents the results of one week of observations. On all days, the vast majority of the time structure is occupied by the Off state. The largest share during the operation of the shearer is occupied by extraction activity (E) and the smallest share by the Idle (I) state. The maneuvering state is shown in column M. It should be recalled that the recorded data pertain to the period of implementation and testing of the shearer, and therefore, the results may significantly differ from those obtained in the conditions of industrial exploitation of the deposit.

4. Conclusions

The proposed method can serve as a verification tool for progress reported by employees. In a mine setting, the reporting of work through numbers provided by supervisory personnel is still used. The method described in this study enables a comparison of these values with the values calculated based on the analysis of current intensities, and thus, objective values can be obtained.

The results presented here do not include the shearer direction component, as the relationships discovered in the analysis must be further verified as a continuation of this work.

The presented analysis can also be an example of the fact that, in cases when the required process parameters are not available, they can be discovered based on other available parameters, which facilitates further analysis.

The authors believe that there is the need to develop a method to automatically separate groups of observations based on the shape of the histograms. Past considerations lead to determining the zero positions of the first derivative function approximating the quantitative distribution of observations.

The presented method can be implemented in information systems reporting the course of the production process in hard coal mines.

The calculations described in this paper are in line with the authors' earlier works on monitoring the operation of machines. Issues related to identifying deviations from the norms of machine operation are presented in [38]. The subject of monitoring the analysis of the effectiveness of machine use in hard coal exploitation by generating reports enabling the analysis of the degree of machine use is presented in [39].

Author Contributions: Conceptualization, M.K.; methodology, M.K. and R.O.; software, M.K.; investigation, M.K. and R.O.; writing—original draft preparation, M.K. and R.O. Writing—review and editing, M.K. and R.O.; visualization, M.K. and R.O.; supervision, M.K. and R.O. All authors have read and agreed to the published version of the manuscript.

Funding: This research was prepared as part of AGH University of Science and Technology in Poland, scientific subsidy under number: 16.16.100.215.

Institutional Review Board Statement: Not applicable.

Informed Consent Statement: Not applicable.

Data Availability Statement: The data presented in this study are new and have not been previously published.

Acknowledgments: In the main research, we have used the following R libraries: RMySQL, ggplot2, dplyr, and reshape.

Conflicts of Interest: The authors declare no conflict of interest.

References

1. Zhao, T.; Zhang, Z.; Tan, Y.; Shi, C.; Wei, P.; Li, Q. An innovative approach to thin coal seam mining of complex geological conditions by pressure regulation. *Int. J. Rock Mech. Min. Sci.* **2014**, *71*, 249–257. [CrossRef]
2. Lorenz, U. Coal as fuel with good prospects/Węgiel jako paliwo z przyszłością. *Przegląd Gór.* **2005**, *10*, 8–14. (In Polish)
3. Turek, M. Do not leave thin seams/Nie pozostawiajmy pokładów cienkich. *Przegląd Gór.* **2005**, *10*, 15–19. (In Polish)
4. Zhang, L.M. The Current Situation and Developing Trend of Extremely Thin Coal Seam Mining of China. *Appl. Mech. Mater.* **2013**, *295–298*, 2918–2923. [CrossRef]

5. Wang, J. Development and prospect on fully mechanized mining in Chinese coal mines. *Int. J. Coal Sci. Technol.* **2014**, *1*, 253–260. [CrossRef]

6. Mamaikin, O.; Kicki, J.; Salli, S.; Horbatova, V. Coal industry in the context of Ukraine economic security. *Min. Miner. Depos.* **2017**, *11*, 17–22. [CrossRef]

7. Griadushchiy, Y.; Korz, P.; Koval, O.; Bondarenko, V.; Dychkovskiy, R. Advanced experience and direction of mining of thin coal seams in Ukraine. In *Technical, Technological and Economical Aspects of Thin-Seams Coal Mining, International Mining Forum*; Taylor & Francis/Balkema: London, UK, 2007; pp. 1–7.

8. Pavlenko, I.; Salli, V.; Bondarenko, V.; Dychkovskiy, R.; Piwniak, G. Limits to economic viability of extraction of thin coal seams in Ukraine. In *Technical, Technological and Economical Aspects of Thin-Seams Coal Mining, International Mining Forum*; Taylor & Francis/Balkema: London, UK, 2007; pp. 129–132.

9. Korski, J.; Bednarz, R. Low shearer longwall system as an effective option to plough systems/ Kombajnowy system ścianowy jako efektywna alternatywa dla strugów ścianowych. *Mech. Autom. Gór.* **2012**, *9*, 31–38. (In Polish)

10. Chai, W.Z. The lectotype of the thin coal bed synthesis mining equipment and the technological parameter optimizes. *Sci. Technol. Tongmei* **2007**, *2*, 2–4.

11. Zhai, X.; Shao, Q.; Li, B.; Li, X. Shearer mining application to soft thin-seam with hard roof. *Procedia Earth Planet. Sci.* **2009**, *1*, 68–75. [CrossRef]

12. Wang, F.; Tu, S.; Bai, Q. Practice and prospects of fully mechanized mining technology for thin coal seams in China. *J. S. Afr. Inst. Min. Metall.* **2012**, *112*, 161–170.

13. Lööw, J.; Abrahamsson, L.; Johansson, J. Mining 4.0—The Impact of New Technology from a Work Place Perspective. *Min. Metall. Explor.* **2019**, *36*, 701–707. Available online: https://link.springer.com/article/10.1007/s42461-019-00104-9 (accessed on 22 May 2021). [CrossRef]

14. Bezerra, F.; Wainer, J. Anomaly detection algorithms in business process logs. In Proceedings of the Tenth International Conference on Enterprise Information Systems—Volume 2: ICEIS, INSTICC, Barcelona, Spain, 12–16 June 2008; SciTePress: Setúbal, Portugal; pp. 11–18. [CrossRef]

15. Böhmer, K.; Rinderle-Ma, S. Automatic Signature Generation for Anomaly Detection in Business Process Instance Data. In *Enterprise, Business-Process and Information Systems Modeling*; Schmidt, R., Guédria, W., Bider, I., Guerreiro, S., Eds.; Springer International Publishing: Cham, Switzerland, 2016; pp. 196–211.

16. Chen, D.; Panfilenko, D.V.; Khabbazi, M.R.; Sonntag, D. A model-based approach to qualified process automation for anomaly detection and treatment. In Proceedings of the 2016 IEEE 21st International Conference on Emerging Technologies and Factory Automation (ETFA), Berlin, Germany, 6–9 September 2016; pp. 1–8.

17. Nolle, T.; Seeliger, A.; Mühlhäuser, M. *Unsupervised Anomaly Detection in Noisy Business Process Event Logs Using Denoising Autoencoders; Discovery Science*; Calders, T., Ceci, M., Malerba, D., Eds.; Springer International Publishing: Cham, Switzerland, 2016; pp. 442–456.

18. Brzychczy, E.; Gackowiec, P.; Liebetrau, M. Data Analytic Approaches for Mining Process Improvement—Machinery Utilization Use Case. *Resources* **2020**, *9*, 17. [CrossRef]

19. Zhou, J.; Li, X.; Mitri, H.S.; Wang, S.; Wei, W. Identification of large-scale goaf instability in undergroundmine using particle swarm optimization and support vector machine. *Int. J. Min. Sci. Technol.* **2013**, *23*, 701–707. [CrossRef]

20. Hargrave, C.O.; James, C.A.; Ralston, J.C. Infrastructure-based localisation of automated coal mining equipment. *Int. J. Coal Sci. Technol.* **2017**, *4*, 252–261. [CrossRef]

21. Jedliński, Ł.; Gajewski, J. Optimal selection of signal features in the diagnostics of mining head tools condition. *Tunn. Undergr. Space Technol.* **2019**, *84*, 451–460. [CrossRef]

22. Bodlak, M.; Kudełko, J.; Zibrow, A. Machine Learning in predicting the extent of gas and rock outburst. *E3S Web Conf.* **2018**, *71*. [CrossRef]

23. Verma, A.K.; Kishore, K.; Chatterjee, S. Prediction Model of Longwall Powered Support Capacity Using Field Monitored Data of a Longwall Panel and Uncertainty-Based Neural Network. *Geotech. Geol. Eng.* **2016**, *34*, 2033–2052. [CrossRef]

24. Deb, D.; Kumar, A.; Rosha, R.P.S. Forecasting shield pressures at a longwall face using artificial neural networks. *Geotech. Geol. Eng.* **2006**, *24*, 1021–1037. [CrossRef]

25. Mahdevari, S.; Shahriar, K.; Sharifzadeh, M.; Tannant, D.D. Stability prediction of gate roadways in longwall mining using artificial neural networks. *Neural Comput. Appl.* **2017**, *28*, 3537–3555. [CrossRef]

26. Kopacz, M. The impact assessment of quality parameters of coal and waste rock on the value of mining investment projects—Hard coal deposits. *Miner. Resour. Manag.* **2015**, *31*, 161–188. [CrossRef]

27. Jonek-Kowalska, I.; Turek, M. Dependence of total production costs on production and infrastructure parameters in the polish hard coal mining industry. *Energies* **2017**, *10*, 1480. [CrossRef]

28. Qiao, W.; Liu, Q.; Li, X.; Luo, X.; Wan, Y.L. Using data mining techniques to analyze the influencing factor ofunsafe behaviors in Chinese underground coal mines. *Resour. Policy* **2018**, *59*, 210–216. [CrossRef]

29. He, Z.; Wu, Q.; Wen, L.; Fu, G. A process mining approach to improve emergency rescue processes of fatal gas explosion accidents in Chinese coal mines. *Saf. Sci.* **2019**, *111*, 154–166. [CrossRef]

30. FAMUR. Available online: https://famur.com/kompleks-mikrus-1 (accessed on 22 May 2021).

31. Dziura, J. Mikrus Complex—A new technology of extraction low decks/Kompleks Mikrus—Nowa technologia wybierania pokładów niskich. *Masz. Gór.* **2012**, *3*, 3–11. (In Polish)
32. The Comprehensive R Archive Network. Available online: https://cran.r-project.org/ (accessed on 22 May 2021).
33. GGPLOT2. Available online: https://ggplot2.tidyverse.org/ (accessed on 22 May 2021).
34. DPLYR. Available online: https://dplyr.tidyverse.org/ (accessed on 22 May 2021).
35. Package "Runner". Running Operations for Vectors. Available online: https://cran.r-project.org/web/packages/runner/runner.pdf (accessed on 22 May 2021).
36. Replacing NAs with Latest Non-NA Value. Available online: https://stackoverflow.com/questions/7735647/replacing-nas-with-latest-non-na-value (accessed on 22 May 2021).
37. Jaszczuk, M. The influence of the load condition shearer on the possibility of obtaining a high concentration of mining Wpływ stanu obciążenia kombajnu ścianowego na możliwość uzyskania wysokiej koncentracji wydobycia, Gliwice. *Zesz. Nauk. Politech. Śląskiej* **1999**, *240*, 19–25. (In Polish)
38. Kęsek, M. Analysing data with the R programming language to control machine operation/ Analiza danych z wykorzystaniem języka R w kontroli pracy maszyn. *Inżynieria Miner.* **2019**, *1*, 231–235. [CrossRef]
39. Kęsek, M.; Ogrodnik, R. Computer aided monitoring and analysis of machine work/Komputerowe monitorowanie i analiza pracy maszyn. *Mark. Rynek* **2018**, *9*, 383–395. (In Polish)

energies

Article

Experimental Study on Shear Mechanism of Rock-Like Material Containing a Single Non-Persistent Rough Joint

Sayedalireza Fereshtenejad [1], Jineon Kim [2] and Jae-Joon Song [1,*]

[1] Department of Energy Resources Engineering, Research Institute of Energy and Resources, Seoul National University, Seoul 08826, Korea; a.r_fereshtenejad@snu.ac.kr
[2] Department of Energy Resources Engineering, Seoul National University, Seoul 08826, Korea; kjineon@snu.ac.kr
* Correspondence: songjj@snu.ac.kr

Abstract: The geometrical and mechanical properties of non-persistent joints as well as the mechanical behavior of intact rock (rock bridges) are significantly effective in the shear strength of weakness planes containing non-persistent joints. Therefore, comprehensive knowledge of the shear mechanism of both joints and rock bridges is required to assess the shear strength of the planes. In this study, the shear behavior of specimens containing a single non-persistent rough joint is investigated. A novel procedure was used to prepare cast specimens embedding a non-persistent (disc-shaped) rough joint using 3D printing and casting technology, and the shear strength of the specimens was examined through an extensive direct shear testing program under constant normal load (CNL) condition. Three levels for three different variables of the joint roughness, rock bridge ratio, and normal stress were considered, and the effects of these factors on the shear behavior of prepared samples were tested. The experimental results show a clear influence of the three variables on the shear strength of the specimens. The results show that the normal stress applied to the jointed zone of weakness planes is considerable, and thus joint friction contribution should be taken into account during shear strength evaluation. Furthermore, the dilation mechanism of the specimens before and after failure was investigated through a digital image correlation analysis. Finally, a camcorder was used to analyze the location and sequence of the initiated cracks.

Keywords: shear behavior; non-persistent joint; rock bridge ratio; joint roughness; normal stress; digital image correlation

check for updates

Citation: Fereshtenejad, S.; Kim, J.; Song, J.-J. Experimental Study on Shear Mechanism of Rock-Like Material Containing a Single Non-Persistent Rough Joint. *Energies* **2021**, *14*, 987. https://doi.org/10.3390/en14040987

Academic Editor: Krzysztof Skrzypkowski

Received: 8 January 2021
Accepted: 9 February 2021
Published: 13 February 2021

Publisher's Note: MDPI stays neutral with regard to jurisdictional claims in published maps and institutional affiliations.

Copyright: © 2021 by the authors. Licensee MDPI, Basel, Switzerland. This article is an open access article distributed under the terms and conditions of the Creative Commons Attribution (CC BY) license (https://creativecommons.org/licenses/by/4.0/).

1. Introduction

When the engineering dimensions of an investigated site exceed the average joint size in the domain, the joints are surrounded by intact rock (rock bridge) in the rock mass and should be considered as non-persistent. Hence, joint size (persistence) should be precisely measured during the field survey owing to its substantial influence on the rock mass strength [1]. A combined shear plane where failure occurs is usually formed by the interaction of various non-persistent joints. Deng and Zhang [2] and Segall and Pollard [3] noted that the rupture of rock bridges connecting en-echelon joints may lead to the development of faults. In natural faults damage zones, en-echelon fractures, which occur as a unique set of sub parallel fractures, have often been found [4]. A comprehensive understanding of the spatial and geometrical properties of joints in a rock mass reveals various potential failure paths passing through the joints and rock bridges. Not only the bridges but also the joints have a significant effect on the shear behavior of the failure paths [5,6].

There has been considerable experimental and numerical research on the shear behavior of rough persistent rock joints from different points of view [7–29]. However, relatively few studies have investigated the shear mechanism of rock masses with non-persistent joints surrounded by intact rock. In previous experimental studies, edge-notched embedded joints were considered, and artificial materials were preferred to natural rocks for creating specimens containing precise joint configurations with varying parameters. The spatial and geometric properties of joints in a rock mass (joint angle, rock bridge angle, joint size, rock bridge size, joint dispersion, length of overlap, and joint offset) as well as the boundary condition (normal load) and mechanical characteristics of the intact rock and joint surface (internal friction, internal cohesion, and joint friction) were systematically investigated [5,30–33]. In addition, to conduct a quantitative analysis which precisely detects the full failure mechanism of a rock mass with non-persistent joints, different numerical approaches such as the finite element method (FEM), discrete element method (DEM), and boundary element method (BEM), have been employed [31,34–39].

This study investigates the shear behavior of specimens with an embedded non-persistent rough joint using a direct shear test machine under constant normal load (CNL). The joint size/rock bridge ratio and the normal load as well as the joint roughness are the variables whose effects on the shear mechanism of the specimens are studied [40]. This research is different from previous studies for the following reasons.

- The applied specimen preparation techniques of previous studies resulted in specimens containing joints which are edge notched. Thin sheets were located within a casting frame prior to adding mortar and were removed as the mortar hardened. However, in this study, a novel method is applied to create non-persistent joints surrounded by intact material;
- In previous studies, the sheets applied to make joints caused a measurable aperture (gap) between the joint walls, corresponding to the thickness of the sheets. This gap prevents the contact between the joint walls, and consequently joint friction cannot be mobilized during shear process. However, using the casting approach applied in this study, the embedded joints are closed and joint friction is mobilized from the beginning of direct shear tests;
- In most previous experimental research on the shear behavior of rock masses, smooth non-persistent rock joints were modeled and analyzed. However, the specimen preparation procedure here allows one to create non-persistent close joints with different roughness levels along different directions.

2. Experimental Procedure

2.1. Equipment and Experimental Settings

A stiff servo-controlled direct shear test machine was used to perform all of the shear tests in this study (Table 1). Normal and shear displacements were measured using linear variable differential transformers (LVDTs), and strain gauge type load cells were applied to measure the shear loads. The load and displacement data were continuously recorded using a data acquisition system. The tests were conducted under constant normal load (CNL) condition by means of a hydraulic actuator. According to International Society for Rock Mechanics (ISRM) suggested methods, the normal load was continuously increased at a similar rate (0.01 MPa/s) up to the target normal stress (σ_n) [41]. The specimens were sheared up to 20 mm at the shear displacement rate of 0.2 mm/min. A 10 mm gap between the specimen holders was considered following the ISRM standard [41]. Moreover, fracture initiation moments were monitored using a camcorder at a frame rate of 30 fps (Figure 1b).

Table 1. Mechanical properties of 3DP and plaster specimens.

Material Type	UCS (MPa)	Tensile Strength (MPa)	Young's Modulus (GPa)	Poisson's Ratio	Basic Friction Angle (°)	Cohesion (MPa)	Internal Friction Angle (°)	Density (gr/cm^3)
3DP	15.9	2.4	6.16	—	41	—	—	—
Plaster	34	3.85	7.08	0.23	39.3	7.0	40	1.86

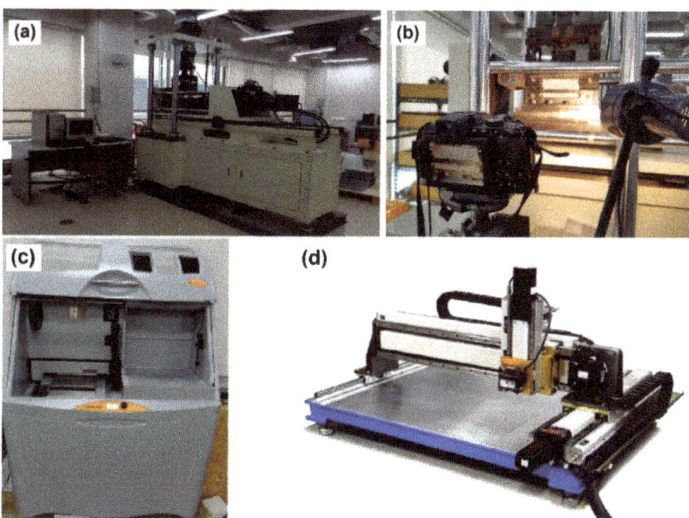

Figure 1. The equipment applied in this study; (**a**) Servo-controlled direct shear test machine, (**b**) Camera (Nikon, COOLPIX P600), (**c**) Zprinter® 450, (**d**) 3D laser profiler.

In the process of mold preparation to make rough joints, a powder-based 3D printer machine (Zprinter® 450) that uses binder jetting technology was used (Figure 1c). Comprehensive information about the functionality of the 3D printer and its applicability to rock mechanics is available in the authors' previous study [42]. In addition, a 3D laser profiler was applied to measure the surface roughness of the 3D printed (3DP) molds, the silicone molds, and the plaster joints applied in this study (Figure 1d). The 3D laser profiler determines the roughness parameter by digitizing the surface of a joint using a laser displacement meter (Kenyence LK-G150). This profiler is composed of a laser displacement meter, a motion control system, and a LabVIEW computer which controls the entire system and exports the measured data. The motion control system (Jeongwon Mechatronics) controls the position of the laser displacement meter using servo-motors which are placed on the x, y, and z axes.

2.2. Materials

An industrial gypsum powder was mixed with water at a mass ratio of 3:1 to create the mortar required to cast plaster specimens. The powder consists of more than 99% bassanite. Three NX-size plaster specimens with a height of 130 mm and three more with a height of 65 mm were prepared. Three uniaxial compressive strength (UCS) tests and three Brazilian tests were carried out on the prepared specimens using MTS 816 system to measure the UCS, Young's modulus, and tensile strength according to the ISRM suggested methods [43]. Moreover, three direct shear tests at different normal stresses of 1 MPa, 1.5 MPa, and 2 MPa were carried out to measure the basic friction angle of the plaster specimens. Table 1 lists the mechanical properties of plaster specimens. VisiJet PXL Core powder and VisiJet® PXL^{TM}$_{Clear}$ binder were used to make 3DP molds. A printing layer thickness of 0.089 mm and a binder saturation level of 120% were selected to print these specimens. Some of the mechanical characteristics of the 3DP molds are listed in Table 1. To create the silicone molds to cast the plaster rough joints, two types of silicone (Molkang), one with viscosity of 6000 MPa·s (Moldmaster (Soft)) and the other one with viscosity of 16,000 MPa·s (Moldmaster (Normal)) were applied.

2.3. Variables and Levels

This study investigates the effects of three variables (rock bridge ratio (ζ), normal stress (σ_n), and joint roughness (Z2)) on the shear behavior of plaster specimens with a single non-persistent joint. Z2 (the root mean square of the first derivative of the profile) is a commonly used statistical parameter applied in this study to quantify the roughness of joint profiles [44]. The roughness levels selected in this study, low (LR), medium (MR), and high (HR), almost cover a wide range of joint roughness in nature. Here, the rock bridge ratio (ζ) is defined as the ratio of the bridged area to the weakness plane area. As listed in Table 2, three different levels were selected for each factor. As the loading capacity of the applied direct shear test machine in shear direction was limited, 2 MPa was selected as the maximum value of the normal stress. The normal stress condition of this study is similar to the condition that exists at shallow depth. This normal stress condition was frequently selected in previous studies [20,23,45].

Table 2. Selected levels for the factors affecting the shear behavior of specimens with a non-persistent rough joint.

Factors (Variables)	Levels		
	1	2	3
Joint roughness	JRC = 6.6 Smooth, nearly planar	JRC = 11.7 Smooth undulating	JRC = 17.6 Rough undulating
Rock bridge ratio (ζ)/Joint size (mm)	0.71/80	0.59/95	0.45/110
Normal stress (MPa)	1	1.5	2

2.4. Specimen Preparation Procedure

2.4.1. Rough Joint Specimens (J Specimens)

Three different levels of roughness, denoted here as low, medium, and high, were selected to make 27 disc-shaped joint specimens to investigate the shear behavior of disc-shaped rough joints of three sizes under three different normal stress levels. Nine different surfaces for a combination of two factors (joint size, joint roughness) at three levels were required. To make rough plaster joints, nine silicone molds were created using the following procedure. First, a Matlab script was used to generate isotropic artificial rough surfaces with the given parameters (root mean square roughness, fractal dimension, size, and resolution) in the point cloud data format [46]. The code is based on simulating the surface topography/roughness by means of fractals. It uses the Fourier concept (specifically the power spectral density) for surface generation. Afterwards, the point cloud data files were converted to the STL format, and the STL files were then modified and 3D printed. Oil-based paint was uniformly sprayed onto rough surfaces of the 3DP molds for easily release from the final silicone molds (Figure 2a). Finally, to cast the final disc-shaped silicone molds, the 3DP molds were placed in steel molds located on a leveling table. A 10 mm layer of a durable silicone with relatively high viscosity was poured onto the rough surface of the 3DP molds, and a thinner layer (5 mm) of low viscosity silicone was poured onto the cured first layer to make a smooth and flat bottom (Figure 2b). Each plaster cylinder with a rough surface made by applying the silicone and steel molds (Figure 2c) was then encapsulated in the same casting material with a higher gypsum powder to water ratio (3.55:1) to be secured in each half of the specimen holder (Figure 2d).

All of the plaster disc-shaped joints were cast using these nine silicone molds. However, the roughness of the generated surface in Matlab gradually degraded during the molding process. As the final goal was to create three plaster joints of various sizes (three sizes) with the same roughness, continuous attempts to generate rough surfaces (by Matlab) were made until the final goal was reached. The roughness value of each joint was measured considering several sampling profiles parallel to the shear direction. The distance between two adjacent sampling profiles was set to be 0.5 mm. As the applied roughness parameter ($Z2$) is sensitive to the sampling interval [47], the same sampling interval of 0.5 mm was considered along each sampling profile. The number of points sampled for 80, 95, and 110 mm joints were 7338, 10,336, and 13,882, respectively. $Z2$ values of the sampling profiles were measured for each joint, and the average value was eventually considered as the representative roughness value for the joint. Table 3 lists the measured $Z2$ values of the generated, 3DP, and final plaster joint surfaces for all nine cases. The converted JRC values in the table were calculated using the equation proposed by Tse and Cruden (JRC = 32.2 + 32.47 log Z2) [44]. Figure 3 illustrates the surfaces of the final plaster joints scanned by a 3D laser profiler.

Figure 2. Preparation procedure of joint specimens; (**a**) 3DP mold, (**b**) silicone mold, (**c**) plaster rough joint, (**d**) encapsulated joint.

Table 3. Roughness properties of generated, 3DP, and final plaster joint surfaces.

Joint Diameter (mm)	Roughness Level	Roughness Value (Z_2)		
		Generated Surface	3DP Joint	Final Plaster Joint
80	Low (LR)	0.268	0.162	0.163 (Converted JRC = 6.62)
	Medium (MR)	0.390	0.238	0.235 (Converted JRC = 11.78)
	High (HR)	0.659	0.365	0.354 (Converted JRC = 17.56)
95	Low (LR)	0.267	0.161	0.163 (Converted JRC = 6.62)
	Medium (MR)	0.460	0.238	0.234 (Converted JRC = 11.72)
	High (HR)	0.675	0.364	0.356 (Converted JRC = 17.64)
110	Low (LR)	0.267	0.161	0.163 (Converted JRC = 6.62)
	Medium (MR)	0.435	0.239	0.232 (Converted JRC = 11.6)
	High (HR)	0.693	0.364	0.358 (Converted JRC = 17.71)

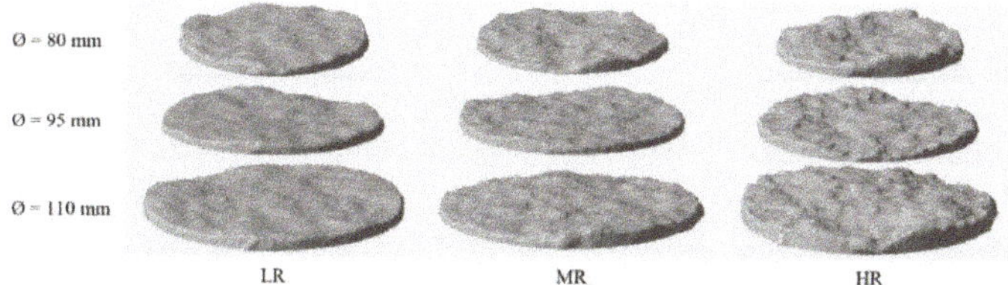

Figure 3. Surfaces of final plaster joints at three levels of roughness.

2.4.2. Specimens Containing a Single Non-Persistent Open Joint (Br specimens)

Several rectangular cuboid plaster specimens (length: 150 mm, width: 115 mm, height: 130 mm) with a single non-persistent open joint were cast to investigate the shear behavior of rock bridge. The preparation process of Br specimens is illustrated in Figure 4a. A detailed explanation about the sample preparation process of Br specimens is explained in the authors' previous study [48].

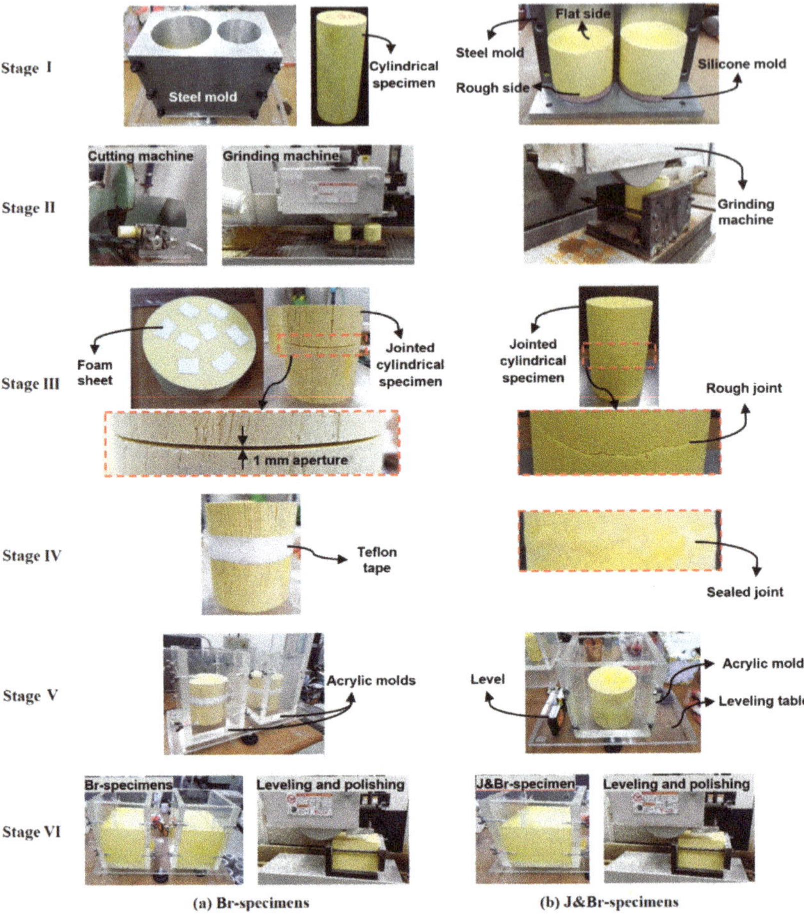

Figure 4. Step-by-step preparation process of (**a**) Br specimens and (**b**) J&Br specimens.

2.4.3. Specimens Containing a Single Non-Persistent Rough Joint (J&Br specimens)

Twenty-seven rectangular cuboid plaster specimens (length: 150 mm, width: 115 mm, height: 130 mm) embedding a single non-persistent rough joint were cast to investigate the simultaneous shear behavior of joint and rock bridge. The specimen preparation process of J&Br specimens is quite similar to that applied for the Br specimens and consists of the following stages (Figure 4b).

1. Two cylindrical plaster specimens each of which has one rough side and one flat side are initially cast using a steel mold and a pair of silicone molds. The rough sides of the cylindrical specimens represent the upper and lower walls of a joint;
2. The other side of each cylindrical specimen (flat side) is smoothed by a grinding machine;
3. The rough sides of the prepared cylindrical specimens are placed on each other to make a cylindrical specimen containing an interlocked rough joint in the middle;
4. To prevent the seepage of plaster mortar into the rough joint, the joint is plastered around the circumference with a very thin layer of slightly cured and sticky mortar;
5. The cylinder with a sealed rough joint is then put in a hexahedron acrylic mold which is placed on a leveling table;
6. The space around the jointed cylinder is filled with plaster mortar, and the sides of the cured specimen are leveled and smoothed utilizing a grinding machine.

The preparation process of one Br specimen or one J&Br specimen requires around eight hours. All of the specimens were tested seven days after the preparation process, during which they were kept at room temperature. As mentioned above, the casting process of Br specimens and J&Br specimens consists of two main steps. First, a jointed cylindrical part is cast, and once it solidifies, the surrounding rock bridge part is cast. Therefore, this inevitable interruption during the casting process may cause a weak surface around the cylindrical specimen. To investigate the possible effect of these weak surfaces on the results, all the specimens made for this study were carefully observed during and after the experiments. No crack was found to be initiated from or propagated toward those weak surfaces. Therefore, the experimental results of this study were not affected by the weak surfaces created during the specimen preparation process. It is important to note that the tensile resistance of the applied Teflon tape is insignificant under small deformations, hence it provides no resistance against the normal and shear deformation of open joints in Br specimens.

3. Experimental Results

The simultaneous contributions of the friction mobilized by the joint and the cohesion of the rock bridge to the shear resistance of weakness planes with non-persistent joints are highly complex. Finding the distribution of the applied normal load on jointed and bridged zones is a key factor to determine the shear strength of the planes. In some previous analytical studies [49], the normal stresses distributed on the jointed and bridged zones of a weakness plane were assumed to be identical and equal to the nominal normal stress (average normal stress applied to the entire area of the weakness plane), $\sigma_n^{nominal}$. In previous experimental research, the load was only applied to the bridged zone of weakness planes due to applied specimen preparation procedures; as a result, the joint roughness contribution to the shear strength was inconsiderable. As the normal stiffness of the rock bridge, K_n^{Br}, and the normal stiffness of the embedded joint, K_n^{J}, typically differ, the amount of normal load applied on each zone is proportional not only to its area on the weakness plane but also to its normal stiffness. Knowing the normal stiffness of the joint and rock bridge determines the amount of normal load distributed on the jointed and bridged zones of a weakness plane when only the normal load is applied. However, it provides no precise information about the normal stress distribution regime when both normal and shear loads are applied (direct shear test condition). Therefore, determination of normal stresses on the jointed and bridged zones of weakness planes (σ_n^{J} and σ_n^{Br}) subjected to the direct shear test is very complicated. The experimental strategy of this study is adopted to investigate the normal load distribution regime and the effects of joint roughness, normal load, and rock bridge ratio on the shear behavior of rock mass embedding a joint, and is categorized into two cases. Below, the experimental results obtained for each case are reported.

3.1. Case I

In case I experiments, the applied normal load (N_L) is entirely distributed over the bridged zone of the weakness plane (A_{Br}). Therefore, the normal stress applied onto the bridged zone is higher than the nominal normal stress ($\sigma_n^{J} = 0$, $\sigma_n^{Br} = \frac{N_L}{A_{Br}} > \sigma_n^{nominal}$), and the shear strength of the specimen is solely dominated by the shear resistance of the rock bridge. Several Br specimens containing a single joint of three different sizes ($\varnothing = 80$ mm, $\varnothing = 95$ mm, and $\varnothing = 110$ mm) were prepared and subjected to direct shear tests under three different normal stress levels (1, 1.5, and 2 MPa). As the joints are open, the shear test results reflect only the resistance of the bridged zone of the weakness planes. Figure 5 shows the shear tests results. As expected, the shear strength is higher for the cases with greater rock bridge ratios subjected to higher normal stresses. The values of the peak shear displacement and peak dilation for the tested samples are illustrated in Figure 6a,b, respectively. As shown, the specimens with greater rock bridge ratios subjected to higher normal stresses were sheared for longer distances before failure, and those with greater rock bridge ratios subjected to lower normal stresses experienced more dilation at the failure moment.

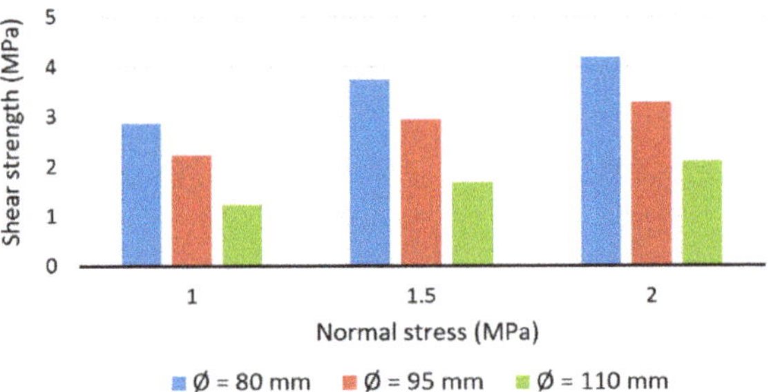

Figure 5. Shear strength of Br specimens (case I: $\sigma_n^J = 0$).

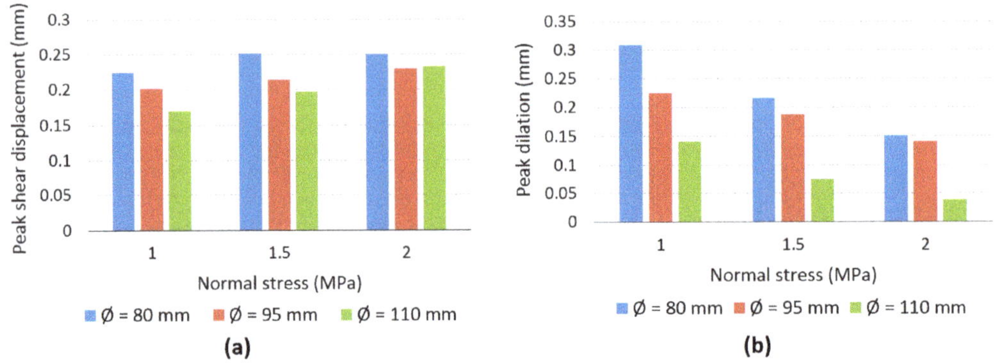

Figure 6. (a) Peak shear displacement and (b) peak dilation values for Br specimens subjected to direct shear test condition (case I: $\sigma_n^J = 0$).

3.2. Case II

In this section, the experimental results of direct shear tests carried out on J&Br specimens are provided. These results reflect a complicated mechanism when a weakness plane is under direct shear test condition with the shear resistance depending on the simultaneous mobilization of joint friction and intact rock cohesion. Figure 7 shows the shear strength of the J&Br specimens with a single embedded joint of three different sizes ($\varnothing = 80$ mm, $\varnothing = 95$ mm, and $\varnothing = 110$ mm) and three different roughness levels (LR, MR, and HR) when subjected to three normal stress levels (1, 1.5, 2 MPa). As shown in the figure, the normal stress and rock bridge ratio have strong effects on the shear strength of the J&Br specimens. The shear strength of the specimens increases by when the rock bridge ratio and applied normal stress increase. Although the overall trend shows a direct relationship between the joint roughness and the shear strength of J&Br specimens, the effect of the joint roughness on the shear strength is less than that of rock bridge ratio or normal stress. Figures 8 and 9 depict the peak shear displacement and peak dilation of the J&Br specimens, respectively. The overall results show that the specimens with greater rock bridge ratios experienced greater peak shear displacement when undergoing greater normal stresses (Figure 8). Figure 9 indicates that J&Br specimens with greater rock bridge ratios experienced greater peak dilation when undergoing lower normal stresses.

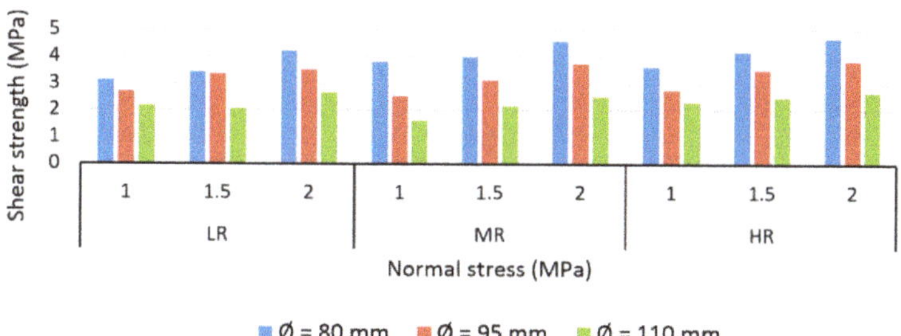

Figure 7. Shear strength of J&Br specimens embedding single joint with different sizes and roughness under different normal stress levels.

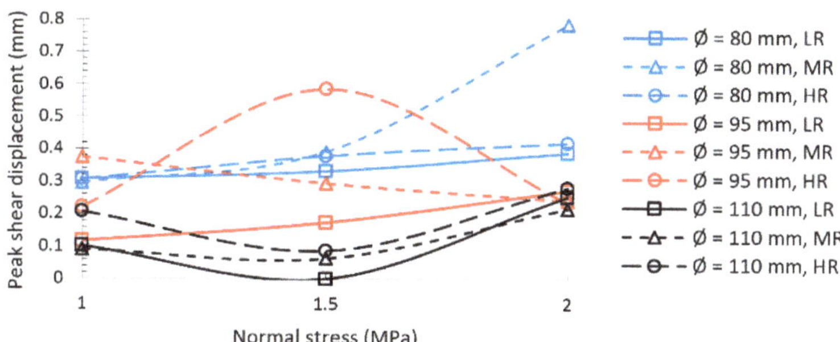

Figure 8. Peak shear displacement of J&Br specimens embedding a single joint with different sizes and roughness under different normal stress levels.

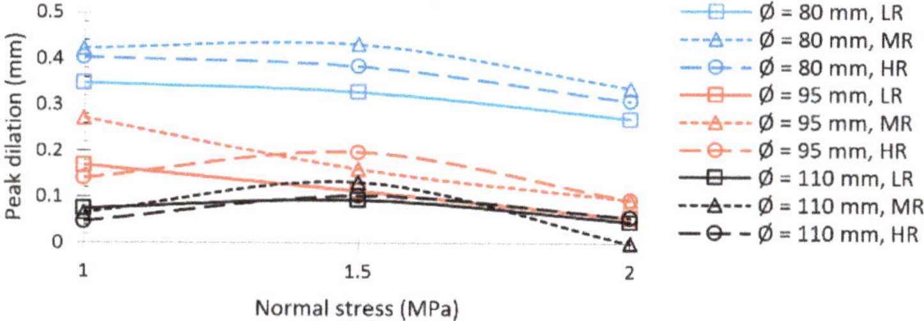

Figure 9. Peak dilation of J&Br specimens embedding a single joint with different sizes and roughness under different normal stress levels.

Five J&Br specimens were subjected to uniaxial compressive loading to study the effects of rock bridge ratio and joint roughness on the normal stiffness. Figure 10 shows the normal stress-normal deformation curves of the J&Br specimens. The results clearly demonstrate that the normal stiffness varies with the normal stress in the initial non-linear stage and remains constant afterwards. The non-linear stage implies the closure behavior of the embedded joint under uniaxial loading and the subsequent linear stage denotes the normal deformation of J&Br specimens when the embedded joint is closed [50]. The linear stage in fact shows the normal deformation of an intact sample subjected to uniaxial loading, hence the curves in this stage are almost parallel. As shown, the normal deformation is greater for the J&Br specimens with wider joints. The normal stress-normal deformation curves of the J&Br specimens embedding a 95 mm diameter joint show that the roughness of embedded joint has no systematic effect on the normal deformation of J&Br specimens subjected to uniaxial compressive loading. The shear stiffness of the

J&Br specimens subjected to direct shear test condition were investigated from their shear stress–shear displacement curves. The shear stiffness values were measured from the linear part of the curves before the peak shear stress (Table 4). The results show that rock bridge ratio, normal stress, and roughness of embedded joints have no systematic effect on the shear stiffness of the J&Br specimens. The residual shear strength of the J&Br specimens was also measured from the shear stress–shear displacement curves (Table 4). The results clearly show that the residual shear strength is greater when J&Br specimens are subjected to greater normal stress levels. In most of the cases, the residual shear strength is greater for the J&Br specimens embedding a rougher joint. The effect of roughness on the residual shear strength is more pronounced when the rock bridge ratio of J&Br specimens is smaller. Moreover, the values of cohesion and friction angle of J&Br specimens are provided in Table 4. The results show that the cohesion increases when the rock bridge ratio is increased. Neither Rock bridge ratio nor joint roughness has a systematic effect on the friction angle of J&Br specimens.

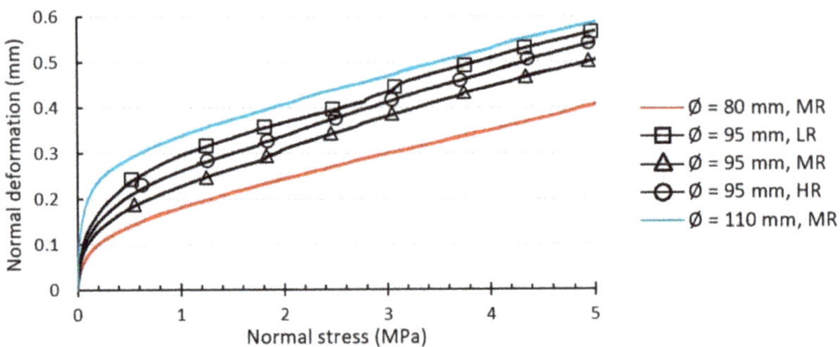

Figure 10. Normal stress-normal deformation curves of J&Br specimens subjected to uniaxial compressive loading.

Table 4. Shear stiffness and residual shear strength of J&Br specimens subjected to direct shear test condition.

Joint Diameter (mm)	Roughness Level	Shear Stiffness (MPa/mm)			Residual Shear Strength (MPa)			Cohesion (MPa)	Friction Angle (°)
		σ_n (MPa) 1	1.5	2	1	1.5	2		
80	LR	7.09	11.47	15.5	1.59	1.74	2.21	1.98	47.1
	MR	14.83	9.42	7.69	1.67	1.85	2.17	2.96	37.9
	HR	16.11	12.8	13.62	1.56	1.99	2.1	2.58	46.7
95	LR	25.19	24.24	7.05	1.44	1.75	2.39	1.98	38.9
	MR	8.57	13.21	10.8	1.51	1.8	2.4	1.3	50.7
	HR	12.16	10.75	15.07	1.7	2.22	2.67	1.75	47.3
110	LR	7.66	6.56	8.97	1.43	1.75	2.16	1.57	25.8
	MR	15.48	9.5	12.44	1.52	1.84	2.36	0.76	41.7
	HR	8.51	12.7	8.44	1.72	2	2.68	1.95	20.0

4. Discussion

4.1. Joint Friction Contribution

The actual shear mechanism of the specimens with a non-persistent rough joint and the true physical interaction between the bridged and jointed zones of the weakness plane during shearing process occur for the case II experiments. Hence, the shear strength of the specimens obtained from the experiments of case I was compared to those obtained from the case II experiments. Figure 11a–c show the shear strengths obtained from case I and case II experiments when the joint diameters are 80 mm, 95 mm, and 110 mm, respectively. The comparison clearly shows that the maximum shear load required to break J&Br specimens is underestimated when the friction contribution of the jointed zone is ignored. The difference between the shear strength obtained from case I and case II experiments (subtracting the values of the red bars from those of the blue bars in Figure 11) is illustrated in Figure 12. As shown in most of the cases, the greater differential shear strength is for the specimens with wider joints. This shows that the involvement of the embedded joint in the shear strength is more pronounced when a wider portion of the plane of weakness is discontinuous. The normal stress and the joint roughness do not appear to have any constant influence on the differential shear strength. Overall, the results of experiments show that the portion of applied normal load distributed to the jointed zone of weakness plane is not negligible and therefore joint friction contribution should be considered when evaluating the shear strength of the planes.

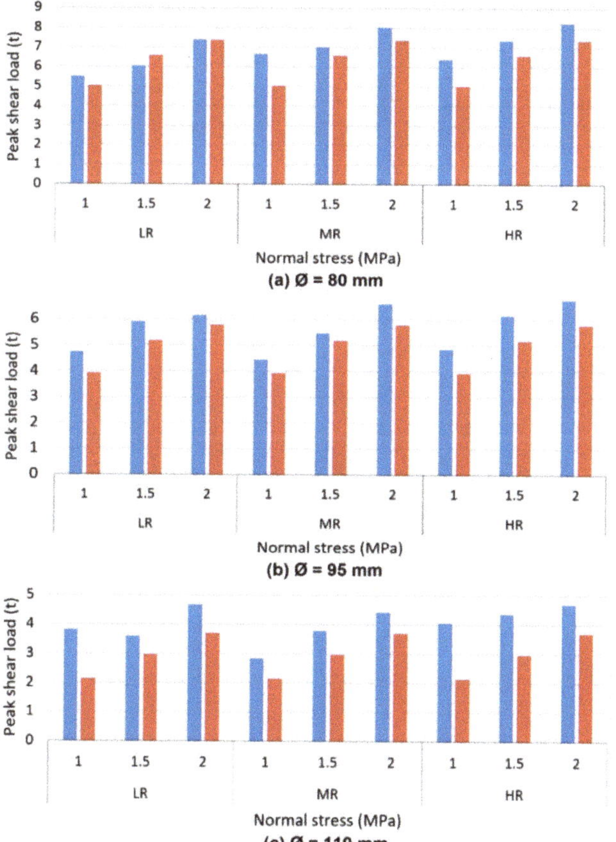

Figure 11. Comparison of the shear strength of the specimens containing a non-persistent rough joint obtained from case II experiments (blue bars) with the ones from case I experiments (red bars) when the joint diameter is (**a**) 80, (**b**) 95, and (**c**) 110 mm.

Joint friction mobilization is another significant factor in the evaluation of the shear strength of weakness planes. A total of 27 J specimens with three different joint sizes ($\varnothing$ = 80 mm, $\varnothing$ = 95 mm, and $\varnothing$ = 110 mm) and three different roughness (LR, MR, and HR) were prepared and subjected to direct shear tests under three different normal stress levels (1, 1.5, and 2 MPa). Figure 13 illustrates the peak shear displacements of J specimens and those of J&Br specimens. As shown, the shear displacement required for J specimens to reach their ultimate shear strength is much greater than the peak shear displacement of the J&Br specimens. This means that the friction of jointed zone of J&Br specimens is partially mobilized at the moment of failure. Moreover, the results of Figure 13 show that the peak shear displacement of J&Br specimens is less than 1 mm. As mentioned, previous specimen preparation technique (application of rough sheets with 1 mm thickness to make non-persistent joints) limits us to make joints with a certain amount of aperture. As the peak shear displacement of the J&Br specimens is smaller than the horizontal gap between the joint walls, joint friction is not mobilized at failure moment. Therefore, the proposed specimen preparation procedure of this study is suggested to investigate the shear behavior of specimens with non-persistent joints, knowing that the friction contribution of embedded joints is considerable.

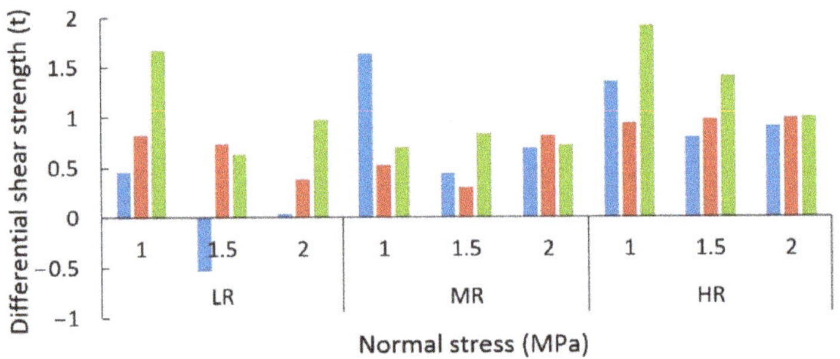

Figure 12. The difference between peak shear loads obtained from case I and case II experiments.

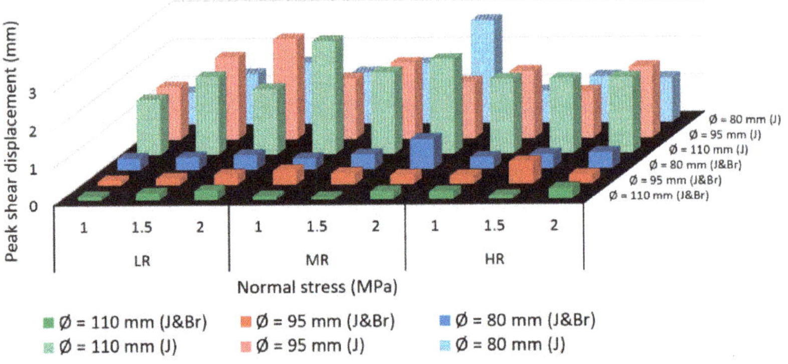

Figure 13. Comparison of peak shear displacements of J and J&Br specimens.

4.2. Dilation Mechanism

Figure 14 shows the shear stress/dilation-shear displacement curves of the J&Br specimens subjected to the direct shear test condition. As shown in the figure, the dilation curves consist of three phases. The first represents the dilation of the specimens before the moment of failure. This phase begins with the initiation of the test and ends at the point corresponding to the peak shear stress, denoted by the dash-dotted line arrows in Figure 14. In the first phase, all of the J&Br specimens experienced a certain amount of dilation, mostly controlled by the normal load applied to them. The shear stress drops sharply at the moment of failure and reaches a certain point (toe). The second phase of dilation starts at the point corresponding to the peak shear stress and ceases at the point corresponding to the toe, denoted by the dashed line arrows. In this phase, the specimens dilate due to crack propagation and coalescence. This dilation is followed by compression for some specimens (e.g., Ø = 95 mm, MR, 2 MPa) because the cracks are closed after coalescence. The third phase starts immediately after the second phase and expresses the dilation behavior of the failure planes created after the specimens were broken into two halves in the second phase. The experimental results show that the specimens subjected to higher normal stresses underwent less amount of dilation, and those with rougher and smaller joints experienced a greater amount of dilation during the last phase of dilation.

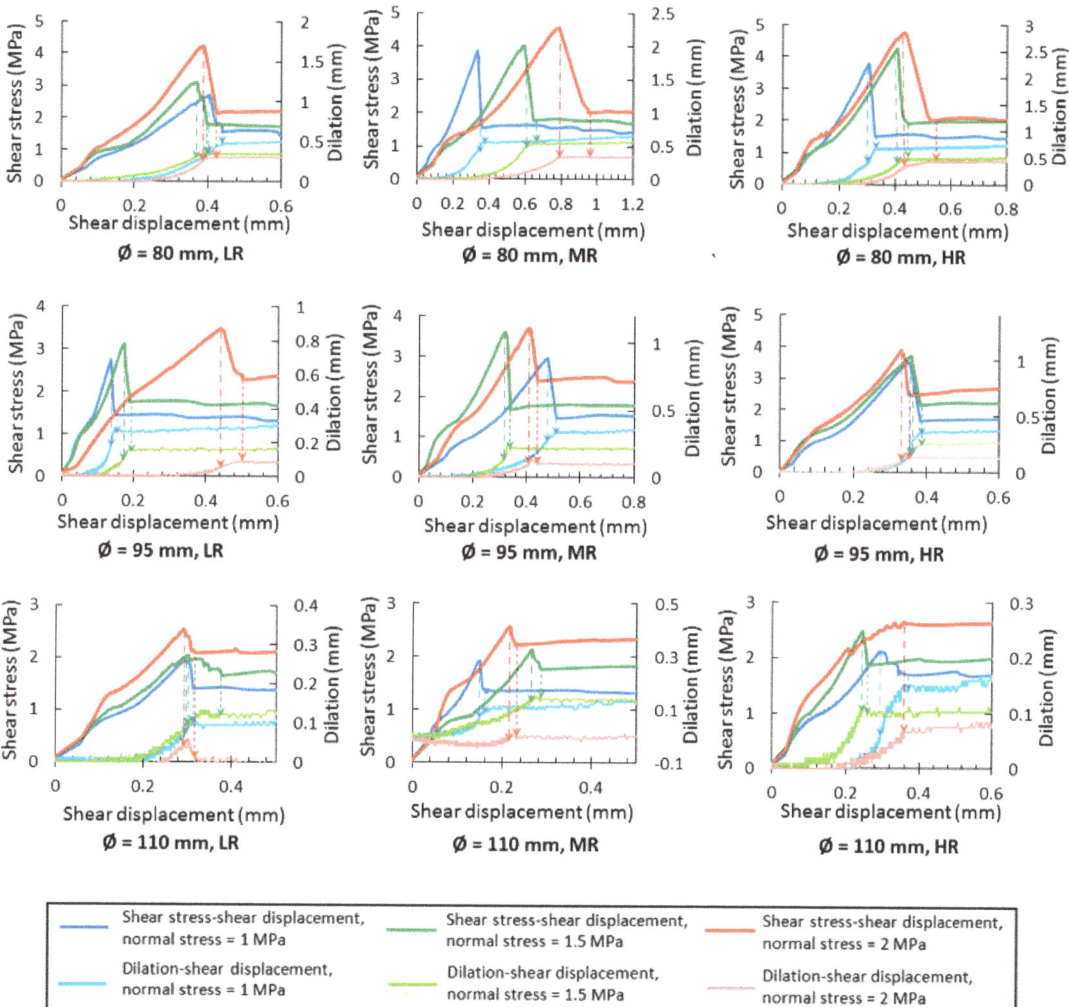

Figure 14. Dilation mechanism of J&Br specimens considering their shear stress–shear displacement curves.

The rough jointed zone of the weakness planes causes dilation to a very limited extent while a J&Br specimen is sheared. However, a large portion of the peak dilation of the J&Br specimens (Figure 9) is expected to be caused by torque exerted due to the existing gap between the specimen holders. To investigate the cause of the dilation, the displacement field on one J&Br specimen (Ø = 95 mm, HLR, 2 MPa) was analyzed using the DIC technique. Figure 15a illustrates the specimen on which some points whose vertical displacements are desired are marked. The central line in the figure depicts the intersection of the shear plane where the joint is located and the outer boundary of the specimen. The vertical displacements of ten points located at the top, center, and bottom of the specimen were investigated (Figure 15a). Figure 15b illustrates the vertical displacements of the points before the failure of the specimen. As shown in this figure, the vertical displacements of the points on the right side of the specimen (Tr, Cr, and Br) are greater than those on the middle (Tm, Tm1, Cm, and Bm), and the vertical displacements of the points on the middle are greater than those on the left side (Tl, Cl, and Bl). This information simply reveals the rotation of the specimen along the axis of rotation, and this rotation causes dilation during the shear test. The points located on the top, center, and bottom parts of the left/right side of the specimen had similar upward displacements (Tl, Cl, and Bl or Tr, Cr, and Br). However, the points located at the middle-top (Tm and Tm1) zone were displaced slightly more than those located at the middle-center (Cm) and middle-bottom (Bm) zones. This relative vertical displacement at the middle of the specimen reflects the initiation of a crack which propagated between Cm and Tm1.

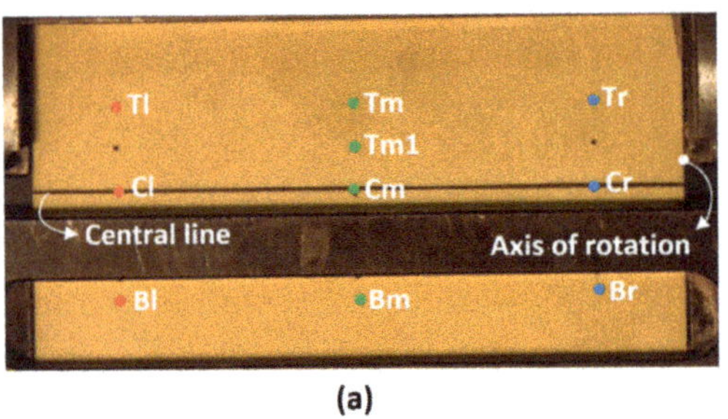

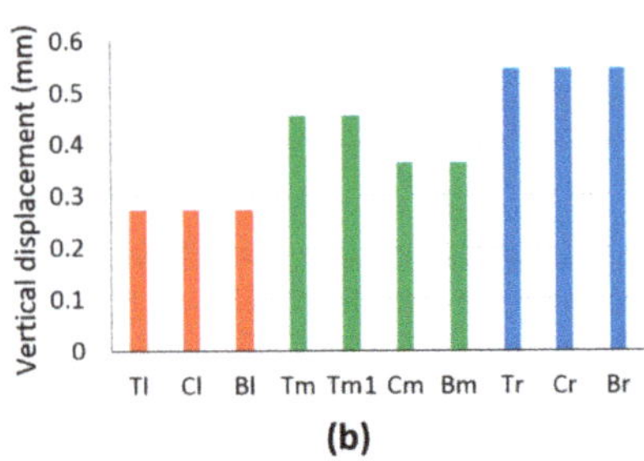

Figure 15. (**a**) The location of the points selected for DIC analysis, (**b**) Upward displacement of the points.

4.3. Cracking Analysis

In most previous experimental and numerical studies of the failure mechanisms of specimens containing edge-notched non-persistent (open) joints under direct shear test condition, the tensile mode was more frequently reported as the cause of the initiation and propagation of cracks inside a rock bridge [5,30,31,35,39]. All of the J&Br specimens were monitored using a camcorder while they were subjected to the direct shear test condition, and the cracking process was studied extensively. Figure 16 indicates that the shear load is applied to the lower part of the specimens from the left side and that the normal load is exerted downward over them. The vertical movement of lower shear box is restricted and the upper shear box is only allowed to move vertically. As shown in the figure, cracks initiate and propagate from three different zones: the left side, the center, and the right side of the specimens.

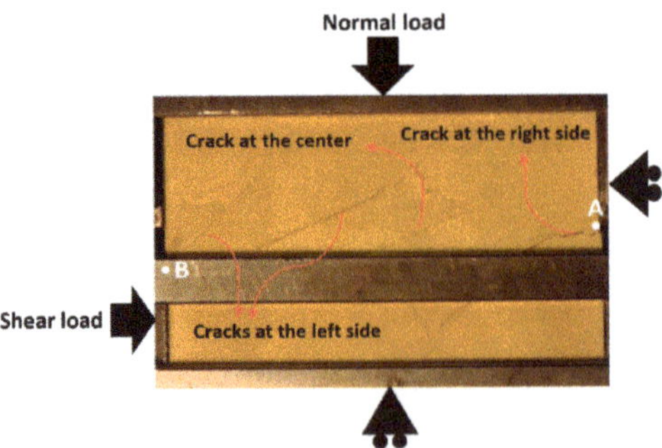

Figure 16. Crack initiation and propagation zones in J&Br specimens under direct shear test condition.

Figure 17 shows the initiation moment and the sequence of cracks on the shear stress–shear displacement curves (by the ×, +, and ◯ symbols) leading to the failure of J&Br specimens subjected to the direct shear test condition. The sequence of initiation of cracks is indicated in the legends of the graphs in Figure 17. The locations of cracks which initiated at the same time are indicated in parentheses, and the star symbol beside a cracking zone indicates that the crack initiated and propagated explosively through the zone. As shown, the first crack initiated in the vicinity of point A (the axis of rotation) and propagated toward the circumference of the embedded joint and then to the lower part of the specimen. For the cases with smaller joints (Ø = 80 mm), the first cracks most commonly initiated at the center of the specimens. The first cracks most commonly initiated before the failure of the specimens and did not have any significant impact on the path of the shear stress–shear displacement curves. The cracks which initiated at the central zone did not affect the curves. Finally, the last crack, connecting the left side of the joint circumference to point B, eventually results in a rupture with the embedded joint and the right-side crack. The final cracks always coincide with the peak shear stress and dramatically reduce the shear resistance of the specimens. The J&Br specimens with greater rock bridge ratios (Ø = 80 mm and Ø = 95 mm) mostly experienced an explosive and violent rupture. However, the final cracks propagated smoothly through the specimens containing a wider joint (Ø = 110 mm). The normal stress and joint roughness have no clear effects on the crack initiation moment and the sequence of the cracks.

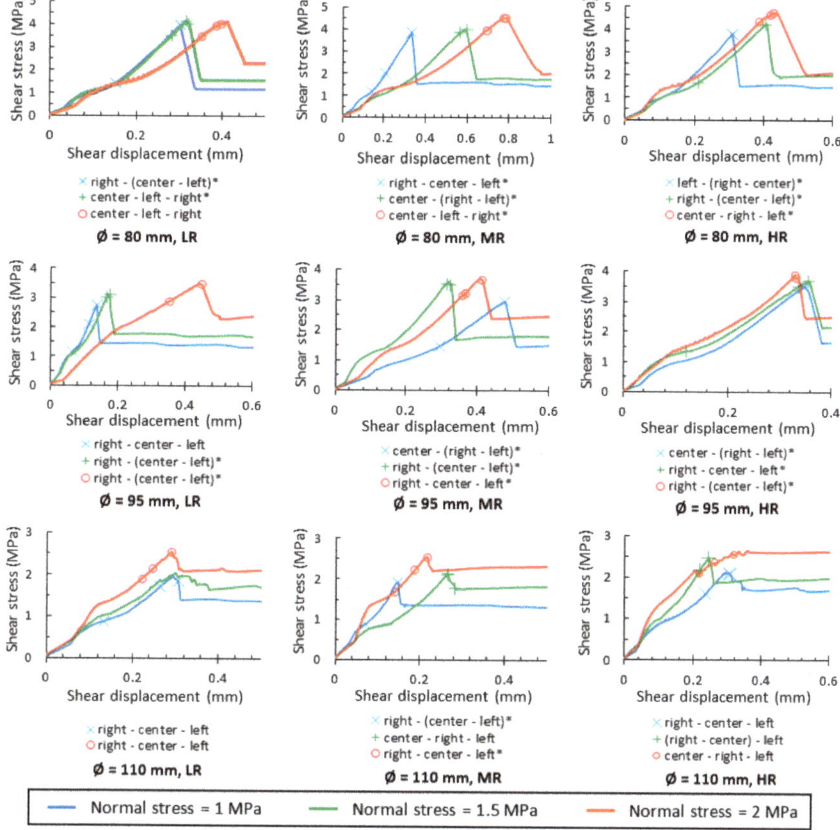

Figure 17. Crack initiation moments and the sequence of the cracks that lead to the failure of J&Br specimens containing a single joint with different sizes (Ø = 80 mm, Ø = 95 mm, Ø = 110 mm) and roughness (LR, MR, and HR) subjected to different normal stresses (1 MPa, 1.5 MPa, and 2 MPa).

5. Conclusions

A novel specimen preparation approach was applied to create rock-like specimens containing a non-persistent rough joint surrounded by intact material from all sides. Applying the casting method, the embedded joints are closed and interlocked, so their friction is mobilized once the specimens are subjected to the direct shear test condition. Unlike previously applied specimen preparation approaches, the proposed method enables the creation of embedded joints with different roughness levels along various directions.

In this study, the effects of three variables, namely rock bridge ratio, joint roughness, and applied normal stress on the shear behavior of rock-like specimens with non-persistent rough joints were experimentally investigated. The results of the experiments clearly demonstrated that all of the variables have impacts on the shear behavior of the specimens to different extents. The effect of rock bridge ratio and the normal stress on the shear strength of the specimens were found to be relatively more influential than that of joint roughness. The overall analysis of the peak shear displacement results revealed that the specimens with greater rock bridge ratios subjected to greater normal stresses undergo more shear displacement before the moment of failure. Moreover, the specimens with greater rock bridge ratios experience greater peak dilation when undergoing lower normal stresses.

The shear strength of the specimens with an open joint (Br specimens) and a single closed joint (J&Br specimens) were experimentally measured and the results were compared. The comparison revealed that the normal stress is partially applied to the jointed zone of the weakness planes, and the contribution of joint friction to the shear strength of the planes is not negligible. Moreover, the impact of joint roughness on the shear strength of J&Br specimens confirms that a fraction of normal load is applied to the embedded joint.

Three distinct phases of dilation were detected for specimens containing a non-persistent rough joint. The first phase demonstrates the dilation of the specimens before the failure point. A DIC analysis revealed that the dilation in this phase is mostly due to the rotation of the specimens as a result of the inevitable existing gap between the specimen holders of the direct shear test machine. In the second phase, dilation due to crack propagation and coalescence is followed by compression because of the closure of the enforced failure plane. Finally, the third phase of dilation complies with the dilation mechanism of a persistent rough joint. The walls of the enforced failure plane created after the failure of the specimens slide over each other in this phase.

The cracking process of the specimens containing a non-persistent rough joint was monitored with a camcorder. In most cases, the first crack was initiated at the axis of rotation and propagated to the circumference of the embedded joint and the last crack propagated from the other side of the joint circumference, eventually resulting in a rupture. Both cracks created a rough enforced shear plane with the embedded joint. It is important to note that the cracks that initiated and propagated before the final crack did not significantly alter the path of the shear stress–shear displacement curves. Most specimens with a smaller joint ($\varnothing$ = 80 mm and $\varnothing$ = 95 mm) were explosively broken whilst the cracks that initiated in the specimens with a wider embedded joint ($\varnothing$ = 110 mm) smoothly propagated. The experimental results show that the joint roughness had no clear effect on the crack initiation moment and the sequence of the cracks.

Author Contributions: Conceptualization, S.F.; Formal analysis, S.F. and J.K.; Funding acquisition, J.-J.S.; Investigation, S.F., J.K. and J.-J.S.; Methodology, S.F. and J.-J.S.; Project administration, J.-J.S.; Supervision, J.-J.S.; Visualization, S.F. and J.K.; Writing—Original draft, S.F.; Writing—Review and editing, J.-J.S. and S.F. All authors have read and agreed to the published version of the manuscript.

Funding: This work was supported by a grant from the Human Resources Development program (No. 20204010600250) of the Korea Institute of Energy Technology Evaluation and Planning (KETEP), funded by the Ministry of Trade, Industry, and Energy of the Korean Government.

Institutional Review Board Statement: Not applicable.

Informed Consent Statement: Not applicable.

Acknowledgments: The authors are grateful to Jiwon Choi for her assistance during the experimental work at Rock Mechanics and Rock Engineering laboratory of Seoul National University.

Conflicts of Interest: The authors declare no conflict of interest.

References

1. Song, J.-J. Distribution-Free Method for Estimating Size Distribution and Volumetric Frequency of Rock Joints. *Int. J. Rock Mech. Min. Sci.* **2009**, *46*, 748–760. [CrossRef]
2. Deng, Q.; Zhang, P. Research on the Geometry of Shear Fracture Zones. *J. Geophys. Res. Solid Earth* **1984**, *89*, 5699–5710. [CrossRef]
3. Segall, P.; Pollard, D.D. Nucleation and Growth of Strike Slip Faults in Granite. *J. Geophys. Res. Solid Earth* **1983**, *88*, 555–568. [CrossRef]
4. Cheng, Y.; Wong, L.N.Y.; Zou, C. Experimental Study on the Formation of Faults from En-Echelon Fractures in Carrara Marble. *Eng. Geol.* **2015**, *195*, 312–326. [CrossRef]
5. Gehle, C.; Kutter, H.K. Breakage and Shear Behaviour of Intermittent Rock Joints. *Int. J. Rock Mech. Min. Sci.* **2003**, *40*, 687–700. [CrossRef]
6. Tuckey, Z.; Stead, D. Improvements to Field and Remote Sensing Methods for Mapping Discontinuity Persistence and Intact Rock Bridges in Rock Slopes. *Eng. Geol.* **2016**, *208*, 136–153. [CrossRef]
7. Barton, N. Shear Strength Criteria for Rock, Rock Joints, Rockfill and Rock Masses: Problems and Some Solutions. *J. Rock Mech. Geotech. Eng.* **2013**, *5*, 249–261. [CrossRef]
8. Barton, N. The Shear Strength of Rock and Rock Joints. *Int. J. Rock Mech. Min. Sci. Geomech. Abstr.* **1976**, *13*, 255–279. [CrossRef]
9. Barton, N. A Relationship between Joint Roughness and Joint Shear Strength. In Proceedings of the Rock Fracture International Symposium on Rock Mechanics, Nancy, France, 4–6 October 1971; pp. 1–8.
10. Barton, N.R. A Model Study of Rock-Joint Deformation. *Int. J. Rock Mech. Min. Sci. Geomech. Abstr.* **1972**, *9*, 579–582. [CrossRef]
11. Barton, N.; Choubey, V. The Shear Strength of Rock Joints in Theory and Practice. *Rock Mech.* **1977**, *10*, 1–54. [CrossRef]
12. Beer, G.; Poulsen, B.A. Efficient Numerical Modelling of Faulted Rock Using the Boundary Element Method. *Int. J. Rock Mech. Min. Sci. Geomech. Abstr.* **1994**, *31*, 485–506. [CrossRef]
13. Fox, D.J.; KaŇa, D.D.; Hsiung, S.M. Influence of Interface Roughness on Dynamic Shear Behavior in Jointed Rock. *Int. J. Rock Mech. Min. Sci.* **1998**, *35*, 923–940. [CrossRef]
14. Ladanyi, B.; Archambault, G. Simulation of Shear Behavior of a Jointed Rock Mass. In Proceedings of the the the 11th U.S. Symposium on Rock Mechanics (USRMS), Berkeley, CA, USA, 1 January 1969.
15. Ge, Y.; Kulatilake, P.H.S.W.; Tang, H.; Xiong, C. Investigation of natural rock joint roughness. *Comput. Geotech.* **2014**, *55*, 290–305. [CrossRef]
16. Ge, Y.; Tang, H.; Ez Eldin, M.A.M.; Wang, L.; Wu, Q.; Xiong, C. Evolution process of natural rock joint roughness during direct shear tests. *Int. J. Geomech.* **2017**, *17*, E4016013. [CrossRef]
17. Grasselli, G.; Egger, P. Constitutive law for the shear strength of rock joints based on three-dimensional surface parameters. *Int. J. Rock Mech. Min. Sci.* **2003**, *40*, 25–40. [CrossRef]
18. Jiang, Y.; Li, B.; Tanabashi, Y. Estimating the relation between surface roughness and mechanical properties of rock joints. *Int. J. Rock Mech. Min. Sci.* **2006**, *43*, 837–846. [CrossRef]
19. Lee, H.S.; Park, Y.J.; Cho, T.F.; You, K.H. Influence of Asperity Degradation on the Mechanical Behavior of Rough Rock Joints under Cyclic Shear Loading. *Int. J. Rock Mech. Min. Sci.* **2001**, *38*, 967–980. [CrossRef]
20. Lee, Y.-K.; Park, J.-W.; Song, J.-J. Model for the Shear Behavior of Rock Joints under CNL and CNS Conditions. *Int. J. Rock Mech. Min. Sci.* **2014**, *70*, 252–263. [CrossRef]
21. Liu, R.; Lou, S.; Li, X.; Han, G.; Jiang, Y. Anisotropic Surface Roughness and Shear Behaviors of Rough-Walled Plaster Joints under Constant Normal Load and Constant Normal Stiffness Conditions. *J. Rock Mech. Geotech. Eng.* **2020**, *12*, 338–352. [CrossRef]
22. Maksimović, M. The Shear Strength Components of a Rough Rock Joint. *Int. J. Rock Mech. Min. Sci. Geomech. Abstr.* **1996**, *33*, 769–783. [CrossRef]
23. Park, J.-W.; Song, J.-J. Numerical Simulation of a Direct Shear Test on a Rock Joint Using a Bonded-Particle Model. *Int. J. Rock Mech. Min. Sci.* **2009**, *46*, 1315–1328. [CrossRef]
24. Patton, F.D. Multiple Modes of Shear Failure in Rock. In Proceedings of the 1st ISRM Congress, Lisbon, Portugal, 1 January 1966.
25. Prassetyo, S.H.; Gutierrez, M.; Barton, N. Nonlinear Shear Behavior of Rock Joints Using a Linearized Implementation of the Barton–Bandis Model. *J. Rock Mech. Geotech. Eng.* **2017**, *9*, 671–682. [CrossRef]
26. Seidel, J.P.; Haberfield, C.M. A Theoretical Model for Rock Joints Subjected to Constant Normal Stiffness Direct Shear. *Int. J. Rock Mech. Min. Sci.* **2002**, *39*, 539–553. [CrossRef]
27. Zhang, Q.; Li, X.; Bai, B.; Hu, H. The Shear Behavior of Sandstone Joints under Different Fluid and Temperature Conditions. *Eng. Geol.* **2019**, *257*, 105143. [CrossRef]
28. Zhao, J. Joint Surface Matching and Shear Strength Part A: Joint Matching Coefficient (JMC). *Int. J. Rock Mech. Min. Sci.* **1997**, *34*, 173–178. [CrossRef]
29. Zhao, J. Joint Surface Matching and Shear Strength Part B: JRC-JMC Shear Strength Criterion. *Int. J. Rock Mech. Min. Sci.* **1997**, *34*, 179–185. [CrossRef]

30. Ghazvinian, A.; Nikudel, M.R.; Sarfarazi, V. Effect of Rock Bridge Continuity and Area on Shear Behavior of Joints. In Proceedings of the International Society for Rock Mechanics and Rock Engineering, Lisbon, Portugal, 1 January 2007.

31. Savilahti, T.; Nordlund, E.; Stephansson, O. Shear Box Testing and Modeling of Joint Bridge. In Proceedings of the International Symposium on Rock Joints, Loen, Norway, 4 June 1990; pp. 295–300.

32. Shaunik, D.; Singh, M. Strength Behaviour of a Model Rock Intersected by Non-Persistent Joint. *J. Rock Mech. Geotech. Eng.* **2019**, *11*, 1243–1255. [CrossRef]

33. Wong, R.H.C.; Leung, W.L.; Wang, S.W. Shear Strength Studies on Rock-like Models Containing Arrayed Open Joints. In Proceedings of the the 38th U.S. Symposium on Rock Mechanics (USRMS), Washington, DC, USA, 1 January 2001.

34. Elmo, D.; Donati, D.; Stead, D. Challenges in the Characterisation of Intact Rock Bridges in Rock Slopes. *Eng. Geol.* **2018**, *245*, 81–96. [CrossRef]

35. Ghazvinian, A.; Sarfarazi, V.; Schubert, W.; Blumel, M. A Study of the Failure Mechanism of Planar Non-Persistent Open Joints Using PFC2D. *Rock Mech. Rock Eng.* **2012**, *45*, 677–693. [CrossRef]

36. Jiang, M.; Liu, J.; Crosta, G.B.; Li, T. DEM Analysis of the Effect of Joint Geometry on the Shear Behavior of Rocks. *Comptes Rendus Mécanique* **2017**, *345*, 779–796. [CrossRef]

37. Romer, C.; Ferentinou, M. Numerical Investigations of Rock Bridge Effect on Open Pit Slope Stability. *J. Rock Mech. Geotech. Eng.* **2019**, *11*, 1184–1200. [CrossRef]

38. Shang, J.; Zhao, Z.; Hu, J.; Handley, K. 3D Particle-Based DEM Investigation into the Shear Behaviour of Incipient Rock Joints with Various Geometries of Rock Bridges. *Rock Mech. Rock Eng.* **2018**, *51*, 3563–3584. [CrossRef]

39. Zhang, H.Q.; Zhao, Z.Y.; Tang, C.A.; Song, L. Numerical Study of Shear Behavior of Intermittent Rock Joints with Different Geometrical Parameters. *Int. J. Rock Mech. Min. Sci.* **2006**, *43*, 802–816. [CrossRef]

40. Fereshtenejad, S. Fundamental Study on Shear Behavior of Non-Persistent Joints. Ph.D. Thesis, Seoul National University, Seoul, Korea, 2020.

41. Muralha, J.; Grasselli, G.; Tatone, B.; Blümel, M.; Chryssanthakis, P.; Yujing, J. ISRM Suggested Method for Laboratory Determination of the Shear Strength of Rock Joints: Revised Version. In *The ISRM Suggested Methods for Rock Characterization, Testing and Monitoring: 2007–2014*; Ulusay, R., Ed.; Springer International Publishing: Cham, Switzerland, 2015; pp. 131–142. ISBN 978-3-319-07713-0.

42. Fereshtenejad, S.; Song, J.-J. Fundamental Study on Applicability of Powder-Based 3D Printer for Physical Modeling in Rock Mechanics. *Rock Mech. Rock Eng.* **2016**, *49*, 2065–2074. [CrossRef]

43. Ulusay, R. (Ed.) *The ISRM Suggested Methods for Rock Characterization, Testing and Monitoring: 2007–2014*; Springer International Publishing: Cham, Switzerland, 2015; ISBN 978-3-319-07712-3.

44. Tse, R.; Cruden, D.M. Estimating Joint Roughness Coefficients. *Int. J. Rock Mech. Min. Sci. Geomech. Abstr.* **1979**, *16*, 303–307. [CrossRef]

45. Lin, H.; Ding, X.; Yong, R.; Xu, W.; Du, S. Effect of Non-Persistent Joints Distribution on Shear Behavior. *Comptes Rendus Mécanique* **2019**, *347*, 477–489. [CrossRef]

46. MathWorks. Available online: https://www.mathworks.com/matlabcentral/fileexchange/60817-surface-generator-artificial-randomly-rough-surfaces (accessed on 7 May 2020).

47. Gao, Y.; Wong, L.N.Y. A Re-Examination of the Joint Roughness Coefficient (*JRC*) and Its Correlation with the Roughness Parameter Z_2. In Proceedings of the 48th U.S. Rock Mechanics/Geomechanics Symposium, Minneapolis, MN, USA, 18 August 2014.

48. Fereshtenejad, S.; Song, J.-J. Applicability of Powder-Based 3D Printing Technology in Shear Behavior Analysis of Rock Mass Containing Non-Persistent Joints. *J. Struct. Geol.* **2021**, *143*, 104251. [CrossRef]

49. Lajtai, E.Z. Shear Strength of Weakness Planes in Rock. *Int. J. Rock Mech. Min. Sci. Geomech. Abstr.* **1969**, *6*, 499–515. [CrossRef]

50. Kulatilake, P.H.S.W.; Shreedharan, S.; Sherizadeh, T.; Shu, B.; Xing, Y.; He, P. Laboratory Estimation of Rock Joint Stiffness and Frictional Parameters. *Geotech. Geol. Eng.* **2016**, *34*, 1723–1735. [CrossRef]

 *energies*

Article

Microgravity Survey to Detect Voids and Loosening Zones in the Vicinity of the Mine Shaft

Slawomir Porzucek * and Monika Loj

Faculty of Geology, Geophysics and Environmental Protection, AGH University of Science and Technology, Mickiewicza 30 Av., 30-059 Kraków, Poland; mloj@agh.edu.pl
* Correspondence: porzucek@agh.edu.pl; Tel.: +48-126172354

Abstract: In mining and post-mining areas, the assessment of the risks to the surface and its infrastructure from the opening or closed mine is of the utmost importance; particular attention should be paid to mine shafts. The risks include the occurrence of undetected voids or loosening zones in the rock mass. Their detection makes it possible to prevent their impact on a mine shaft and surface infrastructure. Geophysical methods, and in particular, a microgravity method lend themselves for the detection of changes in the distribution of masses (i.e., the density) due to voids and loosening zones. The paper presents the results of surface microgravity surveys in the vicinity of three mine shafts: under construction, working, and a liquidated one. Based on the gravity anomalies, the density distribution of the rock mass for all three cases was recognized. The properties of the anomalies allowed to determine which of the identified decreased density zones may pose a threat to the surface infrastructure or a mine shaft. The microgravity survey made inside the working mining shaft provided information on the density of rocks outside the shaft lining, regardless of the type of lining. No significant decrease of density was found, which means that there are no larger voids outside the shaft lining. Nevertheless, at a depth of 42 m in running sands layer, the decreasing density zone was located, which should be controlled. Additionally, measurements in two vertical profiles gave the possibility of directional tracking of density changes outside shaft lining. Such changes were observed on three boundaries of geological layers, with two of them being on the boundary of gypsum and other rocks.

Keywords: geophysics; microgravity; hard coal mine; mine shaft; mining and post-mining area; rock density; voids and loosening zones

check for
updates

Citation: Porzucek, S.; Loj, M. Microgravity Survey to Detect Voids and Loosening Zones in the Vicinity of the Mine Shaft. *Energies* **2021**, *14*, 3021. https://doi.org/10.3390/en14113021

Academic Editor: Adam Smoliński

Received: 26 April 2021
Accepted: 20 May 2021
Published: 23 May 2021

Publisher's Note: MDPI stays neutral with regard to jurisdictional claims in published maps and institutional affiliations.

Copyright: © 2021 by the authors. Licensee MDPI, Basel, Switzerland. This article is an open access article distributed under the terms and conditions of the Creative Commons Attribution (CC BY) license (https://creativecommons.org/licenses/by/4.0/).

1. Introduction

Deep mining is one of the methods for exploiting mineral deposits. Shafts are the most important infrastructure as they play a key role in such mines. They allow not only to excavate the minerals but to transport people and equipment necessary for the mine to operate. Shafts play a vital role in ventilation; they also convey electricity and water. This is why it is important to regularly monitor mine shafts in order to prevent the events that could lead to their damage or, even worse, destruction. Unfortunately, despite all the safety measures and precautions, shafts may become damaged or destroyed with dramatic consequences [1]. The causes may vary but ultimately the failure is usually a result of a human error such as insufficient monitoring, routine, or inappropriate investigation methods.

Many factors may threaten the stability of a mine shaft and the entire infrastructure related to it. One of them is damage to shaft lining, which apart from common causes such as inappropriate exploitation and maintenance of the shaft, may be a result of causes that are described by Lecomte et al. [1]. One of such cause is excessive pressure of water on the shaft lining, which may lead to ruptures (shaft located at Tirphil, New Tredegar, England, 2010). Underground water may also lead to chemical or mechanical suffusion, i.e., sliding

143

of the material outside of the shaft lining, which results in voids and loosening zones; it also leads to tensions in the shaft lining, which may cause fracturing. A similar phenomenon but at a larger scale may occur near the surface, posing a threat to the shaft and the surrounding infrastructure (coal shaft V, Knurow-Szczyglowice colliery in Poland) [2]. Another hazard may be posed by an unfavorable geological structure, as was the case with the collapse of the shaft at the Jinchuan Nickel Mine, Gansu Province, China, 2005 [3,4] or the coal shaft V in Pniowek colliery, Poland, 2007 [5].

When a mine is closed, the shafts are liquidated in a manner that depends on the country and time. Unfortunately, liquidated shafts pose danger to the surrounding area due to the factors indicated above as well as the following [1,6]:

- a collapse of the material filling the shaft (shafts n°8 and n°8 bis, at Noeux-les-Mines, France, 2007),
- a failure of the shaft head (West Midlands Shaft, England, 2000 and shaft III Matylda-East colliery, Swietochlowice, Poland, 2008),
- a failure of deep closure structure located in the shaft galleries (shaft n°2, Vieux Condé, France, 1987),
- a risk of subsidence due to remobilisation of filling material or surface development (coal shaft Nord at Noyant d'Allier, France, 2001),
- a specific focus on surface development (Low Hall n°7, New Zealand Pit at Abrams Lancashire, England, 1945).

Hazards to a shaft and its surrounding areas may be identified using many methods that determine the condition of the shaft lining, its damage or horizontal or vertical dislodgments of the shaft elements, and the surface or surface infrastructure. These methods, however, are limited to detecting certain occurrences and usually do not determine their direct causes, which in the cases indicated above are related to the condition of the rock mass. In order to identify the sources of hazards for the shaft and its infrastructure, geophysical methods should be applied.

Geophysical methods are widely used to identify shallow rock mass, but their application is determined by the shallow geological setting, specific purpose, and the location. In the case of hazards related to mine shafts, the geophysical surveys may be applied to the surface surrounding the shaft and the shaft's interior. Surface surveys can be applied to operating and liquidated shafts.

Theoretically, almost all geophysical methods can be applied for surface surveys, but in practice there are limitations to the applicability of a given method.

Seismic method is characterized by very good vertical and horizontal resolution and is frequently used to identify shallow geological structures and detect voids and loosening zones [7–9]. Unfortunately it requires profile surveying, which renders it inapplicable for investigating mine shafts due to the surface infrastructure of the surveyed areas and the fact that it is impossible to run profiles that would allow for area-wide reconnaissance of the rock mass. Another issue is multiple distortions resulting from the operations of the shaft and its infrastructure. Seismic method can be applied to liquidated shafts, although it is applicable only along profiles and it is not always economical to increase the number of profiles.

Geoelectrical methods, including electrical and electromagnetic (including Ground Penetration Radar) ones, allow for examination of shallow parts of the rock mass [10–12]. However, as is the case with the seismic method, the measurements are taken along profiles, which means that both methods have similar limitations. Additionally, many distortions occur that are a result of buried utilities such as water, electricity, gas, telecommunications, and more. This is why, while geoelectrical methods give good results in undeveloped areas [13,14], they are of limited use in the case of mass rock surrounding the shaft accompanied by the broadly-understood infrastructure.

The limitations of both geophysical methods render a magnetic method useless for the purpose of surveying the rock mass surrounding the shaft that carries any infrastructure. The method is inapplicable also due to many electromagnetic distortions. Outside of urbanized areas, the method can be used to detect buried shafts on the condition that near-surface elements of the shaft contain iron.

The last geophysical method, gravity method, registers the spatial distribution of masses, and consequently, the distribution of density [15]. The advantage of this method is its low susceptibility to external interference. Since the method is based on the common phenomenon of gravity, the only errors come from the observation device, i.e., the gravimeter. Due to the construction of the gravimeter, significant measurement errors come from instrument vibrations, coming from ground vibrations or vibrations caused by strong wind gusts. The appropriate method allows, however, to reduce such errors. For near-surface surveys, a variant of gravity method, i.e., microgravity, is applied. The prefix micro points to low anomalies generated by distortion and small distances between observation points.

Microgravity method is widely used for the detection of objects whose density differs from the density of surrounding forms, which renders it particularly applicable for detection of voids and loosening in the rock mass [16–18], posing a risk to the stability of the shaft and its surrounding infrastructure. As the measurements are taken at observation points, the method is applicable to surveys conducted in the areas covered by infrastructure or urban areas [19–21]. The measurements can be taken inside facilities such as halls and sheds, allowing for more complete coverage of the surveyed area. For this reason, the method can be applied to operating shafts as well as liquidated ones [22,23].

As has already been mentioned, geophysical surveys conducted in a working mine shaft, aimed at examining the rock mass outside of the shaft lining, are a separate issue. Voids and loosening zones outside of the shaft lining pose a particular risk to the shaft as they may lead to damage of the lining. Due to specific construction of the shaft and conditions inside the shaft, the only applicable method that can provide the desired results is the gravity (microgravity) method. As Hammer demonstrated in 1950 [24] and McCulloh confirmed in 1965 [25], this method allows to assess medium density rock mass outside of the shaft lining. On this basis, the loosening zones and voids outside of lining were detected as they lower the average density calculated from observation points located in the shaft. Madej has presented many examples in detail in his work [26].

For the above reasons, the authors present the application of microgravity method to assess risks related to mine shaft, while it is still operating as well after its liquidation. They also present the example where the risk was assessed during the sinking of the shaft. As the method registers the distribution of density in the rock mass, it gives good results in detecting voids and loosening zones in the rock mass. It can also be applied for detecting buried shafts and reconnaissance of rock mass surrounding the identified liquidated mine shafts. It also allows to assess the degree to which the liquidated mine shaft is backfilled in its near-surface part. Before shaft sinking, the seismic methods allows to recognize deeper parts of the rock mass [27], but the microgravity method allows to recognize the shallow part. Unfortunately, due to the lack of shaft infrastructure, there is no possibility to perform a survey inside the shaft during its sinking.

As it is relatively straightforward, microgravity method can be applied for monitoring the safety of shaft during its construction, exploitation, and after its liquidation [26,28,29]. The results can be used to vary mining modellings and simulations [30,31].

2. Materials and Methods

Geophysics is a field of science related to the problems connected with the Earth's structure and the processes that take place in there, using the analysis of natural and unnatural induced physical fields. A gravity method, based on the Earth's natural gravity field, is one of the passive methods. It means that the devices only register the received signal. Gravity surveys are based on measuring the changes in the gravity field, which reflect the changes in the distribution of masses (and, consequently density) in the rock mass. In fact, the measured value is the value of the vertical component of Earth's acceleration (g_M). As it follows from the Newton's Law of Universal Gravitation and Second Newton's Law of Motion [32], acceleration caused by Earth's gravity field is directly proportional to its mass and inversely proportional to the square of the distance. In light of the above the observed changes in gravity, forces may be used for a broadly understood reconnaissance of the distribution of masses in the rock mass.

In applied geophysics, the gravity method is applied to identify a geological structure frequently in order to detect mineral or oil and gas deposits. In engineering, a microgravity method is used to determine the condition of near-surface rock mass, particularly to detect loosening zones and natural and anthropogenic voids [16,18,33]. Gravity method also allows to determine the volume density of the rock mass "in situ".

The change in gravity between observation points can have two primary sources. The first one is related to the distribution of density in the rock mass and the other to the observation point elevation and the terrain topography. While the first cause is the subject of the investigation, the other is undesired and should be eliminated.

For this purpose, the terrain correction δg_T is calculated. It eliminates the gravity impact of the terrain surrounding the observation point. Thanks to this correction, the excess of masses (above the observation point) and the deficit of masses (below this point) are eliminated. In both cases, the vertical component of this impact leads to the reduction of the gravity value, so the correction is always added do observation value. Introducing a correction renders the surrounding terrain flat from the point of view of each measurement point. In the case of geological investigations, the correction is calculated up to 25 km from the observation point and in the case of microgravity surveys the distance of 25 m is usually sufficient [34]. Obviously, if the terrain is not very varied, there is no need to introduce the correction.

The impact of the elevation on the measured gravity is consistent with Newton's Universal Law of Gravitation: The gravity lessens in inverse proportion to the square of the distance between the observation point and the element of the Earth. For this reason, the measured values of gravity must be reduced to some assumed reference level (datum) with some fixed equipotential surface. The basic datum in gravity studies is the surface of the Earth model, i.e., the surface of the oblate ellipsoid.

In order to reduce gravity measurements, corrections are introduced to eliminate individual factors related to the difference in elevation between the physical Earth surface (the observation point) and the datum, as well as the deposition of different rock masses between them.

The first correction is the free-air correction δg_F, which eliminates the effect of elevation difference between the location of the observation point and the datum. Using the Earth model, the average vertical gradient of gravity can be calculated, which is 3.086 nm·s^{-2} per meter. Therefore, the measured value of the gravity force transferred to a datum that is closer to the center of mass (Earth) must be increased.

The gravity impact generated by an infinite slab with a thickness equal to the distance between the observation point and the datum and a certain average density is eliminated with the Bouguer correction δg_B. The value of the Bouguer correction is negative as the gravity impact of the mass of the slab located below the observation point is eliminated.

The sum of all the corrections described above is called Bouguer reduction and it allows to calculate the value of the Bouguer anomaly, which is caused solely by the changes in the density distribution in the Earth's crust. The Bouguer anomaly stands for the difference between the measured gravity reduced to a datum and the normal gravity calculated at that level also called latitude correction.

The normal gravity value g_N is calculated using the International Gravity Formula derived from the Geodetic Reference System 1980 Earth model [35].

$$\Delta g = g_M + \delta g_F + \delta g_B + \delta g_T - g_N \tag{1}$$

Micogravity surveys are frequently carried out in urban areas and in many such cases it is necessary to take into account one more correction, as buildings and other structures located in the vicinity of observation points also have a certain mass. When such masses are close to the observation points they have an impact on the gravity value, in a similar way as terrain relief does. In order to eliminate this impact, the building correction δg_U is calculated [19,20].

In mining areas, microgravity measurements are conducted both at the surface and in mine shafts and mining galleries. Each of the underground workings constitute a mass deficiency that affects the measured gravity values at the observation points in or near them. It is therefore necessary to eliminate this influence by introducing mining correction δg_G corresponding to gravity effects from the shaft (shaft correction) or galleries (gallery correction). New geodetic methods allow you to model the shape of the above objects with very high accuracy [36].

The Bouguer anomaly is a superposition of the gravity impact of all the rock masses in the geologic medium. However, from the point of view of engineering investigations, the most interesting part is the one concerning the changes in density distribution in the most near-surface part of the rock mass, called local anomalies. Anomalies originating from sources outside the surveyed area are called regional anomalies [37]. There are several different analytical methods by which the distribution of regional anomalies is calculated from the Bouguer anomaly distribution. The difference between the Bouguer anomaly and the regional anomaly results in a residual anomaly [38], which is a mathematical approximation of the local anomaly.

The authors of this paper used three methods to calculate regional anomalies, such as approximating the regional field with a low-order polynomial, Butterworth and Gauss filtering. The calculation of the regional field using the polynomial method involves approximating the trend seen in the Bouguer anomaly distribution using the polynomial order 1–4. The filtering methods, on the other hand, are performed in the wavenumber domain by transforming the Bouguer anomalies, usually with Fast Fourier Transformation. The transformation allows for the calculation of the power spectrum, which is used to determine the wavelengths representing the regional and residual anomalies. The wavelength depends on how deep the source of the anomaly is located and how large it is. It provides the basis for filtering methods. In gravity, the Butterworth and Gauss filter is most commonly used to separate both types of anomalies.

The gravity method also allows for the determination of bulk density values "in situ". This task can be performed in two ways: using surface measurements [39,40] and using measurements in boreholes. In the second case, two types of gravity surveys are available, i.e., surveys in small-diameter boreholes [41] and in mine shafts [24].

In order to determine the bulk density of the rock medium surrounding a mine shaft, the "in situ" interval density method is applied, in which the microgravity observations are made inside the mine shaft, in observation points along the vertical profile [26,42].

A regularity discovered by McCulloh [24] underlies the use of the gravity method, in its vertical profiling version. It indicates that the gravity impact from a horizontal, infinite rock layer limited from above and below by observation points is generated by that part of the layer that is adjacent to the measurement profile within a radius equal to five times its thickness. It is known that in the measured values, up to 90% of the information comes from the structure of the geological medium located immediately behind the shaft lining [26].

The gravity values (g_M) recorded in the shaft are influenced by gravity impact of the shaft, its lining, and other shaft bottom and mine workings occurring nearby. For this reason, the mining correction δg_G needs to be applied. The topography of the terrain around the shaft has a similar effect, which sometimes necessitates the introduction of terrain correction δg_T.

On the basis of the measured gravity values with corrections (g), the density ρ_i of the rock slab is determined, with the slab delineated from the top and bottom by the following observation points [24,25]:

$$\rho = 3.682 - 1.193 \cdot (\Delta g / \Delta h) \tag{2}$$

$\Delta g = g_{i+1} - g_i$—gravity difference between a roof and a foot of a slab, $nm \cdot s^{-2}$, $g_i = g_{M,i} + \delta g_{G,i} + \delta g_{T,i}$, $\Delta h = h_{i+1} - h_i$—rock slab thickness, m.

Density error is calculated according to the formula:

$$\delta\rho = |\,1.193 \cdot \delta g / \Delta h\,| \tag{3}$$

δg—mean square measurement error, $nm \cdot s^{-2}$.

Microgravity surveys for geological and engineering purposes use gravimeters that measure relative values of gravity. It is not necessary to determine the absolute value of gravity in these surveys because the key information does not concern the specific value at the observation point, but the change of gravity between the observation points. All gravity measurements taken during gravity surveys are reduced relative to a fixed reference base. The reference base is the point at which the gravity force value has been predetermined. This value can be set arbitrarily or measurements can be referred to a point with an established absolute value of gravity, thus assigning absolute values to the measured values.

Relative gravimeters, due to their construction, are characterized by the occurrence of drift, which means a change of the measured gravity value with time. In order to eliminate the drift, its magnitude is examined with the passage of time on a single assumed point called a drift base. In the case of engineering surveys, one and the same point is used as a drift and reference base at the same time. In order to eliminate the drift effectively, gravity measurements are performed in the system of survey loops, starting and ending at the drift/reference base. In order to increase the quality of results, survey loops are not longer than 1–2 h.

Two procedures are used to obtain high quality gravity measurement results. The first one consists in taking at least two or three measurements of the gravity values at each observation point. If the difference between the obtained results does not exceed $0.05\ nm \cdot s^{-2}$, the average value is calculated, otherwise additional measurements are taken. The other procedure consists of repeating the gravity observations for at least 5% of all observation points, and in the case of microgravity surveys, even at two to four observation points from each survey loop.

The procedures described above are applicable in surface gravity surveys and in surveys carried out in vertical profiles in mine shafts. The difference in both cases is related to the manner of distribution of observation points within the survey area. In the case of surface surveys, the points are usually located at regular intervals in the form of a survey grid. The distance between points depends on the expected depth of the source generating the anomaly. In microgravity surveys, these distances range from 1 to 25 m.

In mine shafts, gravity is measured along vertical profiles located on the edges of the mine cage (a skip) or along the shaft axis. The distance between the observation points depends on the distance between the shaft bunton. A spatial measurement plank is placed on the bunton, thanks to which gravity measurements are not affected by the vibrations of the skip (Figure 1).

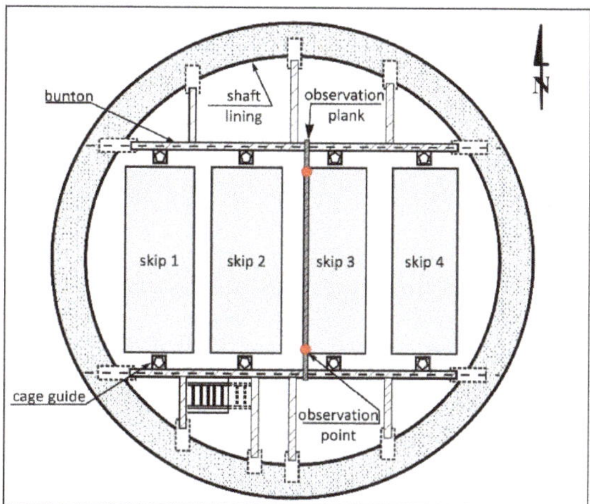

Figure 1. The mine shaft cross-section.

3. Results

Here we would like to present how the microgravity method can give an answer about the state of rock mass around a mine shaft. We have selected three examples of a gravity survey, each from a different stage of mining activity—before, during and after the operations of a mine shaft.

3.1. Working Mine Shaft

One of the main challenges posed before mining operations is ensuring the safety of workers and mining infrastructure. It is particularly important to ensure safe exploitation of mine shafts. In order to identify the causes of mass rock distortions that have an impact on the safety of a mine shaft, the current structure of the rock mass has to be investigated. This can be performed using microgravity method in the form of vertical gravity profiling and surface microgravity method. The application of a surface microgravity survey allows for a reconnaissance of the state of the rock mass surrounding the mine shaft in terms of the existence of possible loosening zones, crack zones, or directions of water migration towards the shaft. Certain issues related to the safety of shafts, such as the state of the shaft lining resulting from the influence of physical phenomena on the shaft, are investigated by means of vertical gravity profiling.

The authors present the application of the above-mentioned methods on the example of research carried out in and around the mine shaft in the area of one of the collieries in Upper Silesia, Poland.

3.1.1. General Geological Setting

The examined shaft is surrounded by rock complexes from three periods at the following depths:

- 0.0–61.8 m—Quaternary,
- 61.8–139.0 m—Neogene,
- from 139.0 m—Carboniferous with coal seams.

The Quaternary sediments lying horizontally and reaching a thickness of 31.5 m are developed in the form of silt over a sandy clay layer and very wet silt. Below there is a 25-m-thick layer of fine sand over medium sand, which occurs as running sand. The last layer of Quaternary sediments is loam with gravel lying at the depth of 56–62 m.

The thickness of the Neogene formations around the shaft is 77 m. In their roof there are 18-m-thick layers of loam and gypsum, below which lies a 30-m-thick layer of sandy loam. The other Neogene formations are layers of sandstone and shale.

The Coal Measures begin at a depth of 139 m and consist of alternating layers of sandstone and shale and coal seams. The direction of extension of the Carboniferous strata is northwestern–southeastern (NW–SE), which dips sharply towards the southwestern (SW).

3.1.2. Microgravity Survey

Surface microgravity surveys were made around the shaft, in a radius of approx. 50 m. Observation points were supposed to be set in a 5×5 grid but only in accessible locations (Figure 2). The measurement was repeated at about 5% of the points and the mean square error stood at 0.07 nm·s^{-2}. The obtained error value is close to the nominal error of the gravimeter and is slightly higher due to ground vibrations caused by the working shaft and the surrounding equipment.

Figure 2. The distribution of Bouguer anomalies (black dots—the observation points).

On the basis of microgravity and geodesy data, Bouguer anomaly values were calculated in accordance with the Formula (1), which provides the starting point for interpretation and subsequent data processing. First, Bouguer anomalies were subjected to Butterworth filtering in order to eliminate unwanted interferences caused by ground vibrations, measurement errors, and geodesic errors. This allowed us to obtain the distribution of Bouguer anomalies shown in Figure 2.

In the presented anomaly distribution, the influence of the gravitational influence of the shaft is clearly visible, manifested by strong, relatively negative values in its vicinity. Since some points were located close to buildings, they undoubtedly also influenced the measured values. The influence of buildings was eliminated using building correction.

Buildings were approximated by a rectangular prism, using geodetic data and information about the materials used in the construction of each building [20].

Mining correction was used to eliminate the influence of the shaft. Mining (shaft) correction was calculated using archival information from the excavation of the shaft. The mine shaft with diameter of 7 m had a brick and concrete lining to a depth of 65 m and a tubing lining below this depth. The different sections of the shaft were approximated by vertical cylinders. The foot and roof of each cylinder corresponded to the position of the rock layers surrounding the shaft pipe. The densities of each cylinder were chosen according to the layers that each cylinder corresponded to. The influence of variable shaft lining was also taken into account in the calculation of the correction. The correction was calculated at all points on the ground surface as well as at observation points inside the shaft (which will be described later).

It follows from the calculations that the total gravitational influence of the shaft and buildings on the ground surface reaches a maximum value of slightly more than $1.1 \text{ nm} \cdot \text{s}^{-2}$ (Figure 3). For a microgravity survey, this value is significant and its disregard would have a significant impact on the interpretation.

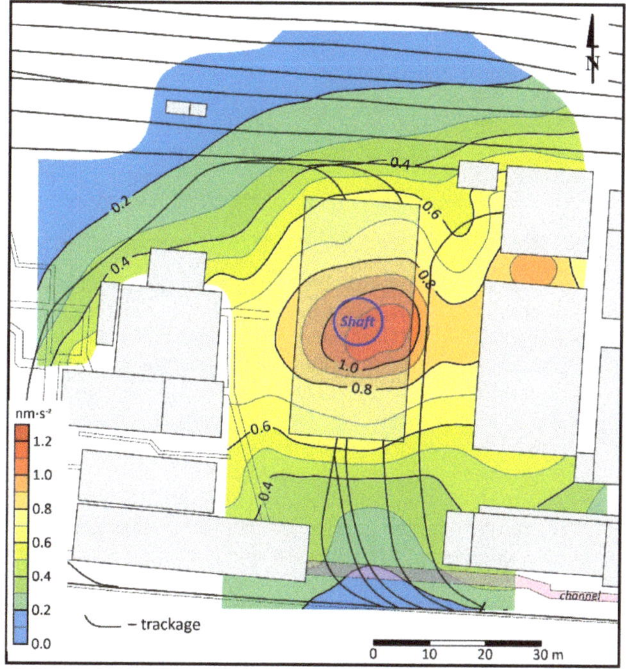

Figure 3. Distribution of the sum of building and mining correction.

After taking into account both gravity corrections above, the distribution of Bouguer anomalies is as shown in Figure 4. In general, the course of the anomalous field is from NW to SE, and the values of the anomalies increase from SW to NE (northeastern), and the total range of Bouguer anomalies is $2.2 \text{ nm} \cdot \text{s}^{-2}$. Thus, the course of the field is closely related to the geological structure described earlier, i.e., the course and dip of the Carboniferous strata.

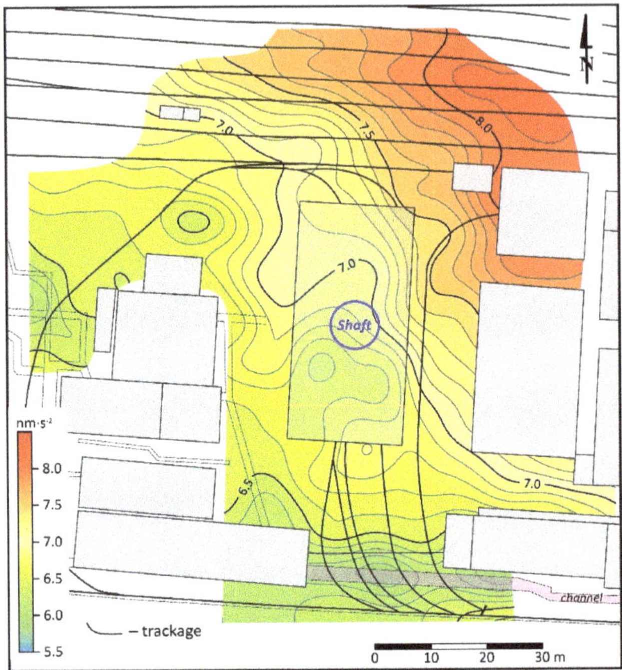

Figure 4. Distribution of Bouguer anomalies with building and mining correction.

In the obtained distribution of Bouguer anomalies compared to the general trend of changes in Bouguer anomalies, local changes of small horizontal extent and small amplitude (in relation to the whole field) can also be observed. These local anomalies provide information on the condition of the rock medium around the shaft. In order to separate them, a regional anomaly corresponding to the influence of the general geological structure was calculated. The regional anomaly was approximated using the surface polynomial order 2, which allowed the calculation of residual anomalies (Figure 5). The residual anomalies reflect the actual local anomalies, but are only an approximation of them.

In the obtained distribution of residual anomalies, the most visible is a sequence of relatively negative gravity anomalies P located in the southern part of the image. It is characterized by a small horizontal extant but significant amplitude. Its course coincides with that of a drainage channel located shallowly underground.

A relatively negative gravity anomaly Q was recorded in the direct vicinity of the shaft. This anomaly is a culmination of a zone of lower gravity values, which runs in the northern direction and joins a sequence of negative anomalies stretching East–West. Within the zone, two small anomalies with amplitudes larger than their surroundings R_1 and R_2 are also observed. The small amplitude and small horizontal extant of the entire zone of relatively negative gravity anomalies indicate that they are related to near-surface local density changes and do not affect the safety of the shaft. The anomalies P and R_2 are associated with railroad infrastructure located in the vicinity of the shaft. The anomaly R_1 occurs in an area devoid of any technical utilities. Therefore, it is caused by a local decrease in the density of the rock medium. This decrease is small and occurs in a limited area. Due to the relatively large distance of this anomaly from the shaft, it does not pose a threat to the investigated object.

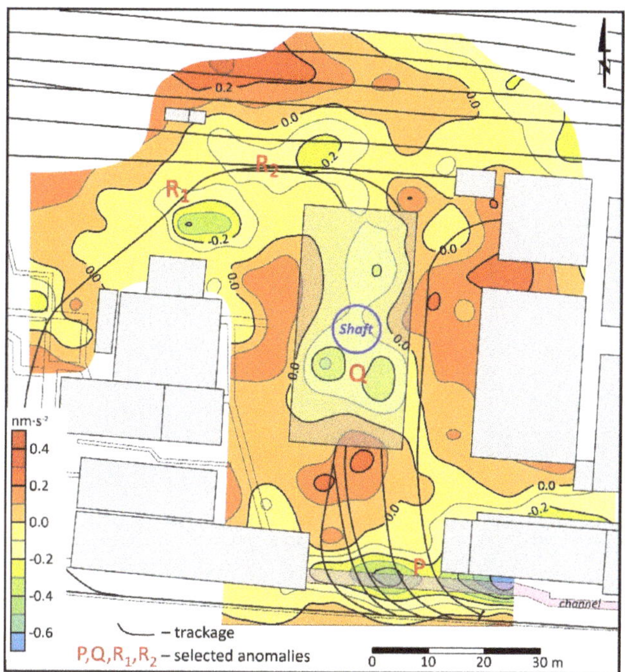

Figure 5. Distribution of the residual anomalies.

3.1.3. Gravity Vertical Profiling

Gravity surveys of the rock medium behind the shaft lining were carried out inside materials and personnel transport shaft. Vertical gravity profiling was performed along two vertical measuring profiles located on the northern and southern side of the shaft (Figure 1). Observations of gravity changes with depth were started from the depth of 6 m below the pit bank (outset) to the depth of 96 m. The gravity measurements were performed according to the rule that the gravimeter should be located in the same position relative to the shaft axis. The average vertical distance between successive buntons was 6 m.

The basis for interpretation of results of a vertical gravity profiling in a mine shaft is the vertical distribution of Bouguer anomalies, taking into account relevant gravity corrections, among which the most important is the shaft correction (a variant of mining correction), which was described earlier. The analysis of the Bouguer anomaly distributions along the north and south profiles shows that the shapes of both distributions are very similar to each other (Figure 6a–c), as they both reflect general changes in gravity due to depth.

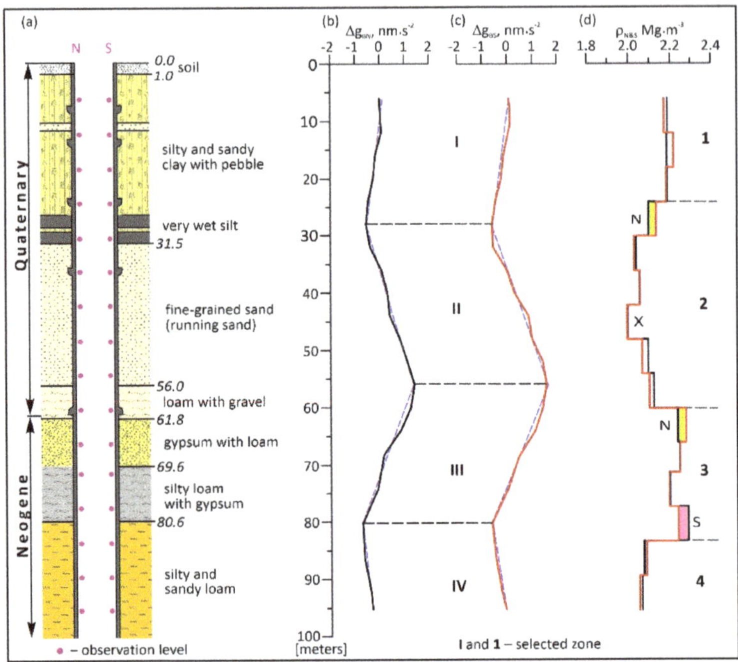

Figure 6. The results of vertical gravity profiling: (**a**) lithology, (**b,c**) "Bouguer anomaly", (**d**) densigram.

Four almost linear variations can be distinguished on both distributions, which correspond to four rock complexes (I–IV) that vary in density. The first density boundary occurs at the depth of 28 m. It separates a predominantly clay complex (I) located higher from the very wet sand located below (II). The next density boundary lies at a depth of 56 m and corresponds to the floor of the above-mentioned sands. The third density boundary lies at the depth of 80 m and marks the bottom of the loam and gypsum (III). Below the depth of 83 m, there is loam of lower density than the gypsum complex located above.

On the background of the general trends of the Bouguer anomalies, local changes in gravity are visible, which correspond to local changes in density within the described complexes I–IV. These changes can be observed in the distribution of interval densities calculated for both profiles (Figure 6d). The thickness of each interval is the distance between the observation points, which was about 6 m. For each interval density, an error was calculated that ranged from 0.002 Mg·m⁻³ to 0.024 Mg·m⁻³, with most of them not exceeding 0.01·Mg·m⁻³ (Figure 7).

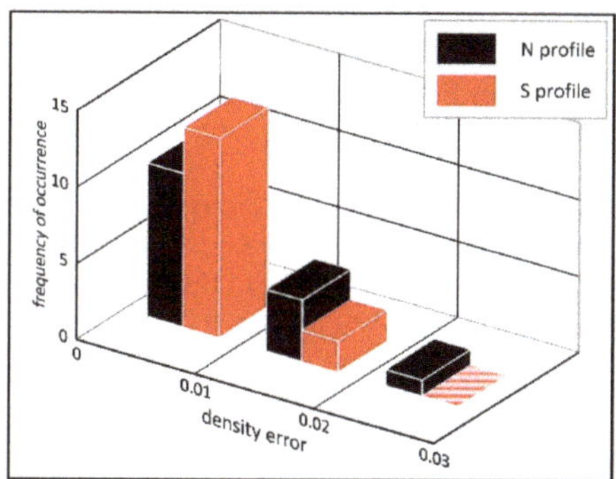

Figure 7. Frequency of density error occurrence.

On the densigram (Figure 6d), we can also distinguish four zones with similar density values. It should be noted, however, that the position of borders of these zones differs slightly from the position of borders determined on the basis of "Bouguer anomaly". This has to do with the fact that density is calculated for constant intervals imposed by the distance between buntons, while boundaries on the densigram are determined on the basis of interval levels. Similarly, density interval boundaries are related to lithological boundaries. Sometimes an interval falls on a lithological boundary, making the calculated density the average of the two boundary layers.

Thus, the first zone (1) reaching the depth of about 24 m is characterized by nearly constant density with an average value of 2.19 $Mg \cdot m^{-3}$ and corresponds to the layer of silty clays. The second zone (2) extends to the depth of 60 m, i.e., to the border of the quaternary and is not as homogeneous in terms of density. In both profiles, interval densities vary between 2.00 and 2.11 $Mg \cdot m^{-3}$. They may represent facial changes within the fine and medium sand layer. There is only one interval density (X) at the depth of 42 m that has a noticeably lower density than the average density of the zone (2). It is clearly visible one both vertical profiles, and this indicates that loosening exists around the shaft. It is possible because the density is in the running sand layer. Just below the quaternary boundary, the third zone (3) is located with the varied density of between 2.20 and 2.28 $Mg \cdot m^{-3}$. It is composed of loam and gypsum layers, i.e., it is heterogeneous in terms of density. Below, at the end of the measurement profile lies the fourth zone (4) with silty and sandy loam with nearly uniform density of 2.08 $Mg \cdot m^{-3}$.

On the background of the described zones, some horizontal differentiation can also be observed in the northern-southern (N–S) direction. The first differentiation can be seen at the depth of 24–30 m, where a decrease of density can be observed on the northern side. This zone is located within the layer of very wet silt and may be related to leaching of rock material beyond the shaft lining. Within the Neogene formations, two intervals with horizontal density variation are observed, the first one under the quaternary–Neogene boundary (60–66 m) within the gypsum layer (lower density on the northern side) and the second one in the interval at a depth of 78–84 m, at the boundary of silty loam with gypsum and silty loam (lower density on the southern side). It seems that both density variations can be related to the occurrence of minor karst in these intervals. There are also other density differences visible in the densigram, but their magnitude falls within the error limits and there are no grounds to treat it as a loosening zone.

In general, the calculated density distributions indicate that there are no significant loosening zones in the surrounding rock mass in the studied shaft section that could threaten its safety.

3.2. Closed Mine Shaft

A microgravity study was conducted on a plot of land occupied by the Fire Department in Boguszów Gorce, Lower Silesia, Poland. The purpose of the study was to determine the safety of the land surface and urban infrastructure in the area. This was due to the fact that after one of the November rainfalls, two sinkholes with a small horizontal extant were formed, located on the opposite sides of a garage building. These sinkholes were eliminated as they were filled in to make way for fire trucks.

Between the 14th and 19th centuries, silver ore was mined in the town. At the distance of several tens of meters from the sinkholes, there was an old mine shaft "Ludwig", which, according to reports, was liquidated by covering it with a concrete slab and is now probably located under a building [43]. Therefore, it was necessary to investigate if there was a connection between the sinkholes, the mine shaft and shallow exploitation of silver ores.

3.2.1. General Geological Setting

The major part of the town lies on steeply sloping strata of Upper Carboniferous system, Silesian series, Namurian, and Westphalian stage (according to the regional stratigraphy of northwest Europe)—Figure 8. According to Polish stratigraphy, in the research area there are three rock formations: Wałbrzych (Namurian A-B substages), Biały Kamień (Namurian C and some part of Westphalian A substages) and Żalcerz (Westphalian substage). Lithologically, all of them are formed almost in the same way: quartz puddingstone, sandstone, mudstone, claystone, and hard coal. They are bordered on the north by a large rhyolite laccolith. These formations are cut by NW–SE extensional faults with opposite slopes to the Carboniferous strata.

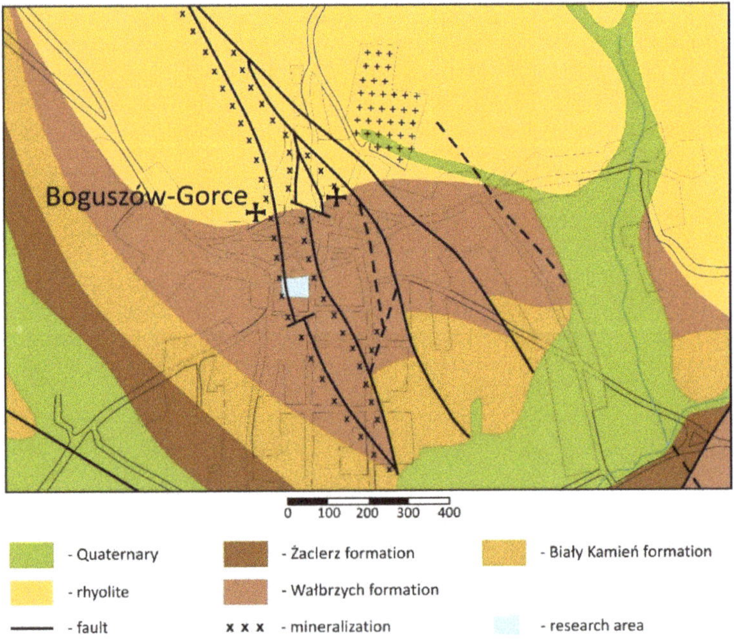

Figure 8. Simplified geological setting in Boguszów-Gorce area.

In the fault zones, in the near-surface part of the rock mass, mineralization of silver ore occurs, while barite is observed at a greater depth. There are four zones in the city area. Two of them and partly the third one are mineralized, and the extreme southern one runs through the surveyed area.

3.2.2. A Brief History of Mining in the Studied Area

The history of Boguszów-Gorce is closely connected with silver mining. The beginnings of exploitation date back to the 14th century, but an important date is 1529 when the "Wags mit Gott" mine was build [43]. The mining history of the town is a story of ups and downs, largely dependent on the wars affecting the area. The end of the 16th century saw the end of the first stage of the town's development, followed by stagnation, and as late as in the second half of the 17th century, timid attempts were made to reactivate mining. The exploitation was carried out through mining galleries with a cross-section of about 1.2 × 0.8 m, and the small lighting shafts had a square cross-section of 1.1 × 1.1 m. It was not until the early 18th century, after the discovery of a new rich vein, that mining flourished again, but by the end of that century it had practically died out. In the second half of the century, attempts were made to reactivate mining, among others by digging a new shaft "Ludwig", but the project was discontinued as soon as 1865. The "Ludwig" shaft was 132 m deep and the shallowest exploitation level was located at a depth of approximately 40 m.

3.2.3. Microgravity Survey

Microgravity survey, aimed at identifying possible loosening zones and voids in the near-surface part of the rock mass, were carried out in an area of approximately 65 × 70 m (Figure 8). This area included both sinkholes and the area adjacent to the decommissioned "Ludwig" shaft, and the measurements were made in a grid of 2.5 × 2.5 m, and twice as densely near the sinkholes. The number of points was limited by buildings and slopes.

Just because of the buildings and escarpments, it was necessary to apply the building and terrain correction. On the basis of the elevation of the land surface around and in the studied area, the terrain correction was calculated assuming an average density of $2\,\text{Mg}\cdot\text{m}^{-3}$. On the basis of maps and geodetic measurements, buildings were mapped and building correction was calculated. The total result of both corrections is shown in Figure 9.

Figure 9. Distribution of a sum of building and terrain corrections.

The maximum value of corrections was about $1.2\,\text{nm}\cdot\text{s}^{-2}$, reaching the highest values at the points located near buildings and slopes. It is worth noting that these corrections significantly impact on the measured gravity value and, consequently, need further interpretation. At the next stage, Bouguer anomaly values were calculated, which were then filtered with Butterworth formula to remove small, random errors. The distribution of the obtained values is shown in Figure 10.

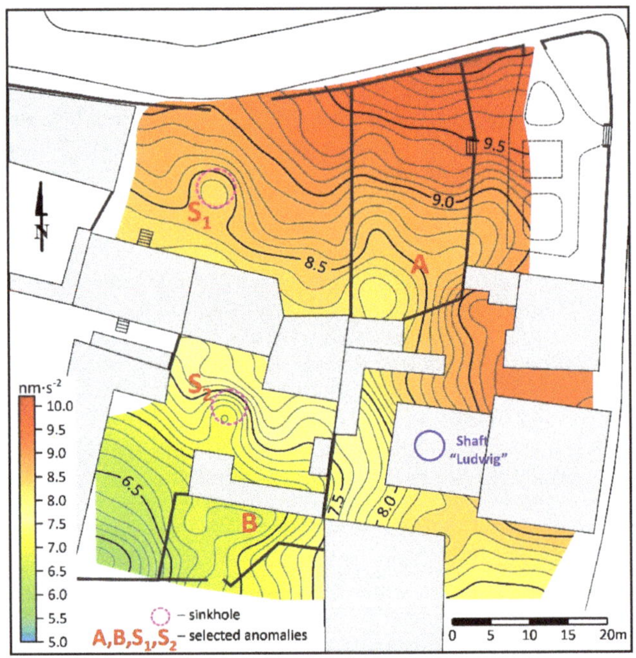

Figure 10. Distribution of Bouguer anomalies with corrections.

The general course of the contours on the Bouguer anomaly distribution is from NW to SE, which is consistent with the extension of the geologic strata of the studied area. The steeply southward collapsing Carboniferous strata cause a significant horizontal gradient, so the Bouguer anomaly values vary by about 5 nm·s^{-2} in such a small area. The two sinkholes that generate the two anomalies S1 and S2 are very prominent in the distribution. Both anomalies are characterized by relatively negative values, which indicates the scarcity of masses in these areas and thus decreased density. Two other anomalies A and B, also with reduced values relative to the surroundings, can be easily distinguished in the distribution. Like the earlier anomalies, these anomalies also indicate mass shortages in the near-surface part of the rock mass.

There is a strong horizontal gradient in the studied area that can mask other anomalies and makes it difficult to define the boundaries described above. For this reason, it was necessary to eliminate this gradient as a regional factor from the large geological structures, which are the Carboniferous strata (Figure 8). Different variants of the calculation of the regional field were performed and finally the Gauss low-pass filter was selected. By subtracting the thus approximated regional field from the Bouguer anomaly, the values of residual anomalies were obtained, the distribution of which is presented in Figure 11.

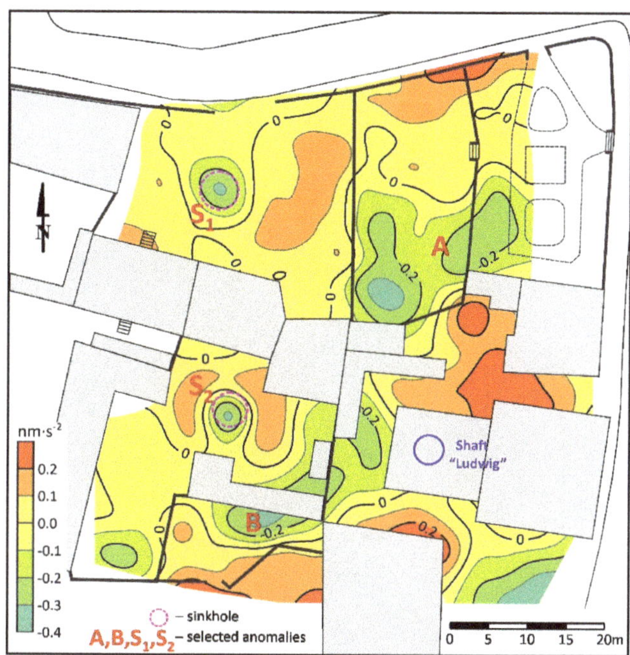

Figure 11. Distribution of residual anomalies.

The boundaries of S1 and S2 are clearly visible on the distribution of residual anomalies. Their horizontal extent is small, almost equal to the area of sinkholes. This means that the density around the sinkholes is not reduced, which indicates that the rock mass was not disturbed there, i.e., there are no cracks and fractures. It can be assumed that the sinkholes were formed by vertical collapses of rock masses only in their area. The collapses most likely occurred only in the area of the shafts, which are the remnants of shallow silver ore mining. Analyzing the simplified geological map (Figure 7), it can be noted that the line connecting the two small shafts coincides with the course of the fault in this area, which contains silver ore mineralization. Thus, it is highly probable that the small shafts define the course of an unknown adit, and were used to illuminate it or exploit the ore.

The separation of the residual anomalies allowed for determination of the horizontal extent of the boundaries of anomalies A and B described earlier (Figure 11). The decomposition shows that anomaly B is larger than it would appear from the Bouguer anomaly distribution and continues in the NE direction. It merges with anomaly A, whose horizontal extent is now more clearly visible.

The distribution of anomaly A does not indicate that it is related with the decommissioned "Ludwig" shaft, which cannot be excluded in the case of anomaly B. Nevertheless, it seems more likely that the sequence of anomalies A and B has a common origin, which may have two causes. The first cause may be a sequence of old mining that has affected the shallow part of the rock mass and thus affected its density. The second cause may be a near-surface loosening of the rock mass caused by water flow along the land surface sloping in a nearly southerly direction. Anomalies A and B lie on the plots with different elevations, separated by retaining walls. Thus, there are grounds for the occurrence of a suffosion phenomenon leading to a decrease in density and thus the formation of anomalies.

3.3. The Shaft under Construction

Microgravity surveys in the area surrounding the drilling site of the new mine shaft were carried out to examine the shallow parts of the rock mass and identify the areas of lower density, which might pose a threat to the shaft itself and the future infrastructure around it. The research was prompted by the discovery of old mining cavities at a depth of approx. 23 m, which indicates that at this depth, previously hard coal was mined in the seam with a thickness of approx. 1.1 m (Figure 12a). The mounds consisting of rock debris, shale, and hard coal were found at a depth of 20 m.

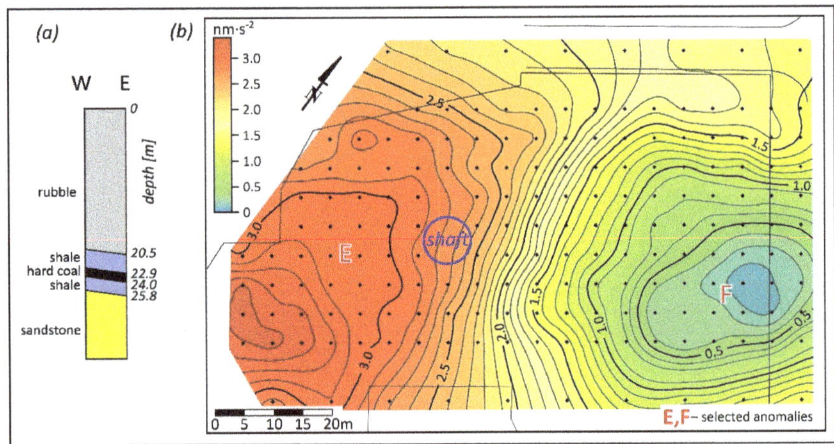

Figure 12. The results of survey: (**a**) main shaft lithology and (**b**) distribution of Bouguer anomalies.

Microgravity measurements were performed around the shaft in the area of 95 × 60 m, in a regular grid of 5 × 5 m, and additionally, in accessible places, points located every 10 m were added on the outside. The studied area was virtually flat, therefore, there was no need to calculate a terrain correction. However, it was necessary to calculate the correction eliminating the gravity impact of the shaft. The distribution of Bouguer anomalies after the correction is shown in Figure 12b. The analysis of the obtained distribution reveals two opposing anomalies: relatively positive E and a relatively negative F. Due to the horizontal extent of both anomalies relative to the entire studied area and the amplitude of both anomalies, it is not possible to calculate the regional anomaly, and both anomalies should be treated as local. Anomaly E, being relatively positive, is not interesting from the point of view of the task at hand. However, the relatively negative anomaly F with an amplitude of about 1.5 nm·s⁻² indicates a significant mass shortage in its area. Based on the information about the occurrence of old workings, it should be believed that its cause is related to the exploitation of shallow coal seam. Given that the dip of the strata is about 9°, it is possible that the anomaly reflects shallow, uncontrolled hard coal mining. Nevertheless, the reduced density of the rock mass poses a threat to the shaft and future infrastructure, so it should be further investigated by drilling.

4. Discussion

The presented examples of surveys demonstrate that the microgravity method can be used in an area that is not easy to measure, the mining area, and, particularly, in the close vicinity of the mine shaft. Despite the shaft and its infrastructure, this method allows to take measurements in the areas where other geophysical methods cannot be applied. What makes it possible are formulas that allow for calculation of corrections that eliminate the gravity impact of the shaft, the underground infrastructure, and the buildings on the surface [19]. The negative anomalies obtained from the processed measurements are always associated with decreased density in the rock mass. Some of these anomalies are easy to interpret as they are related to the existence of identified underground infrastructure, such as an underground channel observed in the gravity distribution near an operating shaft. The other negative anomalies, on the other hand, are due to the presence of voids and loosening zones in the rock mass. Due to the complex geomechanical processes occurring in the rock mass, in most cases, it is difficult to answer unambiguously whether an anomaly is caused solely by a loosening zone, a void, or a void and a loosening zone above it [31,44]. However the analysis of the parameters of the anomaly, especially its amplitude and horizontal extent, allows to determine whether its source can pose a threat to the surface and objects in its vicinity [36]. What is more, an analysis of the knowable shallow geological structure and landforms may point to the anomalies being linked to the hydrogeological conditions in the studied area [45,46].

The presented examples of surface surveys clearly demonstrate how microgravity method can be applied to detect risks related to a mine shaft. Each case was resolved in a different manner. The study in the area of the liquidated shaft "Ludwig" confirmed that the shaft had been liquidated properly and the risk related to the shallow silver ore goafs. The surveys carried out in the vicinity of the working shaft showed that during the surveys there were no density changes that could threaten the stability of the shaft. However, the course of the negative anomaly of (presently) low amplitude in the vicinity of the shaft should be regularly monitored [28,29], as it may be related to leaching, potentially leading to further loss of density. The case of the shaft under construction is an example of the situation where high amplitude indicates the existence of voids in the rock mass, which could threaten the shaft infrastructure and must be investigated further by drilling.

An additional advantage of the microgravity method is that it can be used inside the mine shaft and it is the only geophysical method that can be used when the mine shaft lining is made of steel. As shown in the example described above, the method makes it possible to determine the density of the rock mass outside the shaft lining. The recorded interval density values correspond to the lithological structure around the shaft and its variation [24,26]. The results obtained in this way can have three explanations. The first is the case in which the density values within each geological layer vary slightly and are related to facial changes. This indicates a solid rock mass behind the shaft lining. The second one is the occurrence of zones with significantly decreased density, indicating the existence of voids outside the shaft lining, which may pose a threat to it. The third one, an example of which is described in the article, is a situation where density changes correspond to lithological structure, however, there are small but visible zones of decreased density [26,42]. Additionally, taking measurements at opposing points relative to the center of the shaft, on one level, allowed us to identify on which side such a void was located [26]. Currently, the observed changes do not pose a threat to the mine shaft, however, their presence, especially in the running sand layers, requires monitoring. Linking density information from the measurements taken in the initial part of the shaft and those from the surface gives a more complete picture of the rock mass, which allows for better identification of possible threats to the shaft related to decreased density in the rock mass.

5. Conclusions

The safety problems related to mine shafts have three aspects, that is the safety while sinking the shaft, the safety of using the shaft, and the safety of the area after its liquidation. These are complex issues and they cannot be resolved with the use of a single method. Each method contributes to the safety of the mine shaft by providing new information. In the paper, the authors presented the application of one of the geophysical methods, that is the microgravity method to solve some problems related to all the three aspects. The advantage of this method is that it is non-invasive, not very susceptible to external interference, and it makes it possible to take measurements on the surface and inside the mine shaft. Additionally, what speaks in its favor is easy application. Microgravity method allows to trace the distribution of masses in the rock mass, which is the distribution of density. Voids and loosening zones pose a safety risk for the mine shaft and its surrounding infrastructure, so the method based on the detection of mass (density) lends itself for identification of a risk of this type.

The article presents examples of microgravity survey for each of the above-mentioned cases. Research has shown that the method can be applied not only in the areas of liquidated mine, but also in areas of acting mines, where the measurement conditions are difficult or very difficult.

The presented results of the surface survey showed how important the selection of the appropriate methodology for measurements and subsequent interpretation is. The correct separation of useful anomalies enabled a qualitative determination of the degree of threat of shaft and the surface around it from the voids and loosening zones in the rock mass.

The results of the microgravity survey inside the mining shaft confirmed the possibility of using the method to determine the density of rocks outside the shaft lining. The calculated density allowed to separate depth intervals with a decreased density, which proves the existence of voids or loosening zones. In addition, performing research in more than one vertical profile, their horizontal position can be specified.

It should be emphasized that the unquestionable advantage of the presented research method is the fact that a threat may be detected from the emergence of noticeable changes in infrastructure and on the surface of the area, as well as in the shaft lining. Thus, by performing the measuring in time intervals, it is possible to monitor density changes in the rock mass.

Author Contributions: Both authors equally contributed to the data analysis and the simulation, the results analysis, the writing, and review. Both authors have read and agreed to the published version of the manuscript.

Funding: This research received no external funding.

Institutional Review Board Statement: Not applicable.

Informed Consent Statement: Not applicable.

Data Availability Statement: Not applicable.

Conflicts of Interest: The authors declare no conflict of interest.

Abbreviations

ρ	bulk density, $\mathrm{Mg \cdot m^{-3}}$
g_M	measurement value of gravity, $\mathrm{nm \cdot s^{-2}}$
δg_F	free-air correction, $\mathrm{nm \cdot s^{-2}}$
δg_B	Bouguer correction, $\mathrm{nm \cdot s^{-2}}$
g_N	gravity normal value, $\mathrm{nm \cdot s^{-2}}$
δg_T	terrain correction, $\mathrm{nm \cdot s^{-2}}$
δg_G	mining correction, $\mathrm{nm \cdot s^{-2}}$
g	gravity value with corrections, $\mathrm{nm \cdot s^{-2}}$

References

1. Lecomte, A.; Salmon, R.; Yang, W.; Marshall, A.; Purvis, M.; Prusek, S.; Bock, S.; Gajda, L.; Dziura, J.; Niharra, A.M. Case studies and analysis of mine shafts incidents in Europe. In Proceedings of the 3rd International Conference on Shaft Design and Construction (SDC 2012), London, UK, 22–28 April 2012. Available online: https://hal-ineris.archives-ouvertes.fr/ineris-00973661/document (accessed on 21 May 2021).
2. Bobek, R.; Śledź, T.; Twardokęs, J.; Ratajczak, A.; Głuch, P. Problemy stateczności obudowy szybów w świetle doświadczeń KWK Knurów–Szczygłowice. *Zesz. Nauk. Igsmie Pan* **2016**, *94*, 41–52.
3. Ma, F.; Deng, Q.; Cunningham, D.; Yuan, R.; Zhao, H. Vertical shaft collapse at the Jinchuan Nickel Mine, Gansu Province, China: Analysis of contributing factors and causal mechanisms. *Environ. Earth Sci.* **2013**, *69*, 21–28. [CrossRef]
4. Sun, Q.; Ma, F.; Guo, J.; Li, G.; Feng, X. Deformation Failure Mechanism of Deep Vertical Shaft in Jinchuan Mining Area. *Sustainability* **2020**, *12*, 2226. [CrossRef]
5. Tor, A.; Jakubów, A.; Tobiczyk, S. Zagrożenia Powstałe w Wyniku Uszkodzenia Lunety Wentylacyjnej Szybu V w Jastrzębskiej Spółce Węglowej S.A. KWK "Pniówek" w Pawłowicach. In Proceedings of the conference 13th Warsztaty Górnicze, Baranów Sandomierski, Poland, 18–20 June 2008; pp. 47–56.

6. Longoni, L.; Papini, M.; Brambilla, D.; Arosio, D.; Zanzi, L. The risk of collapse in abandoned mine sites: The issue of data uncertainty. *Open Geosci.* **2016**, *8*, 246–258. [CrossRef]
7. Cardarelli, E.; Marrone, C.; Orlando, L. Evaluation of tunnel stability using integrated geophysical methods. *J. Appl. Geophys.* **2003**, *52*, 93–102. [CrossRef]
8. Grandjean, G.; Leparoux, D. The potential of seismic methods for detecting cavities and buried objects: Experimentation at a test site. *J. Appl. Geophys.* **2004**, *56*, 93–106. [CrossRef]
9. Dec, J. High resolution seismic investigations for the determination of water flow directions during sulphur deposits exploitation. *Acta Geophys.* **2010**, *58*, 5–14. [CrossRef]
10. Van Schoor, M. Detection of sinkholes using 2D electrical receptivity imaging. *J. Appl. Geophys.* **2002**, *50*, 393–399. [CrossRef]
11. Beres, M.; Luetscher, M.; Olivier, R. Integration of ground-penetrating radar and microgravity methods to map shallow caves. *J. Appl. Geophys.* **2001**, *46*, 249–262. [CrossRef]
12. Golebiowski, T.; Porzucek, S.; Pasierb, B. Ambiguities in geophysical interpretation during fracture detection—Case study from a limestone quarry (Lower Silesia Region, Poland). *Near Surf. Geophys.* **2016**, *14*, 371–384. [CrossRef]
13. Balkaya, Ç.; Göktürkler, G.; Erhan, Z.; Ekinci, Y.L. Exploration for a cave by magnetic and electrical resistivity surveys: Ayvacık Sinkhole example, Bozdağ, İzmir (Western Turkey). *Geophysics* **2012**, *77*, B135–B146. [CrossRef]
14. Klityński, W.; Oryński, S.; Dinh, N. Application of the conductive method in the engineering geology: Ruczaj district in Kraków, Poland, as a case study. *Acta Geophys.* **2019**, *67*, 1921–1931. [CrossRef]
15. Torge, W. *Gravity*; Walter de Gruyter: Berlin, Germany; New York, NY, USA, 1989.
16. Bishop, I.; Styles, P.; Emsley, S.J.; Ferguson, N.S. The detection of cavities using the microgravity technique: Case histories from mining and karstic environments. *Geol. Soc. Eng. Geol. Spec.* **1997**, *12*, 153–166. [CrossRef]
17. Styles, P.; Toon, S.; Thomas, E.; Skittrall, M. Microgravity as a tool for the detection, characterization and prediction of geohazard posed by abandoned mining cavities. *First Brake* **2006**, *24*, 51–60. [CrossRef]
18. Porzucek, S. Microgravity survey in the area of the former shallow coal mining. In Proceedings of the 13th International Multidisciplinary Scientific GeoConference SGEM, Albena, Bulgary, 16–22 June 2013; Volume 2, pp. 823–830.
19. Yu, D. The influence of buildings on urban gravity surveys. *J. Environ. Eng. Geophys.* **2014**, *19*, 157–164. [CrossRef]
20. Loj, M.; Porzucek, S. Detailed analysis of the gravitational effects caused by the buildings in microgravity survey. *Acta Geophys.* **2019**, *67*, 1799–1807. [CrossRef]
21. Pringle, J.K.; Stimpson, I.G.; Toon, S.M.; Caunt, S.; Lane, V.S.; Husband, C.R.; Jones, G.M.; Cassidy, N.J.; Styles, P. Geophysical characterization of derelict coalmine workings and mineshaft detection: A case study from Shrewsbury, United Kingdom. *Near Surf. Geophys.* **2008**, *6*, 185–194. [CrossRef]
22. McCann, D.M.; Jackson, P.D.; Culshaw, M.G. The use of geophysical surveying methods in the detection of natural cavities and mineshafts. *Q. J. Eng. Geol.* **1987**, *20*, 59–73. [CrossRef]
23. Porzucek, S.; Madej, J. Detection of near-surface geological heterogeneity at Starunia palaeontological site and vicinity based on microgravity survey. *Ann. Soc. Geol. Pol.* **2009**, *79*, 365–374.
24. Hammer, H. Density determinations by underground gravity measurement. *Geophysics* **1950**, *15*, 585–731. [CrossRef]
25. McCulloh, T.H. A confirmation by gravity measurements of an underground density profile on care densities. *Geophysics* **1965**, *30*, 1108–1132. [CrossRef]
26. Madej, J. Gravity surveys for assessing rock mass condition around a mine shaft. *Acta Geophys.* **2017**, *65*, 465–479. [CrossRef]
27. Kamiński, P. Optimization Directions for Monitoring of Ground Freezing Process for Grzegorz Shaft Sinking. In *Computational Optimization Techniques and Applications*; Sarfraz, M., Karim, S., Eds.; IntechOpen: London, UK, 2021.
28. Pringle, J.K.; Styles, P.; Howell, C.P.; Branston, M.W.; Furner, R.; Toon, S.M. Long-term time-lapse microgravity and geotechnical monitoring of relict salt mines, Marston, Cheshire, UK. *Geophysics* **2012**, *77*, B287–B294. [CrossRef]
29. Loj, M. Microgravity monitoring discontinuous terrain deformation in a selected area of shallow coal extraction. In Proceedings of the 14th International Multidisciplinary Scientific GeoConference SGEM, Albena, Bulgary, 17–26 June 2014; Volume 1, pp. 521–528.
30. Dychkovskyi, R.; Falshtynskyi, V.; Ruskykh, V.; Cabana, E.; Kosobokov, O. A modern vision of simulation modelling in mining and near mining activity. In Proceedings of the E3S Web of Conferences, Shenzhen, China, 26–28 November 2018; Volume 60.
31. Sun, X.; Li, G.; Zhao, C.; Liu, Y.; Miao, C. Investigation of Deep Mine Shaft Stability in Alternating Hard and Soft Rock Strata Using Three-Dimensional Numerical Modeling. *Processes* **2019**, *7*, 2. [CrossRef]
32. Newton, I. *Philosophiae Naturalis Principia Mathematica*; Printed by Joseph Streater; Royal Society: London, UK, 1687.
33. Emsley, S.J.; Bishop, I. Application of the microgravity technique to cavity location in the investigations for major civil engineering works. *Geol. Soc. Eng. Geol. Spec.* **1997**, *12*, 183–192. [CrossRef]
34. Porzucek, S. *Loosenings and Cracks Detection in Rock Mass Located Above Anthropogenic Voids Using Microgravity Method*; AGH UST Press: Kraków, Poland, 2013.
35. Moritz, H. Geodetic Reference System 1980. *B Geod.* **1984**, *58*, 388–398. [CrossRef]
36. Lipecki, T.; Jaśkowski, W.; Gruszczyński, W.; Matwij, K.; Matwij, W.; Ulmaniec, P. Inventory of the geometric condition of inanimate nature reserve Crystal Caves in "Wieliczka" Salt Mine. *Acta Geod. Geophys.* **2016**, *51*, 257–272. [CrossRef]
37. Nettleton, L.L. Regionals, Residuals, and Structures. *Geophysics* **1954**, *19*, 10–22. [CrossRef]
38. Skeels, D.C. What is residual gravity? *Geophysics* **1967**, *32*, 872–876. [CrossRef]

39. Nettleton, L.L. *Geophysical Prospecting for Oil*; Mc Graw-Hill Book Co.: New York, NY, USA, 1940.
40. Parasnis, D.S. A study of rock densities in the English Midlands. *Geophys. Suppl. Mon. Not. R. Astron. Soc.* **1952**, *6*, 252–271. [CrossRef]
41. LaFehr, T.R. Rock density from borehole gravity surveys. *Geophysics* **1983**, *48*, 341–356. [CrossRef]
42. Fajklewicz, Z.; Jakiel, K.; Madej, J.; Radomiński, J. The Gravity Logging in the Mining Shaft for the Detection of Geological Heterogeneties Bihind the Shaft-Lining. In Proceedings of the 35th International geophysical Simposium, Sofia, Bulgary, 22–26 April 1990; Volume 2, pp. 392–462.
43. Wójcik, D.; Krzyżanowski, K. Kopalnia "Wags mit Gott" (Śmiało z Bogiem). In *Dzieje Górnictwa—Element Europejskiego Dziedzictwa Kultury*; Zagożdżon, P.P., Madziarz, M., Eds.; Oficyna Wydawnicza Politechniki Wrocławskiej: Wrocław, Poland, 2009; Volume 2, pp. 335–344.
44. Fajklewicz, Z. Origin of the anomalies of gravity and its vertical gradient over cavities in brittle rock. *Geophys. Prospect.* **1986**, *34*, 1233–1254. [CrossRef]
45. Lee, L.V.; Kudaibergenova, S.S. The Study of Suffusion-karst Phenomena by Using Complex of Geophysical Methods. In Proceedings of the 14th Conference Engineering and Mining Geophysics, Almaty, Kazakhstan, 23–27 April 2018; pp. 1–11.
46. D'Obyrn, K.; Kamiński, P.; Motyka, J. Influence of Hydrogeological Investigation's Accuracy on Technology of Shaft Sinking and Design of Shaft Lining—Case Study from Southern Poland. *Energies* **2021**, *14*, 2050. [CrossRef]

energies

Article

Fault Sealing Evaluation of a Strike-Slip Fault Based on Normal Stress: A Case Study from Eastern Junggar Basin, NW China

Jie Ji [1,2,*], Kongyou Wu [1,2,*], Yangwen Pei [1,2], Wenjian Guo [3], Yin Liu [1,2] and Tianran Li [1,2]

[1] School of Geosciences, China University of Petroleum, Qingdao 266580, China; peiyangwen@upc.edu.cn (Y.P.); liuyin@upc.edu.cn (Y.L.); b17010010@s.upc.edu.cn (T.L.)
[2] Key Laboratory of Deep Oil & Gas, China University of Petroleum (East China), Qingdao 266580, China
[3] Research Institute of Exploration and Development, Xinjiang Oilfield Company, PetroChina, Karamay 834000, China; guowj@petrochina.com.cn
*   Correspondence: seasonchina@sina.com (J.J.); wukongyou@upc.edu.cn (K.W.)

Abstract: The sealing of a fault zone has been a focus for geological studies in the past decades. The majority of previous studies have focused on the extensional regimes, where the displacement pressure difference between fault rock and reservoir was used to evaluate the fault-sealing property from the basic principle of fault sealing. When considering the displacement pressure difference, the impact of gravity on the fault rock was considered, whereas the impact of horizontal stress was ignored. In this study, we utilize the displacement pressure difference as an index to evaluate the sealing capacity of strike-slip faults, in which both the impacts of gravity and horizontal stress on the fault rocks are all integrated. By calculating the values of σ_H/σ_V and σ_h/σ_V in the vicinity of fault planes, the coefficient K of compaction impacts on fault rocks between normal stress to fault planes and gravity was then determined. By revealing the quantitative relationship between the displacement pressure of rocks, burial depth and clay content, the displacement pressure difference between fault rocks and reservoirs were calculated. The results suggest that the sealing capacity of a strike-slip fault is not only related to the magnitude of normal stress to the fault plane, but also to the stress regime. The clay content is also an important factor controlling the sealing capacity of strike-slip faults.

Keywords: normal stress of fault plane; strike slip fault; sealing of faults; displacement pressure; numerical simulation of stress

Citation: Ji, J.; Wu, K.; Pei, Y.; Guo, W.; Liu, Y.; Li, T. Fault Sealing Evaluation of a Strike-Slip Fault Based on Normal Stress: A Case Study from Eastern Junggar Basin, NW China. *Energies* **2021**, *14*, 1468. https://doi.org/10.3390/en14051468

Academic Editor:
Krzysztof Skrzypkowski

Received: 8 February 2021
Accepted: 2 March 2021
Published: 8 March 2021

Publisher's Note: MDPI stays neutral with regard to jurisdictional claims in published maps and institutional affiliations.

Copyright: © 2021 by the authors. Licensee MDPI, Basel, Switzerland. This article is an open access article distributed under the terms and conditions of the Creative Commons Attribution (CC BY) license (https://creativecommons.org/licenses/by/4.0/).

1. Introduction

In the process of subsurface oil and gas migration and accumulation, faults can be barriers to hydrocarbon accumulation [1–6]. As the sealing capacity of faults determines the effectiveness of hydrocarbon traps, the sealing property of fault has been a concern of many studies [7–12]. Many qualitative and quantitative approaches have been proposed to evaluate fault seal potential. In the middle of the 20th century, the research on fault sealing mainly focused on mudstone smear and lithology juxtaposition by considering the juxtaposition relationship between reservoir and non-reservoir [13] and the mechanism of capillary sealing [14–16]. Englder (1974) studied the formation mode of fault gouge by observing the structural characteristics of clastic materials within fault zones, suggesting that fault gouge is only distributed in the vicinity of shear fractures and large sliding surfaces, whereas fault rocks are distributed throughout the whole fault zone. In recent decades, several algorithms have been proposed to evaluate the fault sealing properties quantitatively, either based on the continuity of clay smears or average clay content within the fault zones, e.g., Clay Smear Potential (CSP) [17,18], Shale Smear Factor (SSF) [19], Shale Gouge Ratio (SGR) [20] and Scaled Shale Gouge Ratio (SSGR) [21]. These algorithms evaluate the fault sealing properties by considering the re-distribution of mudstone/shale beds or the clay/content of the beds in sheared fractures.

However, the continuity of clay smearing is determined by a series of parameters including the sedimentary lithification state, the effective stress, the confining pressure, the strain rate and the mineralogy [22]. This makes the above-mentioned algorithms subject to many limitations in practical applications. In particular, some scholars proposed [23–25] calculating the displacement pressure to evaluate the fault-sealing capacity, suggesting that a fault, with displacement pressure no smaller than the reservoir rocks, can seal the oil and gas laterally.

Smith (1966) thought that if the fault has good sealing properties, the fault zone must have high displacement pressure, or there would be a displacement pressure difference between the hanging wall and footwall rocks of the fault. Based on Smith's research, Lyu (2009) quantitatively analyzed the fault sealing by using the difference between the displacement pressure of fault rock and reservoir and considered that the displacement pressure of fault rock depends on the clay content of fault rock and the normal stress on the fault plane. Fu (2014) improved the algorithm and evaluated the vertical sealing ability of the fault based on the displacement pressure difference between fault rock and reservoir; Lei (2019) evaluated the relationship between the SGR value of fault rock and its sealing capacity based on the displacement pressure difference between fault rock and reservoir.

By considering the displacement pressure difference between fault rocks and reservoir rocks, many studies were conducted to quantitatively evaluate the fault-sealing capacity of extensional basins, e.g., the Bohai Bay basin in eastern China, the Yinggehai basin in south China and the Termit basin in Niger [25–27]. However, due to the weak horizontal compressive stress in the extensional basins in these above study areas, these studies considered the impact of gravity on fault zones but without the effect of horizontal lateral compression.

However, in the contractional regimes, horizontal stress plays important control in the fault zone evolution and fault rock development. Therefore, when evaluating the sealing capacity of fault zones using displacement pressure calculation, not only the gravity but also the horizontal stress should be considered. In this study, we selected a contractional basin, the Junggar Basin in NW China, as a template to evaluate fault sealing capacity by integrating both gravity and horizontal stress. A new method is proposed in this study to evaluate the sealing behavior of strike-slip faults developed in the Dongdaohaizi sag in the eastern Junggar basin. Based on the mechanism of fault sealing, the displacement pressure difference between fault rock and reservoir is taken as an index to quantitatively evaluate the sealing strength of the strike-slip fault. The normal stress of vertical stress and horizontal stress on the fault plane is calculated synthetically. The values of σ_H/σ_V and σ_h/σ_V near the fault plane are calculated using Formation Micro Image (FMI) logging data and conventional logging data, and the relationship between comprehensive compaction of fault rock and vertical compaction is deduced. The orientation of σ_H near the fault is also determined by FMI logging data. Based on mercury intrusion experiment data in the study area, a functional relationship was established; the displacement pressure difference between fault rocks and reservoir was calculated to realize the sealing evaluation of a strike-slip fault. This study provides some new understanding of the evolution of fault sealing in the Dongdaohaizi sag. The evaluation result of fault sealing is confirmed by the actual production situation.

2. Geological Setting

Tectonically, the Junggar Basin, located in the junction area of the Kazakhstan, Siberian and Tarim plates, is an important element within the famous Central Asian Orogenic Belt (CAOB) (Figure 1a) [28–32]. The Junggar Basin is constrained by a number of orogenic belts, including the Altai-kelameili mountains to the northeast, the Zaire-Hala'alate mountains to the northwest, and the Yilinheibergen-Bogda mountains to the South (Figure 1b) [33].

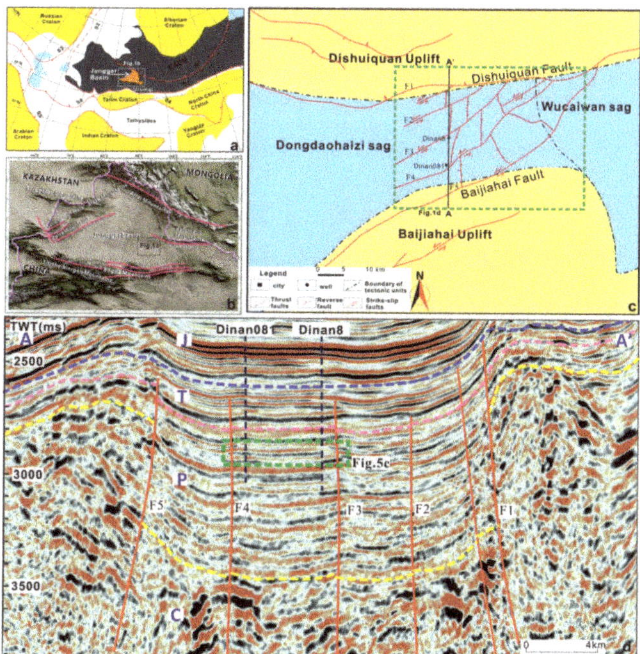

Figure 1. Maps showing the tectonics and geological settings. (**a**) Inset map of the Junggar Basin (modified from Xu et al., 2015); (**b**) Satellite image of the Junggar Basin, with study area marked in the black rectangle; (**c**) Structural subdivision together with fault systems; (**d**) The north–south seismic section A-A', showing the distribution of the strata and the major faults.

Geophysical data suggest that the Junggar Basin is a complex composite basin that has endured multiple polycyclic tectonic events [34–37]. Previous studies indicate that the basin formed in the Late Carboniferous due to the collision and amalgamation of the CAOB [38,39]. The Junggar Basin entered into an intracontinental depression stage in the Mesozoic due to contraction from the northwest and northeast [35,40], forming and reactivating the internal structures within the Junggar basin [38,41]. The reverse boundary faults of the Junggar Basin gradually stopped its slip [37,42] from Late Jurassic to Early Neogene. From the Neogene to the Quaternary, being affected by the collision between the Indian and Eurasian plates, a rejuvenated foreland basin was developed in the Junggar area [35,36]. Due to polycyclic tectonic movements that happened in the Junggar basin, a series of strike-slip faults were developed in the northwest, southeast and northeast Junggar Basin [43–46].

The green rectangle in Figure 1c demonstrates the detailed position of this study, which covers the east portion of the Dongdaohaizi sag, together with the east end of the Wucaiwang sag, the south end of the Dishuiquan uplift and the north end of the Baijiahai uplift. As presented in Figure 1c, complex strike-slip fault systems were developed in our study area, including F1 (the Dishuiquan fault), F2, F3, F4 and F5 (the Baijiahai fault). The fault systems are primarily NE–SW striking, which is sub-perpendicular to the NWW–SEE trending Kelameili strike-slip fault to the northeast. Figure 1d is a north–south trending seismic profile, indicating the development of strike-slip fault systems, including F1, F2, F3, F4 and F5. These faults, with high dip angles, offset the Carboniferous, Permian and Triassic sediments from the deeper to the upper section. The faults F1 (the Dishuiquan fault) and F5 (the Baijiahai fault) present a fault throw of higher than 500 ms, whereas the faults F2, F3 and F4 present very limited fault throw.

The petroleum production data suggest that there are abundant hydrocarbon resources in our study area. In particular, in the vicinity of the faults F3 and F4, there are important petroleum findings in the production wells Dinan8 and Dinan081 (Figure 1c,d). The faults F3 and F4 may play important control on hydrocarbon migration and accumulation here. Therefore, in this study, the fault F3 and F4 were selected to conduct fault sealing evaluation to reveal their sealing capacity and controlling parameters.

3. Methods

3.1. Calculation of Clay Content of Fault Rocks

Fault rocks can be developed in a fault zone between the hanging wall and footwall of the fault during the formation or reactivation of the fault [47,48]. Previous studies suggested that the clay content of fault rocks are important factors controlling the displacement pressure of a fault zone [5,6,49].

As the clay of fault rocks derives from the strata of both the hanging wall and the footwall, the clay content of a fault zone is highly related to both fault throw and clay contents of stratigraphy in two walls. Normally, smaller fault throw and higher clay contents of the stratigraphy may result in higher clay content of fault rocks [24,50].

The clay content of fault rocks can be calculated using Equation (1) proposed by Fristad (1997).

$$V_{fr} = (\sum_{i=1}^{n} h_i \times V_{sh}^i)/D \tag{1}$$

In the above equation, V_{fr} represents shale content of fault rock, h_i represents the thickness of stratum i, V_{sh}^i represents clay content of stratum i, n represents the number of strata sliding through the study point, D represents fault throw.

3.2. Calculation of In Situ Stress of a Fault Plane

It is widely accepted that the in situ stress of a fault plane can be calculated using well-logging data [51–53]. Specifically, by integrating imaging logging data, conventional logging data and dipole acoustic logging data, the maximum horizontal stress (σ_H), (σ_h), (σ_V) curves can be calculated. According to the numerical relationship between σ_H, σ_V and σ_h, three types of regimes can be classified [54,55], including normal fault regime ($\sigma_V > \sigma_H > \sigma_h$, Figure 2a), strike-slip fault regime ($\sigma_H > \sigma_V > \sigma_h$, Figure 2b) and reverse fault regime ($\sigma_H > \sigma_h > \sigma_V$, Figure 2c). The orientation of induced fractures can be interpreted using the imaging logging data, which then can be used to determine the orientation of maximum horizontal stress of a fault plane [56–58].

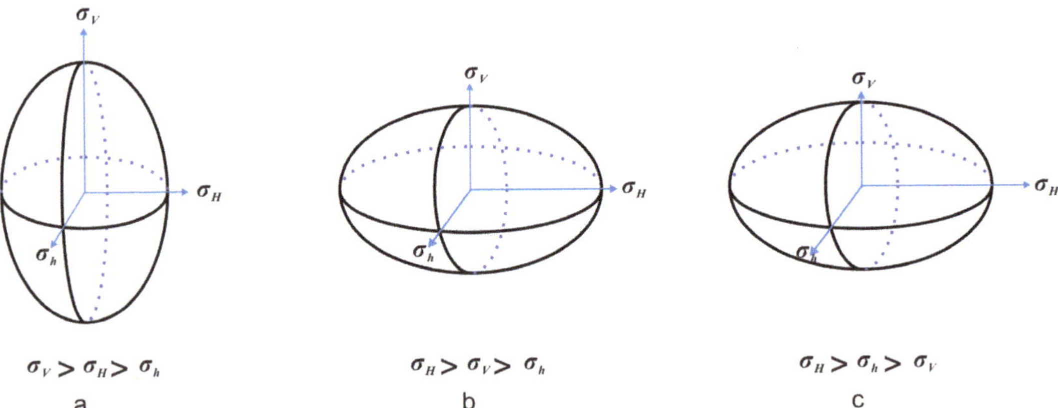

Figure 2. Present in situ stress state of the fault plane; (**a**) Normal fault stress regime (NF); (**b**) Strike-slip fault regime (SS); (**c**) Reverse fault stress regime (RF).

3.3. Calculation of Equivalent Buried Depth of Fault Rock

Fault rocks are compressed by the component that is perpendicular to the fault plane of both vertical stress and horizontal stress [59,60]. The magnitude of vertical stress can be calculated by integration of density logs, and the horizontal stress can be calculated indirectly by the functional relationship between horizontal stress and vertical stress [58,61,62].

The coordinate system was established (Figure 3a) to calculate normal stress (σ_T) on a fault plane (in grey) by integrating vertical stress (σ_V), maximum horizontal stress (σ_H) and minimum horizontal stress (σ_h). The x-axis, y-axis and z-axis are parallel to the maximum horizontal stress σ_H, the minimum horizontal stress σ_h and the vertical stress σ_V.

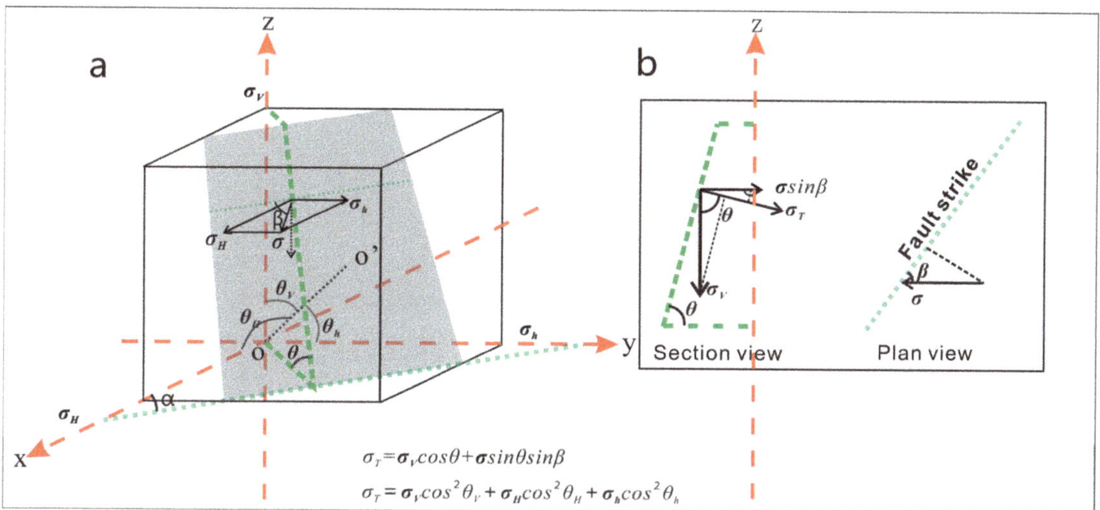

Figure 3. Schematic diagram of depth conversion of fault zone and stress state regime of cross-section. (oo′ is the normal of the fault plane; θ_V, θ_H and θ_h are the angles between σ_V, σ_H, σ_h and the normal out of the fault plane, respectively). (**a**) A description of the stress on the fault plane from a space view. (**b**) A description of the stress on the fault plane from a sectional view.

By integrating the components of σ_V, σ_H and σ_h perpendicular to the fault plane, the resultant normal stress (σ_T) acting on the fault plane can be calculated (Figure 3a) using the following equation:

$$\sigma_T = \sigma_V \times (\cos\theta_V)^2 + \sigma_H \times (\cos\theta_H)^2 + \sigma_V \times (\cos\theta_h)^2 \tag{2}$$

The ratio between normal stress (σ_T) and vertical stress (σ_V) is set as K (i.e., σ_T/σ_V) thus,

$$\sigma_T = K \times \sigma_V \tag{3}$$

in which θ represents fault dip angle, and α represents the sharp angle between the fault strike and the maximum horizontal stress σ_H (Figure 3a,b).

According to the principle of equivalent formation lithostatic pressure (e.g., Lyu et al., 2009; Fu et al., 2014; Lyu et al., 2016), the equivalent buried depth of fault rock Z_h can be calculated (Equation (4)) as:

$$Z_h = K \times Z_f \times \cos\theta \tag{4}$$

in which Z_f represents burial depth of fault rock, θ represents fault dip angle.

3.4. Calculation of Displacement Pressure

Accurate calculation of the displacement pressure of fault rocks and reservoir rocks is of importance in fault sealing evaluation [63,64]. In this study, 357 rock samples were collected from the Dongdaohaizi sag. Based on petrophysical testing results of these samples, a functional relationship between the rock displacement pressures, depth of burial and clay content was constructed (Equation (5)), indicating a positive correlation between the displacement pressure and the product of the clay content and the burial depth [25–27,49].

$$P_d = f(Z, V_{sh}) \tag{5}$$

in which P_d represents the displacement pressure, Z represents the buried depth and V_{sh} represents the clay content.

By mathematical regression of the outer envelope curve for all samples, we obtained Equation (6) to evaluate the largest sealing ability of rock. The equation for the study area is Equation (6).

$$\begin{cases} P_d = 0.5104 \times 10^{\frac{6\times10^{-4}x}{\ln 10}} \\ x = Z \times V_{sh} \end{cases} \tag{6}$$

Equation (6) shows that the product of burial depth and shale content of rock has an exponential relationship with the displacement pressure of rock (Figure 4).

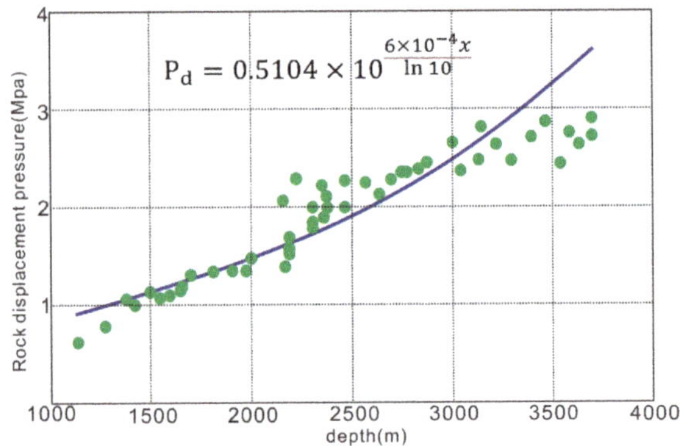

Figure 4. The diagram showing the relationship between displacement pressure of rock with burial depth and clay content of Dongdaohaizi Sag.

3.5. Calculation of Displacement Pressure Difference

By integrating Equations (1)–(5), the displacement pressure difference ($P_{d(f-r)}$) between fault rock and reservoir rock can be calculated as:

$$P_{d(f-r)} = f(Z_h, V_{fr}) - f(Z_r, V_r) \tag{7}$$

in which Z_r represents buried depth of fault rock and V_r represents the clay content of rock.

4. Results

4.1. The Variation of α

The orientation of the maximum horizontal stress near the present strike-slip fault is anisotropic [65–67]. Calibration of the orientation of σ_H is an important step in calculating the α, which has an impact on the fault sealing. Based on the logging data of the areas, the σ_V, σ_H and σ_h curves were obtained for the different locations (Figure 5). The orientation of the σ_H was determined at each location. On the cross-section, the orientation of the σ_H at points 2–7 km apart is not the same, and it is important to highlight that the orientation of the σ_H of these five positions is different from that of the region. The angle between the

σ_H orientation of the northernmost point and the σ_H orientation of the region can reach 40 degrees (Figure 5a). The orientation difference of the σ_H between the closer points is small, while the orientation difference of the σ_H between the farther points is large. On the longitudinal section, the orientation of the σ_H is different in different depths at the same location (Figure 5b,c). The orientation of the σ_H of adjacent wells Dinan 8 and Dinan 081 are different in plane and section (Figure 5c). The orientation of σ_H at Dinan 8 and Dinan 081 is selected as the orientation of σ_H on the fault plane F3 and F4 in (Figure 5c and Table 1), respectively.

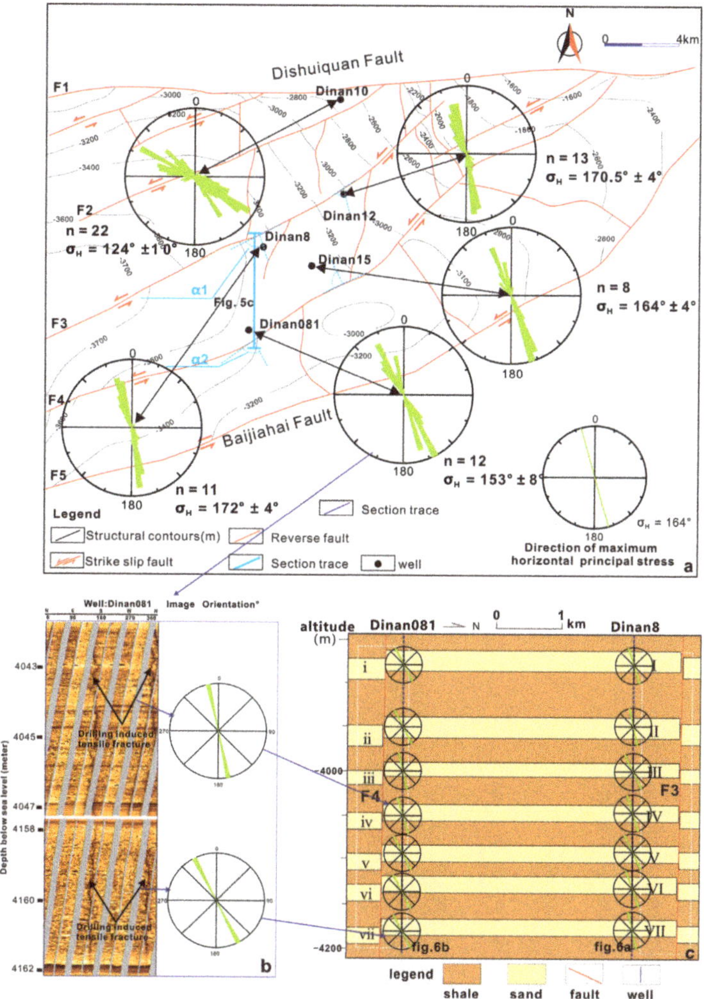

Figure 5. (**a**) The map shows the orientations of the maximum horizontal stress at different sites in our study area. The angle between the orientations of σ_H and the strike of fault F3 is α_1; the angle between the orientations of σ_H and the strike of fault F4 is α_2. The orientation of the regional σ_H is located in the lower right corner (provided by Rock Mechanics Research Center, Xinjiang Oilfield Company, PetroChina, Karamay, China). (**b**) An example of the high-resolution full-bore Formation Micro Image (FMI) from the Borehole Dinan081, showing drilling-induced tensile fractures (ESE–WNW striking) observed at depth of 4042–4047 m and 4158–4162 m (below sea level). (**c**) S–N section showing the maximum horizontal stress orientation of each individual layer.

The angle between the maximum horizontal stress and fault F3 strike is 78 degrees (α_{F3}), and the angle between the maximum horizontal stress and fault F4 strike is 70 degrees (α_{F4}). The change in α angle at F3 has no correlation with depth, the difference between the maximum and minimum value of α angle is 12 degrees, and α angle is less than α_{F3} (Table 1 and Figure 6a). The decrease in α angle relative to α_{F3} results in the decrease of K (Figure 6c), and $P_{d\,(f-r)}$ decreased correspondingly (Figure 6e). There was no correlation between the change in α at F4 and the depth. The difference between the maximum value and the minimum value of α was 14 degrees, and α was greater than α_{F4} (Table 1 and Figure 6b). The increase of α relative to α_{F4} results in the increase of K (Figure 6d) and $P_{d(f-r)}$ correspondingly (Figure 6f).

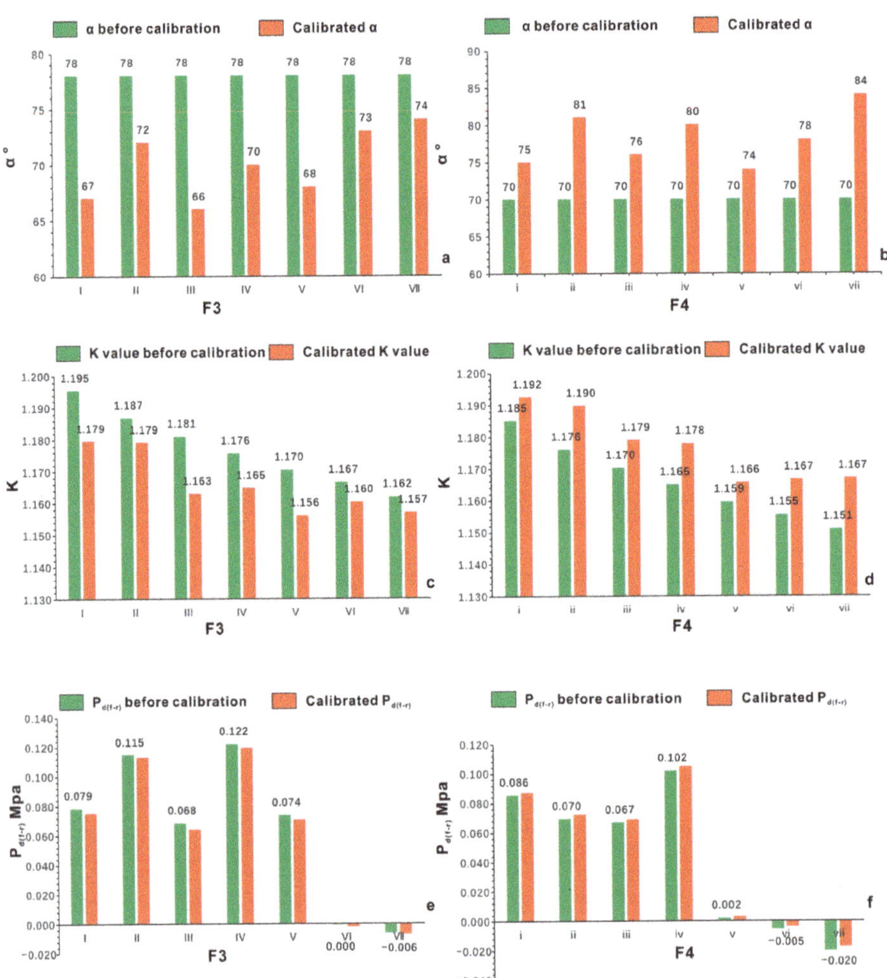

Figure 6. The bar chart showing the change in K value and displacement pressure caused by the change in α. (**a**) Variation in α near fault F3 (α_{F3}). (**b**) Variation in α near fault F4 (α_{F4}). (**c**) Variation in K caused by α_{F3} ($K_{\alpha F3}$). (**d**) Variation of K caused by α_{F4} ($K_{\alpha F4}$). (**e**) The change in displacement pressure difference caused by $K_{\alpha F3}$. (**f**)The change in displacement pressure difference caused by $K_{\alpha F4}$.

Table 1. The known variables (i.e., deep, θ) and calculated results (i.e., σ_H, σ_h, σ_V, α, K) of faults F3 and F4 in the research area.

Fault Number	Reservoir Number	Reservoir Depth	σ_H(psi)	σ_h(psi)	σ_V(psi)	In Situ Stress Regime	α	θ	K
	I	3878	15,715.1	14,193.04	13,804.24	RF	67	78	1.1795
	II	3950	15,943.8	14,361.17	14,098	RF	72	78	1.1791
	III	4000	16,102.6	14,477.92	14,302	RF	66	78	1.1629
F3	IV	4047	16,251.9	14,587.67	14,493.76	RF	70	78	1.1647
	V	4094	16,401.2	14,697.42	14,685.52	RF	68	78	1.1560
	VI	4131	16,518.7	14,783.82	14,836.48	SF	73	78	1.1602
	VII	4176	16,661.7	14,888.9	15,020.08	SF	74	78	1.1569
	i	3885	15,737.4	14,209.38	13,832.8	RF	75	79	1.1925
	ii	3960	15,975.6	14,384.52	14,138.8	RF	81	79	1.1897
	iii	4009	16,131.2	14,498.94	14,338.72	RF	76	79	1.1790
F4	iv	4056	16,280.5	14,608.69	14,530.48	RF	80	79	1.1779
	v	4105	16,436.2	14,723.11	14,730.4	RF	74	79	1.1656
	vi	4142	16,553.7	14,809.51	14,881.36	SF	78	79	1.1665
	vii	4185	16,690.3	14,909.92	15,056.8	SF	84	79	1.1669

The $P_{d(f-r)}$ calculated using α_{F3} is larger than the value using α, but the magnitude of the increase varies from layer to layer (Figure 6e). The $P_{d(f-r)}$ calculated using α_{F4} is lower than the value using α, and the decrease in each layer is not the same (Figure 7f). α was positively correlated with K and $P_{d(f-r)}$ (Figure 7). The accuracy of α angle calculation is positively related to the error.

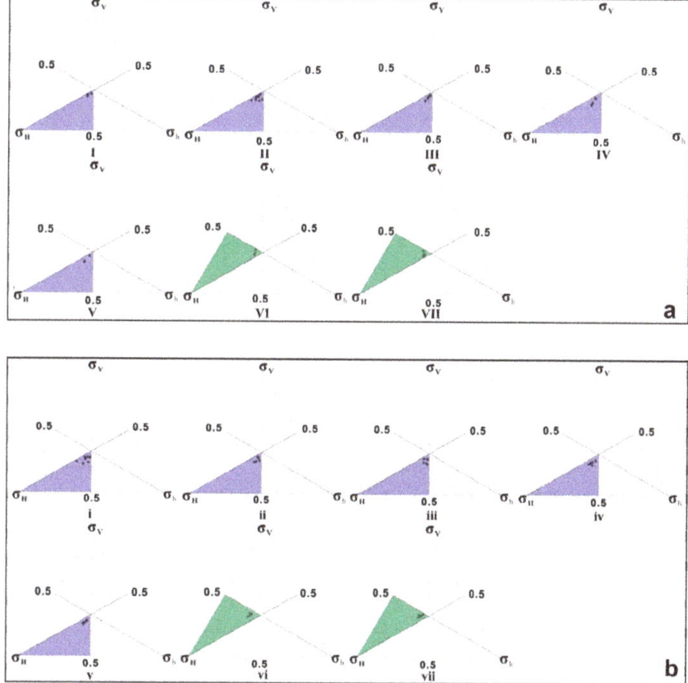

Figure 7. Ternary diagrams showing the stress state near the fault plane of F3 (**a**) and F4 (**b**) (see Figure 5c for detailed settings). The blue shade represents the reverse fault regime, whereas the green shade represents the strike slip regime.

4.2. Stress Regime

The reservoir system and fault system of the fault section are shown in (Figure 5c). The stress values σ_V, σ_H and σ_h of the reservoir are normalized, and ternary diagrams are drawn to reveal the relationship between the σ_V, σ_H and σ_h. The magnitude of σ_h, σ_H and σ_V increases as the burial depth increases.

The dispersion of data points indicates that the ratios of (σ_H/σ_V) and (σ_h/σ_V) are different at different depths. The stress regime of the fault plane corresponding to each reservoir is shown in (Figure 7a,b). The stress regime of the fault plane shows that the upper part (i, ii, iii, iv, v, I, II, III, IV, V) of fault plane F3 and F4 is reverse fault regime, and the lower part (vi, vii, VI, VII) is the strike-slip fault regime. In these ternary diagrams, there is little difference in the distribution of stress regime points of reservoirs under the same stress regime, which indicates that the overall difference of stress regime of each reservoir under the regime stress mode is not too large.

4.3. The Variation of K

If the horizontal stress is not taken into account, and the rock compaction is completely provided by the gravity of the overlying strata; according to (Equation (2)), it can be concluded that the K is equal to $(cos\theta)^2$. Therefore, K is 0.0432 near fault F3 and 0.0364 near fault F4 (Table 1); the corresponding $P_{d(f-r)}$ is shown in (Figure 8c,d). The value of K calculated by combining the normal stress components of superimposed horizontal stress and vertical stress is shown in Table 1 and Figure 8,b. It can be seen that the range of K value increase near F3 fault is 1.1128–1.1363, and the range of K value increase near F4 fault is 1.1292–1.1561.

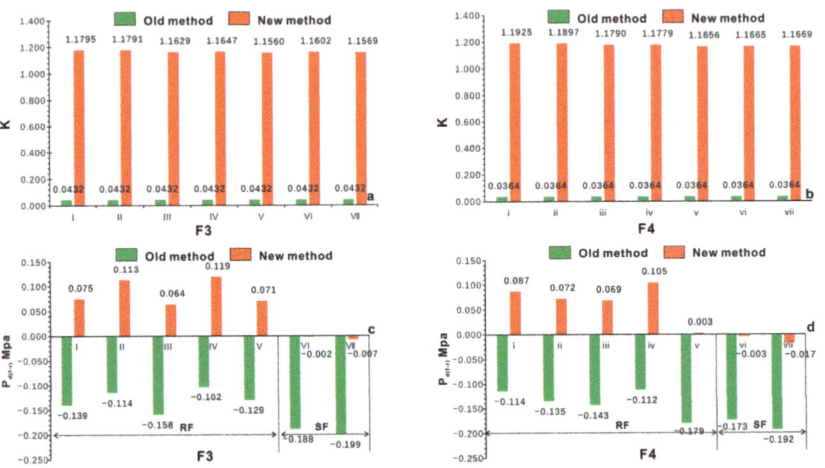

Figure 8. The bar chart showing the change of K value, the old method without considering the horizontal stress, the new method integrating both gravity and horizontal stress. The variation in K value considering the influence of horizontal stress and its influence on the displacement pressure difference. (**a**) Different K values of new and old methods near the F3 fault. (**b**) Different K values of new and old methods near the F3 fault. (**c**) $P_{d(f-r)}$ value of new and old methods near the F3 fault. (**d**) $P_{d(f-r)}$ value of new and old methods near the F4 fault.

The (I–VII) reservoir near the F3 fault corresponds to an increase in $P_{d(f-r)}$ in the range of 0.186–0.227 Mpa (Figure 8c); the (i–vii) reservoir near the F4 fault corresponds to an increase in $P_{d(f-r)}$ in the range of 0.170–0.212 Mpa (Figure 8d).

The results show that the increases in K value and $P_{d(f-r)}$ affected by faults F3 and F4 are different; the increase in K value of the reverse fault stress regime is greater than that of the strike-slip fault (Figure 8a,b); the increase in $P_{d(f-r)}$ value within the depth range of reverse fault stress regime is greater than that of $P_{d(f-r)}$ within the depth range of strike-slip fault stress regime (Figure 8c,d).

There is a positive correlation between K and $P_{d(f-r)}$. Considering the compaction effect of horizontal stress on fault rock, the sealing ability of (F3, F4) fault is obviously increased compared with that without considering horizontal compression. However, the increasing degree of $P_{d(f-r)}$ between fault rock and reservoir is different in different depths (Figure 8c,d).

4.4. Fault Sealing Capacity

As described in Section 3, the displacement pressure difference between fault rocks and reservoir rocks can be calculated using Equation (6). In this research, a fault is not sealed when $P_{d(f-r)} < 0$, whereas a fault is sealed when $P_{d(f-r)} > 0$; the higher the $P_{d(f-r)}$, the better the sealing capacity.

It can be seen from Table 2 that the displacement pressures of reservoir rock in different layers of well Dinan8 are 0.646–0.704 MPa, and the displacement pressures of fault rock calculated by the above method are 0.682–0.706 MPa; the displacement pressures of reservoir rock in different layers of well Dinan081 are 0.630–0.699 MPa, and the displacement pressures of fault rock calculated by the above method are 0.669–0.687 MPa. By comparing the displacement pressures of fault rock and reservoir rock, we found that the displacement pressures of fault rocks in VI,VII, vi, v, vii are all less than that of reservoir rock, so the fault cannot seal hydrocarbons in these formations. Well testing results have proved that these reservoir beds are dry layers or water layers, while the displacement pressures of fault rocks in i, ii, iii, iv, vii, I, II, III, IV, V are all greater than that of reservoir rock; thus the fault can seal hydrocarbons in these reservoir beds. However, the displacement pressure difference between fault rock and reservoir rock is different in different layers, which reflects different sealing capacity. Well testing results have proved that these reservoir beds are all gas layers or oil layers.

Table 2. The known variables (i.e., Reservoir depth, Clay content of fault rock (V_{fr}), Clay content of reservoir (V_r) and calculated results (i.e., Equivalent buried depth of fault rock (Z_h), Displacement pressure of reservoir (P_{dr}), Fault rock pressure and reservoir pressure difference ($P_{d(f-r)}$)) of faults F3 and F4 in the study area.

Fault Number	Reservoir Number	Reservoir Depth	V_{fr}	Z_h	P_{fr}	V_r	P_{dr}	$P_{d(f-r)}$
	I	3878	0.530	482.1	0.682	0.108	0.656	0.025
	II	3950	0.550	506.5	0.692	0.112	0.666	0.026
	III	4000	0.530	491.8	0.686	0.117	0.676	0.010
F3	IV	4047	0.560	523.8	0.699	0.122	0.686	0.013
	V	4094	0.575	541.4	0.706	0.096	0.646	0.060
	VI	4131	0.535	506.9	0.692	0.130	0.704	−0.013
	VII	4176	0.545	520.7	0.698	0.125	0.698	−0.001
	i	3885	0.540	452.0	0.669	0.090	0.630	0.040
	ii	3960	0.545	461.7	0.673	0.111	0.664	0.009
	iii	4009	0.550	470.3	0.677	0.106	0.659	0.018
F4	iv	4056	0.560	482.2	0.682	0.113	0.672	0.010
	v	4105	0.570	494.8	0.687	0.125	0.694	−0.008
	vi	4142	0.530	462.7	0.674	0.120	0.688	−0.014
	vii	4185	0.540	474.5	0.679	0.125	0.699	−0.020

It can be seen from Figure 8c,d that the displacement pressure difference between fault rock and reservoir of F3 fault calculated by this method and previous methods ranges from 0.186 to 0.227 MPa, and that of F4 fault ranges from 0.17 to 0.212 MPa. Because this method considers the influence of horizontal stress comprehensively, the sealing ability of fault calculated by this method is greater than that calculated by previous methods, which indicates that if the influence of horizontal stress on the sealing ability of fault is ignored, the evaluation result will be lower than the actual value. If the old method is adopted, the $P_{d(f-r)}$ of I–VII and i–vii reservoirs are less than zero, which indicates that the oil and gas of I–VII and i–vii reservoirs cannot be sealed, which is seriously inconsistent with the actual situation (Figure 8c,d). In this case, the evaluation result of the new method is consistent with the actual oil and water distribution, which can prove that this method is scientific and feasible.

5. Discussion

5.1. Influence of Different Stress Regime

The σ_h, σ_V and σ_H varies with the depth of the fault plane [68,69], and the σ_h, σ_V and σ_H stress curves near the fault plane will change correspondingly with the change of geometry and kinematics characteristics of faults [53,70]. The ratios of σ_H/σ_V and σ_h/σ_V will change according to the change of stress regime. According to (Equation (3)), it is obvious that with the change in depth, the stress regime determines the magnitude of normal stress σ_T on the fault plane.

It is assumed that the plane is smooth and that the dip and the dip angle are constant (Figure 9a,b); the reverse fault stress regime (RF), $(\sigma_H/\sigma_V) > 1$, $(\sigma_h/\sigma_V) > 1$; the stress regime of the strike-slip fault (SF), $(\sigma_H/\sigma_V) > 1$, $(\sigma_h/\sigma_V) < 1$; normal fault stress regime (NF), $(\sigma_H/\sigma_V) < 1$, $(\sigma_h/\sigma_V) < 1$. The calculations show that $K_{RF} > K_{SF} > K_{NF}$ (Equation (2)).

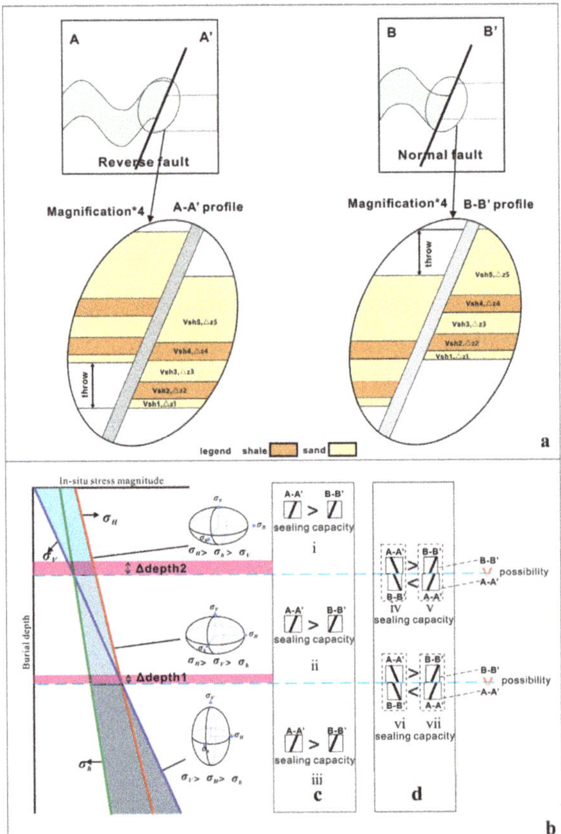

Figure 9. (**a**). The Shale Gouge Ratio (SGR) value of AA′ section is equal to that of BB′ section. In AA′ and BB′ sections, the dip angle and strike of faults are the same, and the stratigraphic characteristics are also the same. (**b**) The AA′ and BB′ sections are under different stress regimes. (**c**) The depth and stress regimes of AA′ and BB′ are the same. (**d**) The depths of AA′ and BB′ are different, and the stress regimes are also different.

Therefore, although the sealing capacity of AA' is better than that of BB' under the same stress regime (Figure 9b,c), the sealing capacity of AA' and BB' are different under different stress regimes. When the depth is fixed, the value of K determines the magnitude of normal stress σ_T (Equation (3)). The difference i sealing capacity can be shown by the following equations:

$$P_{d(RF)} = f(K_{RF} \times \cos\theta \times Z_h, V_{fr}) - f(Z_r, V_r) \tag{8}$$

$$P_{d(SF)} = f(K_{SF} \times \cos\theta \times Z_h, V_{fr}) - f(Z_r, V_r) \tag{9}$$

$$P_{d(NF)} = f(K_{NF} \times \cos\theta \times Z_h, V_{fr}) - f(Z_r, V_r) \tag{10}$$

From the results presented, it is clear that the stress regime has a large impact on the fault sealing capacity. Due to the change in stress regime, the sealing capacity of the BB' plane may be greater than that of the AA', which is deeper than it (Figure 9b,d).

According to the equations in this paper, for different values of (σ_H/σ_V) and (σ_h/σ_V) and the same remaining parameters, the sealing capacity of the shallow faults will be greater than that of the deep ones due to the variation of the stress regimes, within the range of variation of the depths of Δdepth1 and Δdepth2. The ranges of Δdepth1 and Δdepth2 are shown in Figure 9b, and their expressions are as follows:

$$\Delta depth1 = (K_{SF} - K_{NF}) \times \Delta Z_f \times \cos\theta \tag{11}$$

$$\Delta depth2 = (K_{RF} - K_{SF}) \times \Delta Z_f \times \cos\theta \tag{12}$$

It can be inferred that the values of Δdepth1 and Δdepth2 are positively correlated with the value of $(K_{SF} - K_{NF})$. Therefore, we can also draw the conclusion that under the condition of different stress mechanisms and shale content, the sealing property of shallow fault may be better than that of deep fault; the sealing property of strike slip fault has no positive correlation with depth. The stress regime at the depth of fault plane plays an important role in fault sealing.

5.2. Influence of the Orientation of the σ_H

The orientations of the σ_H are different near the fault plane [71], the maximum variation range of the direction can reach 90 degrees [55,58,72] and the angle α between the orientation of the σ_H and the fault strike vary greatly in different sections. This will result in large differences in the normal stresses on the fault plane of the strike-slip fault [73,74], changing the values of K and affecting the fault sealing.

In Equation (2), the interval of variation of α is 0–90. The range of K value affected by this variation is $\left[\cos^2\theta + \sin^2\theta \times \frac{\sigma_h}{\sigma_V}, \cos^2\theta + \sin^2\theta \times \frac{\sigma_H}{\sigma_V}\right]$. Assuming that the angle of α is fixed, the greater the θ, the greater the effect on sealing capacity. Because the θ of the strike-slip fault is usually big, the change in sealing capacity caused by the change in angle is much greater than that of fault with small dip angle. Assuming that the fault dip angle θ is constant and the (σ_H/σ_V), (σ_h/σ_V) is fixed, α angle is an important factor to determine the sealing ability of fault and the sealing ability of fault is positively correlated with α angle.

Accurate calculation of the α angle is helpful to improve the accuracy of fault sealing evaluation. It is more accurate to use the angle between the orientation of the σ_H near the fault plane and the strike of the fault, compared with the angle between the orientation of the σ_H in the region and the strike of the fault. However, due to the limitation of the number of imaging logging data near the fault, the calculation of the σ_h direction near the fault is restricted. Because the accuracy of α angle calculation is positively related to the error of fault sealing evaluation results, using abundant imaging logging data or using software to calculate α angle can improve the accuracy of evaluation results.

5.3. Influence of Different Fault Activity Periods

Many oil and gas reservoirs are not controlled by single faults but by two or more faults [49,75]; the evolution of a strike-slip fault system is the overall response of a series of fault activities under the same stress field environment. The change in the fault characteristics affects the change in the normal stress on the other fault planes in the region, and the change in normal stress on the fault planes caused by different faults is different [76]. Comprehensive analysis of the change in normal stress on the fault plane and its causes is of great significance to the study of oil and gas migration.

In this study, the influence of fault activity periodicity on normal stress of fault plane was simulated by 3D move software, and then the influence on fault sealing was indirectly analyzed. When the activity of the F1 fault in the Indochinese phase was enhanced, the normal stress in the blue region of (Figure 10 i, iii, v) decreased, and the normal stress in other areas increased. When the activity of the F5 fault in Yanshanian period increased, the normal stress of fault in the blue region in (Figure 10 ii, iv, vi) decreased; in other areas, the normal stress increases.

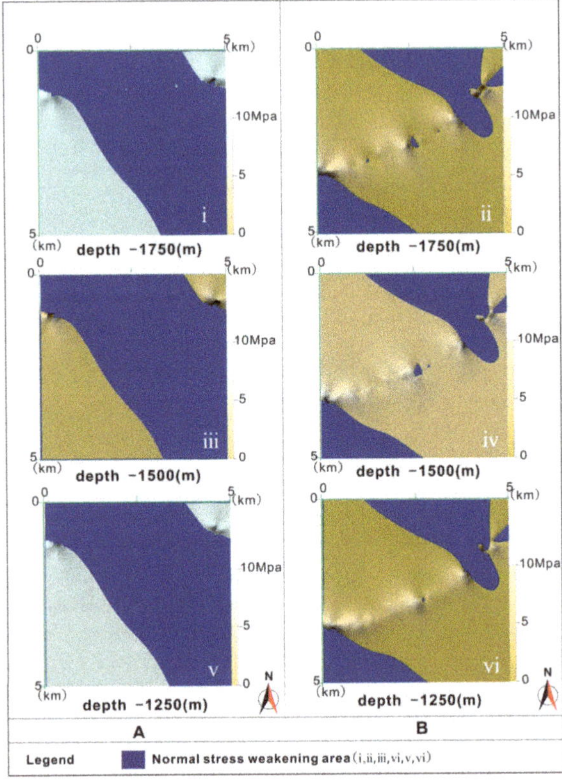

Figure 10. Diagram of normal stress change caused by fault activity sequence. Numerical simulation comparison of plane normal stress distribution caused by activity difference of F1 and F5 faults in different periods. (**A**) are diagrams of normal stress variation at different depths when fault F1 is strongly active. (**B**) are diagrams of normal stress variation at different depths when fault F5 is strongly active. (i), (iii) and (v) are the change diagram of normal stress of −1750 m, −1500 m and −1250 m of the model, respectively when the activity of fault F1 is enhanced. (ii), (iv) and (vi) are the change diagram of normal stress of −1750 m, −1500 m and −1250 m of the model, respectively, when the activity of fault F5 is enhanced. The blue area is the area where the normal stress is relatively weakened. The rest of the area is the area where the normal stress increases relatively.

The sealing capacity of the fault in the normal stress weakening area is weakened, and the sealing capacity of the fault in the normal stress strengthening area is enhanced.

The simulation results show that the activity of faults F1 and F5 and the periods of their activities cause a great difference in the change of normal stress in the region; in Figure 11a, the normal stress in the No. 3 color area is weak and continues to weaken, and the production of the explored oil wells in the area is very low. The normal stress of the fault plane in No.1 color is always strong, where the superimposed areas of No. 2 and No. 4 color regions are also located in this area; the production of oil wells explored in the area is also very low.

Although the success of individual prospects is a function of numerous factors, such as reservoir quality, charge and structure, it is interesting to note that all major fields are located in the No. 5 color area, and the oil well production in the No. 6 green line area is the highest in the study area. The No. 6 green line area is the overlapping area of No. 4 and No. 5 colors, in which the fault transportability is good in the early stage, the fault normal stress is the strongest in the late stage and the sealing is the best, which is most favorable for the preservation of oil and gas reservoirs.

Through the simulation results, we can deduce the conclusion that the intensity and sequence of boundary fault activity control the change in regional normal stress and affect the change of fault sealing capacity; the change law of normal stress on the fault plane in the region caused by it is different, which has an important impact on the migration and accumulation of oil and gas.

According to the variation and superposition of normal stress in the study area simulated by the model, the sealing capacity of faults F2, F3, F4 and F5 can be divided based on the variation in normal stress on the fault plane. The fault sealing in the study area can be divided into three types: high capacity of fault sealing (the normal stress of fault plane is continuously strong), media capacity of fault sealing (the normal stress of fault plane change from weak to strong) and low capacity of fault sealing (the normal stress of fault plane is continuously weak) (Figure 11b).

Therefore, the systematic analysis of fault sealing is helpful to understand the migration and accumulation of oil and gas. It is of great significance for oil and gas exploration and analysis to determine favorable places for prospecting.

5.4. Influence of Uncertain Factors

This method improves the accuracy of the sealing capacity evaluation of slip fault, but there are still many shortcomings. The fault surface is assumed to be smooth, dip and unchanged in tendency, which helps to simplify the calculation but deviates from the actual situation [77]. In this study, the displacement pressure difference between fault rock and reservoir is calculated indirectly by fitting the displacement pressure relationship. However, the accuracy of the fitting equation is limited by the quality and quantity of samples. The uniformity of the distribution of the mass and quantity of samples in different depth ranges will also have a certain impact on the fitting equation. During fault evolution, the relationship between the magnitude of lateral tectonic stresses and the strength of the lateral compaction effect on the faulted rocks is not completely proportional, and a proportionally equivalent treatment method will inevitably lead to errors. Since fault rocks are not conventional sedimentary rocks, the clay content in fault rocks must not be evenly distributed. The current clay content calculation method cannot accurately calculate the clay content of fault rock. Exploring a more practical algorithm and equation, calculating the clay content of fault rock more accurately, increasing the quantity and quality of fitting rock samples and describing the displacement pressure difference between fault rock and reservoir more directly and quantitatively can further improve the accuracy of fault sealing evaluation.

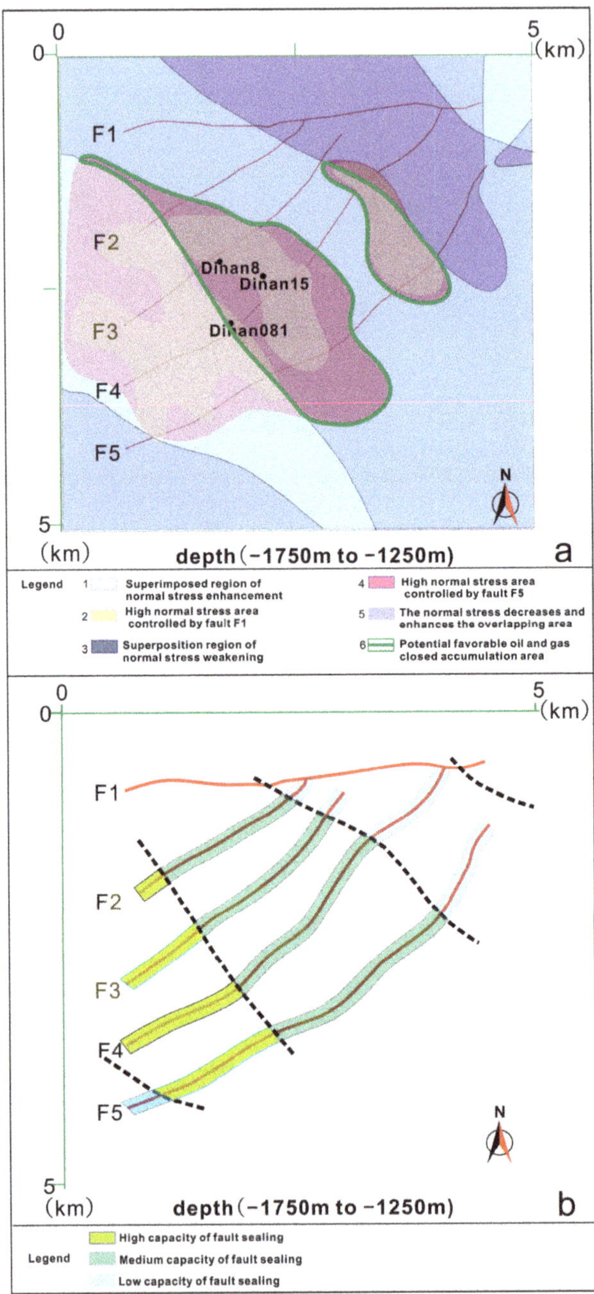

Figure 11. (**a**) Superposition analysis chart of numerical simulation based on normal stress change of the fault plane. (**b**) Schematic diagram of classification and evaluation of fault sealing based on normal stress intensity of fault plane.

6. Conclusions

The stress analysis of logging data was applied to the sealing evaluation of a strike slip fault, which has good applicability and practicability. The research results show that the stress regime corresponding to different depths of fault planes has an important influence on fault sealing. Under the condition of similar buried depth and clay content of fault rock, the fault in reverse fault stress regime has the best sealing performance, followed by strike-slip fault stress regime, and the lowest sealing performance under normal fault stress regime. Comprehensive use of logging stress analysis to evaluate the sealing property of the strike-slip fault can improve the accuracy of fault sealing evaluation and is applied to the evaluation of fault sealing of normal fault and reverse fault.

The normal stress of a fault plane is an important factor to determine the sealing capacity of the fault. In order to accurately calculate the normal stress on the fault plane, it is necessary to comprehensively calculate the normal stress components of the vertical stress and horizontal stress on the fault plane, which is of great significance to the evaluation of fault sealing ability. Neglecting the compaction effect of horizontal stress on fault rock mass will lead to the low evaluation value of fault-sealing performance. The larger the fault dip angle is, the larger the deviation is. For strike-slip faults with large dip angles, this error is not acceptable.

The orientation of the maximum horizontal stress near the fault plane is also a factor affecting the sealing capacity of the fault. Because the orientation of the horizontal stress near the fault is anisotropic, it is not easy to determine its orientation in the case of fewer data. As a result, it is not easy to determine the degree of its influence on fault sealing. Therefore, it brings potential errors for accurate evaluation of fault sealing, and the errors of different positions are not easy to determine.

Author Contributions: Conceptualization, J.J. and K.W.; methodology, J.J. and Y.P.; software, J.J. and Y.P.; validation, W.G.; formal analysis, Y.L.; investigation, T.L.; writing—original draft preparation, J.J. and Y.P.; writing—review and editing, J.J. and K.W. All authors have read and agreed to the published version of the manuscript.

Funding: This research was supported by the "Strategic Priority Research Program of the Chinese Academy of Sciences" (NO: XDA14010301) and "the National Natural Science Foundation of China" (NO: 41872143). The APC was funded by the "Strategic Priority Research Program of the Chinese Academy of Sciences" (NO: XDA14010301).

Acknowledgments: The authors want to thank the Xinjiang Oilfield Company (PetroChina) for providing the original data.

Conflicts of Interest: The authors declare no conflict of interest. The funders had no role in the design of the study; in the collection, analyses, or interpretation of data; in the writing of the manuscript, or in the decision to publish the results.

References

1. Harding, T.P.; Tuminas, A.C. Structural interpretation of hydrocarbon traps sealed by basement normal block faults at stable flank of foredeep basins and at rift basins. *AAPG Bull.* **1989**, *73*, 812–840. [CrossRef]
2. Berg, R.R.; Avery, A.H. Sealing properties of tertiary growth faults, Texas Gulf Coast. *AAPG Bull.* **1995**, *79*, 375–392. [CrossRef]
3. Knipe, R.J. Juxtaposition and seal diagrams to help analyze fault seals in hydrocarbon reservoirs. *AAPG Bull.* **1997**, *81*, 187–195. [CrossRef]
4. Huang, L.; Liu, C.-Y. Evolutionary characteristics of the sags to the east of Tan–Lu Fault Zone, Bohai Bay Basin (China): Implications for hydrocarbon exploration and regional tectonic evolution. *J. Asian Earth Sci.* **2014**, *79*, 275–287. [CrossRef]
5. Pei, Y.W.; Paton, D.A.; Knipe, R.J.; Wu, K.Y. A review of fault sealing behaviour and its evaluation in siliciclastic rocks. *Earth-Sci. Rev.* **2015**, *150*, 121–138. [CrossRef]
6. Xie, L.; Pei, Y.; Li, A.; Wu, K. Implications of meso- to micro-scale deformation for fault sealing capacity: Insights from the Lenghu5 fold-and-thrust belt, Qaidam Basin, NE Tibetan Plateau. *J. Asian Earth Sci.* **2018**, *158*, 336–351. [CrossRef]
7. Allan, U.S. Model for hydrocarbon migration and entrapment within faulted structures. *AAPG Bull.* **1989**, *73*, 803–811. [CrossRef]
8. Gibson, R.G. Fault-zone seals in siliciclastic strata of the Columbus Basin, offshore Trinidad. *AAPG Bull.* **1994**, *78*, 1372–1385. [CrossRef]
9. Yielding, B.F.G. Quantitative fault seal prediction. *AAPG Bull.* **1997**, *81*, 897–917. [CrossRef]

10. Chatterjee, R. Effect of normal faulting on in-situ stress: A case study from Mandapeta Field, Krishna-Godavari basin, India. *Earth Planet. Sci. Lett.* **2008**, *269*, 458–467. [CrossRef]

11. Yielding, G. Using probabilistic shale smear modelling to relate SGR predictions of column height to fault-zone heterogeneity. *Pet. Geosci.* **2012**, *18*, 33–42. [CrossRef]

12. Nicol, A.; Childs, C. Cataclasis and silt smear on normal faults in weakly lithified turbidites. *J. Struct. Geol.* **2018**, *117*, 44–57. [CrossRef]

13. Perkins, H. Fault-closure type fields, Southeast Louisiana. *Gulf Coast Assoc. Geol. Soc. Trans* **1961**, *11*, 177–196.

14. Smith, D.A. Theoretical considerations of sealing and non-sealing faults. *AAPG Bull.* **1966**, *50*, 363–374. [CrossRef]

15. Weber, K.J.; Mandl, G.J.; Pilaar, W.F.; Lehner, B.V.F.; Precious, R.G. The role of faults in hydrocarbon migration and trapping in Nigerian growth fault structures. In Proceedings of the Offshore Technology Conference, Houston, TX, USA, 3–8 January 1978; pp. 2643–2653.

16. Smith, D.A. Sealing and nonsealing faults in Louisiana Gulf Coast salt basin. *AAPG Bull.* **1980**, *64*, 145–172. [CrossRef]

17. Bouvier, J.D.; Sijpesteijn, C.H.; Kluesner, D.F. Three-dimensional seismic interpretation and fault sealing investigations, Nun River Field, Nigeria. *AAPG Bull.* **1989**, *739*, 1397–1414.

18. Fulljames, J.; Zijerveld, L.; Franssen, R. Fault seal processes: Systematic analysis of fault seals over geological and production time scales. *Nor. Petrol. Soc. Spec. Publ.* **1997**, *7*, 51–59. [CrossRef]

19. Lindsay, N.G.; Murphy, F.C.; Walsh, J.J.; Watterson, J.; Flint, S.; Bryant, I. Outcrop studies of shale smears on fault surface. *Int. Assoc. Sedimentol. Spec. Publ.* **1993**, *15*, 113–123.

20. Yielding, G. Shale gouge ratio-calibration by geohistory. *Nor. Petrol. Soc. Spec. Publ.* **2002**, *11*, 1–15. [CrossRef]

21. Çiftçi, N.B.; Giger, S.B.; Clennell, M.B. Three-dimensional structure of experimentally produced clay smears: Implications for fault seal analysis. *AAPG Bull.* **2013**, *97*, 733–757. [CrossRef]

22. Knipe, R.J.; Jones, G.; Fisher, Q.J. Faulting, fault sealing and fluid flow in hydrocarbon reservoirs: An introduction. *Geol. Soc. Lond. Spéc. Publ.* **1998**, *147*. [CrossRef]

23. Lyu, Y.F.; Huang, J.S.; Fu, G. Quantitative study on fault sealing ability in sandstone and mudstone thin interbed. *Acta Petrol. Sin.* **2009**, *30*, 824–829. [CrossRef]

24. Fu, G.; Shi, J.J. Study of ancient displacement pressure of fault rock recovery and its sealing characteristics. *J. China Univ. Min. Technol.* **2013**, *42*, 996–1001. [CrossRef]

25. Lei, C.; Yuan, X.T. Quantitative characterization of fault lateral sealing capacity based on 3-D SGR model-A case from M field, Niger. *Oil Gas Geol.* **2019**, *40*, 1317–1323.

26. Fu, G.; Wang, H.R.; Hu, X.L. Modification and application of fault-reservoir displacement pressure differential method for vertical sealing of faults. *Acta Petrol. Sin.* **2014**, *35*, 687–691.

27. Lyu, Y.; Wang, W.; Hu, X.; Fu, G.; Shi, J.; Wang, C.; Liu, Z.; Jiang, W. Quantitative evaluation method of fault lateral sealing. *Pet. Explor. Dev.* **2016**, *43*, 340–347. [CrossRef]

28. Şengör, A.M.C.; Natal'In, B.A.; Burtman, V.S. Evolution of the altaid tectonic collage and palaeozoic crustal growth in Eurasia. *Nat. Cell Biol.* **1993**, *364*, 299–307. [CrossRef]

29. Windley, B.F.; Alexeiev, D.; Xiao, W.; Kröner, A.; Badarch, G. Tectonic models for accretion of the Central Asian orogenic belt. *J. Geol. Soc.* **2007**, *164*, 31–47. [CrossRef]

30. Chen, S.; Guo, Z.; Pe-Piper, G.; Zhu, B. Late Paleozoic peperites in West Junggar, China, and how they constrain regional tectonic and palaeoenvironmental setting. *Gondwana Res.* **2013**, *23*, 666–681. [CrossRef]

31. Xu, X.; Jiang, N.; Li, X.H.; Wu, C.; Qu, X.; Zhou, G.; Dong, L.H. Spatial-temporal framework for the closure of the Junggar Ocean in central Asia: New SIMS zircon U-Pb ages of the ophiolitic mélange and collisional igneous rocks in the Zhifang area, East Junggar. *J. Asian Earth Sci.* **2015**, *111*, 470–491. [CrossRef]

32. He, D.F.; Zhang, L.; Wu, S.T. Tectonic evolution stages and features of the Junggar Basin. *Oil Gas Geol.* **2018**, *39*, 845–861.

33. Allen, M.B.; Vincent, S.J. Fault reactivation in the Junggar region, northwest China: The role of basement structures during mesozoic-cenozoic compression. *J. Geol. Soc.* **1997**, *154*, 151–155. [CrossRef]

34. Carroll, A.R.; Liang, Y.H.; Graham, S.A.; Xiao, X.C.; Hendrix, M.S.; Chu, J.C.; McKnight, C.L. Tectonics of Eastern Asia and Western Pacific continental margin Junggar Basin, Northwest China: Trapped late paleozoic ocean. *Tectonophysics* **1990**, *181*, 1–14. [CrossRef]

35. Cao, J.; Jin, Z.; Hu, W.; Zhang, Y.; Yao, S.; Wang, X.; Zhang, Y.; Tang, Y. Improved understanding of petroleum migration history in the Hongche fault zone, Northwestern Junggar Basin (northwest China): Constrained by vein-calcite fluid inclusions and trace elements. *Mar. Pet. Geol.* **2010**, *27*, 61–68. [CrossRef]

36. Yang, D.S.; Chen, S.J.; Li, L. Hydrocarbon origins and their pooling characteristics of the Kelameili gas field. *Nat. Gas Ind.* **2012**, *32*, 27–31. [CrossRef]

37. Xiao, W.; Windley, B.F.; Han, C.; Liu, W.; Wan, B.; Zhang, J.; Ao, S.; Zhang, Z.; Song, D. Late Paleozoic to early Triassic multiple roll-back and oroclinal bending of the Mongolia collage in Central Asia. *Earth-Sci. Rev.* **2018**, *186*, 94–128. [CrossRef]

38. Yu, Y.; Wang, X.; Rao, G.; Wang, R. Mesozoic reactivated transpressional structures and multi-stage tectonic deformation along the Hong-Che fault zone in the northwestern Junggar Basin, NW China. *Tectonophysics* **2016**, *679*, 156–168. [CrossRef]

39. Wang, Q.; Shu, L.; Charvet, J.; Faure, M.; Ma, H.; Natal'In, B.; Gao, J.; Kroner, A.; Xiao, W.; Li, J.; et al. Understanding and study perspectives on tectonic evolution and crustal structure of the Paleozoic Chinese Tianshan. *Episodes* **2010**, *33*, 242–266. [CrossRef]

40. Wilhem, C.; Windley, B.F.; Stampfli, G.M. The Altaids of Central Asia: A tectonic and evolutionary innovative review. *Earth-Science Rev.* **2012**, *113*, 303–341. [CrossRef]

41. Wu, K.; Paton, D.; Zha, M. Unconformity structures controlling stratigraphic reservoirs in the north-west margin of Junggar basin, North-west China. *Front. Earth Sci.* **2012**, *7*, 55–64. [CrossRef]

42. Wu, K.Y.; Zha, M.; Wang, X.L.; Qu, J.X.; Chen, X. Further researches on the tectonic evolution and dynamic setting of the Junggar Basin. *Acta Geosci. Sin.* **2005**, *26*, 217–222. [CrossRef]

43. Charvet, J.; Shu, L.; Charvet, L.S. Paleozoic structural and geodynamic evolution of eastern Tianshan, NW China: Welding of the Tarim and Junggar plates. *Episodes* **2007**, *30*, 162–186.

44. Ding, W.L.; Jin, Z.J.; Zhang, Y.J.; Zeng, J.H.; Wang, H.Y. Experimental simulation of faults controlling oil migration and accumulation in the central part of Junggar Basin and its significance for petroleum geology. *Earth Sci.* **2011**, *31*, 73–82. [CrossRef]

45. Zheng, M.; Tian, A.; Yang, T. Structural evolution and hydrocarbon accumulation in the eastern Junggar Basin. *Oil Gas Geol.* **2018**, *39*, 907–916. [CrossRef]

46. Liang, Y.; Zhang, Y.; Chen, S.; Guo, Z.; Tang, W. Controls of a strike-slip fault system on the tectonic inversion of the Mahu depression at the northwestern margin of the Junggar Basin, NW China. *J. Asian Earth Sci.* **2020**, *198*, 104229. [CrossRef]

47. Sibson, R.H. Fault rocks and fault mechanisms. *J. Geol. Soc.* **1977**, *133*, 191–213. [CrossRef]

48. Childs, C.; Manzocchi, T.; Walsh, J.J.; Bonson, C.G.; Nicol, A.; Schöpfer, M.P. A geometric model of fault zone and fault rock thickness variations. *J. Struct. Geol.* **2009**, *31*, 117–127. [CrossRef]

49. Shi, L.; Chu, W.; Deng, S. Catalytic properties of Cu-Co catalysts supported on HNO3-pretreated CNTs for higher-alcohol synthesis. *J. Nat. Gas Chem.* **2011**, *20*, 48–52. [CrossRef]

50. Vrolijk, P.J.; Urai, J.L.; Kettermann, M. Clay smear: Review of mechanisms and applications. *J. Struct. Geol.* **2016**, *86*, 95–152. [CrossRef]

51. Zhou, X.G. The study of fault closure by use of entry pressure and its application in North Tarim. *J. Geomech.* **1997**, *3*, 47–53.

52. Ma, J.H.; Sun, J.M. Calculation of formation stress using logging data. *Well Logging Technol.* **2002**, *26*, 347–351.

53. Liu, G.-L. A novel limiting strain energy strength theory. *Trans. Nonferrous Met. Soc. China* **2009**, *19*, 1651–1662. [CrossRef]

54. Rajabi, M.; Tingay, M.; King, R.; Heidbach, O. Present-day stress orientation in the Clarence-Moreton Basin of New South Wales, Australia: A new high density dataset reveals local stress rotations. *Basin Res.* **2016**, *29*, 622–640. [CrossRef]

55. Ziegler, M.; Rajabi, M.; Heidbach, O.; Hersir, G.P.; Ágústsson, K.; Árnadóttir, S.; Zang, A. The stress pattern of Iceland. *Tectonophysics* **2016**, *674*, 101–113. [CrossRef]

56. Vernik, L.; Zoback, M.D. Estimation of maximum horizontal principal stress magnitude from stress-induced well bore breakouts in the Cajon Pass scientific research borehole. *J. Geophys. Res.* **1993**, *30*. [CrossRef]

57. Yan, S.; Qiao, W. Discussions about In-situ stress calculation of sand shale formations using cross-dipole acoustic logs. *Well Logging Technol.* **2003**, *27*, 122–124. [CrossRef]

58. Rajabi, M.; Tingay, M.; Heidbach, O. The present-day state of tectonic stress in the Darling Basin, Australia: Implications for exploration and production. *Mar. Pet. Geol.* **2016**, *77*, 776–790. [CrossRef]

59. Gao, Z.Y.; Cui, J.G.; Feng, J.R. Modification mechanism of physical properties of deeply-buried sandstone reservoir due to the burial compaction and lateral extrusion in Kuqa Depression. *Geoscience* **2017**, *31*, 302–314.

60. Xia, L.; Liu, Z.; Li, W.; Yu, C.; Zhang, W. Initial porosity and compaction of consolidated sandstone in Hangjin Qi, North Ordos Basin. *J. Pet. Sci. Eng.* **2018**, *166*, 324–336. [CrossRef]

61. Tingay, M.; Hillis, R.; Morley, C.; Swarbrick, R.; Okpere, E. Variation in vertical stress in the Baram Basin, Brunei: Tectonic and geomechanical implications. *Mar. Pet. Geol.* **2003**, *20*, 1201–1212. [CrossRef]

62. Xie, G.A. New method to calculate the maximum and minimum horizontal stress using log data. *Well Logging Technol.* **2005**, *29*, 82–90. [CrossRef]

63. Bretan, P.; Yielding, G. Using buoyancy pressure profiles to assess uncertainty in fault seal calibration. *AAPG Hedberg Ser. Tulsa* **2005**, *2*, 151–162.

64. Yielding, G.; Bretan, P.; Freeman, B. Fault seal calibration: A brief review. *Geol. Soc. Lond. Spéc. Publ.* **2010**, *347*, 243–255. [CrossRef]

65. Mount, V.S.; Suppe, J. Present-day stress orientations adjacent to active strike-slip faults: California and Sumatra. *J. Geophys. Res. Space Phys.* **1992**, *97*, 11995. [CrossRef]

66. Tingay, M.; Bentham, P.; De Feyter, A.; Kellner, A. Present-day stress-field rotations associated with evaporites in the offshore Nile Delta. *GSA Bull.* **2010**, *123*, 1171–1180. [CrossRef]

67. Kingdon, A.; Fellgett, M.W.; Williams, J.D. Use of borehole imaging to improve understanding of the in-situ stress orientation of Central and Northern England and its implications for unconventional hydrocarbon resources. *Mar. Pet. Geol.* **2016**, *73*, 1–20. [CrossRef]

68. Martin, C.; Chandler, N. Stress heterogeneity and geological structures. *Int. J. Rock Mech. Min. Sci. Géoméch. Abstr.* **1993**, *30*, 993–999. [CrossRef]

69. Tang, X.M.; Cheng, N.Y.; Cheng, C.H. Identifying and estimating formation stress from borehole monopole and cross-dipole acoustic measurements. In Proceedings of the SPWLA 40th Annual Logging Symposium, Oslo, Norway, 30 May–3 June 1999.

70. Gudmundsson, A. Effects of Young's modulus on fault displacement. *C. R. Geosci.* **2004**, *336*, 85–92. [CrossRef]

71. Zoback, M.; Barton, C.; Brudy, M.; Castillo, D.; Finkbeiner, T.; Grollimund, B.; Moos, D.; Peska, P.; Ward, C.; Wiprut, D. Determination of stress orientation and magnitude in deep wells. *Int. J. Rock Mech. Min. Sci.* **2003**, *40*, 1049–1076. [CrossRef]
72. Su, S.; Stephansson, O. Effect of a fault on in situ stresses studied by the distinct element method. *Int. J. Rock Mech. Min. Sci.* **1999**, *36*, 1051–1056. [CrossRef]
73. Zoback, M.D.; Mount, V.S.; Suppe, J.; Eaton, J.P.; Healy, J.H.; Oppenheimer, D.; Reasenberg, P.; Jones, L.; Raleigh, C.B.; Wong, I.G.; et al. New evidence on the state of stress of the San Andreas fault system. *Science* **1987**, *238*, 1105–1111. [CrossRef] [PubMed]
74. Müller, B.; Zoback, M.L.; Fuchs, K.; Mastin, L.; Gregersen, S.; Pavoni, N.; Stephansson, O.; Ljunggren, C. Regional patterns of tectonic stress in Europe. *J. Geophys. Res. Space Phys.* **1992**, *97*, 11783–11803. [CrossRef]
75. Wu, K.Y.; Wang, X.L.; Cui, D. Structural characteristics and fluid effects of Nanbaijiantan fault zone. *Coal Geol. Explor.* **2012**, *40*, 5–11.
76. Chen, Y.Q.; Zhou, X.G. Sealing factors of faults and their sealing effects. *Petrol. Explor. Dev.* **2003**, *30*, 38–40. [CrossRef]
77. Antonellini, A.A.M. Effect of Faulting on Fluid Flow in Porous Sandstones: Geometry and Spatial Distribution. *AAPG Bull.* **1995**, *79*, 642–670. [CrossRef]

energies

Article

Influence of Hydrogeological Investigation's Accuracy on Technology of Shaft Sinking and Design of Shaft Lining—Case Study from Southern Poland

Kajetan d'Obyrn [1], Paweł Kamiński [2],* and Jacek Motyka [1]

[1] Faculty of Geology, Geophysics and Environmental Protection, AGH University of Science and Technology, Mickiewicza 30 Av., 30-059 Kraków, Poland; dobyrn@agh.edu.pl (K.d.); motyka@agh.edu.pl (J.M.)
[2] Faculty of Mining and Geoengineering, AGH University of Science and Technology, Mickiewicza 30 Av., 30-059 Kraków, Poland
*   Correspondence: pkamin@agh.edu.pl

Abstract: Accuracy of hydrogeological and geotechnical investigation in place of shaft sinking is a key factor for selection of sinking method and design of the shaft lining. The following work presents the influence of the rising level of accuracy of geological data gathered in the area of shaft sinking in the Silesian Coal Basin and technical projects of shaft lining and technology of its sinking, which have been changing over the years. The initial project of the shaft was repeatedly modified. Each modification eventuated in rising requirements for the shaft lining, such as increasing its thickness or changing concrete class. It has become necessary to use additional methods of reinforcing rock mass around the shaft.

Keywords: mine shaft; hydrogeological investigation; shaft lining design; shaft lining reinforcement

check for **updates**

Citation: d'Obyrn, K.; Kamiński, P.; Motyka, J. Influence of Hydrogeological Investigation's Accuracy on Technology of Shaft Sinking and Design of Shaft Lining—Case Study from Southern Poland. *Energies* **2021**, *14*, 2050. https://doi.org/10.3390/en14082050

Academic Editor: Feng Dai

Received: 4 March 2021
Accepted: 6 April 2021
Published: 7 April 2021

Publisher's Note: MDPI stays neutral with regard to jurisdictional claims in published maps and institutional affiliations.

Copyright: © 2021 by the authors. Licensee MDPI, Basel, Switzerland. This article is an open access article distributed under the terms and conditions of the Creative Commons Attribution (CC BY) license (https://creativecommons.org/licenses/by/4.0/).

1. Introduction

Despite investing in renewable energy sources and development of other energy sources, Polish energy is still coal based. Poland also has extensive resources of coking coal, which is necessary for steel production. These factors cause a need for coal mine operation, because of the coal demand. Economic mine operation requires maintenance, modernization, or elongation of existing mine shafts and even building new. A new mine shaft was designed in the Silesian Coal Basin in southern Poland to raise the effectiveness of a coal mine. The process of the shaft design is presented in the following article. The location of boreholes, with respect to mine workings in coal seam no. 207, geological faults, and borders of water hazard area are presented in Figure 1.

The selection of the shaft sinking method is based on hydrogeological conditions in the chosen place [1]. A particularly important factor is the number of aquifers and their parameters, such as hydraulic contact between them, chemical composition of underground water in different aquifers, and occurrence of fault zones [2]. A common method of geological and hydrogeological investigation is research drilling in the vicinity of the designed shaft [3]. This method is required by Polish law [4] as well as the Polish Standard [5]. The following work presents an example of the evolution of the technical project of a mine shaft, caused by the rising level of accuracy of hydrogeological data collected in two boreholes, drilled at different time intervals as well as the results of mining operations on the mine level.

Boreholes A and B were drilled consecutively in 2007 and 2014. Based on data collected from these wells, the occurrence of Tertiary, Quaternary, Triassic, and Carboniferous formations was specified.

Data gathered in the borehole A showed that Quaternary formation is represented by medium and coarse-grained sand, loamy sand, sandy, silty clay and clay loam, silt, and clay. The bottom part of the quaternary formation is formed by a clay–sand–gravel mixture, rock and loam rubble, and gravel. Quaternary formation is covered with 0.2 m thick layer of sandy soil and humus. The thickness of the Quaternary formation is equal 40.5 m.

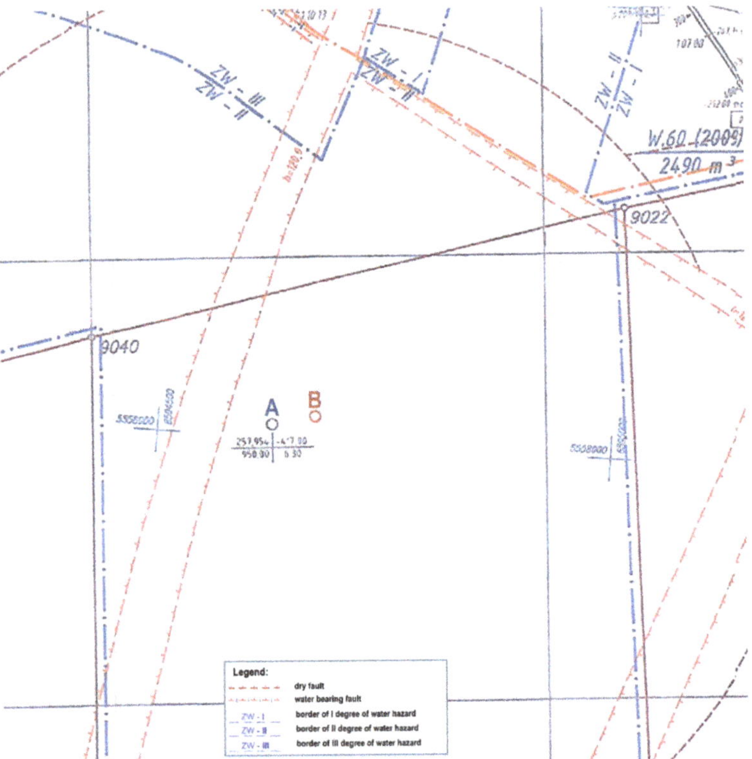

Figure 1. A map of mine workings in the coal seam no. 207 with geological faults, borders of the water hazard area, and boreholes presented.

The tertiary formation consists of clay divided by a layer of micrite limestone at the depth of 94.1–96.7 m and knobby limestone in its bottom part. The thickness of the Tertiary formation is 71.6 m.

The Triassic formation is represented by carbonate rocks—limestone, dolomite, and marl, laying on clay. The total thickness of Triassic formation equals 125.9 m. In both boreholes A and B, Middle (112.1–186.5 m below the surface) and Early Triassic (185.5–238.0 m below the surface) was found.

The Carboniferous formation consists of the Libiąż layers (Westphalian D) and Łaziska layers (Westphalian B and C). Westphalian D, with a thickness of 171.5 m, is represented by thick shoals of coarse-grained, poorly cemented and fractured sandstone containing gravel and pebbles, and fine-grained sandstone delaminated with conglomerate, mudrock, and seams of coal. The thickness of the Westphal C formation is 353.4 m and Westphal D is 187.2 m. They are formed by coarse-grained sandstone with gravel, medium and fine-grained sandstone with mudrock, and several seams of coal.

Tectonics of the designed shaft region, as well as the whole area of the coal deposit, is extremely rich. Numerous faults occur, characterized by throw up to 180 m. Traces of fault lines are located on the N-S, NE-SW, NW-SE, and W-E directions. Seam extent direction is between S-N and NWW-SEE, its dip is oriented to E and NE, and its angle is from 3° to 10°. In the distance of about 350 m on the north from the borehole A, there is a fault of W-E direction and a south-oriented throw with a value between 100 and 140 m. In this borehole, at the depth of 109.80 m, contact between dolomite and limestone breccia was located in the Tertiary formation, which is probably a fault of a 40° dip and a 2.3 m thick zone of breccia. It is also plausible that two caverns filled with clay were drilled through. A geological profile of the rock mass in the vicinity of the designed shaft and boreholes is presented in Figure 2.

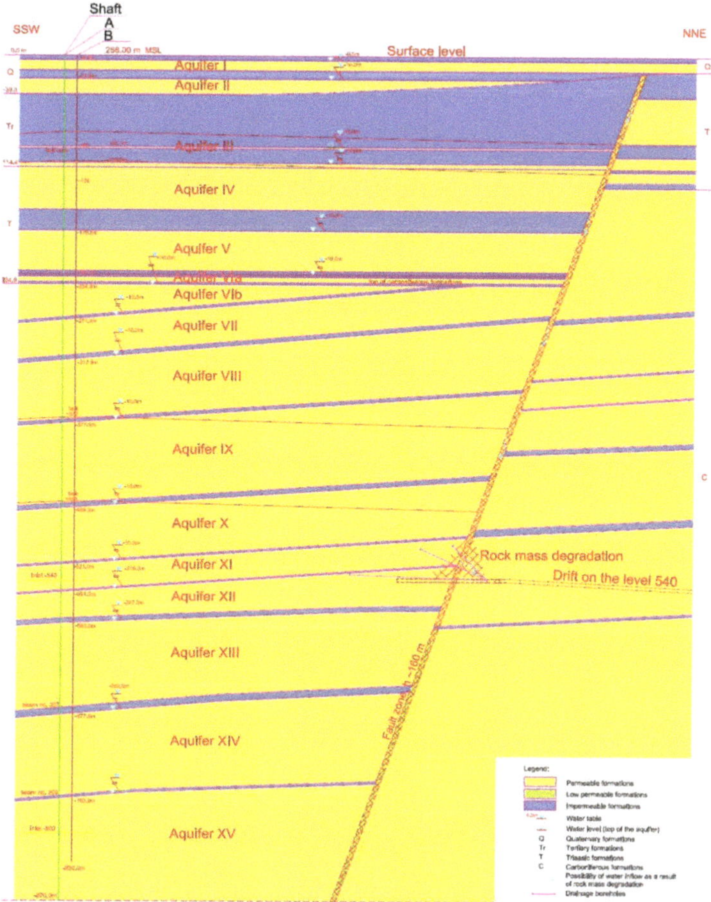

Figure 2. Geological profile in the vicinity of the designed shaft and boreholes.

2. Results

2.1. Borehole A—Stage I

The following hydrogeological tests were conducted in the borehole A:

- Pumping tests,
- Tests with tube samplers at the depths of 550.0–560.0 m, 630.0–635.0 m, 745.0–750 m, 855.0–860.0 m, and 925.0–930.0 m,
- Determination of following parameters in groundwater samples: pH, carbonate hardness, water hardness, total alkalinity, content of the aggressive CO_2, So_4, Mg, Cl, NH_4, NO_3, NO_2, HCO_3, Ca, Fe, K, Na, Mn, SiO_2 ion content, dispersion, dry residue, and assessment of level of water aggressiveness and corrosivity,

- Samples of sandstones layer over the minimum 1.5 m thick coal seams were taken from the drill core and their open porosity, seepage, and chemical composition of pore waters were tested in the laboratory,
- Tests of radioactivity (concentration of radium isotopes) and occurrence of barium ions were conducted for groundwater pumped from a depth of over 500.0 m.

The hydraulic conductivity of aquifers was calculated using the Dupuit (1) formula, and the cone of depression size was determined based on the Sichtard formula (1) using the successive approximation method.

$$k = \left(\frac{R}{3000 * s} \right)^2 = \frac{0,366 * Q * lg\frac{R}{r}}{m * s}, \frac{m}{s} \tag{1}$$

where:

    k—hydraulic conductivity,
    R—depression cone,
    Q—discharge well,
    r—well radius,
    m—aquifer thickness,
    s—drawdown.

Based on the obtained value of the hydraulic conductivity, water inflows (Q) of aquifers was calculated with the so-called "great well" method, using the following formula (2):

$$Q = 2.73 * \frac{k * m * s}{lgR_0 - lgr_0}, \frac{m^2}{s} \tag{2}$$

where:

$$R = 3000 * s * \sqrt{k}, m$$

$$R_0 = R + r, m$$

Pumping tests were carried out in boreholes, without the use of observation (control) boreholes. In such a case, the cone of depression area calculation was done using the empirical Sichtard formula (1), which is widely accepted by hydrogeologists [6,7]. The Sichtard formula provides an approximated value of the cone of depression radius, of which the logarithm is often used in calculations of water inflow, thus its influence on the final result is minimized.

Based on data gathered from a 950-m-deep borehole A, drilled in 2007, 148 different rock layers were defined, as well as four aquifer systems:

- The Quaternary aquifer system at the depth of 4.50–16.00 m below the surface, in which the pressure head is equal to 0.026 MPa.
- The Triassic aquifer system at the depth of 112.10–222.35 m below the surface, in which the pressure head is 0.915 MPa and the average value of the linear slit index is 6.7.

There are two aquifers in Carboniferous formation:

- The Upper Carboniferous aquifer system at the depth of 234.00–516.00 m below the surface, in which the pressure head is 1.225 MPa and the average am value of the linear slit index is 5.4.
- The Lower Carboniferous aquifer system at the depth of 516.00–947.24 m below the surface, in which the pressure head is 3.376 MPa and the average value of the linear slit index is 4.2.

An example of a graph showing the linear slit index is presented in Figure 3. Similar graphs were prepared for the Carboniferous formations of Westphalian B, C, and D.

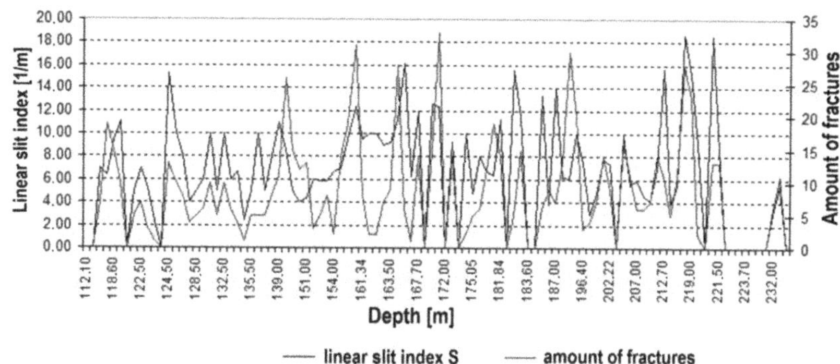

Figure 3. Graph of the average fracture amount per meter in Triassic formation in the borehole A.

Groundwater was also sampled in borehole A on different levels. The analysis of samples confirmed a phenomenon of rising groundwater mineralization with depth (Figure 4.). The value of the total mineralization rises from about 0.35 g/L in overburden formations, through 3.25 g/L at the depth about 560 m, up to 4.9 at the depth of 700 m.

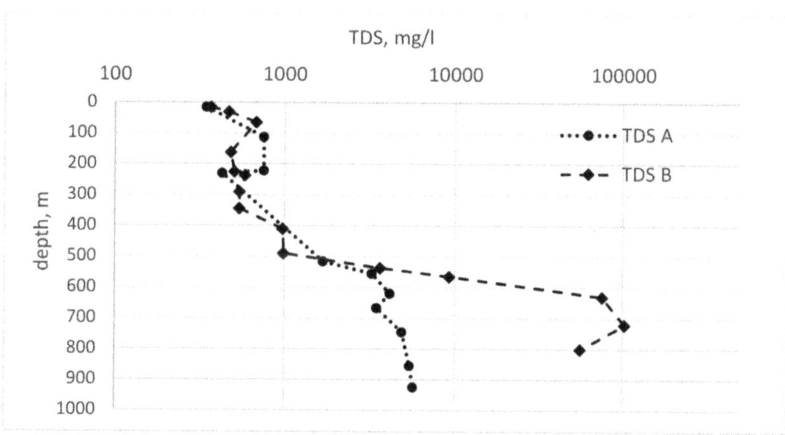

Figure 4. Total dissolved solids (TDS) profile in boreholes A and B.

Key hydrogeological parameters of different formations are shown in Table 1.

Calculations revealed that the maximum expected total water inflow is 14 m^3/min. Considering fractured and cavernous rock mass (Triassic formation) and pressurized groundwater in some aquifers, the possibility of sudden uncontrolled water inflow to the shaft was recognized as a real threat. Depth intervals in the permeable Carboniferous formations in which the average value of the linear slit index is greater than or equal to double the average value of this index for the aquifer were specified, as well as areas of increased values of RQD, which might be an indicator of increased accumulation of water. Data gathered during drilling show a risk of fault zones occurrence on depths of 109.8, 461.8, 462.8, and 570.5 m below the surface.

Based on the profile of the Quaternary formation, a 20m-deep diagram wall was proposed as a primary shaft lining in this section. The diagram wall could be built without lowering the groundwater level. It is a construction element and it is made of adhering sections. The occurrence of the Tertiary loam below the Quaternary formation allowed the shaft at this section to be sunk without using special methods.

Table 1. Results of hydrogeological investigation in boreholes A and B.

Aquifer System/Aquifer	Depth	Water Table		Hydraulic Conductivity k		Water Inflow	
		A	B	A	B	A	B
	m	m	m	m/s	m/s	m^3/min	m^3/min
Q/I	5.5–16.0	1.9	1.55	6.64×10^{-6}	4.09×10^{-6}	0.121	0.065
Q/II	24.2–39.5	dry rocks	10.90	dry rocks	1.02×10^{-5}	-	0.310
Tr/III	94.2–96.2		13.40		3.34×10^{-5}		0.368 [1]
T/IV	111.0–159.4		13.00		1.88×10^{-6}		1.408 [1]
T/V	180.6–222.3	20.6	17.20	1.84×10^{-6}	1.28×10^{-6}	2.80	0.938 [1]
T/VIa	225.2–235.0		132.15		1.16×10^{-6}		0.105 [1]
C/VIb	236.0–269.7		116.90		2.14×10^{-6}		0.848 [1]
C/VII	272.5–309.7		127.50		8.34×10^{-7}		0.463 [1]
C/VIII	314.0–375.1	111.5	130.10	1.05×10^{-6}	1.83×10^{-7}	7.43	0.249 [1]
C/IX	378.6–461.5		127.00		2.40×10^{-7}		0.546 [1]
C/X	466.4–522.0		199.30		2.60×10^{-7}		0.386
C/XI	524.1–550.7		200.20		1.49×10^{-7}		0.120
C/XII	552.4–579.7		217.35		3.82×10^{-7}		0.292
C/XIII	584.7–671.7	178.4	228.40	1.69×10^{-7}	2.66×10^{-8}	3.45	0.103
C/XIV	678.7–762.3		264.00		1.27×10^{-8}		0.057
C/XV	765.0–830.0		174.40		7.48×10^{-8}		0.370
					$\Sigma=$	13.801	5.853

[1] possible sudden water inflows of high amount from faults and fractures zones.

The field and laboratory investigation conducted revealed that Triassic and Carboniferous aquifer systems are characterized with good and very good filtration properties. The Triassic aquifer is of fracture-karstic type, while the Upper Carboniferous aquifer is characterized by a high fracturing level. Both Triassic and Carboniferous formations are fractured. Groundwater in aquifers is under high hydrostatic pressure, while the rock mass has good collector properties. The conducted calculations shows that water inflow from the Triassic formation into the shaft during its sinking exceeds the acceptable value of 0.5 m^3/min. A similar situation is forecasted for the Carboniferous formation. This situation requires prior rock mass sealing.

Hydrogeological conditions determined the design of the shaft lining and shaft sinking method. The value of water inflow into the shaft requires insulating the Upper Carboniferous, Quaternary, and Tertiary aquifer systems by using a water-tight shaft lining. General principles of rock mass sealing [8] were adapted to local conditions. The proposed construction of the shaft lining consists of hydro-insulating screens, using modified loam as a base of hydro-insulating solution [9]. Elements of the hydro-insulation include eight control-injection boreholes, which are located on the circle around the shaft outline at the distance of no less than 4–5 m and no higher than 10–12 m from the outline. In case of discontinuity of the shaft insulation, there is the possibility of supplementary rock mass sealing behind the lining from the shaft heading [10]; however, such a situation should not happen.

Some rock layers of the Carboniferous formation, especially sandstone of the Libiąż layers, is characterized with low values of strength. Fractures in fault zones, high hydrostatic pressure, and a low strength value of rock mass might affect rock falling of the shaft walls and expanding of the shaft heading area. In such situations, a web depth should be reduced to 2–3 m. After exposure of the fractured areas in the shaft heading, shaft walls should be supported with a temporary lining or covered with steel segments, or any other way of preventing rocks from falling down from the walls should be used. For certain rock layers, it is recommended to prevent them from contact with water, because of their tendency for dripping (basing on the test of Skutta [11]) and to reduce their exposure time to a maximum of 4 h.

A concept of shaft lining at this stage comprises the following lining solutions:

- Multilayer lining with hydro insulation:
  - Concrete + concrete panels lining,
  - Concrete + concrete lining,
- Concrete + concrete panels complex lining,
- Single-layer concrete lining with drainage.

The designed panel lining consists of 0.25 m thick concrete panels made of C20/25 class concrete. One of the lining's section comprises C35/45 concrete panels. The primary concrete lining is to be made of C25/30 concrete of a thickness between 0.3 and 0.55 m. The permanent lining is designed as a 0.5–1.1-thick concrete layer from C20/25 to C35/45. The primary column of the complex lining is to be made of 0.5 m-thick (0.9 m at one section) C25/30 class concrete (C35/45 at the 0.9 m-thick section). The permanent lining is to be made of 0.25 m-thick concrete panels of C25/30 class concrete (locally C30/37). The remaining section of the shaft is to be supported with a single-layer lining of C20/25 to C30/37 class concrete with a thickness between 0.75 and 0.85 m.

2.2. Borehole B—Stage II

Field and lab investigations of borehole A revealed difficult hydrogeological and engineering conditions in the profile of the designed shaft. As a proper hydrogeological investigation is crucial for the shaft design, additional research should be conducted [12]. To provide more accurate data of geological conditions, borehole B was drilled in 2014 to the depth of 830 m, at the distance of 13m from the designed shaft. An additional hydrogeological investigation was also related to the proposition of change of the shaft sinking method—from hydro-insulating shields to ground freezing. Similar to the borehole A drilling process, drilling fluid loss zones, tightening, and breakouts in the boreholes were analyzed. In situ tests revealed numerous fracture and tightening zones.

Hydrogeological tests in borehole B consist of drawing water using sludger in case the water inflow is less than 10l/min or a deep-well pump for greater inflow values. Pumping with the deep-well pump was conducted with one to three depressions, increased by about one-third of the lowered water head after the water table stabilization. The conducted research allowed to determine the depth and thickness of aquifers and their hydrostatic pressure and hydraulic conductivity. A scope of conducted tests was similar to the case of borehole A, but sampling density was significantly higher. One hundred and fifty-four geotechnical layers were specified. Basied on the conducted investigation, 16 aquifers were identified, of which 2 were in the Quaternary formation, 1 was in the Tertiary-Triassic formation, 2 were in the Triassic formation, and 10 were in the Carboniferous formation. Calculations of hydraulic conductivity and water inflow for aquifers were conducted using the same formulas as in the case of borehole A, but for 16, not 4 aquifers. The results of these calculations are presented in Table 1.

Comparison of the water mineralization revealed significantly greater values of TDS measured in borehole B. In the overburden, it is about 0.55 g/L. At the depth of 560 m, it is equal to 9.25 g/L, while in borehole A, it was only 3.25 g/L. Similarly, below 600 m, TDS varies between 75 and 103 g/L, while in borehole A, it was between 3.5 and 4.9 g/L. Such differences might be an effect of the fault zone occurrence, which allows for both freshwater descending from the overburden formations and brine ascending from deeper formations. Water flow is additionally stimulated by mine workings drainage in this area.

Protecting the shaft walls with a water-tight shaft lining as presented above in terms of calculated water inflows into the shaft might be very complicated [13]. The occurrence of weak rocks softening in water, together with the necessity for reducing water inflows into the shaft, forced the use of the special shaft sinking method, which is ground freezing, to the depth of 475 m. The depth of freezing is based on rock mass hydraulic conductivity, porosity, RQD, and the Protodyakonov coefficient [14]. Below this depth, a 280 m-long drainage hole is designed for the purpose of drainage in the shaft heading. Water is to be transported with the borehole to the mine drainage system. The designed borehole is protected by a steel casing pipe with a diameter of 150 m, and it is able to discharge water in an amount several times greater than its inflow into the shaft.

Ground freezing to the depth of 475 m is carried out using 40 freezing boreholes. Three additional boreholes were drilled for the purpose of control of the ground freezing process. The cooling capacity of the freezing installation is 4.0 MW. The designed shaft sinking method is realized using the shaft excavator in the frozen section. The drill and blast method is forbidden in this section, because of the risk of freezing boreholes damage done by an explosion. The remaining shaft section is to be sunk using the drill and blast method.

The following figures present elements of freezing installation, including the model of the primary system in Figure 5, the model of the secondary system in Figure 6, the cross section of freezing channel between refrigeration plant and shaft in Figure 7, and the cross section of freezing ring in Figure 8. The primary system includes three freezing units, so-called chillers PAC SAB 283 E eco, in which the freezing medium is ammonia. They are cooled by cooling towers Evapco AT 18-3M14. The refrigeration plant also consists of isolated pipes for medium transfer, a brine tank, a discharge tank, a water treatment station, eight pumps, and an armature. The secondary system is basically a channel comprising pipelines used to transfer coolant from refrigeration plant to freezing boreholes and back. The channel consists of two main parts, one of them is a freezing ring and the other one is a channel between the freezing ring and the refrigeration plant. The freezing channel contains two pipelines with a diameter of 350 mm—inlet and outlet pipelines and one venting pipeline with a diameter of 65 mm. Around the shaft outline, pipelines are circle-shaped. The freezing boreholes' pipes are connected directly to pipelines in the freezing ring.

More accurate results of the hydrogeological survey are reflected in the updated shaft lining design. Construction of the shaft lining is basically the same and consists of a multilayer lining with hydro insulation and a complex lining at the frozen section, and a complex and single-layer lining at the remaining section. However, the parameters of lining elements have changed, particularly concrete class, which in the case of panels is at least C30/37 and for concrete lining is between C30/37 and C40/50 class. In fault zones, reinforced C40/50 class concrete lining is designed. The thickness of the modernized shaft lining is between 0.85 and 1.2 m.

Figure 9 presents the layout of the shaft lining of two shaft sections. On the left, the top section of the shaft is shown, including the diagram wall, panel lining, backfill pipeline channel inlet, and shaft collar. On the right, the multilayer lining, comprising panels and concrete, is presented.

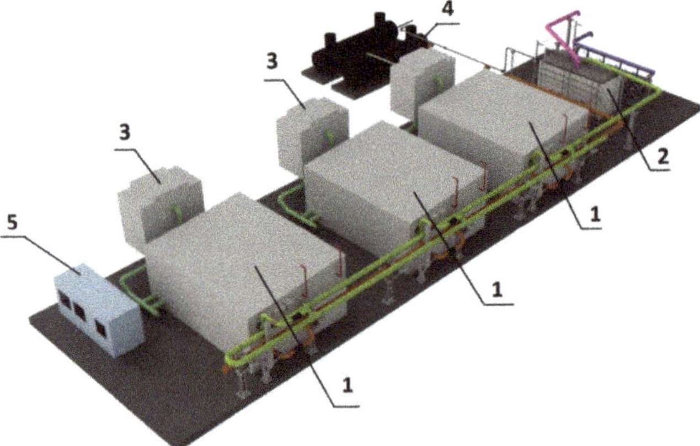

Figure 5. Model of the primary system of freezing installation: 1—cooling aggregates, 2—brine tank, 3—cooling towers, 4—discharge tank, 5—water treatment station.

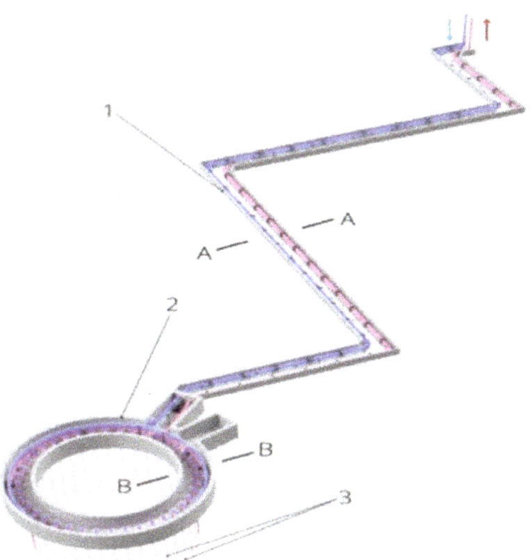

Figure 6. Secondary system of the freezing installation: 1—freezing channel, 2—freezing ring, 3—freezing boreholes.

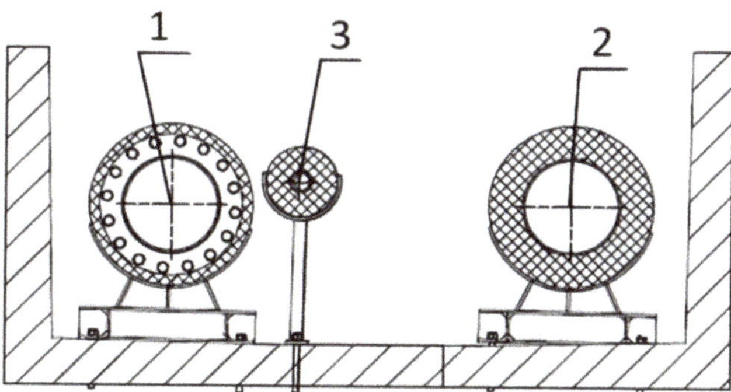

Figure 7. Cross section through the freezing channel (A-A): 1—inlet pipe, 2—outlet pipe, 3—vent pipe.

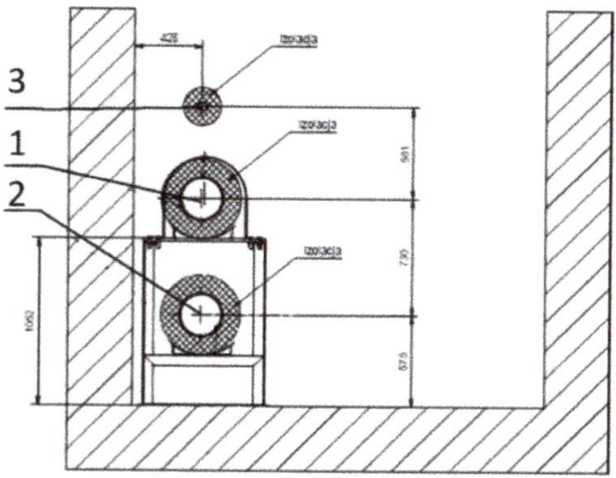

Figure 8. Cross section through the freezing ring (B-B): 1—inlet pipe, 2—outlet pipe, 3—vent pipe.

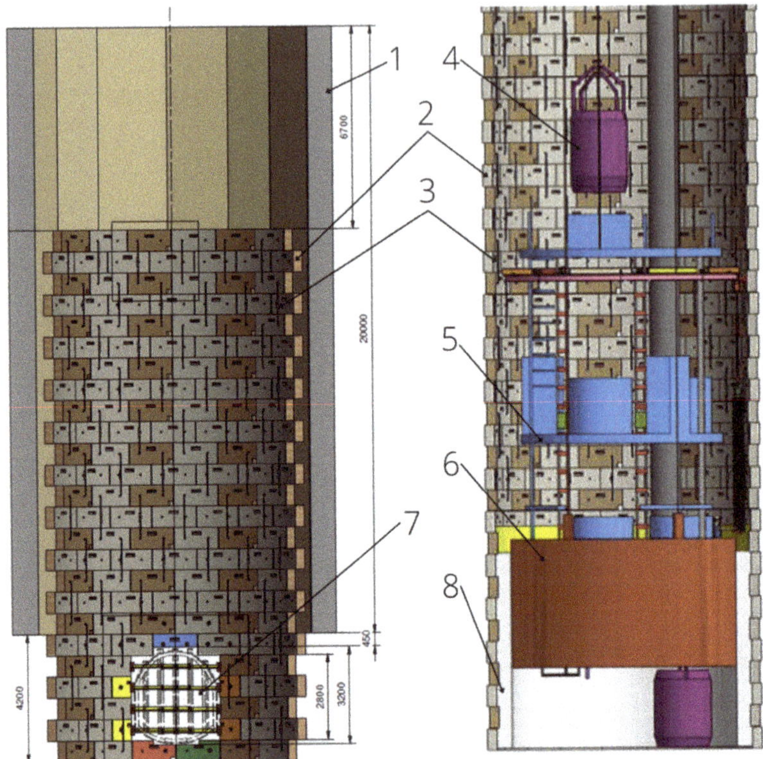

Figure 9. Layout of the shaft lining: 1—diagram wall, 2 –concrete panel (grey—standard panel; brown—panel with gaps for concreting; yellow, blue, red, green, orange—special construction panel), 3—panel fasteners, 4—sinking bucket, 5—suspended stage, 6—steel formwork, 7—inlet of backfill pipeline channel, 8—final concrete lining.

2.3. Mine Workings—Stage III

Since 2002, coal has been excavated from seams no. 207 and 209 on the north from the designed shaft location. Coal seam no. 207 was excavated at the distance of 1.7 km to the north and 1.3 km to the north-east from the borehole B, at depths between 670 and 550 m. The excavation height in this area is up to 4.6m. Coal seam no. 209 was excavated at the distance of 1.7 km to the north-east from the borehole B, and at the depth of 650 m. Seam's thickness is up to 4.3 m. Both coal seams were excavated using the longwall method with caving.

In 2019, the development of roadways at the level of 540 m started. New workings were to connect existing excavations with the shaft. The biggest challenge in this process was a fault zone of about 120 m throw, located about 300 m north of the shaft. Rock mass, in which newly developed workings are made, are characterized with low strength, which is effective in inflows of water and bulk material into workings. Values of these inflows varied in time. Drainage and test boreholes, drilled from mine workings, were tightened or filled up with loose rock material inside the fault zone. Boreholes were drilled from several adjacent excavations and allowed to investigate a 300-m-long part of the fault zone. Water pressure in boreholes did not exceed 1.7 MPa, and groundwater chemistry, based on samples from different boreholes, is typical both for low mineralization seepage water and highly mineralized water (brine), which is characteristic for the Carboniferous formation. The conducted tests revealed a rise of open porosity to about 15.7% and hydraulic conductivity to an average value of 1.9×10^{-6} m/s, several dozens of meters from the fault zone. Water inflow to one of the developed workings reached 2 m^3/min; however, its value decreased over the time. Peak values of water inflow were about 0.6 m^3/min. Water usually contained a significant amount of loam and sand material. The records of the drilling process of boreholes penetrating the fault zones also indicates strong and uneven water accumulation.

Roadways development at the level of 540 m allowed to gather supplementary data about water accumulation of the fault zones and possible connections between aquifer systems in different formations via the fault zone. In addition, data on water pressure in the fault zones and conditions of groundwater flow were collected.

Data on hydrogeological conditions, fault zones, and geotechnical conditions, gathered during the roadways development at the level of 540 m, caused a necessity for another modification of shaft lining construction. New construction of the designed shaft lining has to deal with difficult conditions identified during mining activity at the level of 540 m. The concept of the shaft lining construction did not change. Key differences between previous and new shaft lining construction occur in the area of fault zones and in the application of rock mass injection, which was not considered necessary in previous projects.

Distinctive changes in the original shaft lining project concern fault zones and the shaft section between the depths of 215.4 and 238.0 m, which is a zone of extensive stress. The designed shaft lining in this section comprises steel sections and a concrete layer and is able to transfer loads of 4.6 MPa. In the area of fault zones, a similar solution of a steel and concrete complex shaft lining, able to transfer loads of 4.0 MPa, was designed. In the described sections, concrete panels were replaced with steel sections of different types, depending on the load to transfer. The shaft lining on these sections also consists of q 0.73 m-thick layer of C50/60 class concrete. The layout of the steel sections ring is presented in Figure 10. Orange and blue panels are so-called closing segments, which are assembled in the last step.

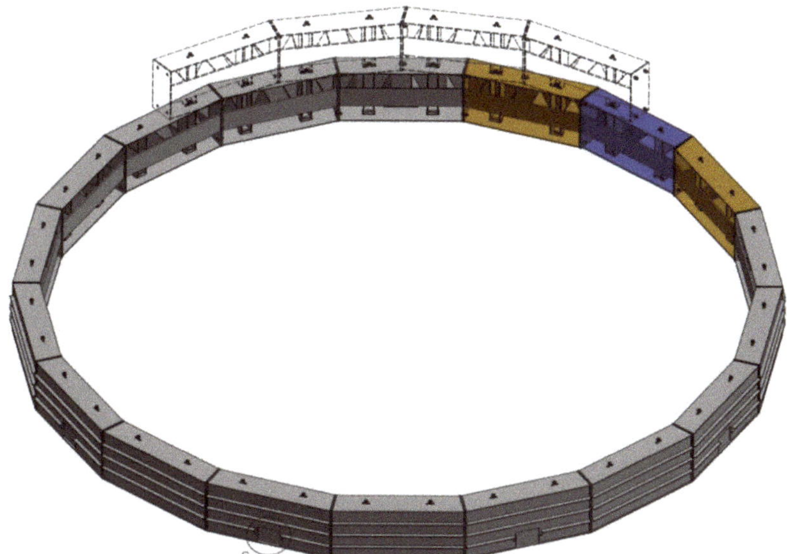

Figure 10. Layout of the steel sections ring (grey panels—typical segments, blue and orange panels—closing segments).

The application of sealing and reinforcing rock mass injection as well as the preliminary injection in different mine shaft sections was also considered necessary. The sealing injection is to be carried out in the vicinity of bottom ends of the freezing boreholes, after the ground defreezing. Injections are to be done at four levels, at depths of 474.90, 476.90,

478.90, and 480.90 m. Rock mass is to be reinforced with a reinforcing injection at the depth of 624.50 m. For the purpose of rock mass injection, eight horizontal boreholes, drilled through the shaft lining, are to be used on each of listed levels. The designed diameter of boreholes is 40 mm.

The necessity for a preliminary rock mass injection is caused by the occurrence of weak and fractured (RQD < 50%) rock mass in a shaft profile over the depth of 473.90 m. In this area, there is a risk of rock falls from shaft walls, which might affect problems with proper shaft lining installation. The decision of using this method of rock mass reinforcement is to be made by the Head of Mining Operations in cooperation with a geologist. A preliminary injection is to be carried out with eight boreholes, drilled from the shaft bottom at an angle between 15 and 30 degrees, along the shaft outline. The maximum length of the injected section must not exceed 8m. The number of boreholes might be changed by a decision of the Head of Mining Operations. The minimum diameter of boreholes is 50 mm. A preliminary sealing and reinforcing injection is to be carried out using expansion heads and polyurethane adhesives.

3. Discussion

Coal exploitation in coal seams no. 207 and 209 and water drainage caused by this process affected the cone of depression occurrence, at a range of about 2.5 km from goaf, therefore reaching the area of designed shaft. It should be noted that the cone of depression size was defined for the layer of permeable sandstones in the roof of excavated coal seams, while the size of the cone of depression and changes in water pressure in other aquifer systems were not defined, because of the lack of geological data.

According to Table 1, the number of hydrogeological tests conducted is significantly higher in the case of borehole B compared to borehole A. Accordingly, hydrogeological parameters obtained in the investigations in borehole A should be considered generalized and average, not related to specific aquifers.

The comparison of forecasted water inflows for boreholes A and B shows that differences between these values are insignificant in the case of overburden formations (Quaternary, Tertiary, and Triassic). However, in the Carboniferous formation, the value of water inflow is three times bigger for borehole A. This difference is an effect of coal excavation in coal seams no. 207 and 209, which changed hydrogeological conditions in this area. Excavation was carried out near the pillar of the designed shaft and caused water drainage of the Carboniferous formation and possibly overlying aquifer systems. A consequence of the drainage is that it also decreases the piezometric pressure in Carboniferous aquifers. According to the investigation carried out in boreholes A and B, the water table in Carboniferous aquifers decreased by about 20 m, while in bottom aquifers, it was up to 88 m. Water drainage caused the development of a cone of depression of significant size, inclined towards goaf. According to lower values of water inflows, obtained in pump tests in borehole B, the calculated values of hydraulic conductivity and forecasted water inflow are also lower. It is notable in aquifers X to XIV, which are under the influence of intensive drainage caused by goaf. It should be noted that despite this situation, the results obtained in calculations based on the data gathered in borehole B are correct, because water relations in this area are not be restored. On the contrary, Carboniferous aquifers are to be drained, which will affect the total drainage of static resources in this area. Inaccuracy in determining water inflows into the shaft are usually caused by a lack of data or errors in forecasting [15], which in turn requires taking into account the specific value of estimation error for the purpose of the shaft lining design.

Modifications of shaft sinking technology and shaft lining design, caused by the growing accuracy of the hydrogeological data, are mostly changes in the construction of the designed lining. Subsequent modifications are connected to rising concrete class or lining thickness, and in some cases, both of these parameters. Additionally, the investigation of fault zones and other areas of increased pressure caused the necessity for the application of complicated, high-strength construction of a shaft lining, primarily made of reinforced concrete and, in the final project, made of steel. This project introduces a steel and concrete complex lining, which consists of a 0.7 m-thick layer of C50/60 concrete class.

Sudden and uncontrolled water inflows into a sunk [16] or operating shaft [17] in the Silesian Coal Basin are extremely rare, but such situations have happened in history. Accurate investigation of rock mass allows one to locate potentially dangerous zones, which can be a threat for shaft stability. To prevent their influence on the shaft, it is necessary to reinforce rock mass in specific areas, e.g., using preliminary or sealing injection. Sealing injection can be very important at the stage of defreezing rock mass in the vicinity of the shaft. The stages of the hydrogeological conditions investigation presented here revealed that shaft design requires an individual approach to this process [18]. It can also be stated that guidelines for shaft design and sinking included in mining regulations and Standards should be treated like general guidance.

4. Conclusions

Shaft sinking is always an expensive venture, which takes many years to complete. The process of the shaft design itself might last several dozens of months and requires significant financial expenses and cooperation of numerous experts in mining, mechanics, civil and electrical engineering, as well as in many other fields, like geology. Proper geological investigation is crucial for the whole process of the shaft design. In the case of an area like the Silesian Coal Basin, where numerous aquifer systems occur, a hydrogeological survey is vitally important.

The example of the designed shaft presented above shows the importance of proper hydrogeological investigation and its impact on the shaft lining design. Subsequent stages of this investigation affect the modifications of shaft lining design, requiring its higher strength, which was done by the application of a higher concrete class or a thicker layer. Ultimately, the application of customized, high-strength constructions was recognized to be necessary.

Obviously, such an intervention in shaft lining construction raises investment cost, as higher-class concrete is consequently more expensive. Similarly, its thicker layer requires higher amounts of concrete, which also raises expenses. Additional reinforcement of the shaft lining or rock mass also incurs extra cost. The necessity for constant modifications of the shaft lining project also raises investment cost, because it requires man hours spent by designers. It is not difficult to imagine a situation where shaft sinking is no longer profitable, because potential gains can no longer cover huge expenditures incurred by the investor, especially taking into account the fluctuations in the resources market and the insecure future of coal-based energy in Europe.

The incurred expenses and their sudden and unexpected growth cannot cover the biggest advantage of proper and accurate hydrogeological survey, which is safety. Incorrect or inaccurate geological survey can lead to a catastrophe, in which the lives and health of people can be in danger.

Author Contributions: Conceptualization, K.d.; methodology, K.d.; data curation, K.d. and P.K.; formal analysis: J.M.; project administration, K.d.; supervision, K.d.; validation, P.K.; writing, P.K. K.d. is responsible for hydrogeological parts, J.M. is responsible for analysis of water threats, P.K. is responsible for engineering elements, i.e., shaft lining and sinking method analysis. All authors have read and agreed to the published version of the manuscript.

Funding: This research received no external funding.

Institutional Review Board Statement: Not applicable.

Informed Consent Statement: Not applicable.

Conflicts of Interest: The authors declare no conflict of interest.

References

1. Kostrz, J. *Górnictwo tom VI, Wykonywanie Wyrobisk część 2, Głębienie Szybów Metodami Specjalnymi*; Wydawnictwo Śląsk: Katowice, Poland, 1964; p. 436. (In Polish)
2. Ledingham, P.; Proughten, A.J.; Saulnier, G.J., Jr. *Science Design for Two Shafts in Phase 1a of the Proposed Rock Characterisation Facility at Sellafield, UK. Characterization and Evaluation of Sites for Deep Geological Disposal of Radioactive Waste in Fractured Rocks, Proceedings from The 3rd ASPO International Seminar Oskarshamn*; Technical Report TR-98-10; Svensk Karnbranslehantering AB: Stockholm, Sweden, 1998; pp. 303–314.
3. Daw, G.P.; Fear, N.J.; Jeffery, R.I.; Pollard, C.A. Hydrogeological Investigations and Ground Treatment for Shaft Sinking at Asfordby New Mine. In Proceedings of the 3rd International Mine Water Congress, Melbourne, Australia, October 1988; pp. 683–692.
4. Ministry of the Energy of Poland. Regulation of the Minister of the Energy of November 23, 2016 of specific requirements of work of underground mines. *J. Laws Pol.* **2017**, *Item 1118*.
5. Polski Komitet Normalizacyjny. *PN-G-05-16:1997 Szyby górnicze—Obudowa—Obciążenia*, 1997.
6. Soni, A.K. Importance of Radius of Influence and its Estimation in a Limestone Quarry. *J. Inst. Eng. India Ser.* **2014**, *96*, 77–83. [CrossRef]
7. Yihdego, Y. Engineering and enviro-management value of radius of influence estimate from mining excavation. *J. Appl. Water Eng. Res.* **2017**, *6*, 329–337. [CrossRef]
8. Daw, G.P.; Pollard, C.A. Grouting for Ground Water Control in Underground Mining. *Int. J. Mine Water* **1986**, *5*, 1–40. [CrossRef]
9. Kipko, E.I.; Kipko Eh, Y.; Williams, R.E. *Integrated Grouting and Hydrogeology of Fractured Rock in the Former USSR*; Society for Mining Metallurgy & Exploration Inc.: Englewood, CO, USA, 1993.
10. Qing, Y.; Kexin, Y.; Jinrong, M.; Hideki, S. Vertical Shaft Support Improvement Studies by Strata Grouting at Aquifer Zone. *Adv. Civ. Eng.* **2018**, *2018*, 5365987. [CrossRef]
11. Skutta, E. Einfache gesteinsmechanische Untersuchungen als Grundlage der Ausbauplannung. *Glückauf* **1962**, *98*, 1455–1461.
12. Greenslade, W.M. *Water Control for Shaft Sinking. Presentation AIME Annual Meeting New Orleans, Louisiana*; Society of Mining Engineers of AIME: Littleton, CO, USA, 1979; pp. 79–138.
13. Kostrz, J. *Głębienie Szybów*, 2nd ed.; Szkoła Eksploatacji Podziemnej: Kraków, Poland, 2014; p. 686. (In Polish)
14. Duda, R. Methods of Determining Rock Mass Freezing Depth for Shaft Sinking in Difficult Hydrogeological and Geotechnical Conditions. *Arch. Min. Sci.* **2014**, *59*, 517–528. [CrossRef]
15. Chudy, K.; Worsa-Kozak, M.; Pikuła, M. Rozwój metod rozpoznania warunków hydrogeologicznych na potrzeby wykonywania pionowych wyrobisk udostępniających złoże—Przykład LGOM. *Prz. Geol.* **2017**, *65*, 1035–1043, (In Polish, with English summary).
16. Czaja, P.; Kamiński, P.; Dyczko, A. Polish Experiences in Handling Water Hazards during Mine Shaft Sinking. In *Mining Techniques—Past, Present and Future 2020*; InTech Open: London, UK, 2020. [CrossRef]
17. Bukowski, P. Water Hazard Assessment in Active Shafts in Upper Silesian Coal Basin Mines. *Mine Water Environ.* **2011**, *30*, 302–311. [CrossRef]
18. Drabek, J.; Sanocki, T.; Wojtaczka, M. Modyfikacja Konstrukcji Obudowy i Technologii Głębienia Szybu w Zawodnionym Górotworze o Niskich Parametrach Wytrzymałościowych. *Inż. Gór.* **2014**, *1*, 49–52. (In Polish)

Article

P-Wave-Only Inversion of Challenging Walkaway VSP Data for Detailed Estimation of Local Anisotropy and Reservoir Parameters: A Case Study of Seismic Processing in Northern Poland

Mateusz Zaręba [1,*], Tomasz Danek [1] and Michał Stefaniuk [2]

[1] Department of Geoinformatics and Applied Computer Science, Faculty of Geology, Geophysics and Environmental Protection, AGH University of Science and Technology, 30-059 Krakow, Poland; tdanek@agh.edu.pl

[2] Department of Fossil Fuels, Faculty of Geology, Geophysics and Environmental Protection, AGH University of Science and Technology, 30-059 Krakow, Poland; stefaniu@agh.edu.pl

[*] Correspondence: zareba@agh.edu.pl or mattzareba@gmail.com

Abstract: In this paper, we present a detailed analysis of walkaway vertical seismic profiling (VSP) data, which can be used to obtain Thomsen parameters using P-wave-only inversion. Data acquisition took place in difficult field conditions, which influenced the quality of the data. Therefore, this paper also shows a seismic data processing scheme that allows the estimation of correct polarization angles despite poor input data quality. Moreover, we showed that it is possible to obtain reliable and detailed values of Thomsen's anisotropy parameters for data that are challenging due to extremely difficult field conditions during acquisition and the presence of an overburden of salt and anhydrite (Zechstein formation). This complex is known for its strong seismic signal-attenuating properties. We designed a special processing workflow with a signal-matching procedure that allows reliable estimation of polarization angles for low-quality data. Additionally, we showed that P-wave-only inversion for the estimation of local anisotropy parameters can be used as valuable additional input for detailed interpretation of geological media, even if anisotropy is relatively low.

Keywords: anisotropy; VSP; unconventional; hydrocarbon; inversion; polarization; signal processing

check for **updates**

Citation: Zaręba, M.; Danek, T.; Stefaniuk, M. P-Wave-Only Inversion of Challenging Walkaway VSP Data for Detailed Estimation of Local Anisotropy and Reservoir Parameters: A Case Study of Seismic Processing in Northern Poland. *Energies* **2021**, *14*, 2061. https://doi.org/10.3390/en14082061

Academic Editor: Yangkang Chen

Received: 11 February 2021
Accepted: 4 April 2021
Published: 8 April 2021

Publisher's Note: MDPI stays neutral with regard to jurisdictional claims in published maps and institutional affiliations.

Copyright: © 2021 by the authors. Licensee MDPI, Basel, Switzerland. This article is an open access article distributed under the terms and conditions of the Creative Commons Attribution (CC BY) license (https://creativecommons.org/licenses/by/4.0/).

1. Introduction

Exploration of dwindling deposits of raw materials requires increasingly more accurate methods of geophysical imaging and accurate interpretation of the obtained results. The role of seismic surveys and maximizing the use of information from seismic wave analysis are now more crucial than ever. The use of VSP walkaway seismic surveys can significantly improve surface seismic information, and it adds value to new geological information [1,2]. In the case of the detailed processing approach, it allows obtaining values of the local anisotropy tensor [3] and provides more accurate information about in situ anisotropy than other types of seismic data surveys [4]. In this paper, we first describe a detailed processing technique which allows the determination of accurate polarization angles for P-wave-only inversion for data acquired in northern Poland, where acquisition was carried out in extremely difficult conditions immediately after a very rainy period. Additional difficulties were related to the examined depth interval (2400–3825 m) in this partially cased well. At that time, well W-1 was one of the deepest wells in Poland and the deepest well for shale gas exploration in the country. Moreover, the target layers were situated below a highly attenuating formation of Zechstein rocks at a depth of approximately 4 km. Both the 3D seismic surface survey and the walkaway VSP data (for some of the shot points) have a low signal-to-noise ratio [5]. The main goal was to perform P-wave-only inversion of this data according to the procedure developed in [4]. The biggest challenge was related to the fact that we already know from the

199

surface seismic survey that this geological medium has very low anisotropy strength, and estimated anisotropy parameters were very low; however, these parameters have a significant impact on seismic processing and migration. This accurate information about local anisotropy could improve reprocessing of seismic data in the future, especially model building for depth migration [6]. The obtained results showed that combining the advanced processing of challenging walkaway VSP data with detailed data analysis of P-wave-only inversion gives additional helpful reservoir information that can be used to make interpretation more accurate when other surveys produce poor quality results and even if anisotropy is relatively low. The walkaway VSP profile was long enough to provide wide angular data coverage with which VSP inversion could be performed. In summary, this work aims to show a processing case study of the first walkaway VSP survey in Poland to determine elastic anisotropy information. The work comprises two practical parts: The first is the processing part; it presents a study that allows a selection of the individual processing sequence in determining the inclinations of the P-wave. The second part is a study showing whether the anisotropy parameters can be obtained using the data from the first part. These two parts provide a complete follow-up that allows maximizing data use, presents unique processing techniques, and increases current knowledge about VSP and anisotropy in general. To compare the results with the lithological characteristic, at the beginning of the article we include a detailed geological description of the Wysin area. This allows for a thorough understanding of the presented case study.

2. Region and Acquisition Characterization

The presented data were acquired as part of a project supporting the development of technologies for shale gas extraction: "Polish Technologies for Shale Gas". The presented survey was conducted by Geofizyka Toruń S.A. (GT Services) on behalf of the Faculty of Geology, Geophysics and Environmental Protection of AGH University of Science and Technology in Kraków. The project was supervised by Professor Michal Stefaniuk from the Department of Fossil Fuels. Data were collected in November 2016 under many unfavorable conditions related to weather and legal matters. Before walkaway VSP acquisition was done, a 3D seismic survey was performed, processed, and interpreted by GT Services for Polish Oil and Gas Company (POGC).

2.1. Description of the Region and Local Geology

The area of the W-1 survey is located in the Kashubian Lake District (in accordance with the nomenclature of Kondracki [7]) in northern Poland on the Eastern European Precambrian Platform (Figure 1). This region has a young glacial character of moraine plateau with numerous frontal moraine hills. The absolute height of those hills is about 160 m asl, but the relative height is no more than 30 m. Many glacial gutters, most of which contain lakes, can be found in this region. The Wierzyca River flows in this region, which contributes to the formation of peat-boggy depressions. Figure 2 shows the chronostratigraphic and lithostratigraphic units in well W-1.

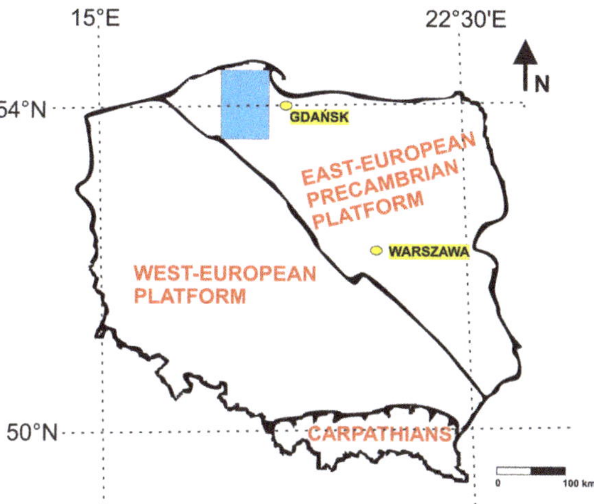

Figure 1. Location of the study area (blue rectangle) in relation to Poland's general tectonic units (see in [9] (modified)).

The analyzed area is in the central-western part of the Baltic Syneclise, at the eastern edge of the Łeba elevation. The two structural complexes of Lower Paleozoic and Permo-Mesozoic rocks occur on the denuded surface of crystalline rocks. Sedimentation begins with the Lower Paleozoic Żarnowiec formation and is built of alluvial sediments, including alluvial fans. Together with Lower and Middle Cambrian rocks, these sediments constitute one big sedimentary complex. Upper Cambrian rocks are generally very eroded. Ordovician and Silurian complexes lie directly on Cambrian rocks and contain many gaps caused by erosion and sedimentation breaks. Upper Permian formations made of Zechstein evaporites lie directly on Silurian sediments above the erosional boundary. The Mesozoic sequence begins with sediments of claystone, mudstone, and sandstone, with carbonate intercalations. Above the Muschelkalk, which consists of marl, dolomite with limestone can be found. The Lower and Middle Jurassic complexes, in which stratigraphic and erosion gaps often occur, are situated directly on Mesozoic deposits. The Cretaceous sediments consist mostly of glauconitic sandstones, marly limestones (Lower Cretaceous), and marls (Upper Cretaceous). Tertiary rock complexes are incomplete and are represented by Miocene formations in the form of marly clays with sand silt layers. Quaternary sediments are represented by a series of glacial sand and gravel sediments, often with numerous pebbles. Three sealing complexes can be found in this region. Two mostly of these complexes consist of silt and claystone with low porosity and permeability (Ludlow and Pridoli). The third sealing complex of the Zechstein formation consists of a salt and anhydrite complex that is approximately 300–400 m thick. Zechstein rocks have an average P-wave velocity of approximately 6000 m/s and are known for attenuating the propagation of seismic signals. The examined zone presented in this paper is below the Zechstein formation. Figure 3 shows the lithostratigraphic well correlation section between the wells, including Well W-1, together with a structural interpretation of seismic data (see in [8] (modified)).

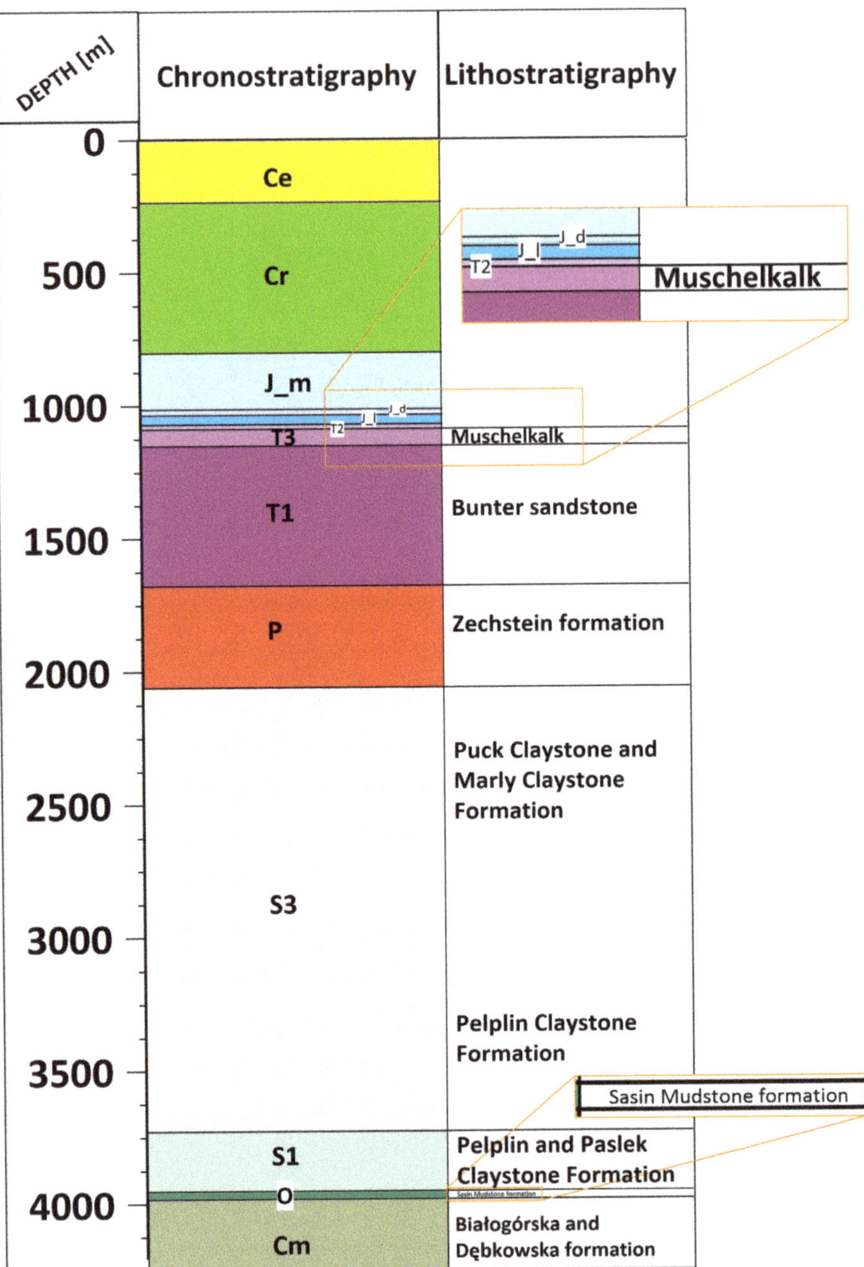

Figure 2. Chronostratigraphic and lithostratigraphic units in well W-1. Cm—Cambrian, O—Ordovician, S1—Lower Silurian, S3—Upper Silurian, P—Permian, T1—Lower Triassic, T2—Middle Triassic, T3—Upper Triassic, J_l—Lias (Early Jurassic), J_d—Dogger (Middle Jurassic), J_m—Malm (Upper Jurassic), Cr—Cretaceous, Ce—Cenozoic (based on core information from Well W-1).

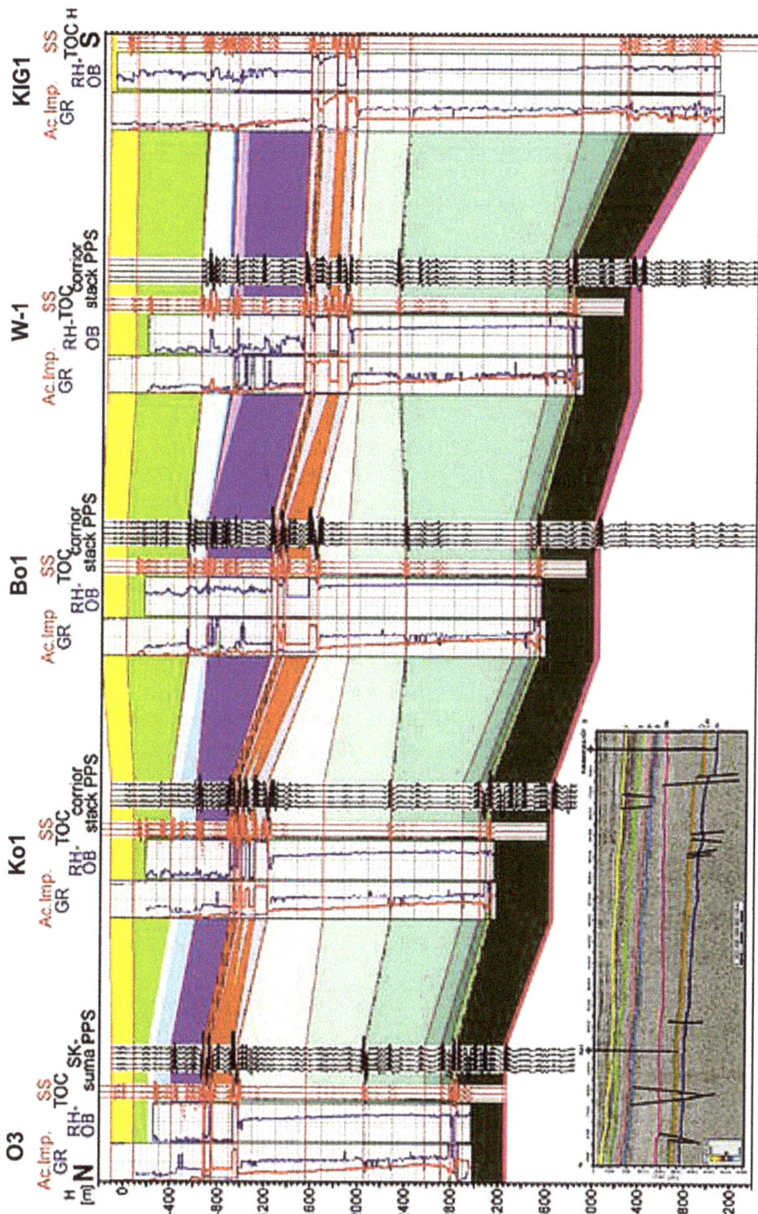

Figure 3. Lithostratigraphic well correlation section between wells O3, Ko1, Bo1, KIG1, and Well W-1, together with a structural interpretation of seismic data (see in [8] (modified)).

2.2. Acquisition Characterization

The survey was carried out in a village in northern Poland. Well W-1 is vertical and has a total depth of 4040 m (2.5 mi); it was drilled in 2013. Two horizontal wells were drilled in 2015 and 201: W2H (total length, including horizontal and vertical sections, is 5450 m) and W3H (total length, including horizontal and vertical sections, is 5540 m).

W2H was the 18th well drilled for shale gas exploration and the 3rd horizontal well drilled in Poland. The hydraulic fracturing and microseismic monitoring had also already been completed in wells W2H and W3H. A 3D seismic survey was carried out in 2013 on an acreage of 76 km^2. The profile of the walkaway VSP shot points was parallel to the horizontal direction of wells W2H and W3H. Figure 4 shows the locations of the VSP walkaway profile and the wells. Acquisition was finished by the end of 2016. In the vertical well, W-1, ninety-six BSR 3C receivers removable tool were used (Oyo Geospace Company) with GeoRes Downhole System Digital Module 3-CH X 24-BIT DIGITIZER (Geospace Technologies). The distance between receivers was 15 m. The cable between receiver sections was BSR type (a total of 97 cables were used). The carrying and transmission cable was a well-logging fiber-optic cable type 6E4FO592 with a maximum of 4200 m. In this study, we present the results from the depth interval between the first receiver at a depth of 2400 m and the ninety-sixth receiver at a depth of 3825 m MDGL (measured depth from ground level). Acquisition was carried out after heavy rainfall between October and November. It was not possible to perform acquisition at a different time, so all challenges had to be overcome at the processing stage. The very soggy ground caused a significant difference in shot point elevation between sweeps. Moreover, the characteristics of the near-surface zone at this point also changed due to compaction or a change in the filling fluids in the pore space. Heavy seismic vibrator trucks almost sank into the soaked ground. The receivers were placed mainly in Wenlock and Upper Silurian rocks. This made it possible to observe and investigate the behavior of seismic signals (anisotropy and P-wave polarization analysis) passing through the Zechstein formation, which is well known for its highly seismic signal-attenuating properties. The interval velocity model used for seismic depth migration and the trajectory of well W-1 is presented in Figure 5 (the Zechstein formation is colored black). The sweep length was 16 s with a frequency range of 6–120 Hz; recording time was 4 s. The distance between the 480 shot points was 25 m; total length across the profile was 12 km. Due to the extremely bad acquisition conditions and the low signal-to-noise ratios, up to 8 sweeps were generated on each shot point.

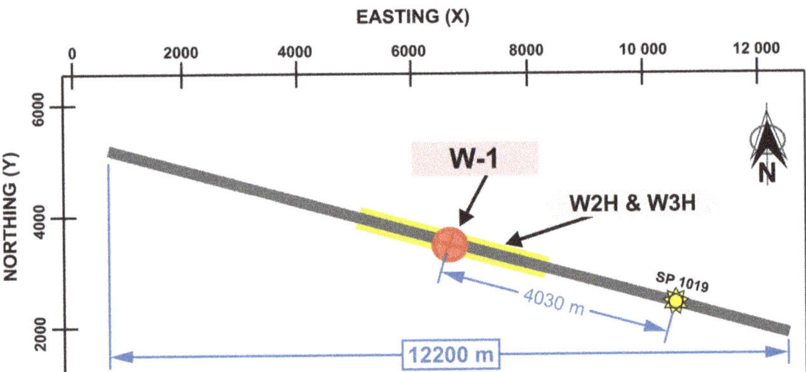

Figure 4. Location map of vertical Well $W - 1$ (**red dot**), horizontal wells W2H and W3H (**yellow line**), with shooting profile line (**gray**) and outermost shot point SP 1019 (**yellow star**) used for this study (see in [5] (modified)).

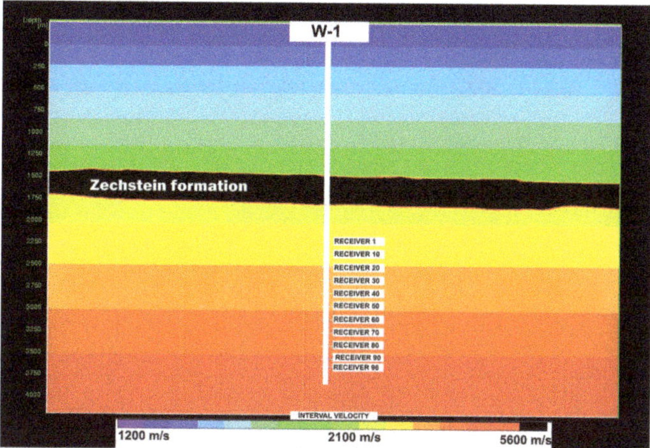

Figure 5. Velocity model for seismic migration requirements, together with the trajectory of well W1 (**white line**) and receivers. The black layer is the Zechstein formation (see in [5] (modified)).

3. Data Processing

In this survey, one of the main challenges was to increase the signal-to-noise ratio of the seismic signal for proper wavefield separation, polarization analysis, and the anisotropy study. All these processes rely on first-break picks. In this example, the low quality of data was caused by several factors:

1. Changes in the near-surface zone between successive sweeps. A core processing step is stacking, which makes it possible to reduce various types of noise and strengthen the useful signal. In this survey, multiple sweeps were performed on each shot point. Due to weather and ground conditions, the near-surface zone changed significantly between sweeps; therefore, we needed to apply a new procedure in the processing sequence before performing vertical stacking.

2. The receivers were located in the depth interval where seismic imaging produces low-quality results due to the properties of the overburden layers. This is a very well-known exploration problem in northern Poland. The Zechstein formation is mostly made of salts, anhydrite, and dolomite. This complex can be described as a shielding screen for seismic signals. This is believed to be the main cause of the failure of hydrocarbon exploration in this region because sub-Zechstein formations cannot be reliably interpreted [10].

3. Artifacts related to acquisition. This group includes weak tool anchoring, which can possibly generate various types of noise (hard to classify) due to pressure and the forces present in the well. The device hung on a cable about 3 km long. Note that noise related to the tool's resonance is often removed by changing its position and moving it up or down. In this case, the tool was in the same place for the whole acquisition period. Furthermore, in many parts of the well there was no casing between the inner and the outer tubes. Ring noise can be generated when a casing is unbounded.

We wanted to examine the best processing workflow for wavefield separation. Typical VSP acquisition consists of 3C receivers that can record two horizontal components (H1 and H2) and one vertical component (V). All components are perpendicular to each other. The correct wavefield separation of the compressional downgoing P-wave on the vertical component and the longitudinal SH and SV waves on the horizontal components is an important seismic processing step. This can be done using the analytical method proposed by Galperin [11], which is based on the directions of the motion of particles. We tested four processing sequences to determine which one allows the best wavefield separation and, consequently, to accurately determine the polarization angles of the longitudinal wave.

This is crucial in the case of P-wave inversion, which needs properly estimated polarization angles [4]. We tested the following workflows:

1. **Workflow 1 (W1)**—component rotation on raw data for each shot; vertical stacking; noise attenuation.
2. **Workflow 2 (W2)**—vertical stacking of raw data; component rotation using stacked data; noise attenuation.
3. **Workflow 3 (W3)**—noise attenuation before vertical stacking; component rotation on each shot separately; vertical stacking.
4. **Workflow 4 (W4)**—noise attenuation; vertical stacking followed by component rotation.

Noise attenuation was always the same and consists of the following procedures:

*   Single time-invariant, zero-phase Ormsby band-pass filter; frequency range: 16–80 Hz, low-cut: 8 Hz, high-cut ramp: 40 Hz.
*   Monofrequency noise attenuation (this type of noise was widely present in these data).
*   Time-space frequency filtering using short-time Fourier transform (STFT) for noise related to tool resonance or surface noise passed via the cable. We used a 200 ms window; the aperture was equal to five traces; threshold value: 3; frequency range: 20–80 Hz.
*   Vertical and horizontal median filtration for high-energy noise attenuation (a general problem for near-offset shot points). Filter parameters were 100 ms for the vertical window, 30 traces for the horizontal dimension, and 50 ms cosine taper zone; no-scaling factor was used.
*   Attenuating energy over first breaks using top mute.

Our processing sequence selection criteria were the lowest polarization angle determination error and the most stable distribution along the whole depth interval. The polarization angle is the angle between the axis of the vertical well (parallel to the receiver's vertical component) and the P-wave incident ray. We will call this the angle inclination; see Figure 6.

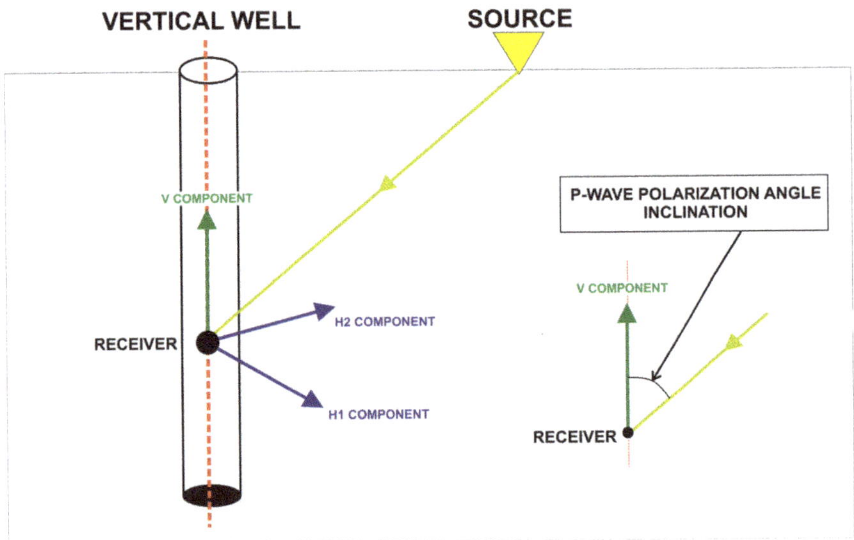

Figure 6. A measurement scheme diagram together with the explanation of the nomenclature used in this paper.

In order to estimate the error of the determination of the polarization angles for each receiver point, the standard deviation of each 5-receiver group was calculated according to Equation (1):

$$\sigma(R_n) = \sqrt{\frac{\sum_n^N (R_n - \overline{R})^2}{N}},$$

(1)

where (R_n) is inclination value for a particular receiver, $\sigma(R_n)$ is the error for a receiver at depth point n, N is a group of 5 receivers, and $\overline{R}$ is the average value of the inclination angle for this group. Then, the simple arithmetic averages of the standard deviations calculated in the previous step were considered as average errors for each shot point using Equation (2):

$$average\ error(SP) = \frac{\sum_n^{N_E} \sigma(R_n))^2}{A_E},$$

(2)

where A_E is the number of estimated angles and $\sigma(R_n)$ is less than 5^0.

In this step, we decided to use workflow 4, which produced the lowest error. In general, the stacking procedure strengthens the useful signal and reduces random noise. In workflow 4, we first performed noise attenuation procedures to reduce noise, which is not random, and to avoid its possible subsequent reinforcement in the stacking process. We had to take this additional step due to weather and ground conditions. The surface zone changed between sweeps, and the seismic vibrator trucks sank into the water-logged ground. The elevation changes were significant. The first and the last sweep in each shot point were performed under different geotechnical conditions. Pore space decreased as a result of compaction; as a consequence, the amount of water filling these pores also changed. To make sure that vertical stacking in this situation would in fact help to reduce unwanted interference and strengthen the useful signal, we decided to add a signal-matching procedure before stacking. We know a priori from 3D seismic surface surveys that anisotropy strength is very low. The observed anisotropy is at the limit of detection by seismic methods (10^{-3} to 10^{-2}). The combination of the great depth and very small values of parameters forced us to obtain the most accurate polarization angles without additional distortions caused by the conditions under which the acquisition was performed. Before we decided to use a time-frequency matching filter, we performed a correlation between records with the calculated reference record (the average of all sweeps on the shot point), see Figure 7.

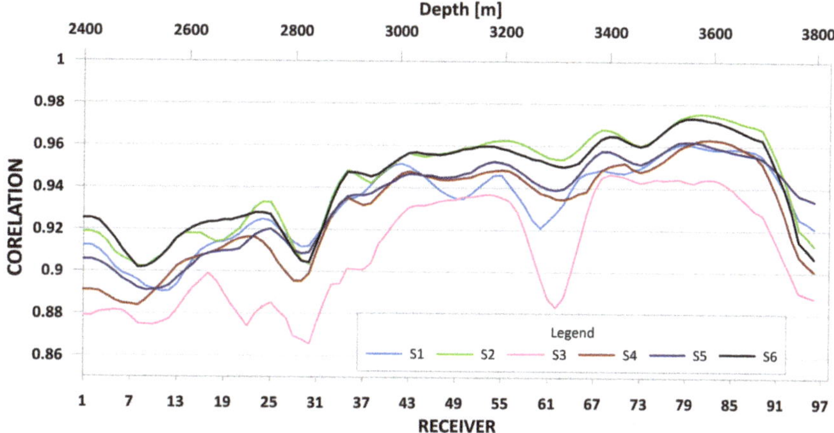

Figure 7. Correlation between records for one shot point. Each line represents the correlation between one sweep and the pilot, calculated as the average of all sweeps on the point (see in [5] (modified)).

At the beginning of the examined depth interval, we can see the lowest correlation for all sweeps. This is an expected effect, as there was no casing between the 9″ and the 7″ tubes in this part. In the rest of the well, where casing is present, we can see significantly higher correlation, except for the 3200–3400 depth interval, where the correlations are weaker. This is probably due to weak anchorage of receivers in this part of the well; however, this is also a change zone between the Ludlow and upper Silurian complexes.

To evaluate this methodology, we plotted estimated inclinations for near-offset SP in Figure 8 (offset 300 m) and for far-offset SP in Figure 9 (4000 m). In both cases, the lowest average error was obtained using workflow 4 with the additional signal-matching procedure. For both offsets, the least optimal workflow was workflow 1 because the Principal Components Method (PCM), which we used for component rotations, needs the first-break times on its input. In this step, the downgoing P-wave energy is used. The window that defines this energy should be centralized around the first-break time. In workflow 1, the first step was first-break picking. We performed hand picking because it was impossible to use an automatic picker due to noise. Performing first-break picking on this step is not very accurate due to the interference between the signal and the noise. These phenomena can lead to changes in the maximum and minimum amplitude position of the signal. This, on the other hand, may cause incorrect determination of the first-break times. Workflow 2 has a slightly lower error than workflow 1 because vertical stacking reduces some random noise; as a result, the first-break picking accuracy can be higher. However, this error increases with the offset. For near-offset shot points, the relative change of average error is not as significant as for shot points located in the far offsets. For the 300 m offset, the average error of workflow 2 is approximately 6% lower compared to workflow 1, while for the 4000 m offset there is a 35% difference. Data processed using workflow 3 have a smaller error compared to data from workflows 2 and 3. The smallest error was obtained using workflow 4. Importantly, adding a matching filter to the workflow with the lowest average error significantly improved the inclination estimation. Compared to workflow 1, the average error reduction for workflow 4 with the signal-matching procedure applied is approximately 71% lower for the near-offset shot point, and it is 83% lower for the far-offset shot point. In the case of original workflow 4, adding the signal-matching procedure resulted in 47% lower error for the near offset and 37% lower error for the far offset. Figure 10 shows a comparison between the S/N ratio for workflow 4 and workflow 4 with the signal-matching procedure with reference to the lithostratigraphic column. We can see a similar phenomenon as in Figure 6: a lower S/N ratio in the depth interval where there is no casing between the inner and the outer tube. The significantly greater value of the S/N ratio can be seen in the Ludlow depth interval. This geological complex is built of marly claystone with SPWI 53%, then marl and claystone on the bottom. Note that the signal-matching procedure improved the S/N ratio, which is important for component rotation and, consequently, also for estimation of inclination angles. Figure 11 shows a plot of average errors (y axis) for workflow 4, and the signal-matching procedure against the offset (x axis). It can be seen that the error increases linearly as the offset increases. However, there are two groups: the first is from the beginning to the offset at approximately 3200 m, and the second group is from the offset at 3200 m to the offset at 4400 m. The first group has a strong positive linear trend and the error varies from 0.3 to 1.2, while the second group shows a negative linear trend. The highest error value is 2 for the smallest offset of this group; as the offset decreases, the error also subsequently decreases to a value of 1.6. However, there are not enough points to fully interpret this phenomenon. The reason for this is probably related to the aforementioned acquisition problems. The final field report states that some far offset points were skipped because the data quality was below the acceptance level and the ground conditions. Figure 12 shows inclination distributions for the different offsets used for further calculations for estimation of anisotropy parameters.

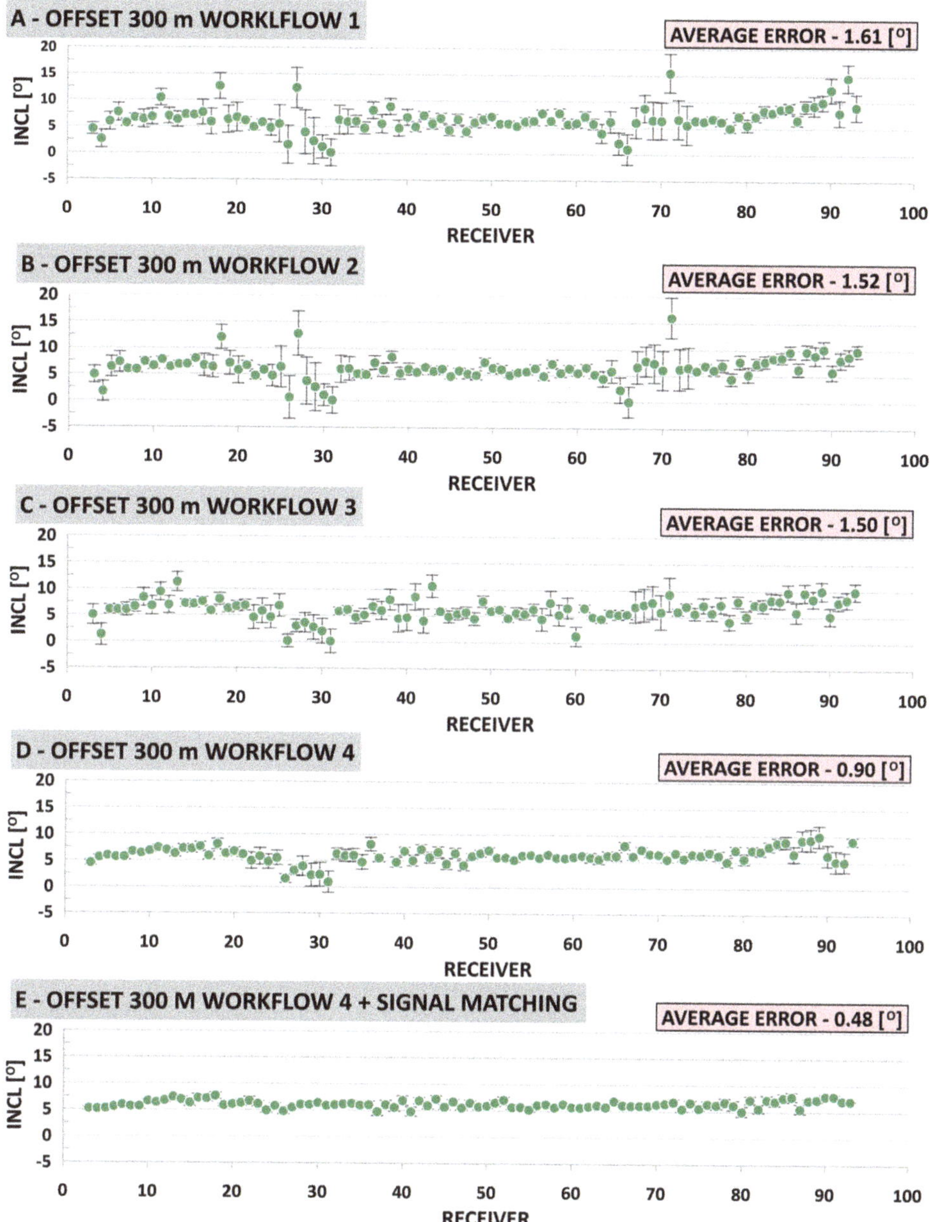

Figure 8. P-wave polarization angles (green dots) with error bars for different workflow schemes for the near-offset shot point (300 m): A is for W1, B for W2, C for W3, D for W4, and E is for W4 with signal-matching procedure added.

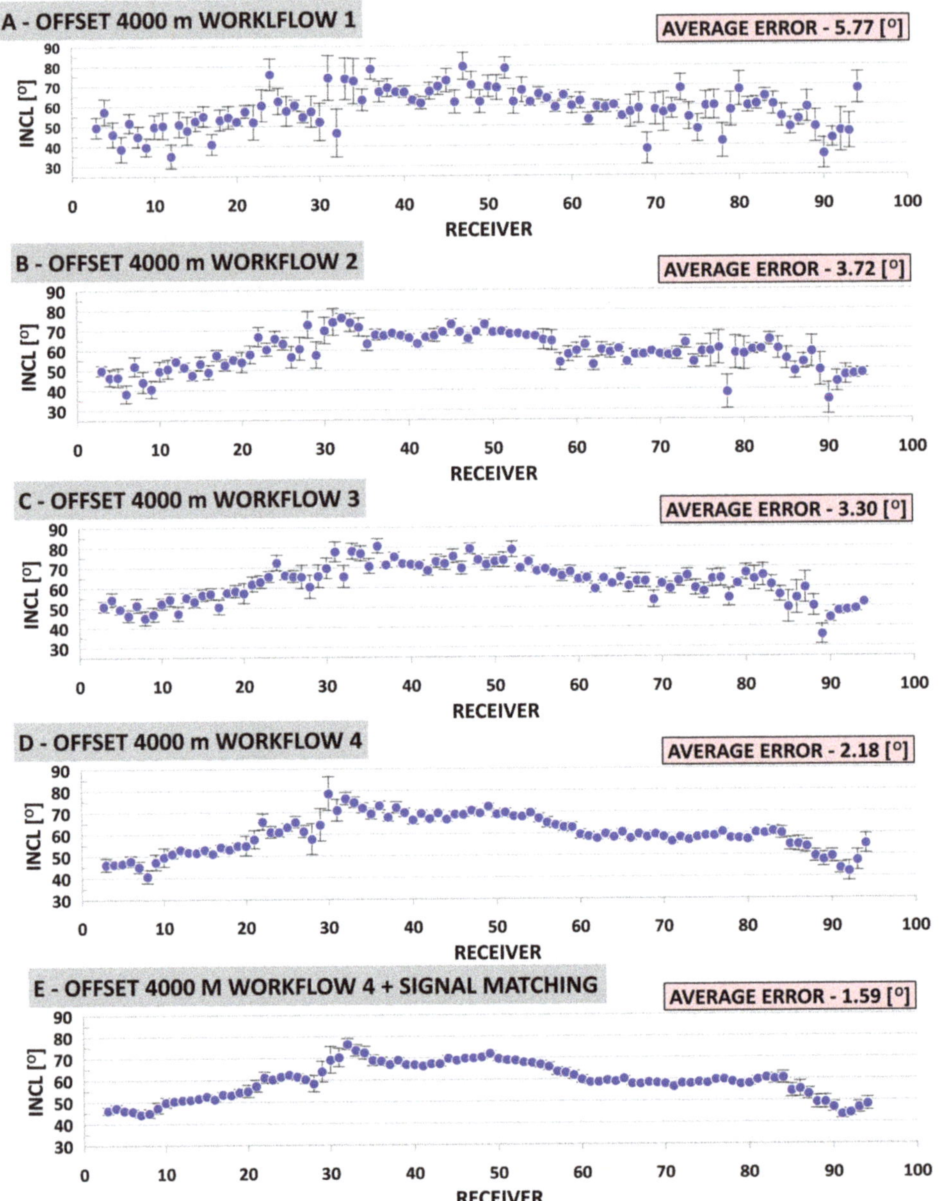

Figure 9. P-wave polarization angles (**dots**) with error bars for different workflow schemes for the far offset shot point (4000 m): A is for W1, B for W2, C for W3, D for W4, and E is for W4 with signal-matching procedure added.

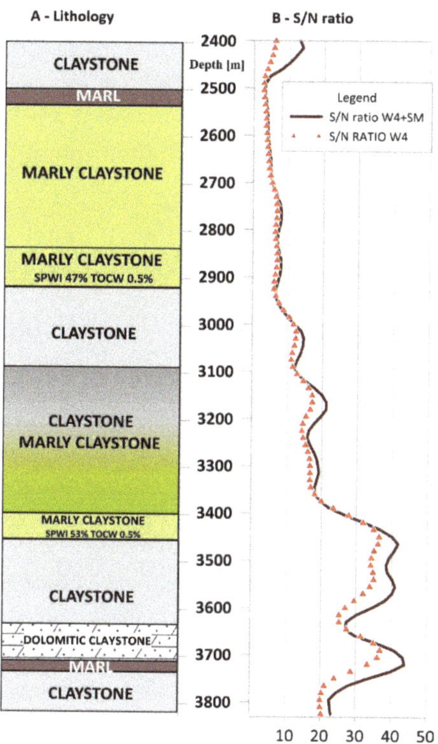

Figure 10. Lithostratigraphic column in Well W-1 with S/N ratio graph for W4 (**red triangles**), and W4 with signal-matching procedure added (**brown solid line**).

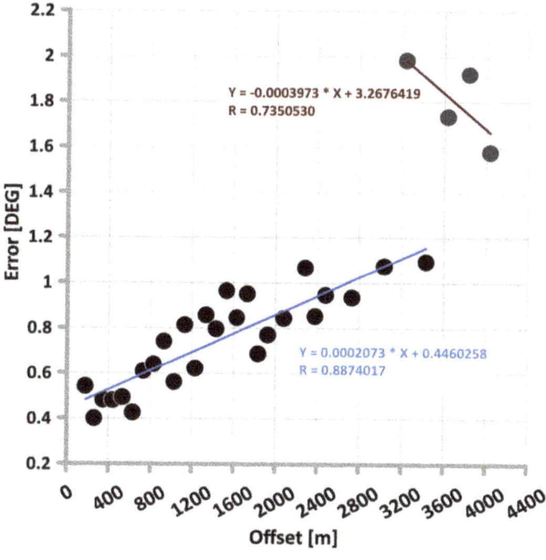

Figure 11. Average polarization angle determination error (y axis) against the offset (x axis) for W4 with signal-matching procedure added (**black dot**) with linear trend line (blue).

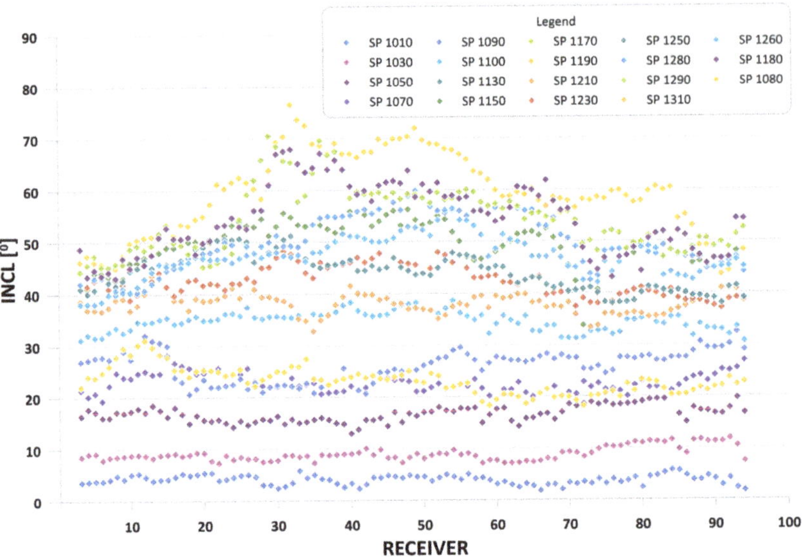

Figure 12. P-wave polarization angle distributions for different shot points (SP 1010 represents the nearest offset, and SP 1310 represents the farthest one).

4. Local Anisotropy Estimation

Several methods allow the determination of anisotropy based on VSP data including slant stack [12], slowness [13], and slowness-polarization method [4]. We used the slowness-polarization method described by Grechka and Mateeva [4] because information about polarizations stabilizes the solution and simplifies the inversion process [4]. Additionally, we wanted to obtain anisotropy parameters at a given depth without entering layer boundaries and any additional assumptions beforehand. Grechka and Mateeva used this methodology for 2D walkaway VSP data gathered in the Gulf of Mexico, where receivers were partially placed inside salt. They proved that estimated in situ anisotropy coefficients can be correlated with lithology and other geophysical methods such as gamma-ray measurement. In the case of vertical transverse isotropy (VTI) media, Grechka and Mateeva speculated that parameter η (Alkhalifah and Tsvankin's anellipticity coefficient [14]) could be slightly less correlated to the lithology than δ (Thomsen's anisotropic coefficient [15]). This is mostly because of the dependence of η on local stress anomalies. Our goal was to show that this methodology can be used even if anisotropy is extremely small. Grechka and Mateeva [4] worked with data acquired in a medium with an anisotropic complex beneath salt and almost isotropic salt. This paper presents the equations necessary to perform the inversion. A detailed mathematical and theoretical description of this method is presented in the appendices of Grechka and Mateeva's paper [4]. The slowness-polarization method starts from the weak-anisotropy approximation. Each time we refer to V_p and V_s in Equations (3)–(19), we mean vertical P and S velocity. In the case of a VTI medium, it can be described by Equation (3) using vertical slowness (q), P-wave inclination angle (α), δ anisotropic coefficient [15], η anellipticity coefficient [14], and V_p velocity. The P-wave slowness polarization vector p component is then described by

$$p_3 \equiv q(\alpha) \simeq \frac{\cos(\alpha)}{V_p} \cdot (1 + \delta_{VSP} \cdot \sin^2(\alpha) + \eta_{VSP} \cdot \sin^4(\alpha)), \tag{3}$$

where α is the inclination angle.

In the case of a VTI medium and measurements in a vertical well, the anisotropic dependence $q(\alpha)$ is controlled by δ_{VSP} (more sensitive to near-offset $q(\alpha)$ variations) and η_{VSP} (more sensitive to far-offset $q(\alpha)$ variations), see Equations (4)–(6):

$$\delta_{VSP} = (f_0 - 1)\delta, \tag{4}$$

$$\eta_{VSP} = (2f_0 - 1)\eta, \tag{5}$$

$$f_0 = \frac{1}{1 - \frac{V_S^2}{V_P^2}}. \tag{6}$$

If one wants to invert directly for η and δ, the p_v ratio has to be known beforehand. The average value of V_S/V_P in Well W-1 is 0.54, therefore the average p_v is 0.29 according to Equation (7):

$$p_v = \frac{V_S^2}{V_P^2}. \tag{7}$$

Using Equation (7), f_0 can be described by

$$f_0 = \frac{1}{1 - p_v} \approx 1 + p_v. \tag{8}$$

The approximation in Equation (8) is used only for the sake of simplification, presenting the theoretical relationships between the parameters η_{VSP}, δ_{VSP}, and η, δ. In fact, we are calculating the f_0 directly in every depth point.

Finally, Equation (3) can be written using p_v in Equation (9):

$$q(\alpha) \simeq \frac{\cos(\alpha)}{V_p} \cdot (1 + p_v\delta \cdot \sin^2(\alpha) + (1 + 2p_v)\eta \cdot \sin^4(\alpha)). \tag{9}$$

To better understand the dependencies of $q(\alpha)$ and η_{VSP} with δ_{VSP} on the offset, an average p_v value of 0.29, and $(1 + 2p_v)$ equal to 1.58, the dependency graph is shown in Figure 13.

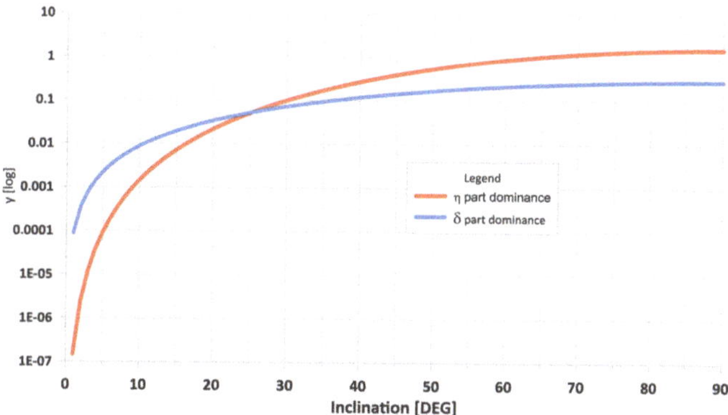

Figure 13. Anisotropic parameters of the offset dependence graph. Axis *x* shows inclination angles. The increase in the inclination angle is related to the increase in the offset.

Figure 13 shows that average polarization angles below 25° are for shot points numbered 1010 to 1180, which correspond to an offset from 0 to 1700 m, and in this range the polarization vector is mostly influenced by the δ part in Equation (9). For offsets from 1700 to 4000 m, the polarization vector is mostly influenced by η (angles over 25°). Thus, it is obvious that for reliable inversion a wide spectrum of polarization angles is needed. The theoretical minimum of δ can be calculated using the following equation ([16]):

$$\delta_{min} = -\frac{1}{2f_0}. \tag{10}$$

which is equal to -0.7034 with an average p_v of 0.28. For the inversion, we need to specify the parameters that will be estimated. These parameters are stored in the vector $\mathbf{k}$:

$$\mathbf{k} = [V_P, \delta_{VSP}, \eta_{VSP}]. \tag{11}$$

Vertical $q^{calc}(k, \alpha)$ is calculated without weak anisotropy approximation, directly from Equations (12)–(15):

$$\mathbf{I} = \begin{bmatrix} \sin(\alpha) \\ 0 \\ \cos(\alpha) \end{bmatrix}, \tag{12}$$

$$\mathbf{p} = \frac{1}{V(\theta)} \begin{bmatrix} \sin(\theta) \\ 0 \\ \cos(\theta) \end{bmatrix}. \tag{13}$$

where $\mathbf{I}$ is polarization vector, $\mathbf{p}$ is slowness vector, θ is polar phase angle, and $V(\theta)$ is the phase velocity of the P-wave. Now, using Christoffel's equation and the Voigt notation for stiffness coefficient c_{ij}, the relation between θ and α can be estimated using Equation (14) (see in [4], Appendix B):

$$\begin{bmatrix} c_{11}\sin^2(\theta) + c_{55}\sin^2(\theta) - V^2 & (c_{13}+c_{55})\sin(\theta)\cos(\theta) \\ (c_{13}+c_{55})\sin(\theta)\cos(\theta) & c_{55}\sin^2(\theta) + c_{33}\sin^2(\theta) - V^2 \end{bmatrix} \times \begin{bmatrix} \sin(\alpha) \\ \cos(\alpha) \end{bmatrix} = 0. \tag{14}$$

When $c_{13} + c_{55}$ is not 0, the phase velocity is equivalent to the same wave mode. Solving the linear equation by eliminating variable V results in the following:

$$(c_{11} - c_{55})tg^2(\theta) + 2((c_{13} + c_{55})ctg(2\alpha)tg(\theta) - (c_{33} - c_{55}) = 0. \tag{15}$$

Considering anisotropic rocks, we know that the equation has two roots because $c_{11} > c_{55}$ and $c_{33} > c_{55}$ in this medium. These roots correspond to the phase angle whose difference between the polarization angle of the P and SV wave is the smallest. The phase velocity is a square root of the Christoffel equation's greatest eigenvalue, therefore $p(\alpha)$ can be calculated using Equation (13). The last step is to minimize the objective function by changing the values of the elements of vector $\mathbf{k}$:

$$F(k) = \sum \frac{[\mathbf{q}^{calc}(\mathbf{k}, \alpha) - \mathbf{q}(\alpha)]^2}{\sigma^2}, \tag{16}$$

where σ^2 describes the errors, which are estimated using the following equation:

$$\sigma^2 = \sigma^2(\mathbf{q}) + (\frac{\sin(\alpha)}{< V_P, log >})^2 \sigma^2(\alpha), \tag{17}$$

where $< V_p, log >$ is the P-wave velocity from the sonic tool. The minimization process of function $F(k)$ starts from an isotropic medium assumption, where vector $\mathbf{k}$ is equal to

$$\mathbf{k} = (V_P, 0, 0). \tag{18}$$

V_P is calculated using

$$V_P = \frac{\sum_n^N \frac{\cos_n(\alpha)}{q_n(\alpha)}}{N}, \tag{19}$$

where N is the number of the considered depth points.

Estimated average P-wave polarization angles for the different shot points along the profile and errors of their estimation (calculated according to Equation (2)) are characterized by a stable linear increase with offset, which is visible in Figure 14. Two areas that deviate from the trend line can be found: one around shot point 1120 (offset around 1230 m) and the other around shot points 1140–1180 (offset 1400–1800 m). We can assume that this is because vibrator trucks did not reach the designated points on the profile. Therefore, they were moved to other locations close to those points but not on the profile line. Another reason could be related to record quality. Besides the observed deviation, we decided to use these data for the anisotropy calculation. In Figure 15, the vertical slowness versus offset graph is presented for all offsets and depth points. Colors represent different depth points. It can be seen that depending on the depth, the difference in elastic properties is obvious. The downward shift along the Y-axis is caused mostly by the velocity increase with depth because Equations (3) and (9) contain the component $\frac{\cos(\alpha)}{V_p}$, by which the whole equation is multiplied. Figure 16 shows a vertical slowness polarization graph for six depth intervals: 2475, 2505, 2700, 2880, 3300, and 3750 m. Green points (dark and light ones) represent all data prepared for the inversion process before validation. Dark green points were used for inversion calculation, and red represents the predicted values. The blue line is calculated according to the assumption that the geological medium is isotropic. For a depth interval of 2880 m, which reaches the top of the marly mudstone layer and is potentially saturated with gas, there is a noticeable shift between the predicted points and the calculated isotropic vertical slowness line from angle 10°. As mentioned before, for inclinations over 25° the η parameter dominates the solution of the slowness-polarization equation. Figure 16 presents raw estimated parameters (δ, η) with no smoothing after inversion. Both parameters reach their absolute maximum at depth point 2880. At depth points 2475 (Figure 16A) and 2505 (Figure 16B), a similar shift can be observed between the isotropic vertical slowness line and the predicted points. The absolute values of the estimated parameters reach their maximum before a rapid change to almost zero. This change can be observed on the border between claystone and the very thin marl layer. Jakobsen and Johansen [17] made a detailed anisotropic approximation in their research on mudrocks. They showed that a negative value of δ is observed in specific kinds of mudrocks. Additionally, they ran a simulation of in situ stress in this kind of rock and examined its impact on Thomsen's parameters, and the values of δ were negative. In Figure 16C, at depth 2700 m the shift between the predicted values and the isotropic curve can be seen; again, it is more noticeable for angles over 10°. This is the center of another claystone layer (bottom 2850 m). We can observe similar values of δ and η as in the first claystone layer. Then (Figure 16D), at depth 2880 in the thin layer of saturated marly claystone, the separation between the isotropic vertical slowness line and the predicted values increases, and the anisotropic parameters have higher absolute values. Going down from depth 2950 to 3400, the thin layer of claystone sits on a thicker layer of claystone and marly claystone. In the clear claystone layer (bottom of the layer at depth 3100 m), δ and η have similar values as in the previous claystone layers. In the layer where claystone transitions into marly claystone, the anisotropic parameters' values become closer to zero. This effect can be observed in Figure 16E, in which the isotopic vertical slowness line crosses the predicted points. A change from negative to positive values in both parameters can also be seen where the marly claystone transitions into dolomitic claystone. In Figure 16F, the theoretical isotropic vertical slowness line passes through the center of the predicted points. Again, this is a very thin marl layer where δ and η are close to zero. These results clearly show that the presented P-wave-only VSP data inversion methodology can be used for lithology studies. Moreover, the estimated anisotropic parameters are reliable and can be used even if the elastic anisotropy of the geological medium is relatively small (on the limit of detection by surface seismic methods). Bandyopadhyay [18] showed that in a laminated geological medium, the sign of Thomsen's parameters can be used as fluid identification, and small values of Thomsen's parameters could be related to the compaction regime.

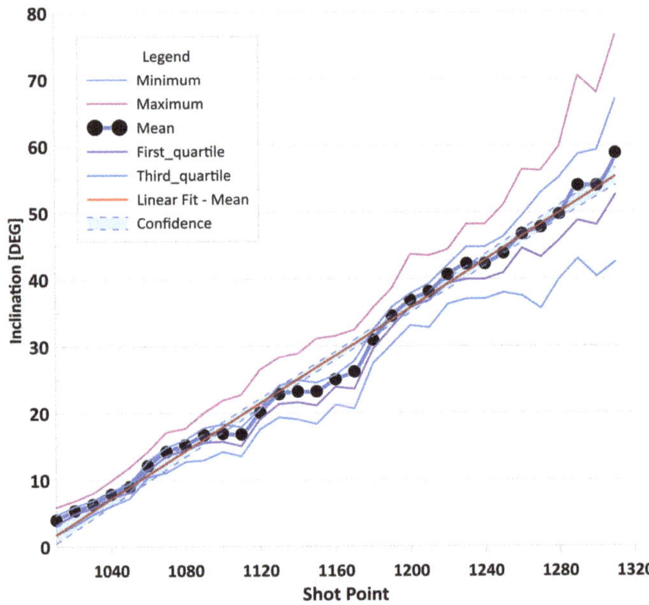

Figure 14. Average P-wave polarization angles for the different shot points along the profile (**black dots**), with trend line (**red**) and basic statistics. Please note that shot point numbers increase with the offset.

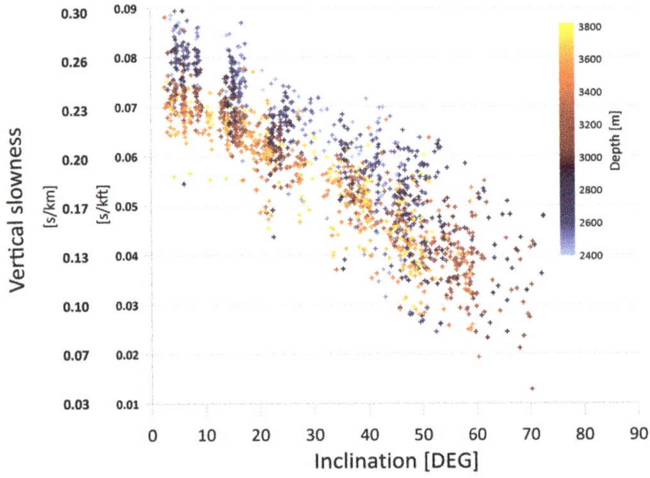

Figure 15. Vertical slowness polarization graph for all offsets and depth points used for calculations.

We also calculated the following parameters: V_r, which is used as a lithology identifier (Equation (20)), and Poisson's ratio using Equation (21). We decided to plot it against the parameters estimated from VSP inversion to analyze its relative change according to the lithology and to compare the sensitivity of these parameters to lithological changes. Thomsen [15] proved that V_r can be used for lithology correlation in rocks with transverse isotropy. Ryan Grigor in 1997 [19] came to similar conclusions. For an isotropic medium, the V_r should directly depend on changes in the Poisson's ratio, which is unique [20]. This factor is not unique for an anisotropic medium [21].

$$V_r = \frac{<V_P, log>}{V_{SX}},\qquad(20)$$

where V_p is sonic P-wave velocity and V_{SX} is sonic wave velocity from horizontal X-component.

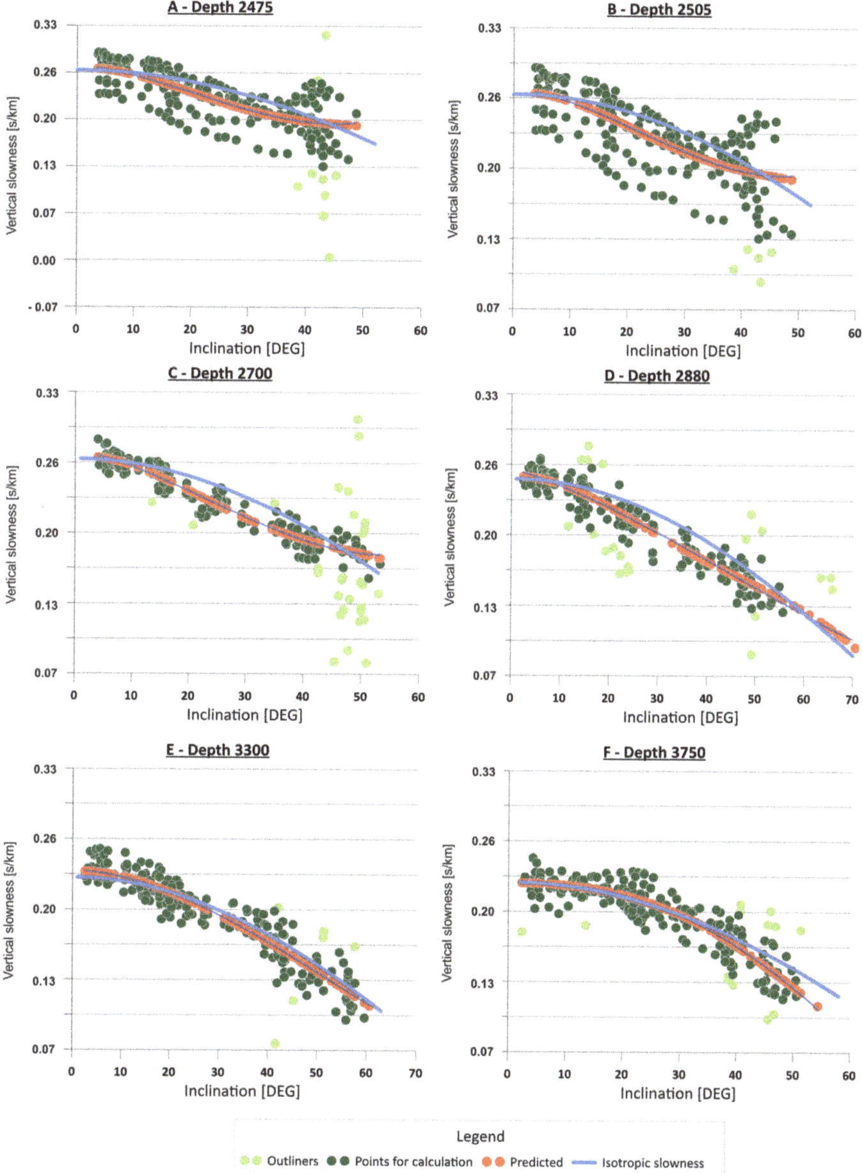

Figure 16. Vertical slowness polarization graph: (**A**) depth 2475 m, (**B**) depth 2505 m, (**C**) depth 2700 m, (**D**) vdepth 2880 m, (**E**) depth 3300 m, (**F**) depth 3750 m. Dark green and light green dots represent all slowness-polarization data points from the VSP survey; dark green dots represent points used for inversion; light green dots are outliers; red dots are points predicted in the inversion process; the blue line is the theoretic isotropic line calculated for average P-wave velocity at this depth.

In the case of well W-1, it does not matter whether V_{SX} or V_{SY} (from the sonic tool) was selected for the calculations as they are almost equal to each other. The average V_{SY}/V_{SX} ratio for the whole investigated depth range is 0.9987.

$$POISSON\ RATIO = \frac{V_r^2 - 2}{2V_r^2 - 2}.$$ (21)

Figure 17 shows sonic V_{SX}, V_{SY}, V_P velocities; δ and η from VSP inversion; and V_r with Poisson ratio. Figure 18 shows the same parameter set averaged to the single value in the layers. In the first marl layer, rapid changes in δ and η are clearly visible, whereas the other parameters remain stable. A similar situation is observed in the thin marl layer located at a depth below 3700 m. Again, there is almost no change in V_{SX} and V_P velocity. A similar situation can be noticed in saturated marly claystone. The relative change in VSP anisotropic parameters is more visible than in V_r. The change in the sonic velocities is noticeable, but their ratio remains almost unchanged. An interesting phenomenon can be observed when claystone transitions into dolomitic claystone, where the signs pf the parameters δ and η change. In Jakobsen and Johansen's [17] laboratory studies on the anisotropy of mudrocks, positive δ was present in the rocks with low porosity, while negative values were related to higher porosity value.

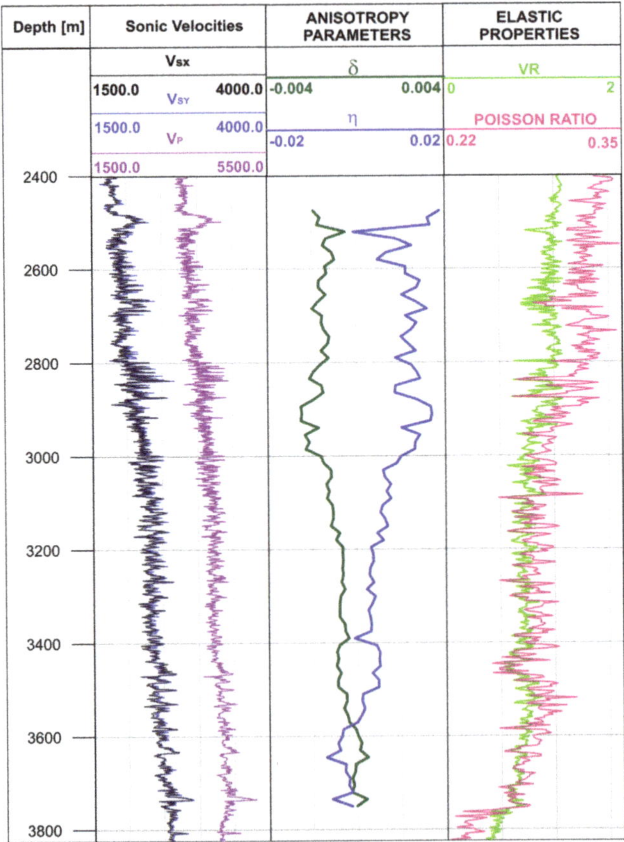

Figure 17. Comparison of sonic V_{SX}, V_{SY}, and V_P velocities (**the first block**); δ and η from VSP inversion (**the second block**); and V_r with Poisson ratio (**the third block**).

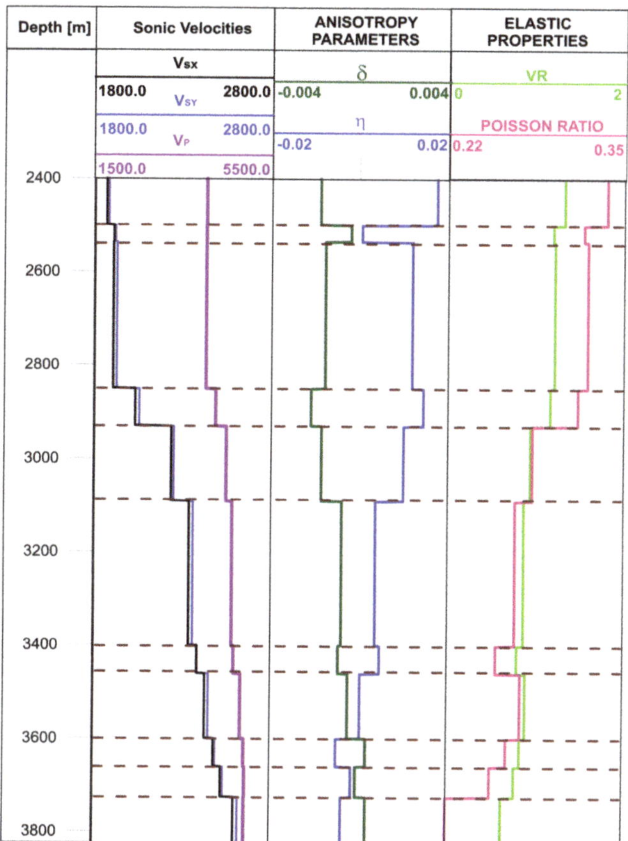

Figure 18. Comparison of averaged lithological layer values of sonic V_{SX}, V_{SY}, and V_P velocities (**the first block**); δ and η from VSP inversion (**the second block**); and V_r with Poisson ratio (**the third block**).

5. Discussion

The P-wave-only inversion of the walkaway VSP data method introduced by Grechka and Mateeva [4] showed good results for local anisotropy estimation on the dataset, with a strong contrast between the near-isotropic salt overburden and sediments with significant anisotropy beneath it. We used this methodology for a very complex survey that was performed in one of the first exploratory wells for shale gas exploration in Poland. We had a priori knowledge about the very low seismic anisotropy strength and the estimated values of Thomsen parameters used for depth migration. Helbig and Thomsen [22] showed the importance of low anisotropy in modern exploration. Accuracy analysis of small seismic anisotropy is crucial to improve surface seismic processing and interpretation. It allows negative effects of noise attenuation to be reduced as well as better coherency and velocity analysis [22]. In this paper, we showed that these parameters can be estimated from a walkaway VSP survey even if the anisotropy strength from the surface seismic survey is calculated to be 2–4%.

These challenging acquisition and workflow tests show that choosing the proper processing scheme is crucial for reliable estimation of polarization angles. We tested four different processing workflows to meet Grechka and Mateeva's [4] criterion for the accuracy of polarization measurements. As the acquisition itself was carried out in problematic conditions that potentially hindered the seismic signal, and the anisotropy of rocks in this place is low, we faced the difficult task of reconstructing the anisotropy parameters. The seismic signal in the case of VSP acquisition goes through the near-surface zone (NSZ) only once; however, in surface seismic surveys the signal travels twice through the NSZ. This zone is one of the most detrimental for seismic signals due to the very low compaction, intensive weathering processes, often-varying rock types, and the complex geology of this zone [23]; however, in this case we observed significant changes in the NSZ zone between sweeps. In this workflow, the first step was noise attenuation, then signal-matching was applied before vertical stacking, and the last step was component rotation; this is the most appropriate method in cases of inclination estimation errors.

The estimated δ and η parameters from walkaway VSP are low: $-0.003 < \delta < 0.001$ and $-0.01 < \eta < 0.02$; however, their relative changes are strongly correlated with lithology. The behavior of δ and η corresponds to other research done for similar rock formations. Estimated anisotropy parameters from VSP show better correlations with lithology than lithology identifier V_r, which was calculated from data from a sonic tool. This is because this sonic tool takes measurements directly in the well, while VSP allows investigation of a larger area whose size is dependent on the offset range. From an offset of approximately 1700 m, parameter η clearly dominates in the solution. Parameter δ dominates in the solution for offset 0–1700 m.

6. Conclusions

This research shows that inversion of P-wave-only data from walkaway VSP surveys can be applied when the geological medium is characterized by very small anisotropy. Additionally, the estimated parameters can be used together with V_r for lithology identification. The sensitivity of VSP anisotropic parameters to lithological changes is higher than the sensitivity of parameters from well logs. In the case of challenging acquisition, when the near-surface zone changes significantly between subsequent sweeps on the same shot point, a special seismic signal processing approach is needed. When workflow 4 is extended with the signal-matching procedure, this allows a significant reduction in the estimation errors of polarization angles.

Author Contributions: Conceptualization, M.Z. and T.D.; methodology, M.Z. and T.D.; validation, M.Z., T.D. and M.S.; formal analysis, M.Z. and T.D.; investigation, M.Z., T.D.; resources, M.S., M.Z. and T.D.; data curation, M.Z. and T.D.; writing—original draft preparation, M.Z.; writing—review and editing, M.Z., T.D. and M.S.; visualization, M.Z.; supervision, T.D.; software, M.Z. and T.D. and M.S.; project administration, M.S. and M.Z. and T.D.; funding acquisition, M.S. All authors have read and agreed to the published version of the manuscript.

Funding: This research was funded by National Center of Research and Development (NCBiR), co-financed by POGC. and Orlen Upstream Sp. z o.o. grant number BG1/GASLUPSEJSM/13 acronym "GASLUPSEJSM".

Acknowledgments: We express our gratitude to POGC for allowing us to use seismic and well-log data acquired as part of the "Polish technologies for shale gas" project. The Walkaway VSP data used in this study were gathered for the Department of Fossil Fuels, a part of the Faculty of Geology, Geophysics and Environmental Protection AGH UST, thanks to the efforts of Michal Stefaniuk, Tomasz Mackowski, and Andrzej Pasternacki. This work and research were done in the context of the following project: "Seismic surveys and their application for the detection of shale gas zones. Selection of optimal acquisition and processing parameters to map the geological structure and distribution of petrophysical and geomechanical parameters of prospective rocks", acronym "GASLUPSEJSM", no. BG1/GASLUPSEJSM/13, a part of Blue Gas I program financed by National Center of Research and Development (NCBiR), co-financed by POGC. and Orlen Upstream Sp. z o.o. We would like to express our sincere thanks to Rafał Kudrewicz from POGC for his valuable remarks.

Conflicts of Interest: The authors declare no conflict of interest.

References

1. Soni, A.; Verschuur, D.J. Full-wavefield migration of vertical seismic profiling data: Using all multiples to extend the illumination area. *Geophys. Prospect.* **2014**, *62*, 740–759. [CrossRef]
2. Steward, R. VSP: An In-Depth Seismic Understanding. *Recorder* **2001**, *26*, 1–13.
3. Dewagan, P.; Grechka, V. Inversion of multicomponent, multiazimuth, walkaway VSP data for the stiffness tensor. *Geophysics* **2003**, *68*, 1022–1031. [CrossRef]
4. Grechka, V.; Mateeva, A. Inversion of P-wave VSP data for local anisotropy: Theory and case study. *Geophysics* **2007**, *72*, 69–79. [CrossRef]
5. Zaręba, M.; Danek, T. VSP polarization angles determination: Wysin-1 processing case study. *Acta Geophys.* **2018**, *66*, 1047–1062. [CrossRef]
6. Hawkins, K.; Leggott, R.; Williams, G.; Kat, H. Addressing anisotropy in 3D pre-stack depth migration: A case study from the Southern North Sea. *Lead. Edge* **2001**, *20*, 528–543. [CrossRef]
7. Kondracki, J. *Regional Geography of Poland*; WN PWN: Warszawa, Poland, 2011.
8. Kasperska, M.; Marzec, P.; Pietsch, K.; Golonka, J. Seismogeological model of the Baltic Basin. In *Opracowanie Map Zasięgu, Biostra-Tygrafia Utworów Dolnego Paleozoiku Oraz Analiza Ewolucji Tektonicznej Przykrawędziowej Strefy Platformy Wschod-Nioeuropejskiej Dla Oceny Rozmieszczenia Niekonwencjon-Alnych Złóż Węglowodorów*; Golonka, J., Bębenek, S., Eds.; Wydawnictwo Arka: Cieszyn, Poland, 2017. (In Polish)
9. Kasperska, M.; Marzec, P.; Pietsch, K.; Golonka, J. Seismo-geological model of the Baltic Basin (Poland). *Ann. Soc. Geol. Pol.* **2019**, *89*, 195–213. [CrossRef]
10. Bajewski, L.; Wilk, A.; Urbaniec, A.; Barton, R. Improving the imaging of under-Zechstein structures based on the reprocessing of 2D seismic in the West Pomerania region. *NG* **2001**, *4*, 195–204. [CrossRef]
11. Galperin, E.I. *The Polarization Method of Seismic Exploration*; Springer: Dordrecht, The Netherlands, 1984.
12. Hake, H. Slant stacking and its significance for anisotropy. *Geophys. Prospect.* **1986**, *34*, 595–608. [CrossRef]
13. Miller, D.; Spencer, C. An exact inversion for anisotropic moduli from phase slowness data. *J. Geophys. Res. Solid Earth* **1994**, *99*, 21651–21657. [CrossRef]
14. Alkhalifah, T.; Tsvankin, I. Velocity analysis for transversely isotropic media. *Geophysics* **1995**, *60*, 1550–1556. [CrossRef]
15. Thomsen, L. Weak elastic anisotropy. *Geophysics* **1986**, *51*, 1954–1966. [CrossRef]
16. Tsvankin, I. *Seismic Signatures and Analysis of Reflection Data in Anisotropic Media*; Pergamon Press: Amsterdam, The Netherlands, 2001.
17. Jakobsen, M.; Johansen, T.A. Anisotropic approximations for mudrocks: A seismic laboratory study. *Geophysics* **2000**, *65*, 1711–1725. [CrossRef]
18. Bandyopadhyay, K. Seismic Anisotropy: Geological Causes and Its Implications to Reservoir Geophysics. Ph.D Thesis, Stanford University, Stanford, CA, USA, 2009.
19. Ryan-Grigor, S. Empirical relationships between transverse isotropy parameters and Vp/Vs: Implications for AVO. *Geophysics* **1997**, *62*, 1942–2156. [CrossRef]
20. Wang, X.Q.; Schubnel, A.; Fortin, J.; David, E.C.; Guéguen, Y.; Ge, H.K. High Vp/Vs ratio: Saturated cracks or anisotropy effects? *Geophys. Res. Lett.* **2012**, *39*, L11307. [CrossRef]
21. Mavko, G.; Mukerji, T.; Dvorkin, J. *The Rock Physics Handbook*; Cambridge University Press: Cambridge, UK, 1998.
22. Helbig, K.; Thomsen, L. 75-plus years of anisotropy in exploration and reservoir seismic: A historical review of concepts and methods. *Geophysics* **2000**, *70*, 9ND–23ND. [CrossRef]
23. Zaręba, M.; Danek, T.; Zając, J. On Including Near-surface Zone Anisotropy for Static Corrections Computation—Polish Carpathians 3D Seismic Processing Case Study. *Geosciences* **2020**, *10*, 66. [CrossRef]

Article

Application of a New Semi-Automatic Algorithm for the Detection of Subsidence Areas in SAR Images on the Example of the Upper Silesian Coal Basin

Maciej Dwornik , Justyna Bała and Anna Franczyk *

Department of Geoinformatics and Applied Computer Science, AGH University of Science and Technology,
al. Mickiewicza 30, 30-059 Kraków, Poland; dwornik@agh.edu.pl (M.D.); jbala@agh.edu.pl (J.B.)
* Correspondence: franczyk@agh.edu.pl

Abstract: The article presents a new method of automatic detection of subsidence troughs caused by underground coal mining. Land subsidence that results from mining leads to considerable damage to subsurface and surface infrastructure such as walls of buildings, road surfaces, and water relations in built-up areas. Within next 30 years, all coal mines are to be closed as part of the transformation of the mining industry in Poland. However, this is not going to solve the problem of subsidence in those areas. Thus, it is necessary to detect and constantly monitor such hazards. One of the techniques used for that purpose is DInSAR (differential interferometry synthetic aperture radar). It makes it possible to monitor land deformation over large areas with high accuracy and very good spatial and temporal resolution. Subsidence, particularly related to mining, usually manifests itself in interferograms in the form of elliptical interferometric fringes. An important issue here is partial or full automation of the subsidence detection process, as manual analysis is time-consuming and unreliable. Most of the proposed trough detection methods (i.e., Hough transform, circlet transform, circular Gabor filters, template recognition) focus on the shape of the troughs. They fail, however, when the interferometric fringes do not have distinct elliptical shapes or are very noisy. The method presented in this article is based on the analysis of the variability of the phase value in a micro-area of a relatively high entropy. The algorithm was tested for differential interferograms form the Upper Silesian Coal Basin (southern Poland). Due to mining, the studied area is particularly prone to various types of subsidence.

Keywords: subsidence detection; image analysis; DInSAR; numerical algorithm

Citation: Dwornik, M.; Bała, J.; Franczyk, A. Application of a New Semi-Automatic Algorithm for the Detection of Subsidence Areas in SAR Images on the Example of the Upper Silesian Coal Basin. *Energies* **2021**, *14*, 3051. https://doi.org/10.3390/en14113051

Academic Editor: Ricardo J. Bessa

Received: 23 April 2021
Accepted: 22 May 2021
Published: 24 May 2021

Publisher's Note: MDPI stays neutral with regard to jurisdictional claims in published maps and institutional affiliations.

Copyright: © 2021 by the authors. Licensee MDPI, Basel, Switzerland. This article is an open access article distributed under the terms and conditions of the Creative Commons Attribution (CC BY) license (https://creativecommons.org/licenses/by/4.0/).

1. Introduction

Differential interferometry synthetic aperture radar (DInSAR) is a well-known, efficient method of monitoring ground deformation on the basis of SAR data from a satellite of an airborne radar [1–3]. It is used to detect vertical ground deformations with high spatial and temporal resolutions and is applied for instance in monitoring of volcanic activity [4], earthquakes [5], or anthropogenic deformations resulting from underground mining [6]. The study is focused on the Upper Silesian Coal Basin (USCB), where subsidence triggered by mining is causing significant damage to the surface and subsurface infrastructure. Every year, coal mines spend even more than a few million euros to repair mining-related damage in a single town or city [7]. Mining has harmful effects on buildings and public infrastructure such as roads, railways, gas and water pipes, power lines, and sewage systems. The damage is expensive to repair, and, in some cases, it can be dangerous for human life and health (e.g., building collapse or gas explosion).

Substantial subsidence of the land and changes to the landform also cause unfavorable changes to the water relations in the area, i.e., disturbances in surface runoff, progressive waterlogging, flooding, and floodplains in subsidence. Flooding is not only caused by the size and distribution of post-mining subsidence but also by hydrotechnical activities undertaken by mines to repair the damage caused [8].

223

Early detection of subsidence allows for taking preventive measures. On a small scale, they could include filling voids or strengthening the site. On a large scale, they could mean relocating people or the infrastructure [9].

Monitoring and studying the environmental effects of mining over large areas is difficult and depends on the size of the area and its location. It is also time-consuming and costly [10]. A far better solution that has been used in recent years is monitoring using DInSAR technology. It can be carried out without ground/field measurements such as surveying, geotechnical engineering, or mining. DInSAR is currently being used by some of the municipalities in the Upper Silesian Coal Basin, located in the impact zone of a coal mine.

The DInSAR technique utilizes two images from the same area that were taken by the same SAR system at different times. First, the images must be co-registered, and then the phase difference ($\Delta\phi$) is calculated for each pixel of the image. In that way, a flattened interferogram is obtained. The digital elevation model (DEM) is used in order to remove the component related to the topography of the studied region from $\Delta\phi$. DEM is converted into a synthetic interferogram that, in turn, is subtracted from the flattened interferogram. As a result, a so-called differential interferogram is obtained. Its phase properties can be directly related to the ground deformation. Subsidence connected among others with mining usually manifests itself in differential interferograms in the form of elliptical interferometric fringes (Figure 1).

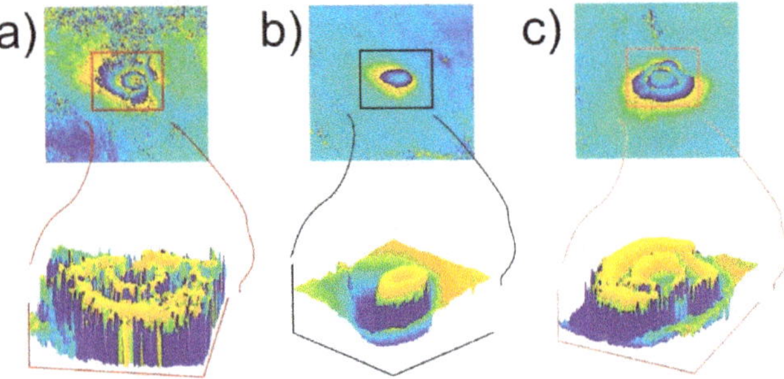

Figure 1. Examples of subsidence troughs visible on interferograms computed for radar images recorded on 4 April 2017 and 16 April 2017 (**a,b**) and 24 December 2017 and 5 January 2018 (**c**) for the area of the Upper Silesian Coal Basin, Poland (Sentinel-1A, descending).

A large number of methods are used to detect subsidence troughs in DInSAR images, namely, Gabor transformation [7], Hough transformation [11], template recognition [12], convolutional neural networks [13], or circlet transform [14]. Each of them have their limitations that render subsidence detection in noisy satellite images inefficient. Most of them can be used to aid subsidence detection in large SAR images but they are insufficient for fully automated work.

The authors propose a new subsidence detection method based on interferogram area searching, where pixels form planes with a slight slope and a relatively high entropy. At the first stage, the effectiveness of the method was tested for 134 interferogram fragments (512 × 512 px) with subsidence troughs. Next, two larger (1400 × 2000 px) areas of the USCB were selected. Due to intensive mining in that area, the subsidence troughs found there are characterized by various extensions and various spatial sizes. As subsidence troughs occur both in urban and rural areas, they can be more or less affected by temporal decorrelation. Such a diversity of the patterns in the interferograms allows for a high-quality verification of the proposed automated method of subsidence detection.

2. Methodology

In interferograms, subsidence troughs take the shape of elliptical areas with a characteristic radial variability of the phase value. The phase value in the areas of subsidence changes monotonically from $-\pi$ to π. After that monotonic increase of the value, a rapid change of the phase value from π to $-\pi$ takes place (Figure 2a). In the interferometric images of the areas without subsidence, the phase value is constant or takes random values in the range of $[-\pi; \pi]$ (Figure 2b).

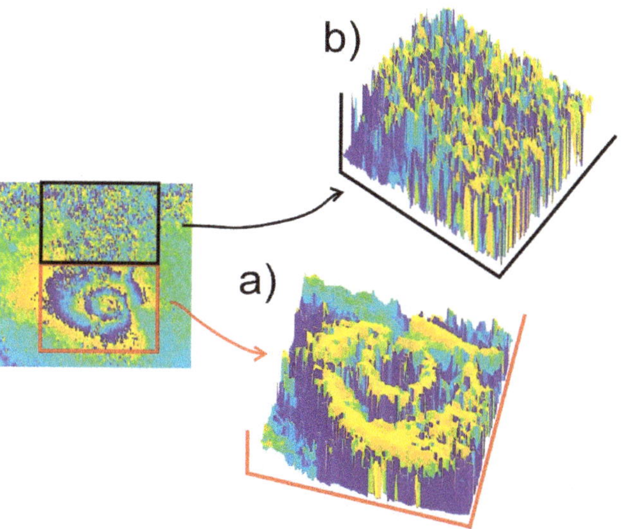

Figure 2. An example of linear monotonic change of the phase value preceded by a step change in its value in interferograms computed from subsidence (**a**) and the random change of the phase value in interferograms characteristic for areas with no subsidence (**b**). The interferogram was computed from radar images taken on 4 April 2017 and 16 April 2017 for the area of Upper Silesia.

Classic methods of subsidence searching utilize the elliptical shape of a trough (Hough). Very often, troughs do not meet that criterion, if at all (e.g., only a part of the trough is visible). The method presented here uses the fact that within a step change, the phase increases linearly from $-\pi$ to π. The other criteria (entropy, standard deviation, or the flatness of the phase histogram) are used to eliminate areas of linear phase change but not related to the subsidence.

The detailed diagram of the algorithm working principle is presented in Figure 3.

First, noise is removed from the image using the median or the adaptive Wiener filters. Next, the image is used as input for a number of simultaneous processes:

1. filtration of the standard deviation of the interferogram;
2. filtration of the image entropy;
3. analysis of the image histogram uniformity;
4. plane searching operation.

The areas with a phase value step change (from π to $-\pi$) are identified by standard deviation filtration procedures and entropy filtration in a square, moving window. In both cases, such areas of the interferogram are searched for where the normalized standard deviation and entropy value is higher than the threshold value.

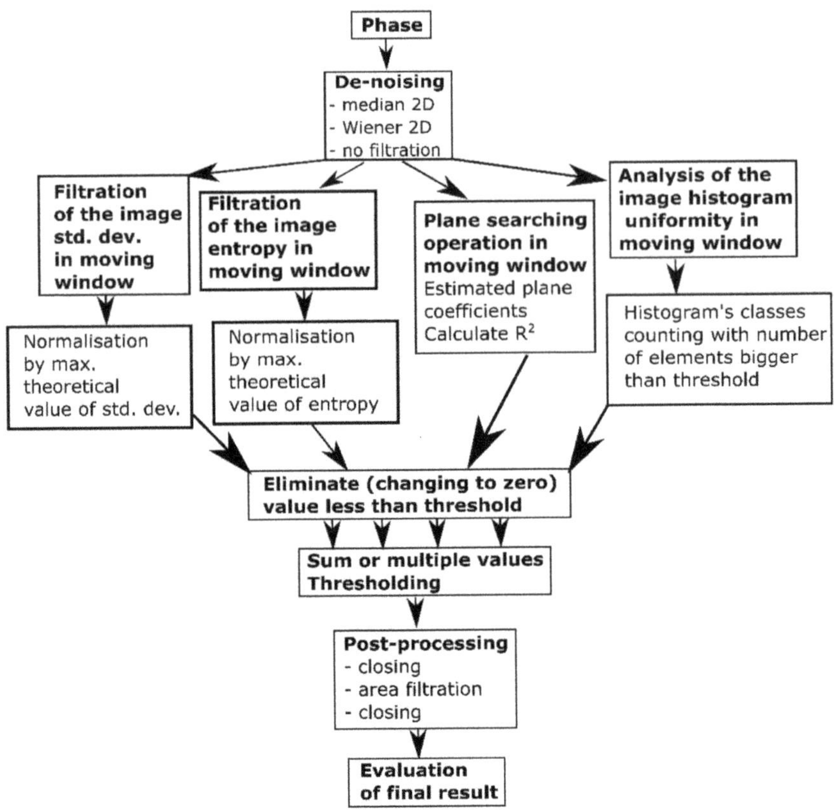

Figure 3. Working principle of the algorithm detecting subsidence troughs in interferometric images.

The analysis of the phase histogram uniformity is carried out in order to eliminate areas where a phase step change occurs that is not preceded by a monotonic increase of the phase value from $-\pi$ do π. This analysis is carried out for a specific subarea of the interferogram and for a specific class number. The subareas whose histograms considerably differ from the uniform distribution do not contain a monotonic change of the phase value. Thus, they are classified as subareas where subsidence troughs do not occur.

The procedure of plane detection is the most important element of the presented algorithm of subsidence detection in interferograms. Its aim is to identify those fragments of an interferogram whose phase values can be approximated by the equation of the plane with a particular slope angle. A fragment of the interferogram is approximated by a plane described by Equation (1)

$$f(x,y) = ax + by + c \tag{1}$$

where (x,y) are the local coordinates of the interferogram. A fragment is classified as an area where a monotonic change of the phase value occurs if the a and b coefficient absolute values of the approximating equation as well as the value of the fitting coefficient R^2 of the approximated plane and the phase values of the analyzed fragment are within the assumed range of values. Figure 4 shows the approximation of three fragments of the interferogram by the plane described by Equation (1).

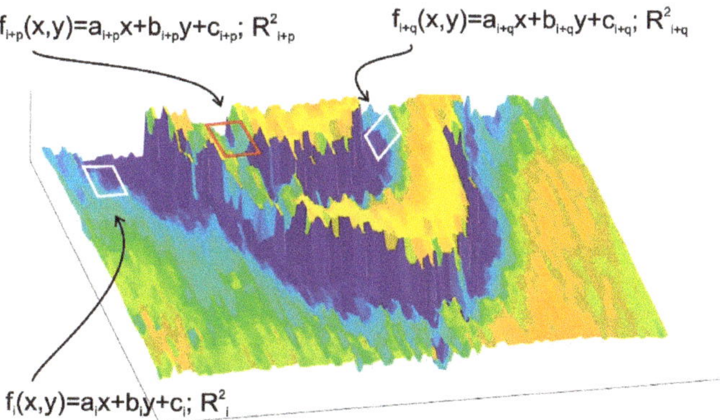

Figure 4. Fitting of interferogram fragments to the plane equation. White squares indicate significant monotonic phase value variability (a, b, and R^2 values high enough). The a, b, and R^2 values of the red square are too low. The presented fragment of the trough was located in the interferogram computed from the radar images recorded on 4 April 2017 and 16 April 2017 for the area of the Upper Silesian Coal Basin, Poland (Sentinel-1A, descending).

The results of all the procedures presented above were taken into account when constructing the output image. In addition, the output image was subjected to final processing that consisted in removing small-area elements and morphological closing. The latter was to remove areas lying closely to one another and to fill small holes in those areas.

3. Experimental Analysis

The algorithm was analyzed using MathWorks MatLAB 2020b software. It was tested for two interferogram sets, which differed in the place where the radar images were taken, the time of their taking, and the size of the areas from which the interferograms were computed. The aim was to find the optimal value of the algorithm parameters, allowing for the detection of all the subsidence troughs occurring in the analyzed sets with the number of false classifications as low as possible. The following parameters determining the efficiency of the algorithm were tested:

- window sizes of median and Wiener filtration (Table 1, columns 2 and 3);
- threshold value adopted at the stage of phase entropy filtration (Table 1, column 4);
- threshold value adopted at the stage of filtration of interferogram phase standard deviation (Table 1, column 5);
- size of the interferogram fragment approximated by Equation (1), fitting threshold value R^2, the weight with which the results of detection of monotonic phase variation areas were taken into account when constructing the resulting map (Table 1, columns 6–8);
- number of histogram classes used at the stage of distribution uniformity analysis (Table 1, column 9);
- resulting map computing method (Table 1, column 10);
- post-processing parameters for removing small areas and morphological closing (Table 1, columns 11 and 12).

The values of the tested parameters for each of the 35 combinations are presented in Table 1.

Table 1. Parameter values adopted at the stage of subsidence detection algorithm tests.

ID	Preprocessing		Thresholds		Plane Filtration			Hist	Sum	Final	
	Med.	Wnr	Entr.	St. dev.	Size	Thresh	Weight	Class	Mult	Thresh	Area
1	7	9	0.2	0.2	13	0.025	0.5	25	S	3	500
2	7	9	0.2	0.2	17	0.025	0.5	25	S	3	500
3	7	9	0.2	0.2	9	0.025	0.5	25	S	3	500
4	7	9	0.2	0.2	11	0.03	0.5	25	S	3	500
5	7	9	0.2	0.2	9	0.03	0.5	25	S	3	500
6	7	9	0.2	0.2	9	0.025	0.5	20	S	3	500
7	7	9	0.2	0.2	9	0.025	0.5	40	S	3	500
8	5	9	0.2	0.2	9	0.025	0.5	25	S	3	500
9	5	9	0.2	0.2	9	0.03	0.5	25	S	3.1	500
10	7	9	0.2	0.2	9	0.03	0.5	25	M	0.25	500
11	7	9	0.2	0.2	9	0.03	0.5	25	M	0.2	500
12	7	9	0.2	0.2	9	0.03	0.5	25	M	0.18	500
13	7	9	0.2	0.2	9	0.03	0.2	25	M	0.18	500
14	7	9	0.2	0.2	9	0.03	0.7	25	M	0.18	500
15	7	9	0.2	0.2	9	0.03	0.7	25	S	3.1	450
16	7	9	0.2	0.2	9	0.03	0.5	25	S	3.1	550
17	7	-	0.2	0.2	9	0.03	0.5	25	S	3	500
18	7	-	0.2	0.2	9	0.03	0.5	25	S	3.2	500
19	7	9	0.2	0.2	13	0.035	0.5	25	S	3	500
20	7	9	0.2	0.2	13	0.035	0.5	40	S	3	500
21	7	9	0.2	0.2	13	0.03	0.5	25	S	3.3	500
22	-	9	0.2	0.2	13	0.03	0.5	40	S	3	500
23	-	9	0.2	0.2	13	0.03	0.5	25	S	3	500
24	7	5	0.2	0.2	13	0.025	0.5	25	S	3	600
25	7	5	0.2	0.2	13	0.025	0.5	25	S	3	500
26	7	9	0.2	0.2	13	0.03	0.5	40	S	3.3	500
27	7	9	0.2	0.2	17	0.03	0.5	40	S	3.3	500
28	7	9	0.2	0.2	15	0.03	0.5	40	M	0.25	550
29	7	9	0.2	0.2	15	0.03	0.5	40	M	0.15	500
30	9	11	0.2	0.2	17	0.03	0.5	40	S	3.3	500
31	-	-	0.2	0.2	9	0.03	0.5	25	S	3.5	500
32	-	-	0.2	0.2	9	0.03	0.5	25	S	3.3	500
33	-	-	0.25	0.25	13	0.03	0.5	25	S	3.3	500
34	-	-	0.25	0.25	13	0.03	0.5	25	M	0.4	550
35	-	-	0.25	0.25	13	0.03	0.5	25	M	0.35	550

3.1. Aplication of the Proposed Algorithm for the First Dataset Covering a Small Area

The algorithm of subsidence detection was analyzed using a test set of 134 images, 512 × 512 px. The images were selected from the interferograms computed from radar images recorded during the Sentinel-1 mission—the first of the European Space Agency missions developed for the Copernicus initiative. The test sets were selected from the Sentinel data products recorded in the years 2017–2018 in the USCB, published under an open access license. The test set comprised interferogram fragments containing as many distinct troughs as possible. An example of three visible troughs from the test set is presented in Figure 1. Each of the images was thoroughly analyzed, which allowed for manual determination of subsidence troughs. The spatial coordinates and the size of 286 such identified troughs in the analyzed test set constituted the so-called reference set. The reference set was used to test the trough detection algorithm. The results of the algorithm—the coordinates of the troughs—were compared with the reference indications, and on that basis, the effectiveness of the algorithm was measured.

The trough detection algorithm was tested not only for its efficiency—the number of detected troughs—but also for false detections, identifying some interferogram parts as troughs. A subsidence trough was considered located correctly if it contained at least a 25-pixel area classified by the algorithm as a subsidence trough. Other interferogram areas classified as troughs that were inconsistent with those determined manually were considered incorrect detections—so-called false alarms. Additional parameters used in the assessment of the algorithm were the number of all detections and the overall area of the interferogram that was classified by the algorithm as subsidence. The algorithm efficiency results for the adopted 35 parameter value combinations (Table 1) are presented in Figure 5.

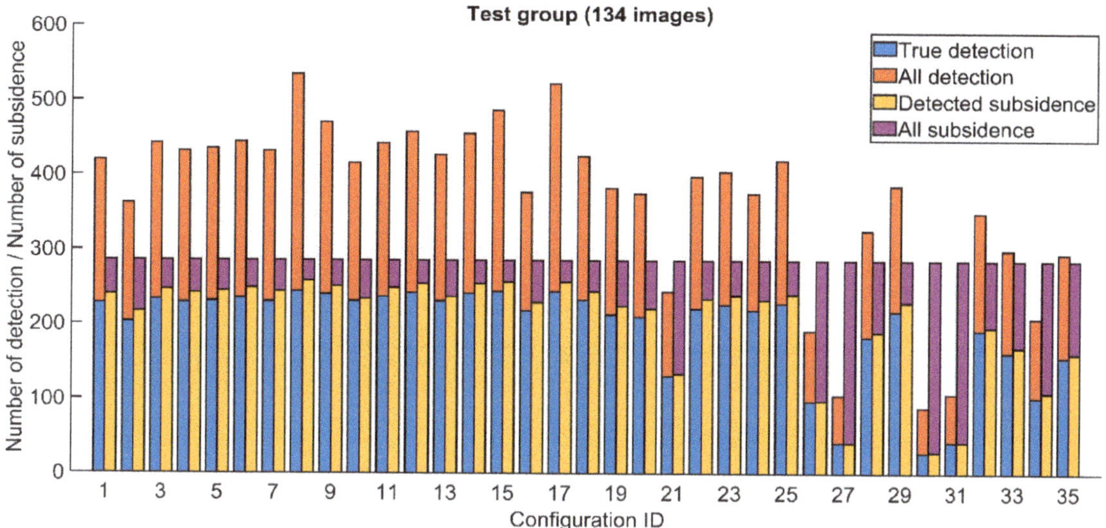

Figure 5. Efficiency of the trough detection algorithm for the 134-element test set for all 35 combinations of parameter values (Table 1).

The presented results show a strong relationship between the parameter values and the efficiency of the algorithm. None of the 35 tested combinations resulted in the detection of all the 286 subsidence troughs. The highest number of troughs was detected using combination 8 but it was at the cost of all detections (535) and a high rate of false alarms (54.4%). In general, for the majority of the tested configurations, it was evident that as the algorithm efficiency increased so did the number of all detections and false detections. The above conclusion does not apply to the cases where the detection number was much lower than the number of existing subsidence areas (particularly configurations 26, 27, 30, and 31). It must be noted that in many cases the number of detected troughs was higher than the number of correct detections of the algorithm. This occurred when subsidence areas were close to one another and the area classified by the algorithm as subsidence matched more than one trough manually indicated in the interferogram.

The efficiency result of the subsidence detection algorithm with an emphasis on the results of particular algorithm operations is presented in Figures 6 and 7.

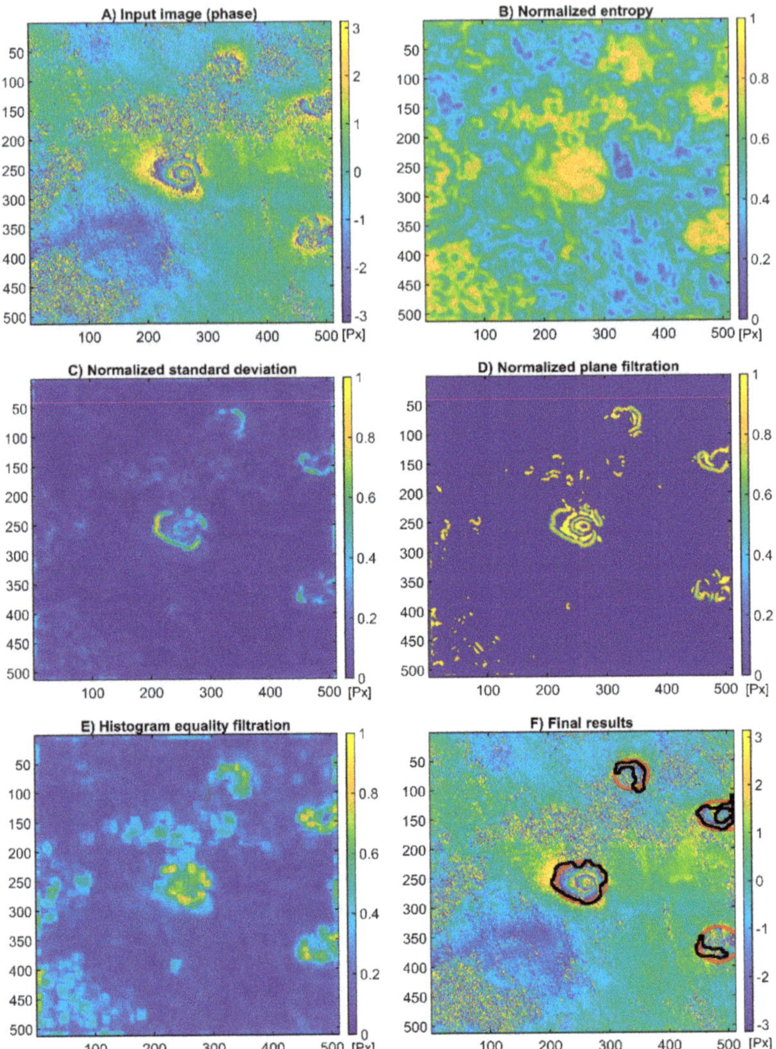

Figure 6. An example of subsidence detection algorithm efficiency with a particular focus on the results of particular algorithm operations. The interferogram computed from the radar images recorded on 4 April 2017 and 16 April 2017 for the area of the Upper Silesian Coal Basin, Poland (Sentinel-1A, descending).

In the subsidence detection, parameter values for combination 14 were used (Table 1), which was characterized by a relatively high detectability (255 detected troughs out of 286 existing ones), an average number of all detections (435), and an average coefficient of false detections (215 out of 456).

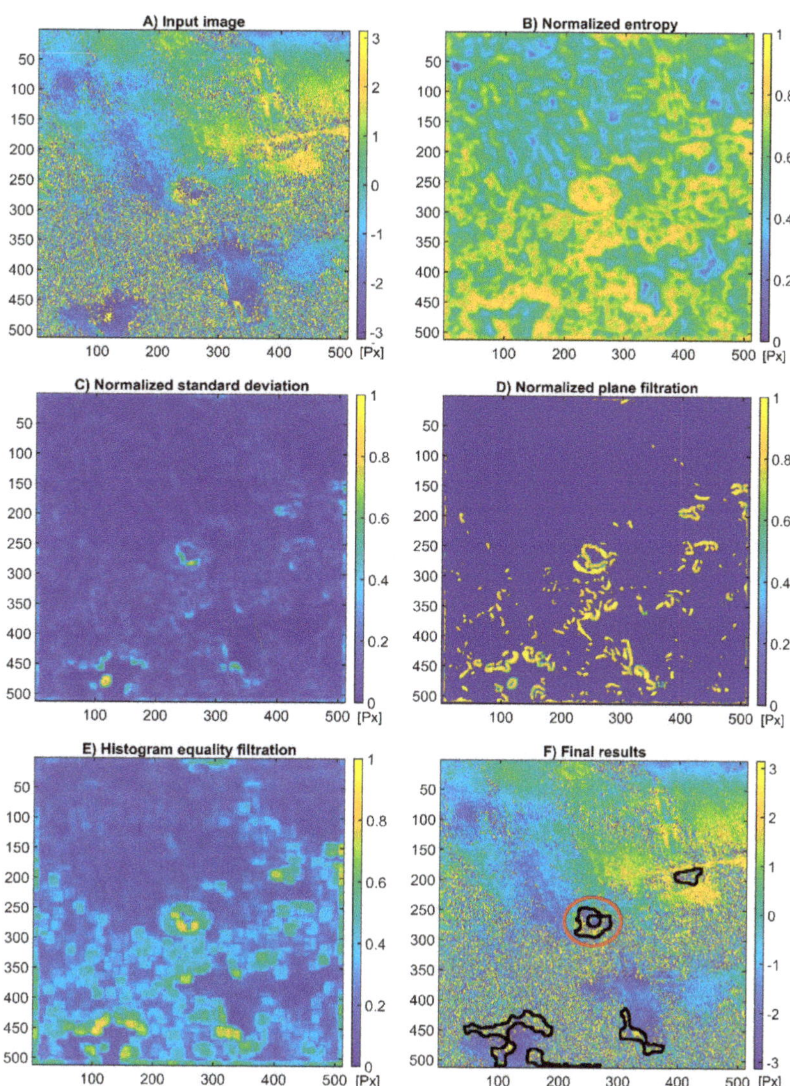

Figure 7. Efficiency of subsidence detection algorithm with a particular focus on the results of its particular operations. The interferogram computed from the radar images recorded on 21 July 2017 and 2 August 2017 for the area of the Upper Silesian Coal Basin, Poland (Sentinel-1A, descending).

Figure 6 shows an interferogram with four subsidence areas of various degrees of development and noise. Those areas were identified manually and marked with a red line (Figure 6F). In the central part of the interferogram, there was a well-developed trough with two clearly marked phase jump lines—phase value changes from $-\pi$ to π. In the east and northeast part, three more troughs were visible, but they were fragmentary and noisy. Figure 6B–E show the results of each of the four algorithm operations. Apart from the true detections, the particular operations also indicated areas without subsidence. The areas incorrectly classified as subsidence troughs were either small or fragmented. Fragmented areas were weakened in the final image construction phase. Single, small areas classified as troughs were removed as a result of the final image postprocessing. The result of the subsidence detection algorithm operations is marked in Figure 6F with a black line.

In the case of very noisy images (Figure 7), the algorithm correctly classified subsidence areas. The algorithm detections contained all true detections as well as false detections (Figure 7). It occurred when uniform areas lay close to noisy areas. In Figure 7A,F, uniform areas with a low phase value were visible in the southern part of the interferogram (vast, blue areas), whereas uniform areas with high phase values, marked with a yellow line, were found in the northeast part.

The incorrect classification was due to the low-pass filtration, which resulted in the formation of areas with a linear phase change at the junction of uniform value zone and the high noise area. These areas can be approximated by a plane equation. Figure 7 shows a detection result for a very noisy interferogram, where apart from the existing trough in that area (red line), four other areas were incorrectly classified as subsidence troughs.

3.2. Aplication of the Proposed Algorithm for the Second Dataset Covering a Large Area

The subsidence detection algorithm was also tested for a test set covering a large area. The tests were carried out on two large (1400 × 2000 px each) interferograms (Area 1 and Area 2) with a high concentration of subsidence troughs. The interferograms were computed from images recorded for the Upper Silesian Coal Basin (USCB). The location of the USCB in Poland is shown in Figure 8.

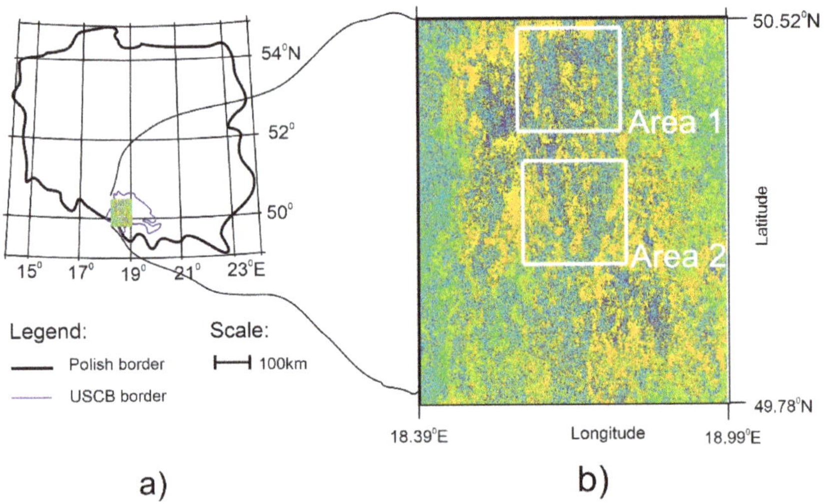

Figure 8. Differential, unwrapped interferogram generated for recordings performed on 10 and 22 October 2016. It was computed from radar images of the Upper Silesian Coal Basin area, Poland (**a**). White squares in the interferogram (**b**) indicate areas with a high concentration of interferometric fringes. These two areas (Area 1 and Area 2) were selected for testing the proposed detection method.

Like for the test set containing a lot of small sections, the test stage was preceded by a manual location of subsidence visible in the interferograms. As a result, 19 subsidence areas were identified and marked. Like for the first set, 35 combinations of the parameter values shown in Table 1 were tested. The results are presented in Figure 9.

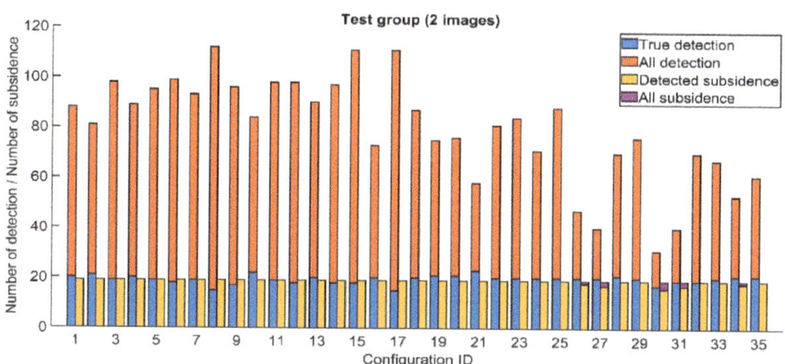

Figure 9. The result of the subsidence detection algorithm operations on a test set covering a large area for 35 parameter combinations.

For the majority of the tested parameter combinations (30 out of 35), all the existing troughs were detected (Table 1). The results of the algorithm efficiency obtained for particular parameter combinations differed in the number of classified areas as subsidence areas and in the number of false detections. The 100% efficiency of the algorithm was also related to the increase in the false classifications. In test set 1, 48% of the detections made by the algorithm were incorrect. In test set 2, that figure was 73%. Figure 10 shows the results of the analysis of particular algorithm operations for parameter configuration 21. When tested with the use of this parameter combination, the algorithm showed 100% detection efficiency, maintaining the lowest number of incorrect detections out of all the analyzed cases (35 out of 58).

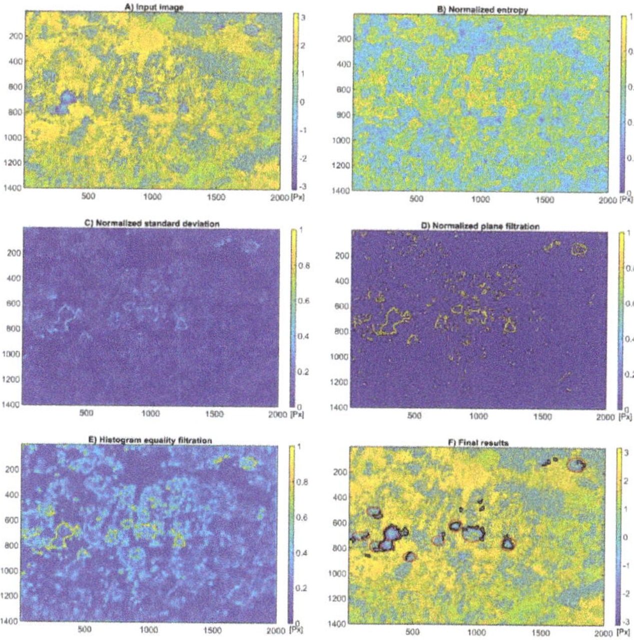

Figure 10. Example of trough detection algorithm with a special focus on the results of particular algorithm operations. The interferogram was computed for radar images recorded on 10 October 2016 and 22 October 2016 for the area of the Upper Silesian Coal Basin, Poland (Sentinel-1A, descending).

In the analyzed image, both manually located troughs and the algorithm-detected troughs were concentrated mainly in the middle lane of the image. Most of the false detections occurred in high-noise parts surrounded by areas of uniform phase values.

4. Discussion

The efficiency of the subsidence detection algorithm depends on the parameters describing its operations. The research carried out in this paper showed that for both test sets there were parameter value combinations that ensured very high efficiency in subsidence detection. One optimal parameter combination that would guarantee high efficiency for both test sets was not found. The comparison of the subsidence detection results for both test sets is shown in Figure 11.

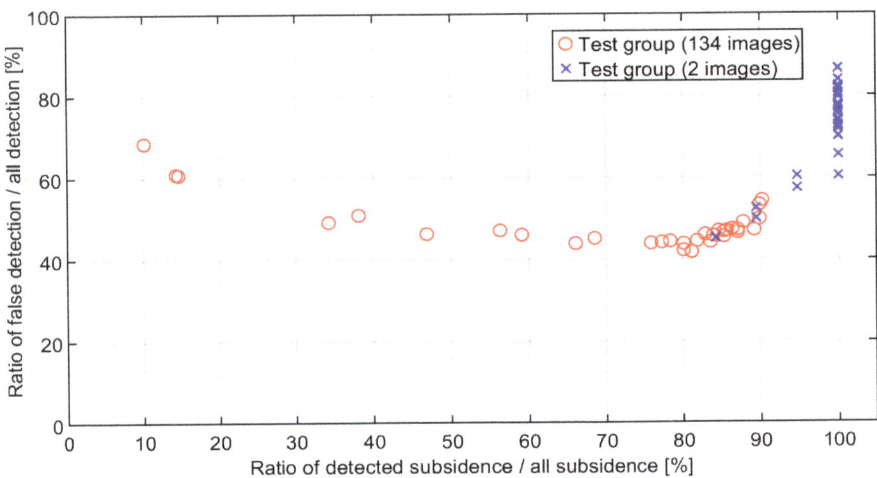

Figure 11. The comparison of the subsidence detection results for both test sets and all the parameter value combinations determining the algorithm efficiency.

The graph shows the relationship between the percentage of the correct classifications of subsidence and the percentage of the incorrect classifications of the interferogram fragments as subsidence areas. The comparison was made for both test sets and for all the parameter combinations that determine the efficiency of the algorithm. The results obtained for both datasets are characterized by a different dynamic of change, which may indicate that both sets were disjoint and there was no one optimal algorithm parameter combination that can be used for any test set that would guarantee satisfactory efficiency in trough detection. The selection of the optimal parameter value combination should be made individually for each area.

As the algorithm efficiency increased so did the percentage of areas that the algorithm incorrectly classified as subsidence. It was observed that the false detections were of the same nature in both test sets. The vast majority of them occurred at the boundaries of noisy parts surrounded by uniform areas with the phase value amounting to $-\pi$ or π. A considerable percentage of false detections also occurred when small areas of a constant phase value were surrounded in the interferogram by noisy areas.

The efficiency of the algorithm was low when troughs were located in noisy interferogram fragments or if the phase change line from $-\pi$ to π was not fully formed.

5. Conclusions and Further Works

The algorithm presented in this article is highly efficient in subsidence detection. It was proved that for both test sets, there was a parameter combination for which the algorithm showed 80–100% subsidence detection efficiency with 40–50% false alarms. The application of the algorithm made it possible to considerably decrease the area that would have had to be controlled manually. It should be emphasized at the same time that due to the cyclical nature of the study and the fact that a subsidence process is long, a false detection in one test does not preclude correct ones in subsequent tests. It particularly applies to high-noise areas that either imitated subsidence or made it impossible to detect it correctly.

What helps to reduce the area for manual analysis is also the fact that both correct detections and false alarms are concentrated in small areas. In such a case, it is vital that the number of detections are not significantly higher than the actual number of subsidence troughs.

The method also showed that its parameters should be adjusted to the parameters of the area. It was proved that thresholding and pre-filtering are of huge importance. In the next stage, the authors plan to create a method of automated parameter selection for particular areas (subareas) and to refine the method of false detection elimination by adding other verification methods such as machine learning algorithms.

Author Contributions: Conceptualization, M.D., A.F., J.B.; methodology, M.D.; software, M.D.; validation, M.D., A.F., J.B.; formal analysis, M.D., A.F., J.B.; investigation, M.D., A.F., J.B.; resources, J.B.; data curation, M.D., A.F., J.B.; writing—original draft preparation, A.F.; writing—review and editing, J.B.; visualization, M.D., A.F.; supervision, J.B.; project administration, A.F.; funding acquisition, M.D., A.F., J.B. All authors have read and agreed to the published version of the manuscript.

Funding: This work was supported by the Faculty of Geology, Geophysics and EnvironmentalProtection, AGH University of Science and Technology, as a part of statutory project.

Institutional Review Board Statement: Not applicable.

Informed Consent Statement: Not applicable.

Data Availability Statement: The data can be accessed upon request from any of the authors.

Conflicts of Interest: The authors declare no conflict of interest.

References

1. Gupta, R.P. *Remote Sensing Geology*, 3rd ed.; Springer: Berlin, Germany, 2017.
2. Sedighi, M.; Arabi, S.; Nankali, H.R.; Amighpey, M.; Tavakoli, F.; Soltanpour, A.; Motagh, M. Subsidence Detection Using InSAR and Geodetic Measurements in the North-West of Iran. In Proceedings of the 2009 Fringe Workshop, Frascati, Italy, 30 November–4 December 2009.
3. Porzycka, S.; Strzelczyk, J. Preliminary results of ground deformations monitoring within mining area of "Prosper-Haniel" coal mine. In Proceedings of the 12th International Multidisciplinary Scientific GeoConferences (SGEM), Albena, Bulgaria, 17–22 June 2012; pp. 895–899.
4. Lu, Z.; Dzurisin, D. *InSAR Imaging of Aleutian Volcanoes: Monitoring a Volcanic Arc from Space*; Springer Science & Business Media: Chichester, UK, 2014.
5. Stramondo, S.; Chini, M.; Bignami, C.; Salvi, S.; Atzori, S. X-, C-, and L-band DInSAR investigation of the April 6, 2009, Abruzzi earthquake. *IEEE Geosci. Remote Sens. Lett.* **2010**, *8*, 49–53. [CrossRef]
6. Nádudvari, A. Using radar interferometry and SBAS technique to detect surface subsidence relating to coal mining in Upper Silesia from 1993–2000 and 2003–2010. *Environ. Socio Econ. Stud.* **2016**, *4*, 24–34. [CrossRef]
7. Porzycka-Strzelczyk, S.; Rotter, P.; Strzelczyk, J. Automatic Detection of Subsidence Troughs in SAR Interferograms Based on Circular Gabor Filters. *IEEE Geosci. Remote Sens. Lett.* **2018**, *15*, 873–876. [CrossRef]
8. Król, Ż.; Mikrut, S.; Gabryszuk, J.; Postek, P.; Mazur, A. Changes in the structure of land use in the areas of mining damage. *Ecol. Eng.* **2015**, *44*, 26–33. [CrossRef]
9. Gamble, J.C.; Gray, R.E. Mine subsidence control. In *General Geology, Encyclopedia of Earth Science*; Springer: Boston, MA, USA, 1988.
10. Kopeć, A.; Trybała, P.; Głąbicki, D.; Buczyńska, A.; Owczarz, K.; Bugajska, N.; Kozińska, P.; Chojwa, M.; Gattner, A. Application of Remote Sensing, GIS and Machine Learning with Geographically Weighted Regression in Assessing the Impact of Hard Coal Mining on the Natural Environment. *Sustainability* **2020**, *12*, 9338. [CrossRef]
11. Klimczak, M.; Bała, J. Application of the Hough Transform for subsidence troughs detection in SAR images. In Proceedings of the 17th International Multidisciplinary Scientific GeoConferences (SGEM), Albena, Bulgaria, 29 June–5 July 2017; pp. 819–826.

12. Huang, J.; Deng, K.; Fan, H.; Lei, S.; Yan, S.; Wang, L. An Improved Adaptive Template Size Pixel-Tracking Method for Monitoring Large-Gradient Mining Subsidence. *J. Sens.* **2017**, *2017*, 3059159. [CrossRef]
13. Rotter, P.; Muron, W. Automatic Detection of Subsidence Troughs in SAR Interferograms based on convolutional neutral networks. *IEEE Geosci. Remote Sens. Lett.* **2021**, *18*, 82–88. [CrossRef]
14. Bała, J.; Dwornik, M.; Franczyk, A. Automatic subsidence troughs detection in SAR interferograms using circlet transform. *Sensors* **2021**, *21*, 1706. [CrossRef] [PubMed]

Article

Application of Hydraulic Backfill for Rockburst Prevention in the Mining Field with Remnant in the Polish Underground Copper Mines

Karolina Adach-Pawelus * and Daniel Pawelus

Faculty of Geoengineering, Mining and Geology, Wroclaw University of Science and Technology, 50-370 Wroclaw, Poland; daniel.pawelus@pwr.edu.pl
* Correspondence: karolina.adach@pwr.edu.pl

Abstract: In the polish underground copper mines owned by KGHM Polska Miedz S.A, various types of room and pillar mining systems are used, mainly with roof deflection, but also with dry and hydraulic backfill. One of the basic problems associated with the exploitation of copper deposits is rockburst hazard. Aa high level of rockburst hazard is caused by mining the ore at great depth in difficult geological and mining conditions, among others, in the vicinity of remnants. The main goal of this study is to investigate how hydraulic backfill improves the geomechanical situation in the mining filed and reduce rockburst risk in the vicinity of remnants. Numerical modeling was conducted for the case study of a mining field where undisturbed ore remnant, 40 m in width, was left behind. To compare the results, simulations were performed for a room and pillar mining system with roof deflection and for a room and pillar mining system with hydraulic backfill. Results of numerical analysis demonstrate that hydraulic backfill can limit rock mass deformation and disintegration in the mining field where remnants have been left. It may also reduce stress concentration inside or in the vicinity of a remnant, increase its stability, as well as prevent and reduce seismic and rockburst hazards. Hydraulic backfill as a local support stabilizes the geomechanical situation in the mining field.

Keywords: hydraulic backfill; remnant; seismic and rockburst hazard; numerical modeling; Polish copper mines

Citation: Adach-Pawelus, K.; Pawelus, D. Application of Hydraulic Backfill for Rockburst Prevention in the Mining Field with Remnant in the Polish Underground Copper Mines. *Energies* **2021**, *14*, 3869. https://doi.org/10.3390/en14133869

Academic Editors: Krzysztof Skrzypkowski and Maxim Tyulenev

Received: 21 May 2021
Accepted: 18 June 2021
Published: 27 June 2021

Publisher's Note: MDPI stays neutral with regard to jurisdictional claims in published maps and institutional affiliations.

Copyright: © 2021 by the authors. Licensee MDPI, Basel, Switzerland. This article is an open access article distributed under the terms and conditions of the Creative Commons Attribution (CC BY) license (https://creativecommons.org/licenses/by/4.0/).

1. Introduction

Seismic and rockburst hazards have represented primary hazards in Polish underground cooper mines since the beginning of ore exploitation in the region. The first strong seismic event, having a magnitude of 2.8, took place on 31 July 1972. With the progress of mining operations, the number and energy of tremors regularly increased and some of them caused rockbursts [1].

Dynamic phenomena occur as a result of rock mass destabilization, leading to the release of potential energy from rocks. Tremors are caused by mining works, which disrupt the original stress state of the rock mass [2,3]. A rockburst is a dynamic phenomenon caused by a mining tremor which leads to sudden and violent destruction or damage of the excavation along with all of the involved consequences [2,3]. Rockbursts are responsible for many mining accidents and damaged excavations. They generate financial losses and disrupt the operational continuity of the mining facility.

The literature knows many classifications of rockbursts (e.g., [4–6]). Essentially, two types of dynamic phenomena are distinguished. The first is directly associated with stopes, while the second is linked to primary tectonic discontinuities [4]. Rockbursts of the first type most frequently occur in the vicinity of excavations. They are the result of both stress redistribution around the excavations and the formation of stress concentration zones. The following may be considered as belonging to this group of rockbursts [6]: strain bursts, pillar bursts, and face bursts. Dynamic phenomena related to tectonic discontinuities include rockbursts caused by fault activation or initiated by a sudden formation and propagation of fractures in undisturbed rock as a result of exceeded shear strength (shear rupture) [6,7]. According to Ortlepp [6], shear rapture is one of the most important mechanisms generating very high-energy mining tremors and major rockbursts. Roof strata fractures occur most often near the edges, old gobs, boundaries between backfill mining and retreat mining, or in the direct vicinity of the mining front.

Both in-situ observations in the Polish underground copper mines and scientific research indicate that a high level of seismic and rockburst hazard is influenced by geological, mining, and organizational factors. The most important geological factors include: increasing depth of mining operations, high in-situ stress in the rock mass, lithology of the rock mass and its geomechanical parameters (very strong lime and anhydrite rocks in the roof layers), as well as tectonics and the thickness of the deposit. The compression strength of roof rocks is very high and therefore they are able to accumulate elastic energy and release it suddenly [1].

The mining factors affecting the dynamic phenomena are: the mining method and its geometry, the roof control method, mining face parameters, concentration of mining works, and whether mining operations are performed under constrained mining conditions [1]. Local stress concentration is the primary mining-related cause of rockbursts, and it can be caused by improperly selected mining systems or by an improper geometry of such a system. Seismic and rockburst hazard is also affected by an excessive concentration of mining works and an increasing scope of mining works performed under constrained conditions, which includes, among other things, mining operations in the vicinity of remnants.

Remnants are undisturbed, typically irregular fragments of the deposit in which mining operations are impossible, technically very difficult, or uneconomical. In-situ observations and scientific research indicate that, in the vicinity of rigid remnants, zones of high stress concentration are located. They cover both the deposit and the rock layers below and above the remnant [8–11]. When the stress values in the remnant exceed the strength of remnant, it may be crushed, and if the conditions are unfavorable, a pillar rockburst may occur [8,12]. Such remnants may also cause the roof layers to collapse above their edges and eventually lead to high-energy seismic events (shear rapture) [8,11,13]. Therefore, it is of great significance to find a way to improve safety of the works in the vicinity of remnants.

Mining practice shows that the roof control method also has an impact on the level of seismic activity recorded in the mining field. Due to the variety of special applications of backfills, there are different types and technologies. For example, cemented hydraulic backfill, which is widely used because it provides high strength and allows the use of waste rock from mining operations, as well as tailings from mineral processing plants as ingredients [14]. Moreover, hydraulic backfill and paste backfill are also used. Meanwhile, Li et al. [15] proposed filling the post-excavation space with waste rocks.

In accordance with the formal regulations currently in force in Poland, copper mines located in the Legnica-Glogow Copper Belt (LGCB), use hydraulic backfill in mining fields where the thickness of the deposit exceeds 7 m [16]. Backfill is also used to control rock mass movement and surface subsidence in order to protect cities or surface facilities located above the excavations.

Hydraulic backfilling is about preparing a backfilling mixture, consisting of granular material (most often backfill sand) and process water and its gravitational hydrotransport to the backfilled excavation (goaf). The mixture flows through the pipeline to the backfilled excavation, where sedimentation of the backfill takes place. The sediment remains in the goaf and constitutes a backfill, and the water through the clarifiers is directed to the mine's drainage system. Preparation of backfilling mixtures in copper mine installations at LGCB, with gravity pipeline hydrotransport and L-shaped geometric pipelines, is carried out by dosing backfilling material and process water. The backfilling technology is characterized by the following parameters:

- Density of backfilling mixture: 1480 kg/m^3
- Flow velocity of the mixture: 6.8 m/s
- Mixture flow rate: 660 m^3/h
- Unit pressure loss: about 1390 Pa/rm
- Average water consumption: 720 m^3
- Average backfilling time: 1.04 h
- Average start-up and rinsing time: 0.90 h

Many years of in-situ observations carried out in the polish copper mines, in the mining fields where hydraulic backfill was used for the liquidation of mined out areas, showed that both seismic activity and rockburst hazard decreased in these fields [17–19].

Scientific research also indicates that the application of backfill can help effectively control the movement deformation of rock layers and reduce stress concentration, thus helping such events as rockbursts as well as increasing the stability of excavations [9,20–29]. The backfilled roof is more stable, its vibrations are at a higher frequency and prove less dangerous, and the duration of the tremor is shorter. At the same time, backfill causes a reduction in the maximum vibration velocities of rock particles (ppv) and in maximum values of their acceleration (ppa) [30].

The development of computer modeling and simulation techniques has significantly expanded research capabilities related to the analysis of seismic phenomena potential in underground mines. Numerical modeling allows for identifying potential rockburst hazard zones, estimating rockburst hazard in every part of the rock mass, even in inaccessible parts, and identifying the conditions under which a dynamic phenomenon may occur [31–33].

In this paper, numerical methods have been used to assess how hydraulic backfill can improve the geomechanical situation in the vicinity of a 40 m wide remnant, which was left behind in a mining field where a room and pillar mining system was used. Numerical simulations performed for the analyzed mining field where hydraulic backfill was applied in the excavations located in the vicinity of the remnant are a continuation of previous research [11,13] conducted for this mining field for the room-and-pillar system with roof deflection. The article includes a comparison of the results of numerical calculations performed for both cases. They can be used to improve the safety and efficiency of underground copper ore mining.

2. Case Study of a Mining Field with Remnant in a Polish Underground Copper Mine

Polish underground copper mines (Lubin, Rudna, and Polkowice-Sieroszowice) are located in the south-west part of Poland and belong to the Legnica-Glogow Copper Belt (LGCB) (Figure 1). The deposit is located in Permian formations, at the contact between dolomite-limestone, sandstone, rotliegend and lower zechstein series. Copper is found associated to sulfides, mainly: chalcocite, bornite and chalcopyrite. The copper deposit is developed in the form of a pseudo-stratum with variable thickness and low inclination (approximately 4°). It is present at a substantial depth of 600 m to 1400 m. A typical lithological cross-section of the LGCB region is characterized by rigid, high-strength rock layers in the roof, capable of accumulating elastic energy, while layers with much lower strength parameters are present in the floor. The deposit is mined with the room and pillar system, mostly with roof deflection, but to improve the geomechanical situation, dry and hydraulic backfill are also used.

2.1. Geological and Mining Conditions of Research Area

The research area in this case study is one of the mining fields located in the Polkowice-Sieroszowice mine (Figure 1). The copper ore deposit is 2.0–2.8 m high, extends in the NW-SE direction, and dips (2–3°) toward NE. The depth of the deposit is approximately 1000 m. It includes mostly sandstone (gray, quartz, and fine-grained sandstone), cupriferous shale (clay and dolomite-clay), as well as dolomite (streaky, dark gray, cryptocrystalline). The roof is built of rock layers being part of the Zechstein carbonaceous series, and the floor is made of grey Rotliegend sandstone. The tectonics in the field is quite poor. One fault is located in the middle part of the mining filed and has an approximately 0.5–1.5 m throw [11].

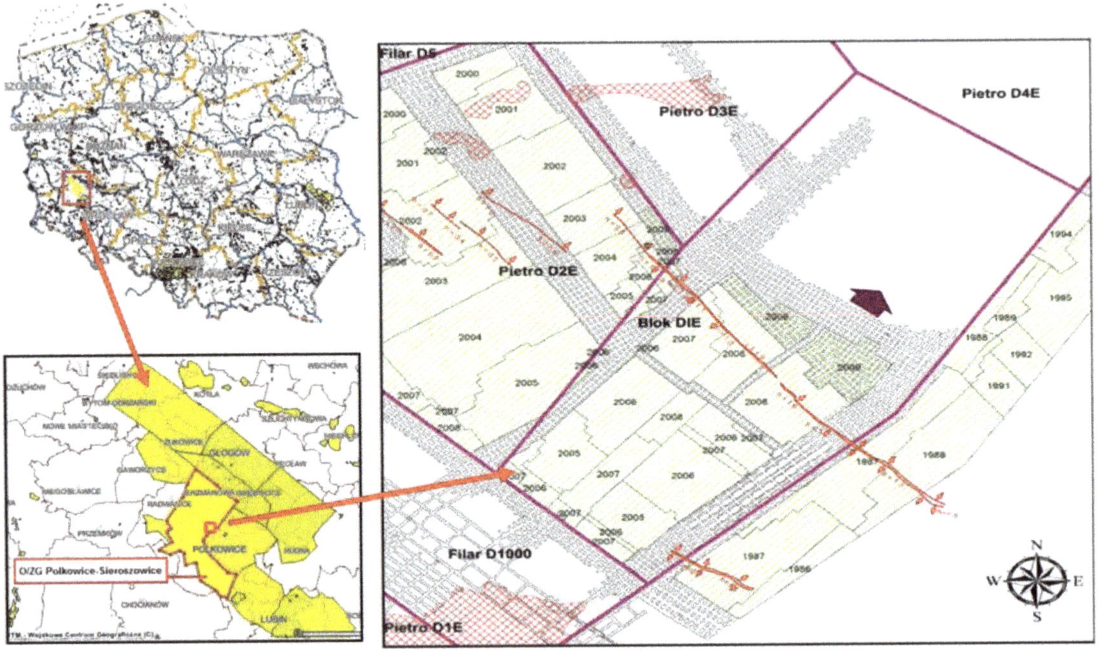

(a)   (b)

Figure 1. Location of the LGCB in Poland, location of the analyzed mining field in the Polkowice-Sieroszowice mine (**a**), analyzed mining field and location of the remnant (**b**).

Mining operations in the analyzed field started in March 2005 using a room-and-pillar system with roof deflection. The deposit was cut with rooms and strips into rectangular technological pillars with basic dimensions of 6×8 m, which were situated perpendicular to the line of the mining front. The excavations were in the shape of an inverted trapezoid, with the roof width equal to 6 m and the inclination angle of sidewalls $10°$. Technological pillars were successively reduced to residual pillars, as the mining front progressed, and left in the mined-out area. In Polish copper mines, the size of technological pillars is selected in such a way that they progressively yield at the stage of separation from the deposit. The width of the working area was generally between 4 and 5 strips (Figure 2) [11].

Exploitation in the analyzed field was carried out in constrained mining conditions. In 2007, because of problems ensuring the roof's stability in the central part and in the right side of the field, a deposit remnant approximately 40×100 m was created, between strip P-38 and strip P-33 (Figure 2) [13].

2.2. Seismic Activity Registered in the Vicinity of the Remnant during Room and Pillar Mining

The analyzed mining field was characterized by a relatively high level of seismicity. Figure 3 presents yearly distribution of the number of mining tremors as well as seismic energy emissions for the period of 2002–2011. The greatest number of seismic events and the highest level of seismic energy emission were recorded in the field in 2006. In December 2006, a very high-energy, class E8 seismic event occurred. It seriously influenced the roof stability in the central part and in the right side of the field. The mining front was being reconstructed with strips P-34 ÷ P-37, but in August 2007 increasing problems with the roof necessitated stopping them, and the progress of the P-38 ÷ P-42 strips began. As a result, a remnant approximately 40 m in width was formed. While restoring the front in the right side of the field, on 13 March 2007, a seismic event having an energy of 10^7 J was recorded. Then, on 4 October 2007, the second event with an energy of 4.4×10^7 J occurred. The epicenters of these events were located in the vicinity of the remnant, while the epicenter of the tremor from October 2007, on the edge of the undisturbed rock remnant. The geophysical mechanism of this event shows rock mass displacement on the reverse fault in the SW direction (towards the gob areas). This surface runs in the direction approximately consistent with the mining front line.

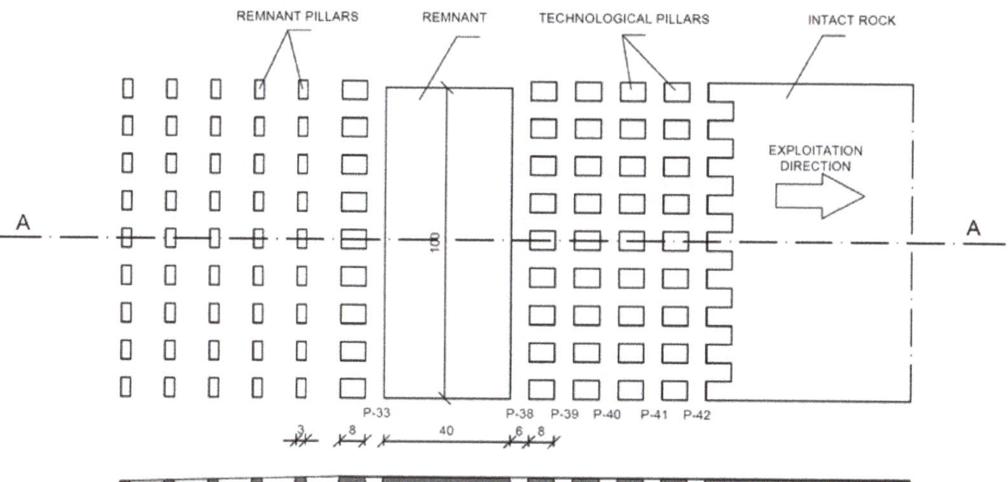

Figure 2. Analyzed mining field (room and pillar mining with roof deflection) and location of the 40 m wide remnant.

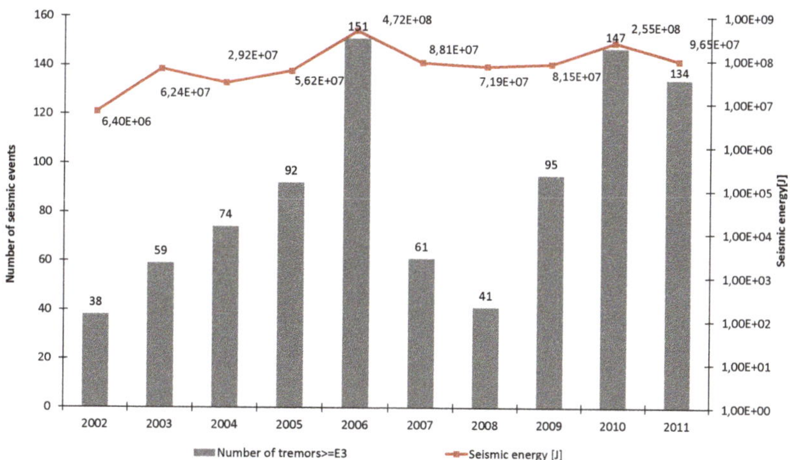

Figure 3. Seismic activity for the period of 2002–2011 in the analyzed field [13].

Moreover, numerical back analysis of the mining front reconstruction, deposit exploitation, and remnant behavior in the field also showed that the roof may have suddenly and violently collapsed at the edge of the rigid remnant, on the side of the field's gob areas, during the reconstruction of the mining front. It may cause a high-energy seismic event induced mainly by the exceeded shear strength [13] (Figure 4). The results of the analysis indicate a significant impact of the mining edge on the occurrence of dynamic phenomena that can be dangerous to the working crew. Therefore, it is of great significance to study the application of hydraulic backfill to reduce the risk and the tendency of seismic events in the working face nearby rock remnants.

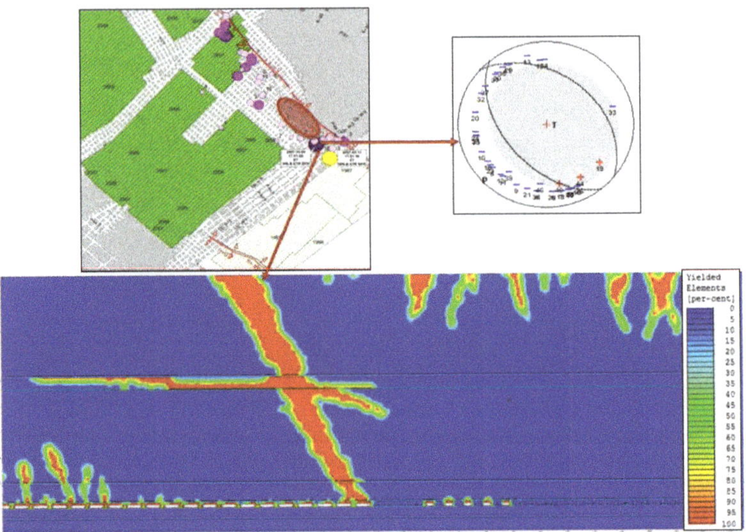

Figure 4. E7 seismic event which occurred in October 2007 at the edge of the created remnant with its probable mechanism [13].

3. Methods: Numerical Simulation of the Application of Hydraulic Backfill in a Mining Field with a Remnant

The numerical model of the analyzed field mined with the room and pillar mining system was created in the RS2 v. 9.0 finite element program in a plane strain state. With the use of adequate computational power, the 2D analysis allowed FEM analyses on a dense mesh, particularly in the vicinity of excavations and deposit remnants. Therefore, the problem could be approached globally, by analyzing a large model and simulating exploitation in the entire field. The model is constructed properly, as the simulation results correspond well to the measurement data related to the convergence of the excavations in the analyzed field.

In the first part of this research [11,13], a back calculation was performed in order to provide the model with a reflection of the actual situation observed in the analyzed mining field. In further calculation steps, the retreat room and pillar mining process was reconstructed in the deposit, based on the actual parameters of the mining system used in the analyzed field (sizes of the technological pillars and the residual pillars, dimensions of the excavation cross-sections, size of the working area, etc.). The numerical model also included a deposit remnant approximately 40 m in width between strips P-33 and P-38, which was formed, at a 460 m front length, by excavating strip P-38 outside the danger zone (31st calculation step). After the remnant of undisturbed rock had been separated, the field was further mined with the use of the room and pillar method with systematic retreat by roof deflection.

In the second part of the research, the use of hydraulic backfill in excavations adjacent to the remnant was proposed as a solution for limiting rock mass deformations (Figure 5).

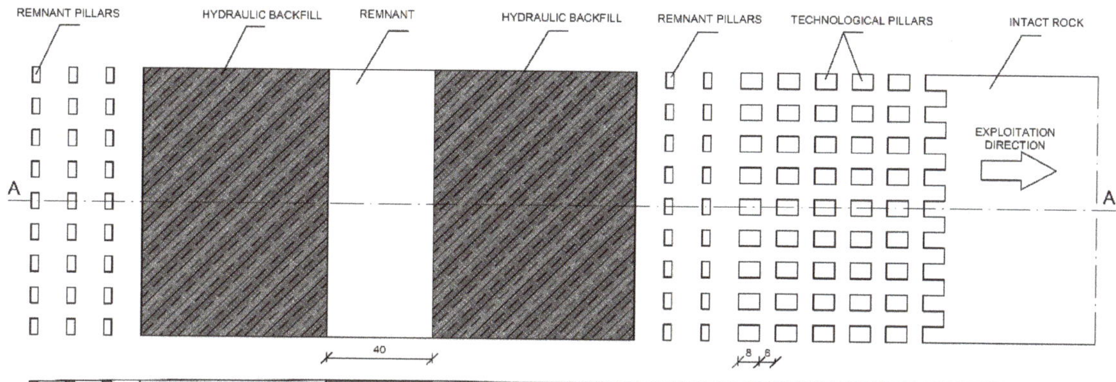

Figure 5. Scheme of room and pillar mining operations with the application of hydraulic backfill in the vicinity of the remnant.

The numerical model reflecting the actual situation observed in the analyzed mining field was modified. In the modified model, stepwise numerical simulations reflected the cutting of the deposit with the room and pillar method with roof deflection until strip P-33 was excavated, when the front length was approximately 460 m (30th step in the numerical model) (Figure 6a). The next (31st) step in the numerical model simulated the hydraulic backfilling of five strips (P-29–P-33) prior to excavating strip P-38 outside the danger zone (Figure 6b). In the next (32nd) step of the numerical model, strip P-38 was excavated to isolate (form) a deposit remnant 40 m in width. Steps 33–37 of the numerical model simulated the excavation of successive strips having a cross-section in the shape of an inverted trapezoid with the width of the excavation roof equal to 6 m and the inclination angle of sidewalls 10°. The rock mass was cut into technological pillars 8 m in width (Figure 6c). When the working area was 5 strips wide, as the front advanced, the mined-out area started to be systematically liquidated with the use of hydraulic backfill. The proposal included using hydraulic backfill to close the mined-out areas on the right side of the remnant, in the area from strip P-38 to strip P-43 (steps 38–42 in the numerical model). In steps 43–65, deposit exploitation in the analyzed field was continued with the use of the room and pillar method with roof deflection (Figure 6d). The rock mass was cut into technological pillars 8 m in width, which were subsequently reduced in size to remnant sizes, and further technological pillars were cut. The assumption concerning the width of the working area in the numerical simulations was 5 strips.

The numerical model was a plate in which the actual geological structure of the analyzed mining field was included (Table 1). Displacement boundary conditions were set in the model:

— Bottom edge of the plate: no vertical displacements
— Side edges of the plate: no horizontal displacements

The upper edge of the plate was loaded with a vertical stress of 17.657 MPa (vertical stress determined on the basis of data from borehole S-294). The self-weight of rock layers was accounted for in the calculations. The virgin horizontal stress was assumed to be equal to the virgin vertical stress (hydrostatic state of initial stresses). In the RS2 v. 9.0 computer program, rock mass was assumed to be homogeneous and isotropic, and the behavior of the rock mass was characterized by an elastic and an elastic–plastic model with softening. The Mohr–Coulomb strength criterion was applied.

As was mentioned before, the numerical model served to reconstruct the actual geological structure of the analyzed mining field. As in the first part of this research, the parameters of the rock mass (for the Coulomb–Mohr criterion) were determined on the basis of laboratory tests of rock samples from boreholes drilled in the analyzed region

(Mo-12 To-2, Mo-12 To-5 and Mo-11 To-3). The Hoek–Brown classification was used. The parameters of the rock mass are presented in Table 1.

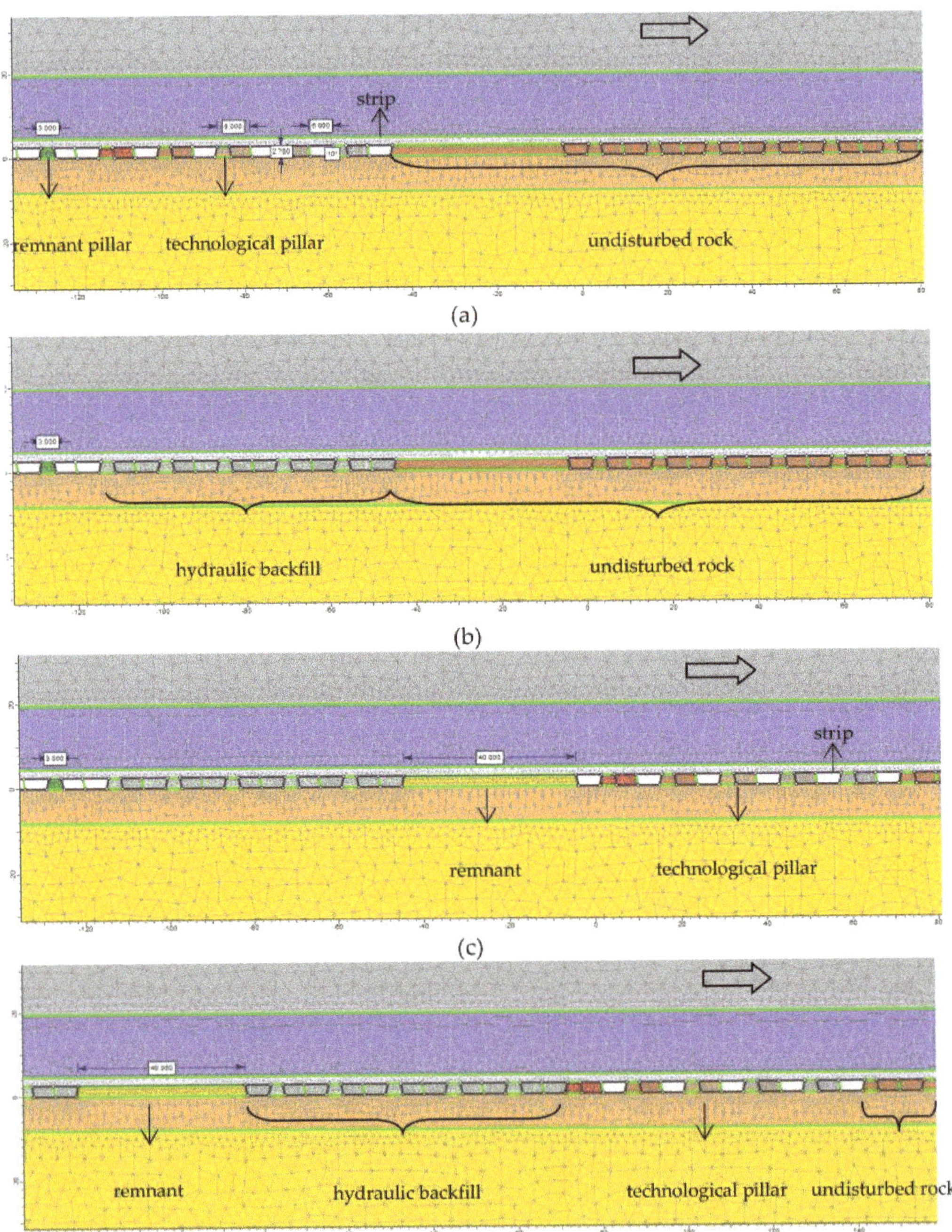

Figure 6. Simulation of mining operations performed in the analyzed field: (**a**) step 30, (**b**) step 31, (**c**) step 37, (**d**) step 42.

Table 1. Rock mass parameters used in the numerical model of the analyzed region [11].

Location	Rock Type	h (m)	E_s (MPa)	v (-)	σ_t (MPa)	c (MPa)	ϕ (°)	c_{res} (MPa)	ϕ_{res} (°)	δ (°)
Roof	Main anhydrite	100.0	41,110	0.24	0.746	6.967	38.66	1.393	36.73	2.00
	Clay-anhydrite breccia	10.0	7100	0.18	0.093	2.507	39.06	0.501	37.11	2.00
	Basic anhydrite	73.0	40,010	0.25	0.765	7.146	38.66	1.429	36.73	2.00
	Calcareous dolomite I	15.0	44,980	0.24	2.933	12.085	39.00	2.417	37.05	2.00
	Calcareous dolomite II	2.0	87,440	0.27	4.715	19.895	39.00	3.979	37.05	2.00
Mined deposit	Mined height	2.7	25,240	0.21	0.825	8.424	39.31	1.350	37.35	2.00
Floor	Quartz sandstone I	8.2	4260	0.15	0.057	1.538	39.06	-	-	-
	Quartz sandstone II	194.5	3220	0.13	0.043	1.160	39.06	-	-	-

The symbols used in the above table are as follows: h—thickness of rock layers, E_s—longitudinal modulus of elasticity, v—Poisson's ratio, σ_t—tensile strength of the rock mass, c—cohesion coefficient, ϕ—internal friction angle, δ—dilatancy angle, c_{res}—residual cohesion coefficient, ϕ_{res}—residual internal friction angle.

The numerical model was built on a finite element mesh with 3-node triangular elements. In order to improve the accuracy of the numerical calculations, mesh density was increased in the central part of the model, in the vicinity of the excavations. The model has 81,864 elements and 41,289 nodes.

The parameters of the hydraulic backfill used in the model were determined iteratively and were based on a practical experience that, under the conditions of the LGCB copper mines, maximum vertical displacements due to the mining of the deposit with the room and pillar system with hydraulic backfill amount to approximately 15–20% of the deposit's thickness [34]. The detailed numerical model for the analyzed region is presented in Figure 7.

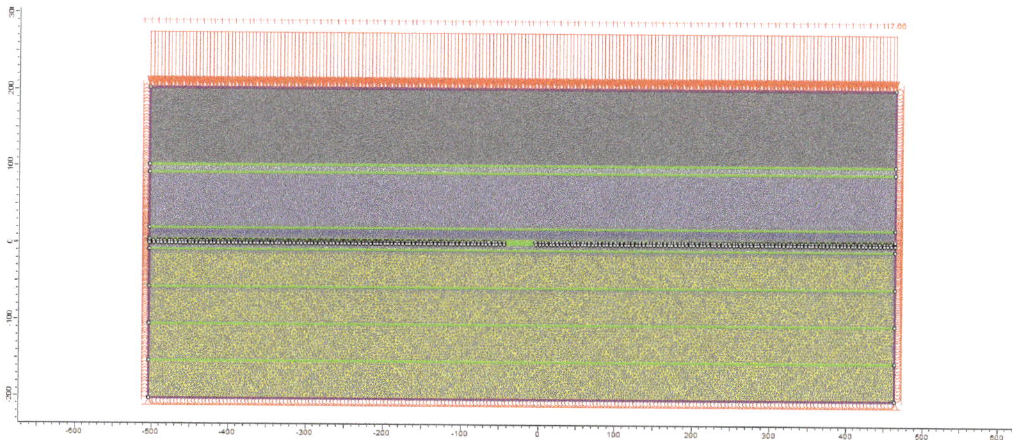

Figure 7. Numerical model for the analyzed region.

The model was validated on the basis of the convergence measurements of excavations driven in the analyzed mining field. The results from the in-situ convergence measurements were compared with the convergence values calculated by means of the numerical simulations. A very good correlation between the numerically calculated convergence and the measurement results may indicate that the model was built correctly (Figure 8).

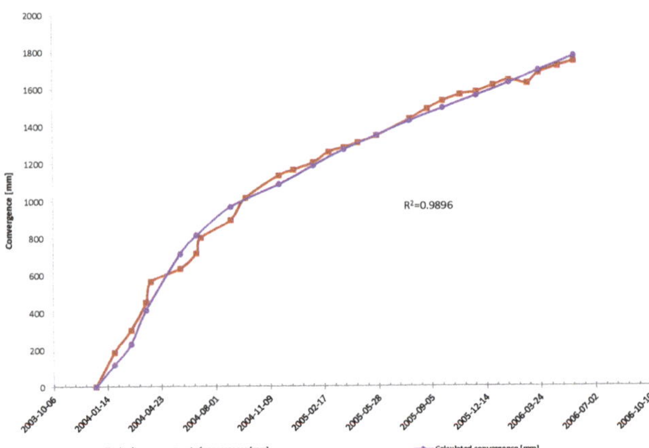

Figure 8. Matching between the results of the in-situ convergence measurements and the convergence values obtained using the numerical model.

4. Discussion

In order to investigate how the hydraulic backfill can improve the geomechanical situation in the vicinity of remnants, vertical stress σ_y distribution and yielded zones have been analyzed. The behavior of the remnant and of the rock mass in its vicinity was analyzed in the successive steps of the room and pillar mining system. The results of the numerical simulations conducted for the case of the mining field in which hydraulic backfill was used in the excavations located in the vicinity of the remnant have been compared with the results of the analysis performed in the previous research for the case of the room and pillar system with roof deflection.

The numerical simulation results for the analyzed mining field demonstrated that the application of hydraulic backfill in the excavations located in the vicinity of the remnant reduces deformations of roof strata generally in the entire mining field and mostly near the remnant.

The results of numerical simulations conducted in the first part of the research for the case of the room and pillar system with roof deflection showed that, in the 35th computational step, during the reconstruction of the mining front, when the fifth strip was created, a yielded zone suddenly (in one computational step) formed in the roof above the edge of the remnant (Figure 9a,b). The transverse line of destruction in the roof was formed near the left edge of the remnant and was inclined at an angle of approximately 60° in the direction of the gob area. These results indicate that a sudden fracturing and collapse of rigid roof layers may occur on the edge of the remnant, mainly due to the exceeded shear strength. This may cause a seismic event with high energy, potentially resulting in a rockburst phenomenon, if appropriate conditions are met [11].

The analysis of the area of the yielded elements numerically calculated in the second part of the research for the case with the room and pillar system with hydraulic backfill in the vicinity of remnant indicates that yielded zones in the field are smaller than in the case with roof deflection. Areas of yielded elements appear in particular above the gobs and they are observed to grow in the successive steps of the simulated room and pillar mining (Figure 10a,b). This indicates a progressing disintegration of roof layers above the mined-out areas. Moreover, in the case with hydraulic backfill, no yielded zone forms suddenly in the roof in the vicinity of the remnant edge over the entire period of mining operations (Figure 10b). A comparison between Figures 9 and 10 also shows that, in both of the analyzed cases, the remnant is stable. Yielded areas inside the remnant occur only near its edges. Their reach increases in the successive steps of the mining operations. After the reconstruction of the entire mining front (five strips on the right side of remnant), the reach of the yielded areas on the remnant edges does not exceed 2 m, while for a front distance of approximately 470 m from the remnant edge the reach of yielded areas reaches a maximum of approximately 3.5 m. During the reconstruction of the mining front, the yielded zones above the strips amount to approximately 2.0–2.9 m.

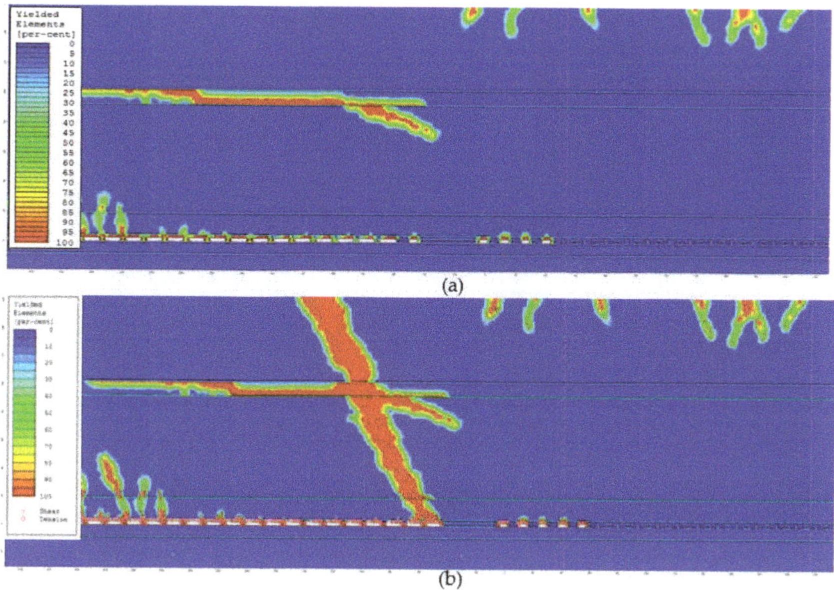

Figure 9. Yielded zones for mining front: (**a**) approximately 50 m from the remnant edge—34th calculation step, (**b**) approximately 60 m from the remnant edge—35th calculation step [11].

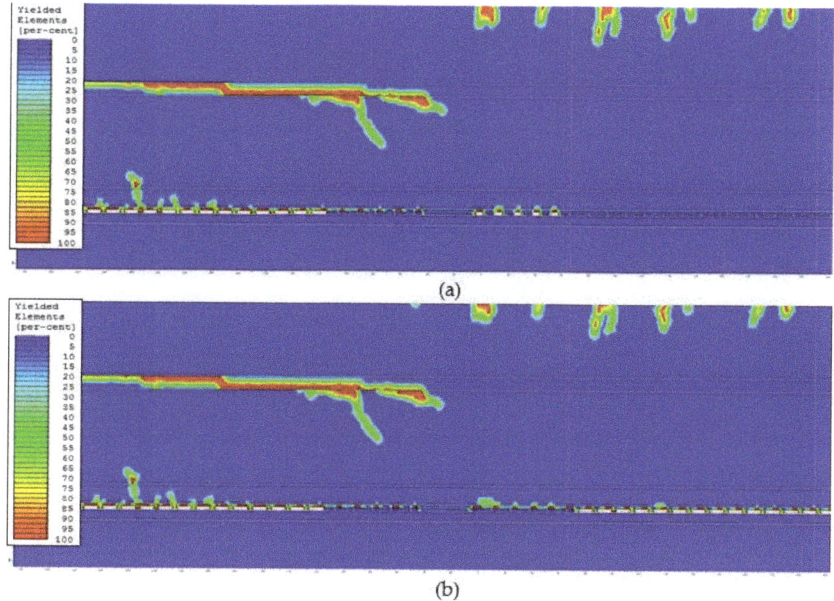

Figure 10. Yielded zones for mining front: (**a**) approximately 60 m from the remnant edge—36th calculation step, (**b**) approximately 470 m from the remnant edge—65th calculation step.

Figures 11–14 show the distribution of vertical stresses σ_y in the analyzed field for the room and pillar system with roof deflection (Figure 11) and with hydraulic backfill (Figures 12 and 13). It can be observed that stress concentration zones occur inside and in the vicinity of the remnant. A comparison between Figures 11 and 12 shows that the violent destruction in the roof above the remnant edge resulted in vertical stress redistributions near the left edge of the remnant and reduced the values of vertical stresses σ_y in the area of the yielded zone.

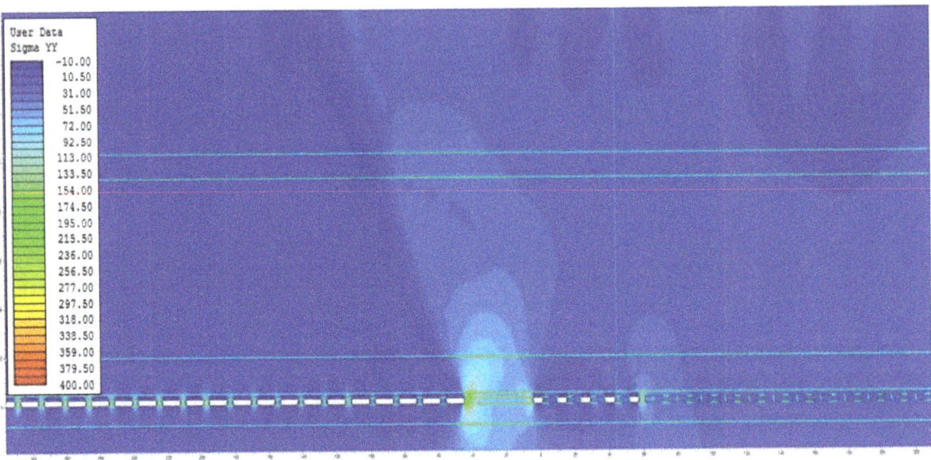

Figure 11. Distribution of vertical stresses σ_y for the front distance approximately 60 m from the remnant edge—35th calculation step [11].

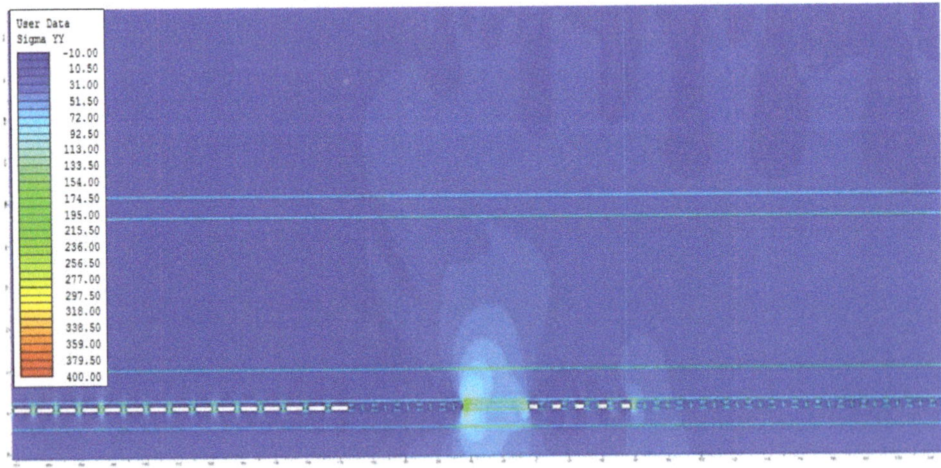

Figure 12. Distribution of vertical stresses σ_y for the front distance approximately 60 m from the remnant edge—36th calculation step.

Analyses of vertical stress σ_y distribution in the roof above the 40 m wide remnant (at distances of 5 m, 10 m, 20 m, and 40 m) for the room and pillar system with hydraulic backfill indicate that this remnant affects the roof layers. The maximum values of vertical stresses σ_y in the roof above the remnant (at distances of 5 m and 10 m) occur at a distance from the edge and are respectively equal to 5 m—120 MPa, 10 m—100 MPa. Their values decrease rapidly towards the level of approximately 20–25 MPa over the hydraulic backfill. With the increasing distance from the remnant, vertical stress σ_y values decrease and tend towards the initial stress state. In the roof layers at a distance of 20 m and 40 m above the remnant, the greatest values of vertical stress σ_y occur above its center (Figure 13). The impact range of the remnant and the values of vertical stresses in its surroundings increase in the successive steps of the simulated mining operations as the front progresses.

A comparison between the values of vertical stress σ_y in the roof 10 m above the remnant for the cases of the room and pillar system with roof deflection and with hydraulic backfill shows that the application of hydraulic backfill reduces the values of vertical stresses in the roof. The values of these stresses near the left edge of the remnant can be noticed to have decreased by a maximum of 20 MPa after the use of hydraulic backfill for the front distance of approximately 470 m from the remnant edge. Moreover, the difference in vertical stress σ_y between the roof deflection case and the hydraulic backfill case increases in the successive steps of the mining operations (Figure 14).

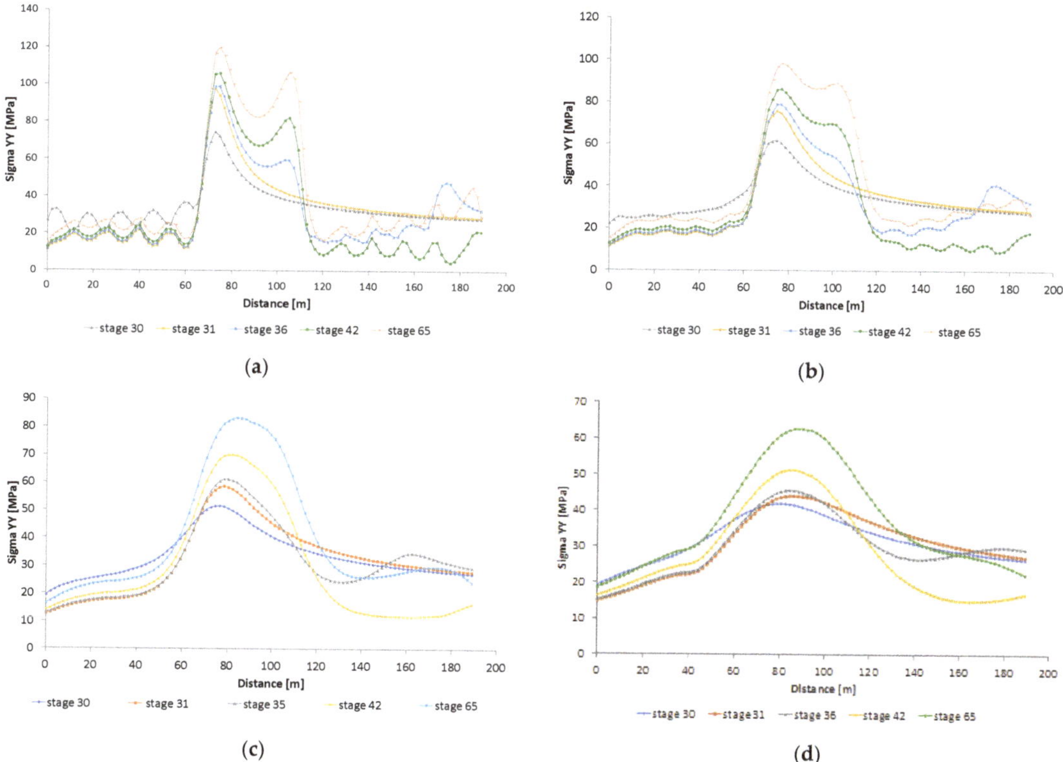

Figure 13. Vertical stress σy in the roof, for vertical distances of: (**a**)—5 m, (**b**)—10 m, (**c**)—20 m and (**d**)—40 m above the remnant.

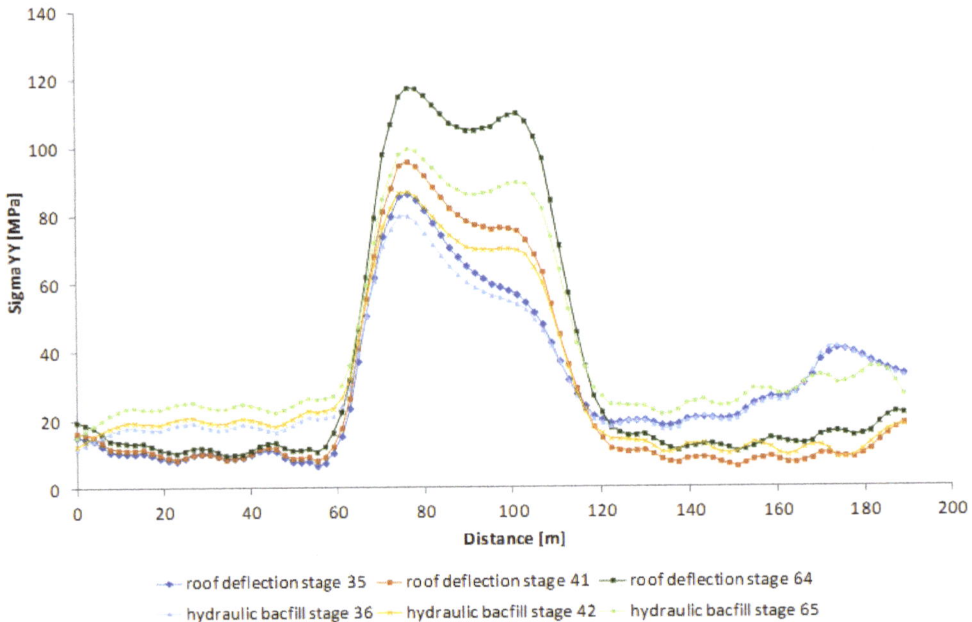

Figure 14. Comparison between vertical stresses σ_y in the roof layer 10 m above the remnant for the roof deflection case and the hydraulic backfill case, for the front distance of approximately: 60 m from the remnant edge—computational steps 35, 36, 146 m from the remnant edge—computational steps 41, 42), 470 m from remnant edge—computational steps 64, 65.

Figure 15 shows a comparison between the values of vertical stress σ_y inside the 40 m wide remnant for the cases of the room and pillar system with roof deflection and with hydraulic backfill. As can be seen, in both cases, the highest values of vertical stresses σ_y occur at a distance of approximately 2 m from its edge. Their values decrease towards the direction of the remnant's center. Moreover, the values of vertical stress σ_y inside the remnant increase in the successive steps of the mining operations. The maximum values of these stresses near the edge as well as in the center of the remnant can be noticed to occur in the case of the room and pillar mining system with roof deflection.

The greatest difference between vertical stress values σ_y inside the remnant in the cases of the room and pillar system with roof deflection and with hydraulic backfill occur on the right side of the remnant and is equal to 50 MPa for the maximum front distance of approximately 470 m from the remnant edge. Similarly, in the center of the remnant, the greatest difference between vertical stress values σ_y is noticed for the front distance of approximately 470 m from the remnant edge and is equal to 15 MPa.

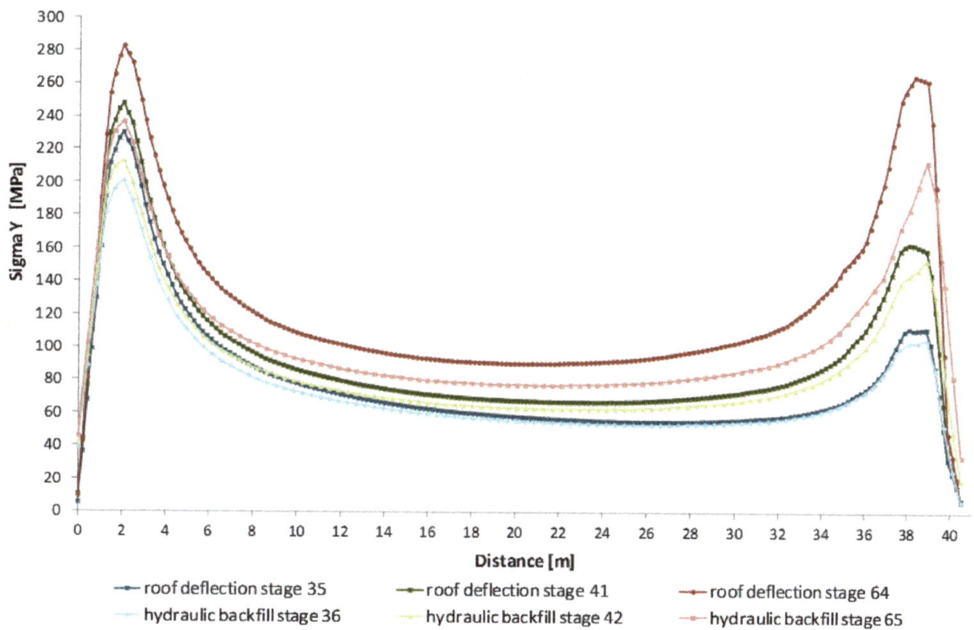

Figure 15. Comparison between vertical stresses σ_y inside the 40 m wide remnant for the roof deflection case and the hydraulic backfill case, for the front distance of approximately: 60 m from the remnant edge—computational steps 35, 36, 146 m from the remnant edge—computational steps 41, 42), 470 m from the remnant edge—computational steps 64, 65.

5. Conclusions

Due to a high level of mining concentration in Polish underground copper mines, mining works are very often carried out in difficult geological and mining conditions, resulting in high seismic and rockburst hazards. Frequent high-energy seismic events pose a threat to crews working underground. Therefore, optimal solutions should be found that will allow the extraction of copper ore deposits as safely and economically as possible. This paper focuses on the problem of the application of hydraulic backfill in excavations located in the vicinity of remnants, in order to improve the safety and efficiency of mining operations. The results of the numerical simulations allowed conducting a qualitative analysis and illustrating the occurrence of certain phenomena in the rock mass, in which copper ore is mined with the room and pillar systems.

The results of the numerical simulations performed for the case study (a mining field located in a Polish underground copper mine where undisturbed ore remnant 40 m in width was left behind) indicate that:

- The application of hydraulic backfill in the vicinity of the remnant strongly improves the geomechanical situation in the mining field, mostly by limiting rock mass deformation and disintegration in the mining field.
- In the case of hydraulic backfill, yielded zones in the field are smaller than in the case of roof deflection. They appear in particular above the gobs and increase in the successive steps of the simulated room and pillar mining.
- When the room and pillar mining system with roof deflection is used, sudden fracturing and collapse of rigid roof layers may occur on the edge of the remnant and cause a high-energy seismic event. In the case with hydraulic backfill, no yielded zone is suddenly formed in the roof in the vicinity of the remnant edge for the entire period of mining operations in the analyzed field.

–   The application of hydraulic backfill also reduces the values of vertical stresses σ_y inside and in the vicinity of the remnant.

The results of the performed analyses indicate that the application of hydraulic backfill in the vicinity of the remnant has a number of very important advantages. It provides progressive roof deflection over the mined-out areas, which eliminates the formation of mining edges. Furthermore, it reduces stress concentrations at the remnant edges and prevents roof layers above edges from collapsing (which may result in high-energy tremors and rockbursts—shear raptures). It also increases the stability of the rock mass in the mining field reduces the stress concentration inside and near the remnant, thus limiting the risk of pillar rockburst. Further research based on numerical calculations showed that increasing the area of the backfill zone near the remnant would further improve stress distribution and reduce rock layer deformations in the mining fields.

The obtained results may facilitate decisions regarding the further development of mining works in areas affected by the presence of remnants and will allow for the verification of mining operation protocols and rockburst prevention methods. The results are also intended not only to improve the safety of mining operations carried out in the mining fields in which leaving a remnant is a necessity, but also to increase the efficiency of copper ore mining.

Further research will include a 3D model of the analyzed field and a comparison of the obtained results with the results of the 2D analysis.

Author Contributions: Conceptualization, K.A.-P.; methodology, K.A.-P. and D.P.; software, K.A.-P.; validation, K.A.-P.; formal analysis, K.A.-P.; investigation, K.A.-P.; resources, D.P.; data curation, D.P.; writing—original draft preparation, K.A.-P.; writing—review and editing, K.A.-P.; visualization, K.A.-P.; supervision, K.A.-P.; project administration, D.P.; funding acquisition, K.A.-P. All authors have read and agreed to the published version of the manuscript.

Funding: The research work was co-founded with the research subsidy of the Polish Ministry of Science and Higher Education granted for 2021.

Data Availability Statement: The data presented in this study are available on request from the corresponding author.

Conflicts of Interest: The author declares no conflict of interest.

References

1. Butra, J.; Kudelko, J. Rockburst hazard evaluation and prevention methods in Polish copper mines. *Cuprum Sci. Tech. J. Ore Min.* **2011**, *4*, 5–20.
2. Cook, N.G.W. Seismicity induced by mining. *Eng. Geol.* **1976**, *10*, 99–122. [CrossRef]
3. Ortlepp, W.D.; Stacey, T.R. Rockburst mechanisms in tunnels and shafts. *Tunn. Undergr. Space Technol.* **1994**, *9*, 59–65. [CrossRef]
4. Gibowicz, S.J.; Kijko, A. *An Introduction to Mining Seismology*; Academic Press: San Diego, CA, USA, 1994.
5. Kaiser, P.K.; McCreath, D.R.; Tannant, D.D. *Canadian Rockburst Support Handbook*; Geomechanics Research Centre: Sudbury, ON, Canada, 1996.
6. Ortlepp, W.D. *Rock Fracture and Rockbursts: An Illustrative Study*; SAIMM Monograph Series M9, 98; The S Af Inst of Min and Metall: Johannesburg, South Africa, 1997.
7. Ortlepp, W.D. Observation of mining-induced faults in an intact rock mass at depth. *Int. J. Rock Mech. Min. Sci.* **2000**, *37*, 23–436. [CrossRef]
8. Salustowicz, A. *Outline of Rock Mass Mechanics*; AGH Science and Technology Press: Krakow, Poland, 2019.
9. Adach-Pawelus, K. Influence of the roof movement control method on the stability of remnant. In *IOP Conference Series—Earth & Environmental Science, Proceedings of the World Multidisciplinary Earth Sciences Symposium, Prague, Czech Republic, 11–15 September 2017*; IOP Publishing Ltd: Bristol, UK, 2017; Volume 95, p. 042022. [CrossRef]
10. Adach-Pawelus, K. The influence of the geology of ore remnant on its behaviour in the mining field. In *Science and Technologies in Geology, Exploration and Mining 1.3, Exploration and Mining, Proceedings of the 18th International Multidisciplinary Scientific GeoConference, Albena, Bulgaria, 2–8 July 2018*; Curran Associates, Inc.: New York, NY, USA, 2018; Volume 18, pp. 1–10.
11. Adach-Pawelus, K.; Butra, J.; Pawelus, D. An attempt at evaluation of the remnant influence on the occurrence of seismic phenomena in a room-and-pillar mining system with roof deflection. *Studia Geotech. Mech.* **2017**, *39*, 3–16. [CrossRef]

12. Lenhardt, W.A.; Hagan, T.O. Observations and possible mechanisms of pillar associated seismicity at great depth. In *SAIMM, Johannesburg, South Africa, Symposium Series S10, Proceedings of the International Deep Mining Conference, Mbabane, Swaziland, 10–12 September 1990*; South African Institute of Mining and Metallurgy: Johannesburg, South Africa, 1990; pp. 1183–1194.
13. Adach-Pawelus, K. Application of seismic monitoring and numerical modelling in the assessment of the possibility of seismic event occurrence in the vicinity of ore remnant. In *IOP Conference Series—Earth & Environmental Science, Proceedings of the World Multidisciplinary Earth Sciences Symposium, Prague, Czech Republic, 3–7 September 2018*; IOP Publishing Ltd: Bristol, UK, 2019; Volume 221, pp. 1–10. [CrossRef]
14. Basarir, H.; Bin, H.; Fourie, A.; Karrech, A.; Elchalakani, M. An adaptive neuro fuzzy inference system to model the uniaxial compressive strength of cemented hydraulic backfill. *Min. Miner. Depos.* **2018**, *12*, 1–12. [CrossRef]
15. Li, M.; Zhang, J.X.; Jiang, H.Q.; Huang, Y.L.; Zhang, Q.A. Thin plate on elastic foundation model of overlying strata for dense solid backfill mining. *J. China Coal. Soc.* **2014**, *39*, 2369–2373.
16. Dz.U. 2017 poz. 1118. Ordinance of the Minister of Energy of 23 November 2016 on Detailed Requirements for Conducting Underground Mining Operations. Available online: http://isap.sejm.gov.pl/isap.nsf/DocDetails.xsp?id=WDU20170001118. (accessed on 23 June 2021). (In Polish)
17. Szpak, M. Impact of mining method cavity caused by the extraction under rockbursts hazard in chosen panels in Rudna—copper mine. *Cuprum Sci. Tech. J. Ore Min.* **2008**, *3*, 15–31. (In Polish)
18. Bartlett, S.; Burgess, H.; Damjanović, B.; Gowans, R.; Lattanzi, C. Technical Report on the Copper-Silver Production Operations of KGHM Polska Miedz, S.A. In the Legnica-Glogow Copper Belt Area of Southwestern Poland; Toronto, ON, Canada. 2013. Available online: https://kghm.com/en/technical-report-copper-silver-production-operations-kghm-polska-miedz-sa-legnica-glogow-copper-belt. (accessed on 23 June 2021).
19. Gogolewska, A.; Junik, P. Seismic hazard related to rate of face advance in Lubin copper ore mine. *Min. Sci.* **2013**, *20*, 87–99. (In Polish) [CrossRef]
20. Pariseau, W.G.; Kealy, C.D. Support potential of hydraulic backfill. In Proceedings of the 14th Symposium on Rock Mechanics, Pennsylvania, PA, USA, 11–14 June 1972; pp. 501–526.
21. Hassani, F.; Ouellet, J.; Servant, S. In situ measurements in a paste backfill: Backfill and rock mass response in the context of rockburst. In Proceedings of the Seventeenth International Mining Congress and Exhibition of Turkey, Ankara, Turkey, 19–22 June 2001; pp. 165–175.
22. Zhang, J.; Zhou, N.; Huang, Y.; Zhang, Q. Impact law of the bulk ratio of backfilling body to overlying strata movement in fully mechanized backfilling mining. *J. Min. Sci.* **2011**, *47*, 73–84. [CrossRef]
23. Xiaojun, F.; Qiming, Z.; Muhammad, A. 3D modelling of the strength effect of backfill-rock on controlling rockburst risk: A Case study. *Arab. J. Geosci.* **2020**. [CrossRef]
24. Li, J.; Yin, Z.Q.; Li, C.M. Waste rock filling in fully mechanized coal mining for goaf-side entry retaining in thin coal seam. *Arab. J. Geosci.* **2019**, *12*, 509. [CrossRef]
25. Feng, J.; Peng, H.; Shuai, G.; Meng, X.; Lixin, L. A roof model and its application in solid backfilling mining. *Int. J. Min. Sci. Technol.* **2017**, *27*, 139–143. [CrossRef]
26. Zhang, J.; Li, B.; Zhou, N.; Zhang, Q. Application of solid backfilling to reduce hard-roof caving and longwall coal face burst potential. *Int. J. Rock Mech. Min. Sci.* **2016**, *88*, 197–205. [CrossRef]
27. Skrzypkowski, K. Decreasing mining losses for the room and pillar method by replacing the inter-room pillars by the construction of wooden cribs filled with waste rocks. *Energies* **2020**, *13*, 3564. [CrossRef]
28. Huang, Y.L.; Zhang, J.X.; Zhang, Q.; Nie, S.J.; An, B.F. Strata movement control due to bulk factor of backfilling body in fully mechanized backfilling mining face. *J. Min. Saf. Eng.* **2012**, *29*, 162–167.
29. Li, M.; Zhang, J.X.; Liu, Z.; Zhao, X.; Huang, P. Mechanical analysis of roof stability under nonlinear compaction of solid backfill body. *Int. J. Min. Sci. Technol.* **2016**, *26*, 863–868. [CrossRef]
30. Hemp, D.A.; Goldbach, O.D. The effect of backfill on ground motion in a stope. In Proceedings of the 3rd Symposium Rockburst and Seismicity in Mines, Kingston, Ontario, 16–18 August 1993.
31. Kaiser, P.K.; Tang, C.A. Numerical simulation of damage accumulation and seismic energy release during brittle rock failure part II: Rib pillar collapse. *Int. J. Rock Mech. Min. Sci.* **1998**, *35*, 123–134. [CrossRef]
32. Wang, J.A.; Park, H.D. Comprehensive prediction of rockburst based on analysis of strain energy in rocks. *Tunn. Undergr. Space Technol.* **2001**, *16*, 49–57. [CrossRef]
33. Zhu, W.C.; Li, Z.H.; Zhu, L.; Tang, C.A. Numerical simulation on rockburst of underground opening triggered by dynamic disturbance. *Tunn. Undergr. Space Technol.* **2010**, *25*, 587–599. [CrossRef]
34. Piechota, S.; Cala, M.; Korzeniowski, W. Evaluation of the value of rock formation strengthening coefficient used for classification of roof in LGOM mines for different methods of gob liquidation. *Min. Rev.* **2006**, *9*, 1–8. (In Polish)

Article

Design and Research on Power Systems and Algorithms for Controlling Electric Underground Mining Machines Powered by Batteries

Artur Kozłowski [1] and Łukasz Bołoz [2,*]

[1] Łukasiewicz Research Network—Institute of Innovative Technologies EMAG, Leopolda 31, 40-189 Katowice, Poland; artur.kozlowski@emag.lukasiewicz.gov.pl

[2] Department of Machinery Engineering and Transport, Faculty of Mechanical Engineering and Robotics, AGH University of Science and Technology, A. Mickiewicza Av. 30, 30-059 Krakow, Poland

* Correspondence: boloz@agh.edu.pl; Tel.: +48-12-617-30-81

Abstract: This article discusses the work that resulted in the development of two battery-powered self-propelled electric mining machines intended for operation in the conditions of a Polish copper ore mine. Currently, the global mining industry is seeing a growing interest in battery-powered electric machines, which are replacing solutions powered by internal combustion engines. The cooperation of Mine Master, Łukasiewicz Research Network—Institute of Innovative Technologies EMAG and AGH University of Science and Technology allowed carrying out a number of works that resulted in the production of two completely new machines. In order to develop the requirements and assumptions for the designed battery-powered propulsion systems, underground tests of the existing combustion machines were carried out. Based on the results of these tests, power supply systems and control algorithms were developed and verified in a virtual environment. Next, a laboratory test stand for validating power supply systems and control algorithms was developed and constructed. The tests were aimed at checking all possible situations in which the battery gets discharged as a result of the machine's ride or operation and when it is charged from the mine's mains or with energy recovered during braking. Simulations of undesirable situations, such as fluctuations in the supply voltage or charging power limitation, were also carried out at the test stand. Positive test results were obtained. Finally, the power supply systems along with control algorithms were implemented and tested in the produced battery-powered machines during operational trials. The power systems and control algorithms are universal enough to be implemented in two different types of machines. Both machines were specially designed to substitute diesel machines in the conditions of a Polish ore mine. They are the lowest underground battery-powered drilling and bolting rigs with onboard chargers. The machines can also be charged by external fast battery chargers.

Keywords: battery drive; battery power; supply self-propelled mining machines; electric mining machines; control algorithms

Citation: Kozłowski, A.; Bołoz, Ł. Design and Research on Power Systems and Algorithms for Controlling Electric Underground Mining Machines Powered by Batteries. *Energies* **2021**, *14*, 4060. https://doi.org/10.3390/en14134060

Academic Editors: Krzysztof Skrzypkowski and Valery Vodovozov

Received: 24 May 2021
Accepted: 2 July 2021
Published: 5 July 2021

Publisher's Note: MDPI stays neutral with regard to jurisdictional claims in published maps and institutional affiliations.

Copyright: © 2021 by the authors. Licensee MDPI, Basel, Switzerland. This article is an open access article distributed under the terms and conditions of the Creative Commons Attribution (CC BY) license (https://creativecommons.org/licenses/by/4.0/).

1. Introduction

The concept of working machine drive based on an electric motor or internal combustion engine, i.e., the obtaining of mechanical energy that can be used to drive the carriage system of the machine and its working systems, refers to complete drive units. Both internal combustion engines and electric motors have been known and successfully used for many years in stationary and mobile machines. However, electric motors do not consume oxygen and do not emit exhaust fumes, which has a positive effect on the environment and human health [1]. They do not generate as much noise, and their efficiency reaches the level of more than 90% or even 98%, so they generate much less heat. In addition, the electric drive unit is less complicated [2]. Therefore, wherever mains power can be used, the electric drive wins. In the case of vehicles, mobile machines, or machines working far from the source of power supply, the internal combustion engine is commonly applied. However, if we want to use battery power, which is also referred to as battery drive, additional serious problems arise. The major drawback of battery power supply is the limited range or limited operating time of the machine, especially when we take into account the long charging time. The use of additional electrically powered systems, such as lighting, air conditioning, or a mechatronic system, further shortens the working time. The key issue in the case of such machines is, therefore, to manage and control the battery condition as well as to optimize battery supervision systems [3]. In the case of underground mining, the operating time reduction due to low temperatures is not the only problem. Apart from the technical aspects, economic considerations related to the cost of purchase and battery operation are also extremely important [4,5].

Very important factors related to the underground mining of metal ores are the costs involved in the ventilation of workings and the negative impact of substances produced during combustion of liquid fuels by the machines on the health of the working staff. In order to reduce costs and improve the working conditions of the workers, it is advisable to replace diesel engines with electric motors powered by the mains and battery. Such a tendency is observed in many countries around the world, with Canada being the precursor [2,6]. This applies in particular to self-propelled mining machines, such as drilling and bolting rigs, LHD loaders, and haul trucks. It should be emphasized, however, that the user expects electrically driven machines to have the same parameters and functional properties as the ones powered by internal combustion engines. This is a serious challenge because battery power still remains a new issue despite the fact that the electric drive is known and widely used also in mining machines. In addition, it should be noted that the difficulty results mainly from the specific conditions of underground mining and related requirements.

The electric drive of mains-powered vehicles and working machines have been used in industry for many years. It is applied not only in trams, trains, stationary machines, cranes or gantries, but also in those working machines that move in a limited area or at a low speed, or in machines that are transported to various workplaces and can be powered by electrical energy supplied by a cable or pantograph, so the output or time of operation are practically irrelevant. Selected examples include the Metso Nordberg NW 80 semi-mobile crushing set, the John Deere autonomous farm tractor with a cable reel, the CAT 7495 cable excavator, the Hitachi ZE85 excavator, or the Hitachi EH4000AC3 dump truck, which is powered by a pantograph. These are, of course, only selected examples of electric working machines. In addition, in underground mining, electric machines powered from the mine's mains are used wherever it is possible.

Of course, mobile machines pose a challenge, as opposed to stationary machines, such as belt conveyors, fans, or even transfer points, i.e., grids. Typical electric and mains-powered underground mining machines are also longwall shearers and roadheaders as well as narrow-gauge loaders, which operate in a limited area. Typical mobile electric machines include haul trucks and loaders, which are powered by mains and are therefore equipped with a cable reel. Examples include the Aramine L150E mini loader for narrow excavations, the Sandvik LH514E loader, the Epiroc Scooptram EST1030 loader, and the Philips HC12BE dump truck. Battery-powered electric working machines are widely used in a variety of industries. The best known are passenger cars, trucks, or increasingly used electric buses, but in the case of these machines, we have practically only a chassis without demanding working systems, and the working conditions are completely different. However, among heavy working machines that perform various processes, there are also battery-operated versions, especially excavators or agricultural machines, including tractors.

In underground mining, due to very difficult working conditions and high requirements, battery-powered machines began to be designed and used relatively late, mostly in the last three years. Currently, several manufacturers have selected solutions of battery-powered machines that are designed to work in various conditions:

- Aramine—L140B miniloader [7]

- Artisan vehicles (since 2019, it has belonged to Sandvik) A4 and A10 loaders, Z50 haul truck [8]
- Epiroc (until 2018 Atlas Copco)—Boomer M2 and E2 drilling rigs, Boltec M and E bolting rigs, Scooptram ST14 loader, Minetruck MT42 haul truck, Easer L, Simba M4, M6, and E7 raiseboring rigs [9]
- Komatsu (do 2017 Joy Global)—several models of haul trucks [10]
- MacLean Engineering—more than 20 models of auxiliary machines [11]
- Normet—a few auxiliary machines [12]
- Phillips Machine—haul trucks [13]
- Sandvik—LH518B loader, DD422iE drill rig, LZ101LE bulldozer [14]

Global companies Komatsu and CAT have announced that in the nearest time, they are going to roll out battery-powered machines, such as drilling and bolting rigs [15], as well as the LHD loader [16].

In the case of drilling, bolting, and auxiliary rigs, the working process is for some time carried out in one place; therefore, the most common solutions are usually applied to charge the battery at the workstation. On the other hand, LHD haul trucks and loaders are in motion most of the time, hence the use of quick battery replacement systems combined with quick charging or quick charging without replacing the battery. Additionally, braking energy, which recharges the batteries, is quite often recovered for all types of battery-operated machines [2]. Braking energy recovery is applied not only in self-propelled mining machines on a wheel-tire chassis but also in cog-wheel railways used in coal mining [17].

The results of simulation tests, however, show the advantage of fast charging, especially in long-term operation. In one of the studies [18], the authors have demonstrated that in the case of five-year operation, fast charging is more advantageous from the point of view of the efficiency and operating costs of machines. It is therefore advisable to use an onboard charger, with the possibility of charging from the mains at the workplace and the function of braking energy recovery. In addition, the machine can be charged with external high-power chargers.

Designing machines for underground mining necessitates the use of modern methods that make it possible to meet the requirements of users, taking into account extremely difficult working conditions [19,20]. The battery power supply of machines is an additional design and economic challenge [21]. Other challenges in the design of modern machines intended for underground operation include aspects related to occupational health and safety, among others the problem of excessive noise, which is increasingly often discussed [22], but also the rapidly developing robotization and automation [23] as well as digitization [24,25].

The works carried out in cooperation with Mine Master, Łukasiewicz Research Network—Institute of Innovative Technologies EMAG and AGH University of Science were used in the design of two self-propelled mining machines, namely a drilling rig and a bolting rig. Self-propelled drilling and bolting rigs perform drilling and bolting, which belong to basic cutting processes applied in the room and pillar mining system. In the case of these machines, the above-mentioned processes determine the working system's power demand. The bolting process is necessary and allows controlling the excavation stability [26,27]. Drilling is performed in the case of both drilling and bolting rigs. The energy consumption and efficiency of the drilling process depend on many factors, especially on the drilling method, hole diameter, type and condition of the tool, as well as the physical and mechanical properties of the mined rock [28,29].

Both machines were manufactured and tested. Bench and operational tests were carried out. The tests, which were mainly focused on the functional properties of battery power supply, confirmed the correctness of the developed power systems and operation algorithms. Technical data and additional information can be found in the catalog on the website of the manufacturer (Mine Master Spółka z o.o., Poland, Wilków) [30]. It should be clearly emphasized that these are the first machines in the world designed for low headings.

The Roof Master RM 1.8KE bolting rig is 1.8 m high, whereas the Face Master FM 1.7LE drilling rig has a height of 1.65 m (Figure 1). They can be maneuvered in workings having a width of 4.5 m in the room and pillar mining system. It should be emphasized that both machines have been equipped with onboard battery chargers, which enable direct charging from the mine's mains. The aforementioned solutions of the competing companies require excavations with dimensions considerably exceeding 2 or even 3 m.

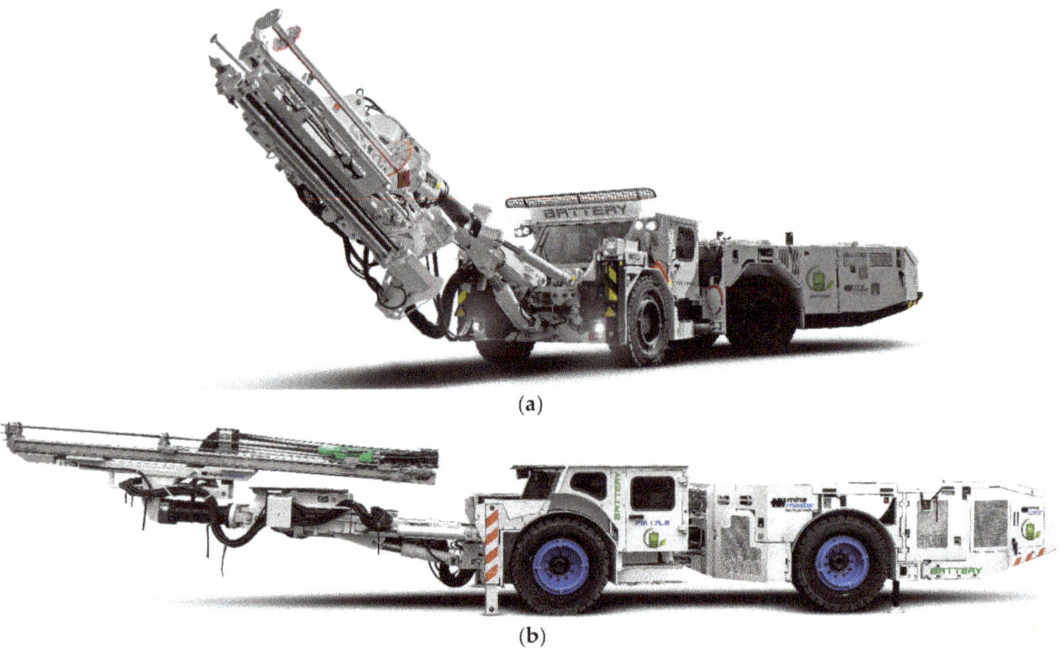

(a)

(b)

Figure 1. Mine Master battery-powered electric self-propelled mining machines: (**a**) Roof Master RM 1.8KE bolting rig, (**b**) Face Master FM 1.7LE drilling rig.

The subject of mining battery machines has been widely discussed in recent years at various conferences and in scientific journals. Manufacturers are offering more and more new solutions to battery-powered mining machines that are designed for specific operating conditions [31]. Due to the necessity of charging, these machines must be adapted to the power grid of the future user. Detailed information on the settings and parameters of power systems and control algorithms still remains the "know-how" of their manufacturers. General control algorithms for a typical cycle of discharging and charging a battery can be found in various variants in numerous studies. [32,33]. Especially in recent years, research has been conducted on modern regulators and artificial intelligence [34,35]. There are also algorithms that take into account other systems, for example, hybrid, charging with solar panels or wind turbines [36,37].

The main problem was to design machines specially for a Polish ore mine that could substitute currently operated diesel machines. However, battery-powered machines are different, and it was necessary to carry out underground tests in the mine to specify detailed requirements concerning power systems and control algorithms. One of the biggest challenges was to design very low machines equipped with onboard chargers and the possibility of using external fast chargers. As stated before, both machines are considerably lower than 2 m, which is a significant achievement compared to competitor products.

The above-quoted selected items relate to machines working overground, especially vehicles. Studies on underground battery-powered low machines, in particular those regarding power systems and control algorithms, are unavailable. The existing research and experiences concern battery-powered solutions applied only in overground machines, so they are not useful in the discussed case. Due to various reasons, only some types of batteries, power systems, and components can be used in underground mine conditions. Batteries pose the biggest problem. It is possible to use many different batteries, such as lead acid (VRLA), nickel cadmium (NiCd), lithium iron phosphate (LFP), lithium nickel manganese cobalt (NMC), lithium nickel cobalt aluminum (NCA), and sodium nickel chloride (Na-NiCl$_2$). Sodium nickel chloride (Na-NiCl$_2$) batteries are the best solution for many reasons, including in particular the risk of explosion and fire, operating temperature, ventilation, and durability. In the case of other components, the manufacturers' recommendations regarding operating conditions are applied [2]. This is an important issue in underground mining machines, which needs to be further explored.

2. Materials and Methods

For each newly designed underground mining machine, it is necessary to define requirements based on the working conditions and the user's expectations. This is particularly important in the case of a completely new solution, such as battery-powered self-propelled electric mining machines. The working conditions and user's requirements are different for each mine. Therefore, the first stage of works involved carrying out underground research, which consisted of recording selected parameters of machines. The recorded and analyzed data, as well as the collected comments and requirements of the future machine user, were used to develop assumptions and guidelines for the designed power system along with control algorithms. Next, power systems and control algorithms were developed and checked in simulation tests in a PLC environment. During the research, the properties of various types of batteries, as well as the structure and parameters of the systems managing their operation, were considered. The control algorithms developed as part of the described tests take into account the requirements resulting from the machine operation schedule with the distinction of basic cycles such as standby, ride, operation, and the adopted concept of the drive system. In order to carry out further investigations, it was necessary to design and construct a laboratory stand allowing for the simulation of the machine load and battery discharge, mains battery charging, and braking energy recovery. Tests were performed for all possible combinations of parameters.

The analysis was carried out for typical cases and situations where the braking energy was recovered for a fully charged battery or the battery was charged during the work of the machine operational systems. The performed validation confirmed the correctness of the developed power systems and control algorithms that were used in the self-propelled drilling rig and self-propelled bolting rig. The operational tests of the manufactured machines also proved the correctness of the power systems and control algorithms.

3. Underground Tests of Machines with a Combustion Engine

Underground tests were carried out for two internal combustion machines, the operational parameters of which were similar to the planned battery-powered machines. The Roof Master RM 1.8 bolting rig (Figure 2a) and the Face Master FM 1.7 drilling rig were used for tests. The tests were conducted in the KGHM Polska Miedź S.A. mine O/ZG Polkowice-Sieroszowice. The parameters of the drive system were recorded during the machine ride, while the parameters of the working process and pressure in the hydraulic system of the working parts were recorded during operation (Figure 2b). The data were recorded during the drilling and bolting process as well as during rides to other workstations and to the repair chamber.

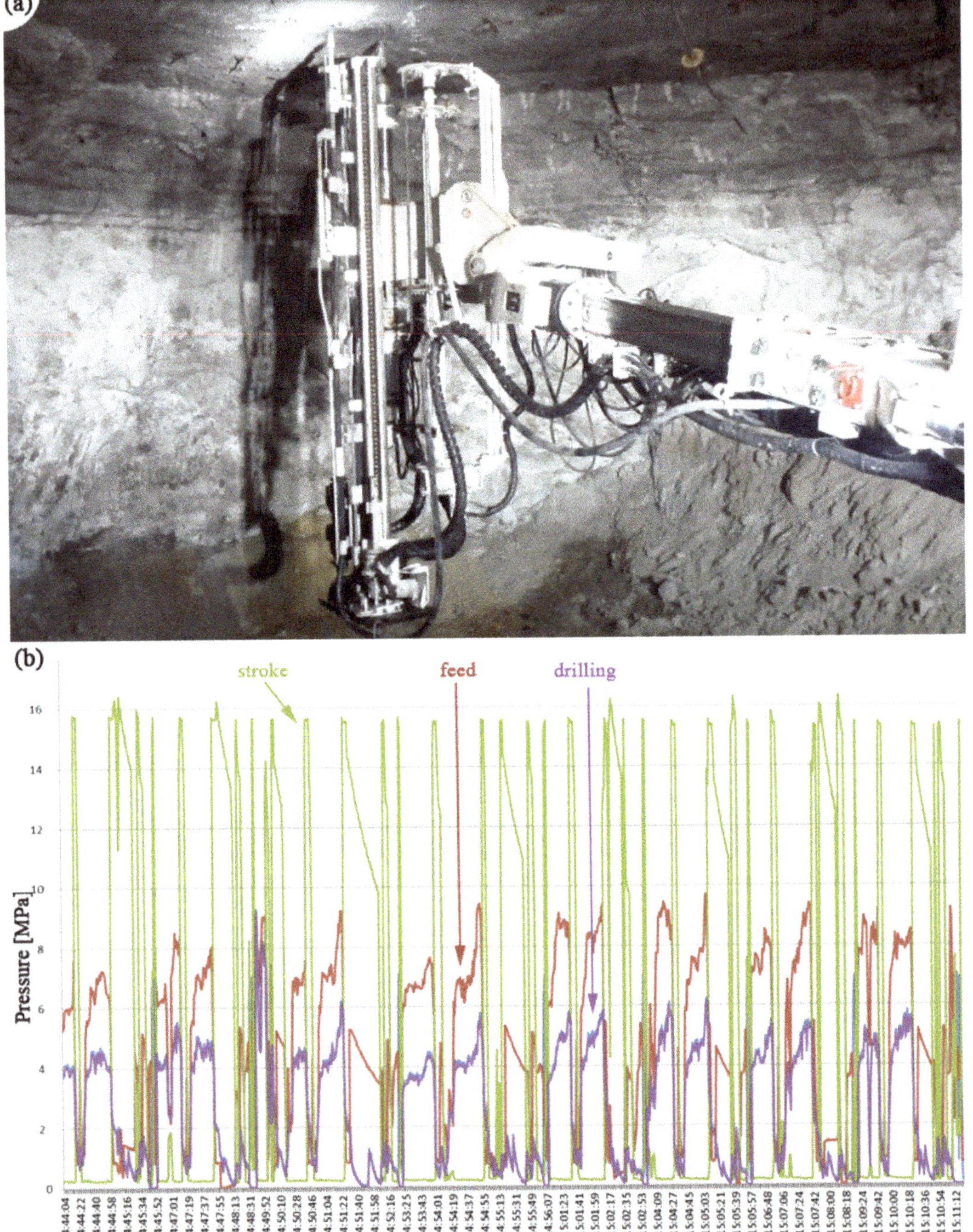

Figure 2. Underground tests in the KGHM mine: (**a**) Roof Master RM 1.8 diesel bolting rig, (**b**) recorded pressure courses in the turret hydraulic system.

Figure 2 presents a selected photo from the tests and an example of a graph of pressure courses versus time.

Based on the analysis of the test results and the user's requirements, the following was determined:

- Typical schedules of machines operation;
- Actual parameters of the supply network;
- Speed of movement of machines, taking into account the slope and quality of the ground;
- Profile of the route covered by the machines;
- Resistance to movement for various routes and floor conditions;
- Power demand for individual phases and operating conditions.

The performed analyses allowed developing a detailed concept of a self-propelled vehicle with an equivalent electric drive powered by batteries.

4. Power System of the Drilling and Bolting Rig

One of the most important systems in the designed machine is the electric power system. It is the electric power supply system that the mobility of the machine, and, in consequence, its working parameters and performance characteristics, depend on. The main electric circuit (high-current) has been designed to have the same configuration for the drilling and bolting rig. The only difference is in the size of the main engine, battery, or hydraulic systems. Based on the assumptions resulting from the functionality of the machines in the combustion version, a power system was designed, the schematic diagram of which is presented in Figure 3.

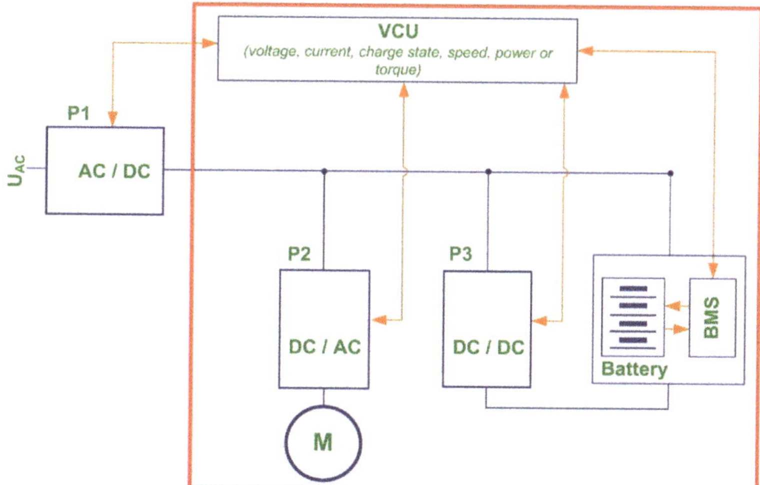

Figure 3. Schematic diagram of the machine power system.

The system is composed of three converters P1, P2, and P3. The P1 converter is used to connect the AC (alternative current) network to the DC (direct current) bus on the machine. For standard network parameters, the rectifier will work in the passive mode, which consists of rectifying the voltage. Improvement in the active mode is achieved by installing an LC (inductor-capacitor) filter on the inverter input. The P3 converter is used to charge the battery, while the P2 inverter is used to supply and regulate the main drive motor M (controller). The control is effected by the VCU (vehicle control unit), which contains a PLC (programmable logic controller) with a program for controlling the machine.

The P2 converter is used during the ride. The P3 converter connected to the battery uses the DC/DC application and operates in the VCM (voltage control mode) with the option of maintaining the voltage on the DC bus side. The P2 converter supplying the drive motor operates in the DC/AC mode at that time. The diagram of the system during the machine's ride has been shown in Figure 3 (red frame). During the machine's ride, the engine draws power only from the battery. A significant advantage is the energy return to the battery during the downhill ride.

Battery charging and operation take place in accordance with the system shown in Figure 3. Owing to the battery charger incorporated in the machine, it is charged directly from the mine's mains. The work of both machines in the excavation, i.e., drilling and bolting, is always carried out when the machine is connected to the mine's mains. In practice, however, voltage drops may occur in the power supply station, which results in a break in the machine's operation. To prevent these interruptions, the possibility of supplying the hydraulic system with a battery has been provided in the system. In the case of very difficult conditions occurring in the process of hard rock drilling, it is possible to use battery power with increased power on the main motor.

According to the assumptions, the ride is battery-powered. The power consumed by the motor is the power from the battery, decreased by losses in the converters. The power flow during the machine ride is shown in Figure 4 (red line).

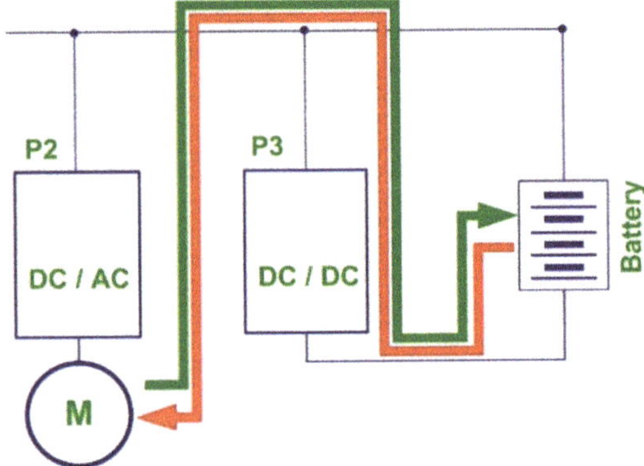

Figure 4. Diagram of power flow during the machine ride (red) and braking (green).

During the ride, energy will always flow from the battery to the main motor. In the analyzed cases, the losses from the converters were not taken into consideration when developing the power flow diagrams. The losses are converted into heat and carried away by the cooling system.

Braking is an important element of the machine's ride, which affects its safety. The main role in this process is played by dynamic braking with the option of main motor braking with energy return to the battery. Battery charging must always finish with a certain reserve so that the battery has some capacity to receive the braking energy. The power flow during braking is shown in Figure 4 (green line). Power flows from the motor to the battery.

The battery is charged near the power supply station or in the repair chamber. Charging takes place via the P1 and P3 converters. Charging energy from the mains is stored in the battery. Figure 5 (green line) shows power flow during charging.

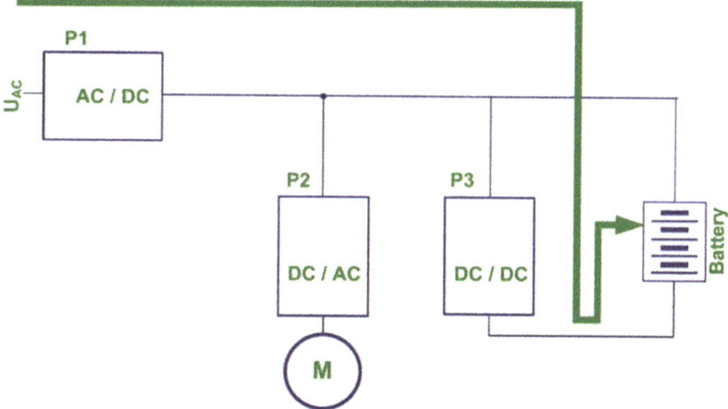

Figure 5. Diagram of power flow during charging of the machine.

After the machine has arrived at the workstation, it is possible to recharge the battery in the time between the drilling of subsequent holes or between the installation of subsequent bolts, or during the auxiliary processes. This allows using the machine's working time more efficiently. The master control system controls the total current drawn from the mains. The power flow has been shown in Figure 5 (green line).

5. Control Algorithms

The next stage involved developing algorithms for controlling the electric power system during the ride and operation of the machine, taking into consideration various combinations. Due to the similar structure of the power electric circuits of the drilling and bolting rigs, common algorithms were developed for both types of machines.

The P1, P2, and P3 converters were configured depending on the selected operating mode. The appropriate configuration ensures the correct operation of the machine and allows applying appropriate limitations, which guarantees the safe operation of the powered equipment. Particular attention was paid to the configuration of converters cooperating with the bank of batteries. The supervision system was configured to prevent deep discharge of the battery and to ensure optimal charging conditions set by the battery management system (BMS). Depending on the mode (ride or operation), the relevant converters are blocked or configured. After the appropriate machine operating mode has been selected, the configuration of the equipment is performed automatically by the PLC controller, which communicates by means of CAN (controller area network) transmission bus.

5.1. Algorithms for Switching on the Supply Voltage

The supply voltage is switched on according to two algorithms, depending on the mode (operation or ride). The main supply voltage is the DC bus voltage for battery operation and the AC voltage when the machine is in the operating mode (drilling or bolting). Before the main voltages are switched on, the auxiliary voltage supplying the circuits of PLC controllers and protection systems is switched on. The converters are configured before switching on the main power.

The algorithm controlling the process of switching on the voltage during the machine ride is shown in Figure 6. During the machine ride, a bank of batteries is used as a power source. The unit checks all the systems one by one and prepares them for operation according to the control algorithm. In this mode, the machine must not be connected to an *AC* power source via cable, the power cord must not be unwound, and the power supply system must function properly. Before the ride, the battery power system must be prepared.

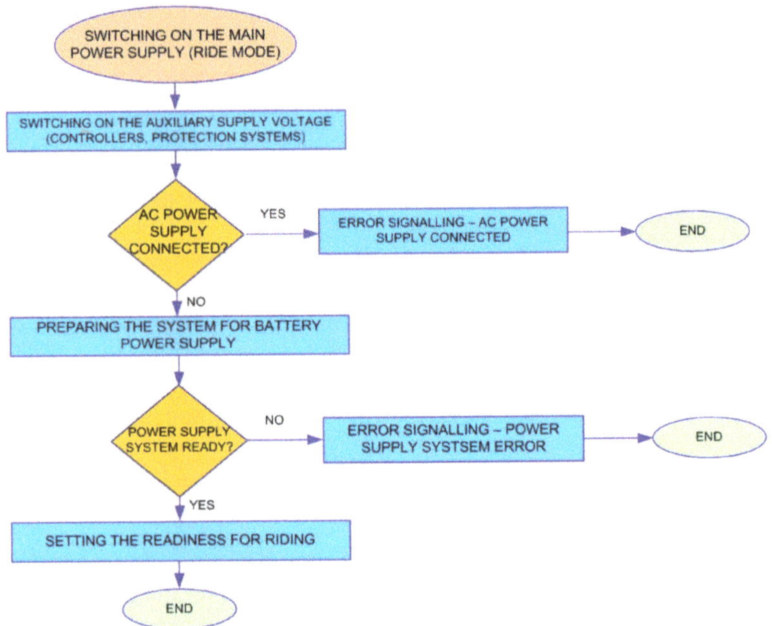

Figure 6. Algorithm for the procedure of switching on the supply voltage in the ride mode.

The algorithm controlling the process of switching on the voltage in the operating mode has been shown in Figure 7. In the operating mode, after switching on the auxiliary voltage, the converters (P1, P2, and P3) are configured for work. While the machine is in operation (dr illing, bolting), the mine's AC network is used as a power source. As in the case of machine rides, the power system is first prepared, and its correct operation is checked. In the next step, the battery is connected, and the P1 and P3 converters are connected to the circuit. The electrical system is now ready for operation.

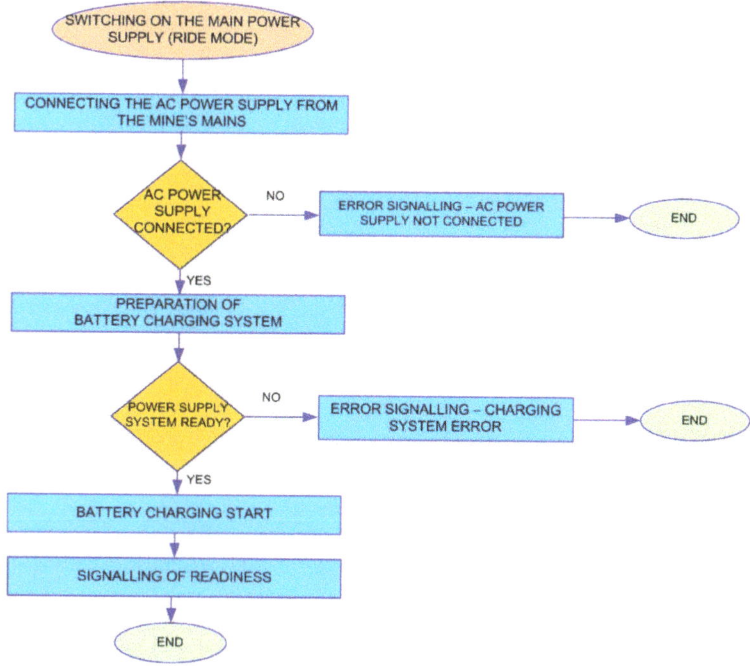

Figure 7. Algorithm for the procedure of switching on the supply voltage in the operating mode.

5.2. Algorithms for Controlling the Electric Power System of the Machine

The algorithm controlling the electric power system is also selected depending on the operation or ride mode. The algorithm during the machine's ride has been shown in Figure 8. In the ride mode, after switching on the voltage, the electric system is ready for riding. The converters (P2, P3) are configured for the ride mode. Before starting a ride, the machine control system checks the state of charge of the batteries. Riding is not possible when the battery is completely discharged. After receiving the command to ride, the control system starts the electric motor of the ride drive. During the ride, the master control system in the PLC continuously monitors the depth of battery discharge. The system informs the operator about the battery discharge status on an ongoing basis. If the critical discharge is reached, the machine must be stopped; otherwise, the battery management system (BMS) will disconnect the battery. Riding cannot be continued unless the battery has been recharged to the minimum value. In most cases, braking of the machine is accompanied by energy return to the battery. In an emergency, when the bank of batteries has been disconnected or the motor is stopped due to a failure of the inverter, the master control system initiates emergency braking using mechanical brakes. The ride (switching on the main motor) is continued when the system is ready for riding, i.e., after the cause of the emergency stop has ceased and the errors are deleted.

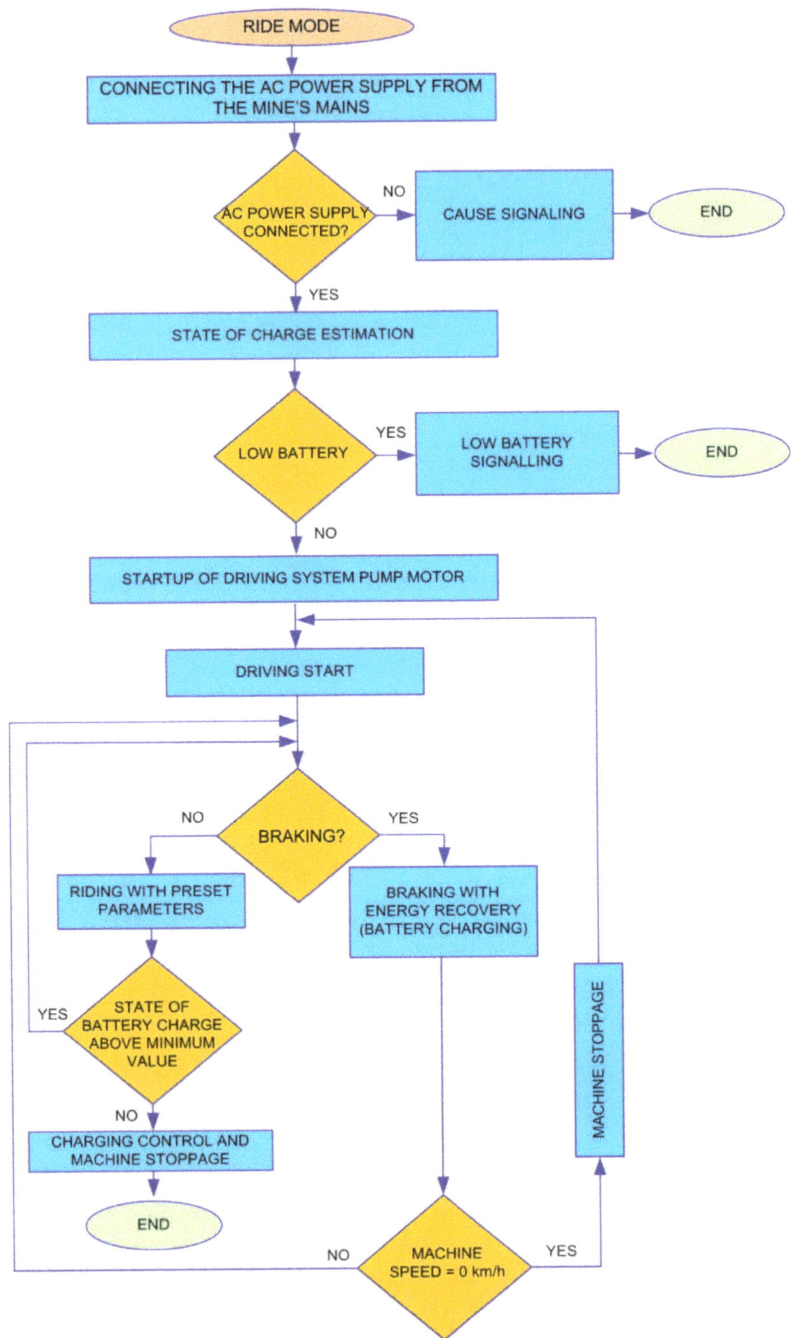

Figure 8. Algorithm controlling the electric power system during the ride.

The algorithm for controlling the electric power system during operation is shown in Figure 9. After switching on the voltage in accordance with the procedure described above, the electrical system is ready for operation. The converters (P1, P2, and P3) are configured for operation. In the operating mode, the state of charge of the battery is constantly monitored by the controller of the master control system. After the work request has been received by the master control system, the motor of the operating system pump is turned on. After the supply voltage has been switched on, the condition of the insulation resistance is constantly monitored. If a decrease in resistance is detected, the supply voltage of the machine must be switched off.

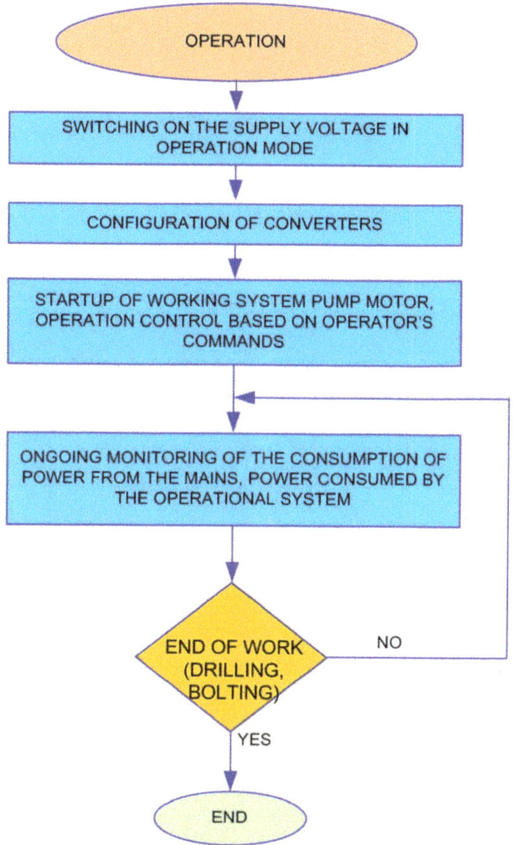

Figure 9. Algorithm controlling the electric power system during operation.

6. Validation Tests

Laboratory validation tests were carried out for the developed power systems and control algorithms. The tests of the electric drive system of the self-propelled drilling rig and the self-propelled bolting rig in the riding and operation mode were conducted at a special stand in the measuring system shown in Figure 10. The system uses a test platform for PMSM (permanent-magnet synchronous motor), equipped with two synchronous motors coupled with each other, an AC/AC frequency converter with an active rectifier, enabling energy return to the supply network, a PLC logic controller, a switching and measuring apparatus and an operator's panel with a display. The stand with the motors and the operator's desk is shown in Figure 11.

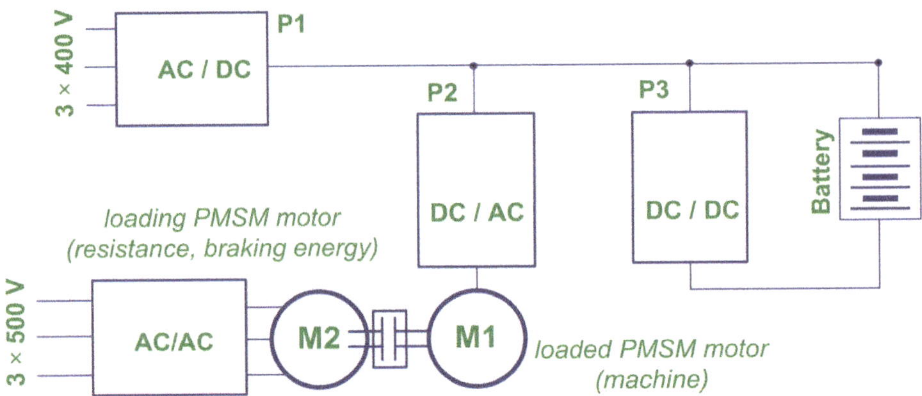

Figure 10. Measuring system for validation tests of the electric drive in the riding and operation mode.

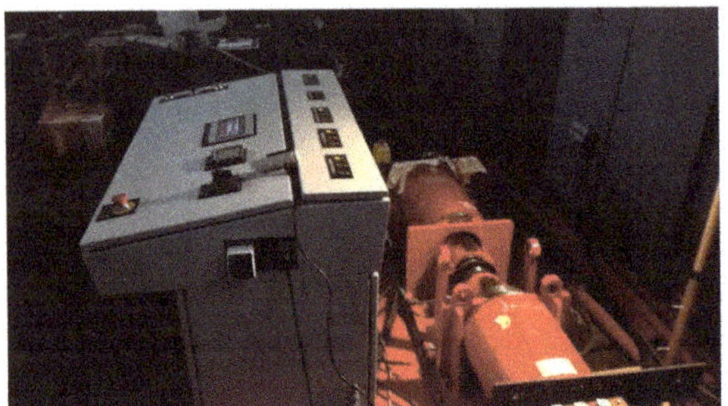

Figure 11. Stand for testing the drive system.

The basic element of the system subjected to validation was the DC busbar. The AC/DC converter (P1 diode rectifier), supplied with the voltage of 3 × 500 V from an inductive regulator, is connected to the bus. The P3 DC/DC converter connects the bus with the bank of batteries. The P2 DC/AC converter controls the main drive motor, i.e., the M1 motor.

In order to check the behavior of the system during mains voltage fluctuations, the AC/DC converter (diode rectifier) was supplied with the voltage of 3 × 500 V from an inductive regulator, allowing for changes of this voltage within a range wider than the expected tolerance range of $0.85 \div 1.2\ U_N$. The DC bus can be powered during standby/operation and, depending on the connection to the mains, from the AC/DC converter or via the DC/DC converter from the main bank of batteries.

Due to the need to check the behavior of the system at different battery voltage values and during mains voltage fluctuations, the stand was equipped with the M2 motor coupled with the M1 machine motor, which was the loading machine (Figure 10). The main motor load generated by the M2 motor can be regulated by the AC/AC converter controlling the motor, and the value and direction of the torque for any value and direction of rotational speed can be fully adjusted, from positive to negative values. This means that the M2 motor can both load and drive the M1 motor, which allows representing (simulating) the downhill ride and, therefore, motor braking. When driving in a flat area or on a slope, the M1 machine works as a drive motor, drawing energy via the DC/AC converter and the DC bus from the bank of batteries. During drilling or bolting, the vehicle is connected to a 3 × 500 V power supply network, and the M1 machine also functions as a drive motor but draws energy via the DC/AC converter and the DC bus, in this case, from this power supply network. The electric energy generated in the load motor is returned to the supply network via the AC/AC converter. This is the most typical type of operation of the vehicle drive system.

The test of discharging the battery corresponded to long riding with a constant load in the real system. Additionally, it was checked how the voltage at the battery terminals changed as the depth of discharge, and the load increased. The tests were conducted using a battery consisting of six modules connected in series with the following parameters:

- Rated voltage: 499 V;
- Rated capacity: 58.8 Ah;
- Rated energy: 30 kWh;
  Max. voltage: 562 V;
- Nominal voltage: 390 V;
- Max. dynamic discharge current: 120 A;
- Max. dynamic charge current: 60 A;
- Continuous discharge power: 60 kW;
- Continuous charge power: 30 kW.

The battery was discharged by the loading drive in the following system: battery > P3 converter > P2 converter > M1 machine drive motor M1 > M2 loading motor > energy return to the mains 3 × 500 V.

The battery was discharged at approximately 30 kW (half the discharge power) to 35% available energy. Further discharge was blocked by the battery management system (BMS). The conducted tests allowed obtaining battery discharge characteristics (Figures 12–14), which enable determining the battery discharge rate, the voltage drop in the state of charge and time function, as well as the amount of load in the state of charge function. A non-linear fitting was performed for voltage changes and a linear fitting for load change so as to find formulas describing these characteristics. An almost perfect match of the functions was achieved. In all the cases, the R-square (COD—coefficient of determination) value is significantly above 0.99. The formulas and matching factors are given in the charts. During the tests, the voltage changed from 504 to 486 V. Reducing the state of charge to 35% decreased the voltage by more than 3%. During battery discharging from 100% to 35%, a load of 37 Ah was consumed. The state of 35% charge was reached after nearly a 42 min operation. The function formula allows calculating that the battery will be completely discharged after 135 min.

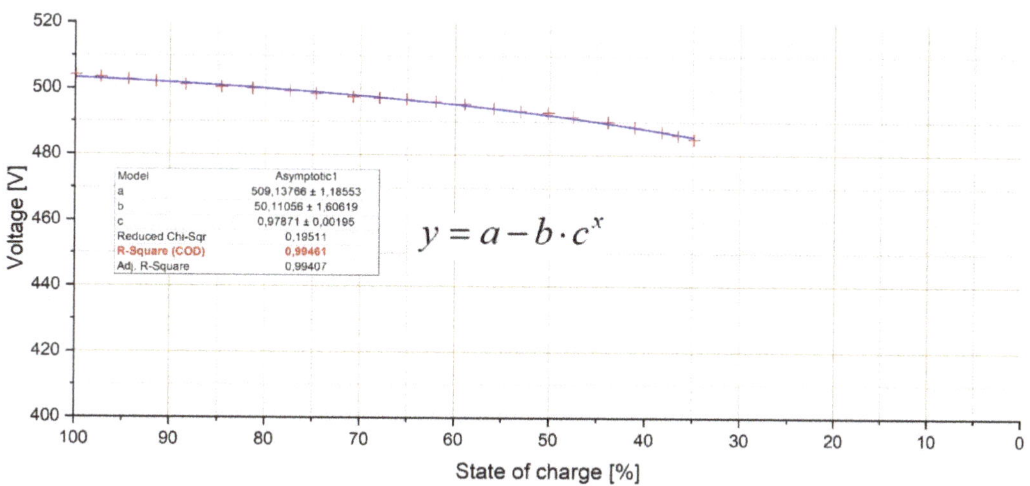

$$y = a - b \cdot c^x$$

Figure 12. Battery discharge—battery voltage as a function of the state of charge.

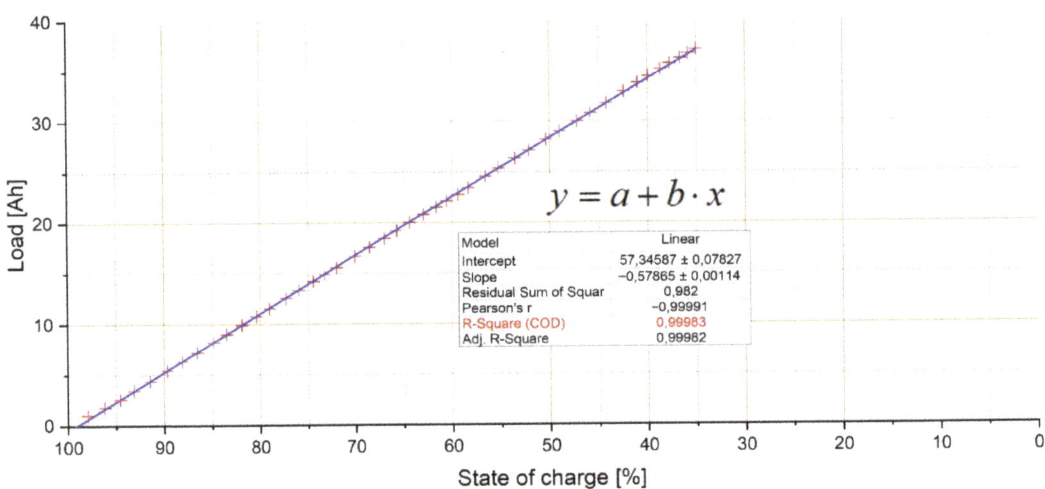

Figure 13. Battery discharge—load as a function of the state of charge.

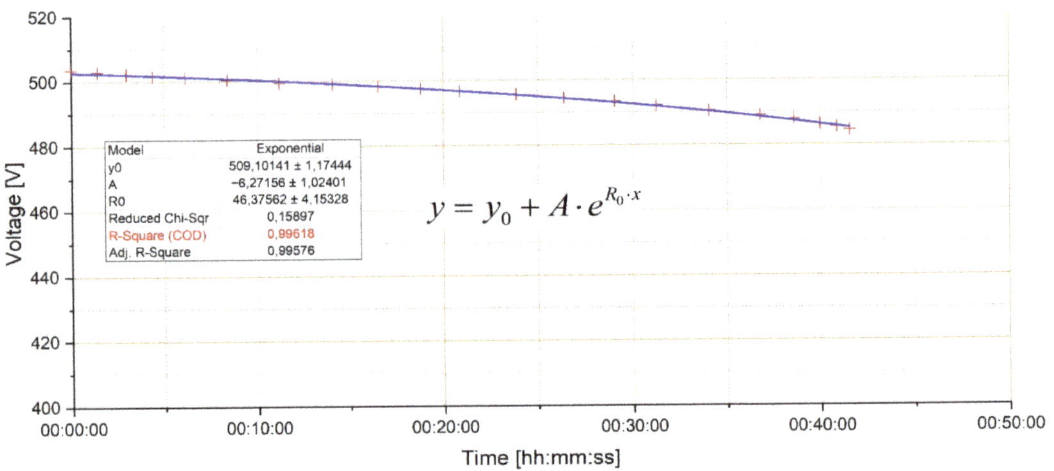

Figure 14. Battery discharge—battery voltage versus discharge time.

Next, tests of charging the bank of batteries were carried out. The *P3* charging converter enables full control over the charging current and voltage. However, the basic condition for obtaining the full required charging dynamics is that the voltage of the *DC* bus must be greater than the voltage of the battery.

Charging tests were carried out in two modes, in accordance with the presented configuration:

- Battery voltage setting mode (energy storage) with the setting of maximum charging current limitation;
- Charging current setting mode with the setting of maximum voltage limitation on the battery.

For the battery voltage setting mode, the preset voltage was the voltage of the fully charged battery: $U_{zad} = U_{bat.max} = 540$ V, and the maximum charging current $I_{max.ład.} = 60$ A was set as the charging current limitation. In this mode, charging with the current set by the battery management system (*BMS*) was continued until the battery was fully charged (100%). The state of charge is indicated by the *BMS* of the battery used for battery tests.

For the charging current setting mode, the preset current was maximum charging current $I_{zad} = I_{max.load} = 60$ A, and the maximum battery voltage $U_{bat.max} = 540$ V was set as voltage limitation. The charging current in this mode was also set by the *BMS* battery management system up to a level of 100%.

Based on the research, it was found that both modes were identical in this case. This is due to the fact that charging parameters are set by the battery management system (*BMS*), which is the master unit for the charging converter. Therefore, the values of the charging parameters set in the converter are treated each time as acceptable parameters, i.e., limit parameters, which the *BMS* does not exceed.

Figure 15 shows the course of the battery charging current. To reduce the current fluctuations around the mean value, it is recommended that a choke with a larger inductance should be used. The inductance of the choke used in the tests was 0.3 mH.

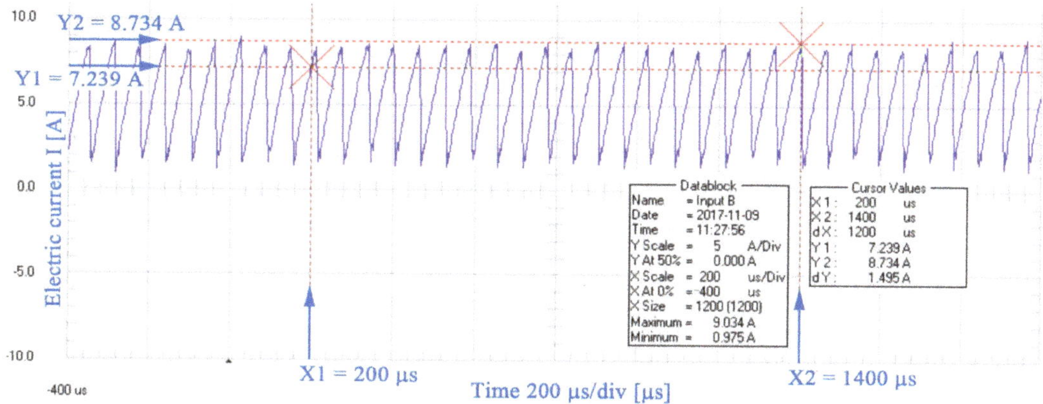

Figure 15. Battery charging current waveform—average battery charging current 5A.

Next, dynamic tests were conducted with battery power and motor load in accordance with the developed system. The dynamic tests involved wide-range changes of the drive motor's loading torque, from the positive to negative nominal value. Therefore, the *M2* motor can both load and drive the *M1* motor, which simulates riding with recovery due to motor braking. During the simulation of a ride in a flat area or on a hill, the *M1* motor works as a drive that draws energy through the *DC/AC* converter and the *DC* bus from the bank of batteries. During the battery discharge, there is no possibility of limiting the current drawn from the battery by the *P3* converter. However, the current can be limited by the inverter supplying the motor. The tests of the work of the inverter powering the drive motor were conducted in the following modes:

- Speed control;
- Torque control;
- Power control.

Speed control—increasing or decreasing the load (torque) during the drive's operation in the speed control mode causes an increase or decrease in the current and power consumed by the drive system while maintaining a constant rotational speed (to the value set in the converter of maximum current/power, beyond which the speed begins to decrease).

Torque control—increasing or decreasing the load (torque) during the drive operation in the torque control mode causes a reduction or increase in the rotational speed until the driving torque and load torque balance is achieved. The current and power consumed by the drive system also change.

Power control—increasing or decreasing the load (torque) during the drive operation in the power control mode causes a reduction or increase in the rotational speed. The current and power consumed by the drive system also change.

The motor braking tests with energy recovery for the battery involved changing the sign of the torque of the machine loading the drive motor, as a result of which it changed its function from loading to driving, and the tested *M1* motor became a generator. The electric power generated in this motor was transmitted through the *P2* converter to the *DC* busbars and, next, through the *P3* converter to the *BA* bank of batteries. The smooth transition of the *M2* machine torque from the load torque to the driving torque resulted in a smooth change of the sign of the power drawn from the bank of batteries to power supplied to this battery in the full range of load torque changes. At this point, it should be noted that the above tests were aimed at verifying the functionality of the proposed electric drive system of the analyzed mining machines and not checking the quantitative relationships, all the more so because the tested system was a model one and did not fully reflect the real target system.

The attempt to limit the power consumed by the drive system was aimed at verifying the possibility of limiting the power consumed by the drive system during drilling or bolting supplied from the mains when the power available from the mains is limited to 65 kW. The test was carried out with mains power supplied through the *P1* converter. As the *P1* converter cannot limit the power consumed, this limitation was effected through the *P2* converter supplying the motor. After the drive motor reached the permissible power (limit value), the drive torque was increased while the rotational speed was reduced. This is an advantageous feature of this type of drive as it prevents the tool from jamming in the rock mass. This is one of the standard limitations in converters working as inverters and can be set in a wide range. The test was conducted for the limitation set at 50% motor rated power with a positive result.

The attempt to limit the drive torque by the inverter supplying the drive motor was aimed at checking the possibility of limiting the torque by the drive system during drilling in order to prevent the tool from jamming in the rock mass. The test was carried out with mains power supplied via the *P1* converter while increasing the load torque above the set limit torque. In addition, in this case, the limitation was effected through the *P2* converter supplying the motor. It is also one of the standard limitations in converters working as inverters and can be set in a wide range. The test was performed for the limit set at 50% rated torque. After exceeding the set torque limit, the speed of the drive motor began to drop rapidly. The result of the driving torque limitation test was positive.

Another attempt to limit the current by the inverter supplying the drive motor was undertaken to check the possibility of limiting the torque by the drive system during operation (drilling) in order to prevent the tool from jamming in the rock mass. The test was carried out with mains power supplied through the *P1* converter. Increasing the load torque causes the current of the drive motor to increase. When the limit current is reached, the load current stabilizes, and the rotational speed is reduced, which should prevent tool jamming. In addition, in this case, the limitation was effected through the *P2* converter supplying the motor. It is also one of the standard limitations in converters working as inverters and can be set at any level. The test was performed for the limitation set at 50% rated current. After exceeding the set current limit, the speed of the drive motor began to drop rapidly. The current limitation test was positive.

7. Conclusions

Wherever machines powered by internal combustion engines are used, attempts are undertaken to replace them with electric equipment. In the case of many working machines and vehicles, the mains supply does not allow them to be fully operational; hence, it is necessary to use a battery power supply. It is crucial for every working machine that its battery-powered version meets the same requirements as the one powered by the combustion engine. Therefore, the requirements in the target place of work in terms of battery power must always be specified. Underground mining is characterized by difficult working conditions and requirements that do not allow implementing solutions applied in other industries. Conditions in underground mines are so diverse that it is not always possible to develop universal solutions and machines.

The aim of the works described in this article was to determine and describe the operating states of the machine as well as to develop power systems and control algorithms that enable work in all operating conditions occurring in the analyzed mine. After the basic operating states of the machine had been described, the algorithms for controlling the electric power system were determined. In simple words, it comes down to the states of charging and discharging the battery under various possible conditions. The above-discussed algorithms for switching on the main voltage, riding, and operation set control over the converters and other elements of the electrical system so as to enable work with electrical parameters corresponding to the rated data of the designed machine. Next, the algorithms were checked during simulation tests and in the real electrical system. The system was based on the currently available elements, but its configuration corresponded to the layout of the designed machines. The developed concept, taking into account the control method, was tested in a model system with properties similar to the actual power supply system of the target machines. Machine operating states, such as riding, braking during ride and operation, were checked and verified. The validation proved the correctness and effectiveness of the power systems and control algorithms, which were then implemented in the designed machines. The developed and tested self-propelled electric bolting rig Roof Master RM 1.8KE and the drilling rig Face Master FM 1.7LE are the best proof that the works were performed properly. These are the first fully electric and battery-powered machines in Poland that have been designed and manufactured for the conditions of KGHM S.A. Battery-powered machines are the inevitable future of mining, which is also being created by Polish research units and Polish companies.

The drilling and bolting rigs are based on the above-presented power systems and control algorithms, which makes them unique on a global scale. The applied power systems and control algorithms are universal enough to be implemented in two different types of machines. Both machines were specially designed to substitute diesel machines in the conditions of a Polish ore mine. They are the lowest underground battery-powered drilling and bolting rigs with onboard chargers. Additionally, the machines can also be charged using external fast battery chargers. Machines offered by other manufacturers are much higher, above 2 m, which is not a challenge. Despite the low height, the presented machines' operation time without charging is similar to that of competitor products. Their limitation lies in the fact that they can work successfully only in approximate conditions in terms of user requirements and underground power grid parameters. However, some modifications can be made to adapt the presented machines to different conditions.

The next step is to develop power systems and control algorithms. There are two major directions. The first is to increase the working time without charging, whereas the second is to enable charging not only in gaps between drilling or bolting but also at the time of machine operation. There are also some general problems to be solved, for example, how to reduce the consumption of energy needed for riding as well as drilling and bolting processes.

Author Contributions: Analysis of literature and state of knowledge, Ł.B.; plan and methodology of researches, A.K.; power supply system and algorithms, A.K. and Ł.B.; implementation of research, A.K. and Ł.B.; elaboration of research results, A.K.; analysis of research results, Ł.B.; analysis and interpretation of results, A.K. and Ł.B.; elaboration of paper, Ł.B. All authors have read and agreed to the published version of the manuscript.

Funding: The research was conducted with the support of the statutory work project number POIR.01.01.01-00-D011/16, Nowa generacja modułowych maszyn, wiercącej i kotwiącej, z napędami bateryjnymi, przeznaczonych do pracy w podziemnych kopal-niach rud miedzi i surowców mineral-nych" ("New generation of battery-powered modular drilling and bolting machines intended for work in underground mines of copper ores and mineral resources").

Institutional Review Board Statement: Not applicable.

Informed Consent Statement: Not applicable.

Data Availability Statement: Data are contained within the article.

Acknowledgments: Many thanks to Mine Master Spółka z o.o. for providing materials and cooperation resulting in the creation of modern battery-powered machines. Thanks to KGHM Polska Miedź S.A. for enabling and supporting the underground tests.

Conflicts of Interest: The authors declare no conflict of interest.

References

1. Fugiel, A.; Burchart-orol, D.; Czaplicka-Kolarz, K.; Smoliński, A. Environmental impact and damage categories caused by air pollution emissions from mining and quarrying sectorsof European countries. *J. Clean. Prod.* **2017**, *143*, 159–168. [CrossRef]
2. Bołoz, Ł. Światowe Trendy w Rozwoju Podziemnych Maszyn Górniczych z Napędami Bateryjnymi. In Proceedings of the VI Science Symposium: Rozwój i Eksploatacja Maszyn Górnictwa Podziemnego Surowców Mineralnych-Efekty Współpracy Przemysłu i Nauki, Online Symposium, 23 April 2021.
3. Guo, Y.J.; Yang, Z.L.; Liu, K.L.; Zhang, Y.H.; Feng, W. A compact and optimized neural network approach for battery state-of-charge estimation of energy storage system. *Energies* **2021**, *219*, 119529. [CrossRef]
4. Miao, Y.; Hynan, P.; Jouanne, A.; Yokochi, A. Current Li-Ion Battery Technologies in Electric Vehicles and Opportunities for Advancements. *Energies* **2019**, *12*, 1074. [CrossRef]
5. Wentker, M.; Greenwood, M.; Leker, J. A Bottom-Up Approach to Lithium-Ion Battery Cost Modeling with a Focus on Cathode Active Materials. *Energies* **2019**, *12*, 504. [CrossRef]
6. The 100% Electric Mine-Converting the Biggest Diesels to Electric. Available online: https://www.nrcan.gc.ca/science-and-data/funding-partnerships/funding-opportunities/current-investments/the-100-electric-mine-converting-the-biggest-diesels-electric/22639 (accessed on 22 March 2021).
7. Aramine–Battery Loader. Available online: https://www.aramine.com/produit/mini-loader-l140b/, (accessed on 24 March 2021).
8. Artisan Vehicles–Battery machines. Available online: https://www.artisanvehicles.com/ (accessed on 24 March 2021).
9. Epiroc–The Zero Emission Fleet. Available online: https://www.epiroc.com/en-gr/innovation-and-technology/zero-emission (accessed on 24 March 2021).
10. Komatsu–Joy Battery Haulers. Available online: https://mining.komatsu/product-details/battery-haulers (accessed on 24 March 2021).
11. MacLean Engineering–EV Series. Available online: https://macleanengineering.com/products/mining/electric-vehicle-series (accessed on 24 March 2021).
12. Normet–Smart Drive. Available online: https://www.normet.com/smartdrive/ (accessed on 24 March 2021).
13. Phillips Machine–Battery Powered Shuttle Cars. Available online: http://phillipsglobal.us/oem-equipment-2/ (accessed on 24 March 2021).
14. Sandvik–Battery Electric Machines. Available online: https://www.rocktechnology.sandvik/en/search/?q=battery&filter=searchsection%3aproducts (accessed on 24 March 2021).
15. Komatsu–Innovation in Hard Rock Mining from Komatsu. Available online: https://mining.komatsu/blog/details/innovation-in-hard-rock-mining-from-komatsu (accessed on 24 March 2021).
16. Cat R1700 XE Features Electric Drive. Available online: https://www.oemoffhighway.com/trends/equipment-launches/mining/press-release/21066468/caterpillar-inc-cat-r1700-xe-features-electric-drive (accessed on 24 March 2021).
17. Polnik, B.; Kaczmarczyk, K.; Niedworok, A.; Baltes, R.; Clausen, E. Energy Recuperation as One of the Factors Improving the Energy Efficiency of Mining Battery Locomotives. *Manag. Syst. Prod. Eng.* **2020**, *28*, 253–258. [CrossRef]
18. Rafi, M.A.H.; Rennie, R.; Larsen, J.; Bauman, J. Investigation of Fast Charging and Battery Swapping Options for Electric Haul Trucks in Underground Mines. In Proceedings of the 2020 IEEE Transportation Electrification Conference & Expo (ITEC), Chicago, IL, USA, 23–26 June 2020; pp. 1081–1087. [CrossRef]
19. Bołoz, Ł.; Castañeda, L.F. Computer-Aided Support for the Rapid Creation of Parametric Models of Milling Units for Longwall Shearers. *Manag. Syst. Prod. Eng.* **2018**, *26*, 193–199. [CrossRef]
20. Bołoz, Ł. Digital Prototyping on the Example of Selected Self-Propelled Mining Machines. *Multidiscip. Asp. Prod. Eng.* **2020**, *3*, 172–183. [CrossRef]
21. Burd, J.T.J.; Moore, E.A.; Ezzat, H.; Kirchain, R.; Roth, R. Improvements in electric vehicle battery technology influence vehicle lightweighting and material substitution decisions. *Appl. Energy* **2020**, *12*. [CrossRef]
22. Biały, W.; Bołoz, Ł.; Sitko, J. Mechanical processing of hard coal as a source of Noise pollution. *Energies* **2021**, *14*, 1332. [CrossRef]
23. Bołoz, Ł.; Biały, W. Automation and Robotization of Underground Mining in Poland. *Appl. Sci.* **2020**, *10*, 7221. [CrossRef]
24. Kozłowski, A.; Wojtas, P. Systemowe podejście do cyfryzacji w procesach technologicznych w górnictwie. *Zesz. Nauk. Inst. Gospod. Surowcami Miner. i Energią Pol. Akad. Nauk.* **2017**, *99*, 47–56.

25. Wojtas, P.; Kozłowski, A.; Wojtas, M. Digitization of Polish mining industry by reducing costs and improving safety and quality of finished product. *Mininig-Inform. Autom. Electr. Eng.* **2017**, *3*, 531. [CrossRef]
26. Skrzypkowski, K. Case Studies of Rock Bolt Support Loads and Rock Mass Monitoring for the Room and Pillar Method in the Legnica-Głogów Copper District in Poland. *Energies* **2020**, *13*, 2998. [CrossRef]
27. Skrzypkowski, K.; Korzeniowski, W.; Zagórski, K.; Zagórska, A. Adjustment of the Yielding System of Mechanical Rock Bolts for Room and Pillar Mining Method in Stratified Rock Mass. *Energies* **2020**, *13*, 2082. [CrossRef]
28. Bołoz, Ł. Interpretation of the results of mechanical rock properties testing with respect to mining methods. *Acta Montan. Slovaca* **2020**, *25*, 81–93. [CrossRef]
29. Kotwica, K.; Małkowski, P. Methods of Mechanical Mining of Compact-Rock—A Comparison of Efficiency and Energy Consumption. *Energies* **2019**, *12*, 3562. [CrossRef]
30. Baterry Electri Rgis, Mine Master Spółka z o.o. Available online: https://www.minemaster.eu/battery-electric-rigs (accessed on 23 March 2021).
31. Weiss, H.; Winkler, T.; Ziegerhofer, H. Large lithium-ion battery-powered electric vehicles—From idea to reality. In Proceedings of the ELEKTRO, Mikulov, Czech Republic, 21–23 May 2018; pp. 1–5. [CrossRef]
32. Kang, T.; Chae, B.; Suh, Y. Control algorithm of bi-directional power flow rapid charging system for electric vehicle using Li-Ion polymer battery. In Proceedings of the 2013 IEEE ECCE Asia Downunder, Melbourne, VIC, Australia, 3–6 June 2013; pp. 499–505. [CrossRef]
33. Chauhan, S.; Singh, B. Grid-interfaced solar PV powered electric vehicle battery system with novel adaptive digital control algorithm. *IET Power Electron.* **2019**, *12*, 3470–3478. [CrossRef]
34. De Luca, F.; Calderaro, V.; Galdi, V. A Fuzzy Logic-Based Control Algorithm for the Recharge/V2G of a Nine-Phase Integrated On-Board Battery Charger. *Electronics* **2020**, *9*, 946. [CrossRef]
35. Zhang, C.-W.; Chen, S.-R.; Gao, H.-B.; Xu, K.-J.; Yang, M.-Y. State of Charge Estimation of Power Battery Using Improved Back Propagation Neural Network. *Batteries* **2018**, *4*, 69. [CrossRef]
36. Notton, G.; Lazarov, V.; Zarkov, Z.; Stoyanov, L. Optimization of Hybrid Systems with Renewable Energy Sources: Trends for Research. In Proceedings of the Conference: Environment Identities and Mediterranean Area, ISEIMA '06, Corte-Ajaccio, France, 9–12 July 2006; pp. 144–149. [CrossRef]
37. Hirech, K.; Melhaoui, M.; Malek, R.; Kassmi, K. A PV System Equipped with a Commercial and Designed MPPT Regulators. In *Proceedings of the Mediterranean Conference on Information & Communication Technologies 2015*; Springer: Cham, Switzerland, 2016. [CrossRef]

Article

Open-Pit Mine Dewatering Based on Water Recirculation—Case Study with Numerical Modelling

Kazimierz Różkowski [1],*, Robert Zdechlik [2] and Wojciech Chudzik [3]

[1] Faculty of Civil Engineering and Resource Management, AGH University of Science and Technology, al. Mickiewicza 30, 30-059 Krakow, Poland
[2] Faculty of Geology, Geophysics and Environmental Protection, AGH University of Science and Technology, al. Mickiewicza 30, 30-059 Krakow, Poland; robert.zdechlik@agh.edu.pl
[3] Lafarge Aggregates and Concrete, Aleje Jerozolimskie 142B, 02-305 Warsaw, Poland; wojciech.chudzik@lafargeholcim.com
* Correspondence: kazik@agh.edu.pl

Abstract: The layout of the dewatering system in open-cast mining must be adapted to mining assumptions and to the size of expected inflows, which, in turn, depend on natural conditions and the operation of other mines and groundwater intakes, affecting the arrangement of the hydrodynamic field. This case study analyses possible dewatering solutions related to a change in the mining drainage system: decommissioning by flooding of a depleted deposit and dewatering of a new one located in the vicinity. As part of numerical modelling, a solution was sought to minimise the environmental impact of drainage. Forecast calculations for two drainage alternatives were made. One of the solutions follows the classic approach: independent dewatering of the new excavation. The second solution assumes the recirculation of waters from dewatering of the new mine through their discharge into a closed and flooded pit located in the vicinity. The results of the forecasts for both variants point to the modification of the hydrodynamic field resulting from expected volumes of inflows and different environmental effects. The use of numerical simulations assisted the selection of the optimal dewatering solution.

Keywords: groundwater; dewatering optimisation; flow model; finite difference method; hydrogeological forecast; water recirculation

check for **updates**

Citation: Różkowski, K.; Zdechlik, R.; Chudzik, W. Open-Pit Mine Dewatering Based on Water Recirculation—Case Study with Numerical Modelling. *Energies* **2021**, *14*, 4576. https://doi.org/10.3390/en14154576

Academic Editors: Maxim Tyulenev and Rajender Gupta

Received: 12 May 2021
Accepted: 26 July 2021
Published: 28 July 2021

Publisher's Note: MDPI stays neutral with regard to jurisdictional claims in published maps and institutional affiliations.

Copyright: © 2021 by the authors. Licensee MDPI, Basel, Switzerland. This article is an open access article distributed under the terms and conditions of the Creative Commons Attribution (CC BY) license (https://creativecommons.org/licenses/by/4.0/).

1. Introduction

The proper exploitation of mineral resources requires the exploitation project and subsequent mining works to be adjusted to the natural conditions identified during the documentation of the deposit. One of the analysed factors includes water conditions (hydrogeological and hydrological), which, in view of the growing environmental awareness, are among the most important ones—both in the course of obtaining a mining license and at the operational stage.

In most cases, the dewatering of open-cast mine workings is necessary to ensure the safety of preparatory works and mining operations. In the case of waterlogging of a section to be mined, a drainage system is constructed, which drains the rock mass up to the depth of the bottom of the planned mining level. As a result, a zone of lowered groundwater pressure (depression cone) is created, sometimes covering the area at a considerable distance from the boundaries of the excavation. The size of inflows depends on a number of factors, including the size of the excavation, the depth of mining and dewatering, hydrogeological parameters of the rock mass, location of the excavation in relation to surface watercourses, flow barriers, or climatic conditions [1]. Therefore, significant amounts of water may have to be pumped out as part of dewatering. The problem is further exacerbated by a high temporal variability of inflows, which also

277

depends on the irregular variability of precipitation, which sometimes may even be short-term in the form of torrential rainfall. That is why it is necessary to select parameters of the used dewatering system properly, which must be able to cope with pumping out increased inflows to ensure operation under specified design conditions. Sample characteristics of inflows to mines are presented, among others, in [2–5].

Mine working dewatering generates additional costs of mine exploitation. The expenses are connected with the purchase of particular components of the dewatering system, their servicing, and, directly, with costs of electricity consumed. Therefore, a proper selection of dewatering system parameters—not only in strictly technical terms (e.g., parameters of pumps) but also in environmental terms (proper spatial and depth location of pumping devices, optimal use of water, and soil conditions by, e.g., water transfer)—is of great importance. When it comes to cost calculation, it becomes increasingly necessary to include the minimisation of negative environmental effects of drainage, as lowering the groundwater level around the excavation significantly can cause, e.g., problems with vegetation, problems with water supply, etc., often determining the possibility of mining works.

In order for large-scale dewatering operations to be carried out in an optimal manner, they should be preceded by a multivariate and multicriteria analysis considering various dewatering scenarios depending on the objective to be achieved. One of the ways of limiting environmental costs of dewatering may include the recirculation of pumped waters, which consists in their reintroduction to the rock mass directly (e.g., by injection) or indirectly (by discharge into, e.g., inactive mine workings). Although the accumulation of water in the reservoir and, consequently, in the rock mass may lead to the intensification of inflows to the operating mine, the total effect, especially the environmental one, despite the necessity of pumping increased amounts of water, may turn out to be more beneficial. The scale of inflow increase will depend on rock mass parameters, distance of the flooded pit and damming water level, and on the quantity of recycled water. An additional advantage of this recirculation method is the use of the existing excavation with a specified capacity, which is first filled with pumped water. The essence of the proposed solution is to achieve the assumed environmental effects and to specify the optimum parameters for the operation of such a combined system.

To make reliable predictions, methods of mathematical modelling of groundwater flow processes should be used. Groundwater flow modelling is considered one of the most accurate methods in hydrogeology related to the analysis of the hydrodynamic field, taking into account various stresses [6–9]. Numerous examples equally demonstrate the usefulness of this method in solving inflow issues in underground [10–14] and open-pit mining [15–21]. The reliability of forecast calculations greatly depends on the correct recognition of the water–soil environment. Methods for the assessment of hydrogeological parameters and their use in mine flooding predictions are presented in [15,22], while the possibilities of using groundwater flow modelling to optimise water abstraction are presented in [23,24].

In practical applications, both the finite difference method—based on simulators of the MODFLOW family [25]—and the finite element method—the FEFLOW programme [26]—are used to model inflows in mining. Simulation calculations may be carried out for steady-state or transient conditions. The calibrated model is used to carry out simulation calculations with the aim to assess the impact of the conducted exploitation on the groundwater environment and to predict changes in the hydrodynamic field as a result of the application of new stresses resulting from the considered changes in the horizontal and depth range of exploitation and dewatering. At the same time, analyses are carried out in terms of the process of flooding workings as a result of rising waters in the rock mass after shutting down the dewatering system.

In the case under consideration, the application of model tests made it possible for the proposed mining solutions to assess the expected environmental effects (associated with both dewatering and flooding of excavations). The ability to compare the results of the forecasts with the actual effect of flooding the workings, which started after conducting this study and has been in progress for over 2 years, is also very valuable.

2. Materials and Methods

2.1. Characteristics of the Problem

Numerical modelling applied to simulate the effects of the designed technical solutions was used to analyse changes in water relations in the surroundings of two adjacent deposits—Radkowice-Podwole and Kowala Mała, located in the vicinity of Kielce in the central southern part of Poland (Figure 1). In the case of the Radkowice-Podwole deposit, the depletion of resources led to a decision to terminate exploitation and commence preparations for water reclamation. On the other hand, exploitation was continued in the Kowala Mała deposit opened in 2010, which is located approximately 700 m to the east of the Radkowice-Podwole deposit. Until now (2021), access and exploitation works have been carried out to the depth corresponding to the ordinate of +230 m above sea level, within the drained rock mass remaining in the range of a depression cone created as a result of dewatering the neighbouring deposit to a level located 40 m below (+190 m above sea level). In the final stage of the exploitation and dewatering of the Radkowice-Podwole deposit, about 27,000 m^3/day of water was drained, as a result of which the excavation had the status of the largest drainage object in the area, exerting significant pressure on the groundwater environment and distinctly lowering the groundwater table across a considerable area.

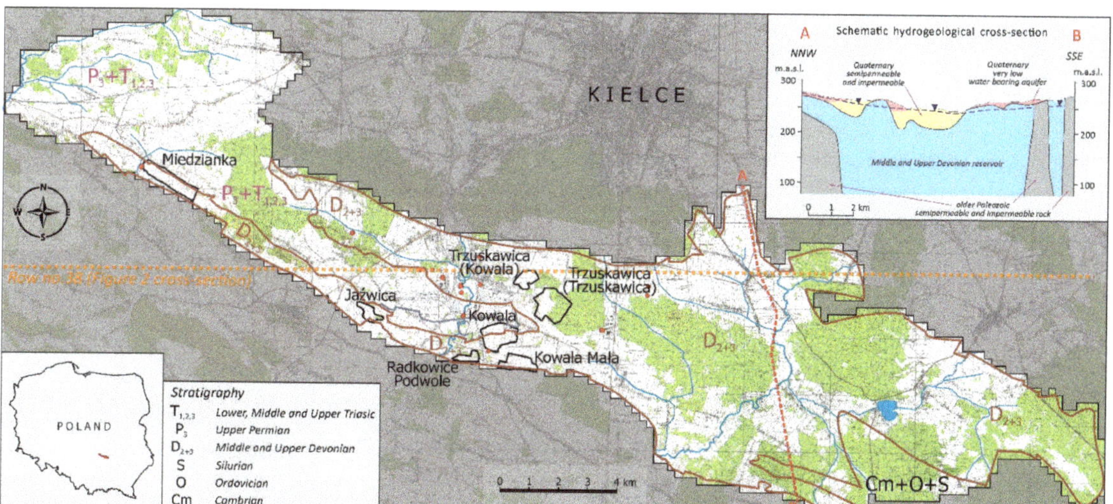

Figure 1. Modelled area of the Gałęzice–Bolechowice–Borków syncline with schematic geological boundaries, location of open pit mines, and exploitation wells marked with red dots (geology and cross-section according to [27]).

Along with the plans of deepening the exploitation of the Kowala Mała deposit and simultaneous termination of the drainage of the Radkowice-Podwole deposit, it became necessary to start draining the Kowala Mała deposit. Taking into account previous drainage conditions, two alternative scenarios were considered. One of the scenarios assumed the application of a "classic" solution, which was to flood the Radkowice pit with water coming from the natural inflow (open-pit liquidated by self-sinking-natural groundwater inflow from the aquifer only) and to dewater the Kowala Mała pit independently. The second scenario also considered flooding the Radkowice pit, but with water from two sources: from natural inflow and piped from the dewatering of the Kowala Mała pit. This assumption, through partial recirculation of water, makes it possible to limit the negative impact on the environment. The analysis of the obtained model prognosis results supported the process of making decisions concerning the designed dewatering system of the Kowala Mała mine.

2.2. Study Area

The Kowala Mała dolomite deposit is one of several neighbouring rock deposits intensively exploited within the boundaries of the Gałęzice–Bolechowice–Borków syncline (Figure 1). The syncline is a structural unit of the Świętokrzyskie Mountains, about 4–6 km wide and 30 km long, filled mostly with permeable Devonian sediments, locally with Carboniferous, Permian and Quaternary cover. The Dyminy and Chęciny anticlines that surround it to the north and south are made up of low-permeability Lower Palaeozoic sediments. In the north-western part, there are Triassic sediments, which are permeable in a part of the profile.

The Gałęzice–Bolechowice–Borków syncline is a productive aquifer system characterised by good water quality. The resources are used by water intakes that supply, among others, a significant part of the nearby city of Kielce (population of about 200,000 inhabitants). Carbonate water-bearing formations form a fissure-karstic aquifer with hydraulic conductivity in the range of $2 \times 10^{-6} \div 9 \times 10^{-4}$ m/s [27]. Due to good hydrogeological parameters, Major Groundwater Reservoir (MGR) No. 418 (Gałęzice–Bolechowice–Borków) was identified within the boundaries of the syncline. According to the national Polish classification [28], an MGR is a geological structure or its fragments with the highest water-bearing and storage capacity in the scale of hydrogeological regions. Within the boundaries of MGR No. 418, covering an area of 132.5 km^2, disposable resources of 63,816 m^3/day were documented [27].

The groundwater of the Devonian aquifer belongs to low mineralised ones with the specific electrolytic conductivity in the range of 550–650 μS/cm. Due to the carbonate environment, it is characterised by a weakly alkaline reaction. With the predominance of ions originating from the dissolution of dolomites and limestones, the water is described to be of a bicarbonate–calcium (HCO_3–Ca) or bicarbonate–calcium–magnesium (HCO_3–Ca–Mg) type. The sulphate concentrations observed throughout the region reach slightly elevated values in the 50–70 mg/dm^3 range. Incidentally, increases in chloride ion concentrations can be observed. Concentrations of inorganic nitrogen compounds remain at low levels. In conclusion, water from the network of observation wells in the vicinity of the pit should be considered of good quality. Unfortunately, water quality analyses are not conducted directly in the flooded pit.

Within the boundaries of the syncline, carbonate sediments—dolomites, limestones, and marls—are mined. Intensive mining works are currently carried out by several mining plants (Figure 1): Miedzianka, Jaźwica, Trzuskawica (two pits), Kowala, Kowala Mała, and until recently also Radkowice. The first one is located at the north-western end of the structure, while the other four are adjacent to each other in the central part of the syncline.

The Radkowice-Podwole dolomite deposit was cut by five excavation levels up to the ordinate of +190 m a.s.l., descending over 60 m below the surface of the ground. Along with the exhaustion of the Radkowice-Podwole deposit, the nearby Kowala Mała deposit was explored, with its resources documented in 2005. Deposit exploitation commenced in 2010 and a few years later it was decided to terminate works within the Radkowice-Podwole deposit. In the Kowala Mała excavation, at particular levels up to the level four inclusive, which correspond to the ordinates: I +270 m a.s.l., II +255 m a.s.l., III +240 m a.s.l., and IV +230 m a.s.l., works were carried out "dry" (without dewatering), as a result of the pit remaining within the depression cone of the neighbouring Radkowice mine. In the second quarter of 2020, works commenced at the next level–V (+215 m a.s.l.). Ultimately, in accordance with the assumptions made on the model, exploitation will be carried out up to +200 m a.s.l.

In the vicinity of the adjacent mines in the central part of the syncline, exploiting raw materials from levels located below the original groundwater table and ongoing dewatering operations led to the formation of depression cones, which over time merged and lowered the water table on a regional scale. In 2016 (for which the hydrodynamic state was reconstructed on the model), the total water abstraction from groundwater intakes reached 6037 m^3/day, with a simultaneous volume of water abstracted by mine drainage systems being almost 12 times higher (about 70,000 m^3/day). The total volume pumped out by intakes and drainage systems of the mines exceeds the disposable resources of the aquifer (63,816 m^3/day) [27], which is a very safe estimation. However, it should be taken into account that a part of the water discharged within the drainage system also originates from mostly forced river infiltration into rock mass. When planning further mining activities in functioning opencast mines, the question of their impact on the water and ground environment is crucial. This aspect may prevail in the decision-making process of appropriate authorities and institutions and may determine the possibilities of further mine exploitation.

2.3. Numerical Model of Hydrogeological Conditions

The numerical model of groundwater flow was prepared using the Processing Modflow programme [29,30]. The geometry of the hydrogeological structure was represented in the model by division into layers for which top and bottom elevations of lithostratigraphic units were determined. Tables containing values of hydraulic conductivity (Table 1) and effective recharge from the infiltration of precipitation were run on the model in a similar way. The nature of changes in pressure or flow in individual cells is defined by determining the so-called boundary condition, which allows stresses affecting groundwater (e.g., exchange with surface water, drainage, etc.) to be mapped.

Table 1. Variation of hydraulic conductivity in different model layers—after model calibration.

Layer	Hydraulic Conductivity (m/Day)			
	Minimum	Average	Median	Maximum
1	2.59×10^{-2}	4.21	2.68	86.06
2	8.54×10^{-4}	1.25×10^{-1}	2.59×10^{-2}	2.59
3	1.10×10^{-3}	4.24	1.55	43.20
4	8.64×10^{-3}	2.75×10^{-1}	8.64×10^{-2}	43.20

The developed model was prepared on the basis of a multilayered hydrogeological model of the Gałęzice–Bolechowice–Borków syncline [31–33], which has been in use for several years [34]. It was updated to include the amount of the discharged mine water from 2016 and the ordinates of drainage in mine workings as of mid-2017. Subsequently, recalibration was carried out using the current inflows to mines and measurements of the water table position. The completed study covered the area of the Gałęzice–Bolechowice–Borków syncline (Figure 1) spanning over 215 km^2. The adopted range of the model makes it possible to study groundwater flow in a regional system, which considers the interaction between groundwater and surface water and local stresses caused by the existing drainage centres related to the extraction of groundwater or mining drainage.

Groundwater recharge results mainly from the infiltration of precipitation and, to a lesser extent, the infiltration of surface water from rivers and streams. Groundwater and surface water are hydraulically connected with each other, which leads to their mutual exchange. Under natural conditions, surface watercourses were drainage zones for groundwater. As a result of the exploitation of groundwater intakes and the dewatering of opencast mines, the original hydrodynamic field saw a transformation. In the vicinity of the dewatered mine workings, the groundwater table dropped considerably, forcing an increased infiltration of water from precipitation and surface watercourses.

The extent of the area covered by the model study was determined on the basis of geological structure and hydrogeological conditions. The boundaries of the model comprised mining pits and groundwater intakes whose activities shape the hydrodynamic field within the considered water-bearing structure. The model is constrained from the south, north, and east by impermeable formations of the Middle Devonian and older Palaeozoic. In the north-western direction, the boundary is based on the course of the Łososina and Czarne Stoki Rivers (Figure 1).

Horizontally, the model study area was discretised based on an irregular grid. A basic division into 250 m × 250 m square cells was adopted. In the central part of the modelled area featuring numerous mine workings and groundwater intakes, the grid was linearly doubled: the objects of larger importance (contours of dewatered pits and intake wells) were mapped on the grid with a block size of 125 m × 125 m. The modelled area was contained in a rectangle comprising of 85 rows and 196 columns, which corresponded to the real size of 17.5 km × 38.25 km.

Vertically, the modelled area was divided into 4 numerical layers (Figure 2). Layer 1 comprises Quaternary water-bearing sediments (mainly sands and gravels in river valleys and glacial deposits; the stratum is discontinuous, locally reaching the thickness of over 50 m). Layer 2 comprises formations occurring in the roof part of the older bedrock, with lower values of filtration parameters (limestones of the lower part of the Upper Devonian, mostly approximately 30 m thick). Layer 3 represents the main aquifer in the studied area (mostly limestones and dolomites of the Middle Devonian; thickness within the model boundaries ranges from approximately 80 to nearly 130 m). Within this layer, groundwater is exploited using deep wells and mining excavations are drained. Layer 4 represents the Devonian part of the aquifer below the primary layer, with lower filtration parameters. For Layer 1, unconfined groundwater flow conditions were assumed. For the remaining layers (2, 3, and 4), mixed confined/unconfined conditions with variable transmissivity were adopted. This way best represents the existing hydrogeological conditions influenced by changes in mining conditions (mine drainage). However, a side effect may be numerical problems associated with rewetting of dry cells during the iterative solution, the minimization of which may require the use of a MODFLOW-NWT solver.

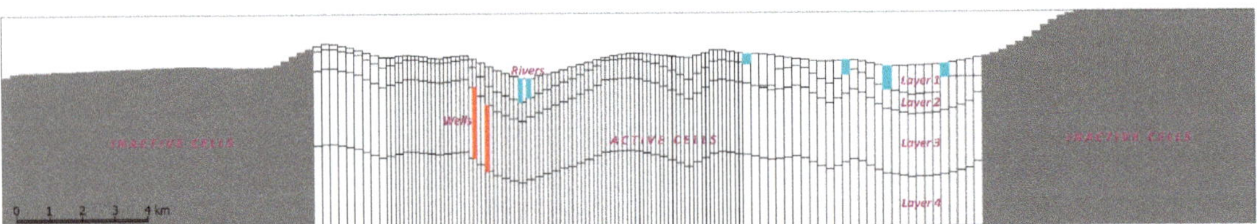

Figure 2. Scheme of vertical discretisation into model layers (example cross-section along row 38, vertical exaggeration—20).

For the contours of mine workings within Layer 3, first-type (Dirichlet) boundary conditions were assumed, allowing inflow to workings of a known drainage ordinate to be simulated. This implementation method serves to correctly represent the functioning dewatering systems. By assuming a known ordinate of the ongoing dewatering (lowest exploitation level), the model was adjusted to actual measurements and observations at the stage of calibration (in this case, the amount of pumped waters). Application of first-type boundary conditions requires control and proper balancing of inflows to mines, obtained in simulations. Alternatively, third-type (Robin) boundary conditions may be used (drain module), but this in turn enforces the necessity of the correct assumption of the values of drain hydraulic conductance.

Forecast calculations were performed under steady-state flow simulation. This research method comes down to the determination of the target state resulting from the assumed stresses, without indicating any intermediate states that characterise the temporal variability of the considered phenomenon. This assumption does not require the use of discretisation of the duration of the considered process and the presented results concern the state after stabilisation of the hydrodynamic field in the simulated system of stresses.

In the table of initial hydraulic heads, corrections were required in blocks with first-type boundary conditions, assumed for Layer 3, corresponding to the contours of dewatered pits. They were assigned values corresponding to the ordinates of the bottom of dewatered levels of active open-cast mines, located within the modelled area, averaged according to the state as of the middle of 2017. In the case of the Radkowice mine, the level of +190 m a.s.l. was assumed.

The values of hydraulic conductivity (horizontal and vertical) corresponding to the modelled aquifers were introduced in zones of cells and then modified during model calibration. Karst phenomena, which are typical of carbonate aquifers, were obviously documented, although their recognition is irregular and only local. Taking into account the regional scale of the study and the adopted degree of discretisation (cell sizes), the model calculations were based on parameter values typical of pore and fracture aquifers. Modifications of filtration parameters were rather limited in some locations where archival studies point to the local presence of karst phenomena.

The distribution of hydraulic conductivity of individual layers obtained after model calibration is presented in the form of synthetic statistics in Table 1.

The average amount of effective recharge from infiltration of atmospheric precipitation reached 4.64×10^{-4} m^3/day/m^2 after model calibration, which is about 27% of the volume of mean annual precipitation recorded for Kielce (629 mm, [35]). Recharge from the infiltration of precipitation was assumed as a second-type boundary condition, with the option of recharge being applied to the highest active cells. The magnitude of the infiltration rate varies depending on the permeability of rocks present in near-surface layers, degree of land development, and location within the range of depression cones of mines and groundwater intakes.

Within Layer 1, the rivers located in the study area were simulated (Figure 3). Due to the limited hydraulic contact between surface water and groundwater, watercourses were simulated with third-type boundary conditions. This way of modelling requires information about the hydraulic conductance of the riverbed (including width and length of watercourse sections) in individual cells and the representation of the hydraulic head. During model calibration, the hydraulic conductance of the riverbed, which determines the contact between surface water and groundwater, was slightly modified.

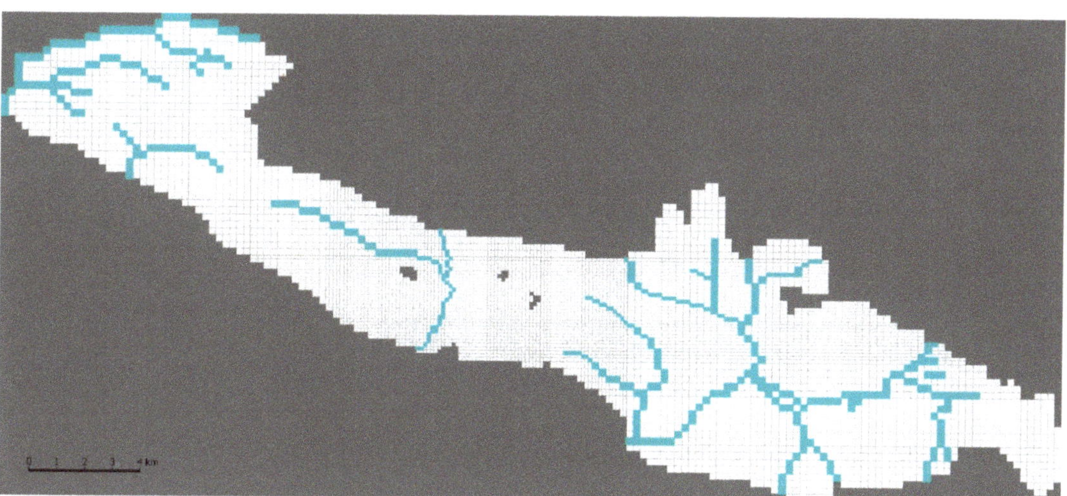

Figure 3. Mapping of rivers within Layer 1 (light blue blocks—rivers with limited hydraulic contact with groundwater, third-type boundary condition).

Intake wells with a constant yield were simulated using second-type boundary conditions. The model included numerous deep wells, from which regular water abstraction was carried out in 2016. Constant well yields, defined as daily averages based on the actual registered abstraction in 2016, were assumed to a total of 6037 m^3/day, of which 1089 m^3/day was within Layer 1 and 4948 m^3/day in the boundaries of Layer 3. In each of the variants, the majority of wells operate under conditions of cooperation, remaining under the hydrodynamic influence of neighbouring intakes or dewatered pits.

3. Results

3.1. Model Calibration

For the numerical computation of groundwater filtration equations, a finite difference method (FDM)-based simulator from the widely used MODFLOW family [8] was used, with MODFLOW-2005 [25] being the most commonly used version at present. The GMG iterative method (solver) was chosen. This method of implementation in the case of an aquifer with significant filtration parameter variability ensures stable numerical calculations [9]. The prepared model was subjected to a calibration process using the results of the authors' own field measurements of the groundwater table depth conducted in mid-2017 in 53 boreholes where no direct water abstraction was carried out (3 boreholes filtered within the Quaternary aquifer and 50 within the Devonian one) and using information on the actual amounts of water discharged as part of dewatering operations performed by individual mines (averaged for 2016). The trial-and-error method [8] was applied to perform the calibration. The original model was previously calibrated to a different hydrodynamic state. Therefore, it can be considered that the calibration was verifying in nature with respect to the earlier version of the model, calibrated to a different state. During the calibration, the spatial distribution of hydraulic conductivity (mainly within Layer 3) was modified and, to a lesser extent, parameters of the riverbed were modified. Only slight changes to the recharge from precipitation were made.

As a result of the calibration, a numerical hydrogeological model was developed, which generates results (water table arrangement and inflows to dewatered excavations) similar to those observed in reality (the so-called Variant 0—the current hydrodynamic state after model calibration). For the adopted calibration points, the differences between the measured water table and that obtained as a result of the interpolation of the resulting matrix distribution from the model were insignificant and, in principle, did not exceed the value of a single meter. The arrangement of the measurement pointed directly in the vicinity of the diagonal of the calibration plot (Figure 4), meaning that there was a good fit of the model response to the actual observations.

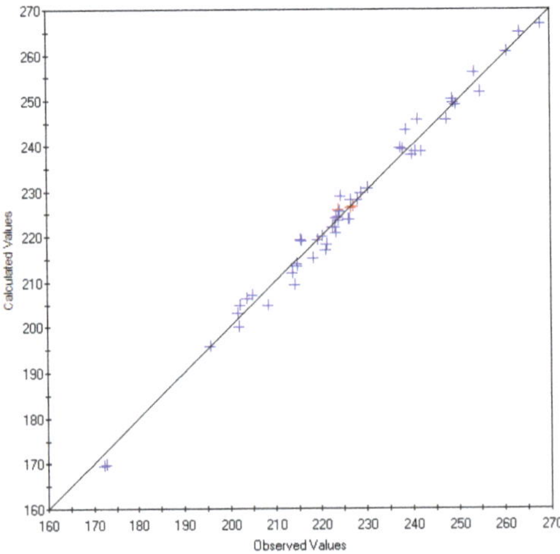

Figure 4. Comparison of the water table ordinates (m a.s.l.) obtained from the model (calculated values) and observed in reality (observed values) for the adopted calibration points (blue—middle Devonian and red—Quaternary).

Considered as the calibration criteria, the error rates were: mean error ME = 0.23 m and mean absolute error MAE = 1.91 m. The normalised value, i.e., related to the amplitude of the water table fluctuation in the considered structure (138 m), of the mean absolute error was calculated to 1.39%. The volumes of inflows to individual pits obtained from the model after the calibration (Q_M) in relation to the observed actual inflows, collected from the operators of the pits (Q_R), differed to a very small degree (Table 2). Absolute differences Q_M–Q_R did not exceed several tens of m^3/day.

Table 2. Magnitudes of inflows to mine workings: actual, obtained after the model calibration, and prognosed in individual variants.

| | Inflow to Open Pit Mine (m^3/Day) | | |
| | Actual | Calibrated | Prognosed | |
Open-Pit Mine	Q_R	Variant 0 (Q_M) Q_M-Q_R	Variant 1 (Q_1) Q_1-Q_M	Variant 2 (Q_2) Q_2-Q_M
Radkowice	26,909.8	26,960.8 +51.0	0.0 −26,960.8	0.0 −26,960.8
Kowala Mała	-	-	6908.6 +6908.6	7934.6 +7934.6
Other open-pit mines	43,482.2	43,438.8 −43.4	44,345.0 +906.2	44,614.4 +1175.5
Total	70,392.0	70,399.6 +7.5	51,253.6 −19,146.0	52,549.2 −17,850.4

The total amount of water circulating in the considered structure for Variant 0 (reconstructed current state) was about 131,000 m^3/day (Table 3). On the inflow side, this value was primarily determined by the effective infiltration of precipitation (about 97,000 m^3/day—74% of total inflow), while the remaining part was formed by the infiltration of waters from surface watercourses (about 34,000 m^3/day). The rivers, mapped within the Quaternary layer, were rather draining in character, changing to infiltrating in the areas where mines had a drainage effect. On the outflow side, the dominant factors included the drainage of open-cast mine workings (about 70,000 m^3/day—53.6% of the entire structure drainage), outflow to surface watercourses (about 55,000 m^3/day), and, to a much lesser extent, exploitation of groundwater intakes (about 6000 m^3/day)—both Quaternary (Layer 1 of the model) and Devonian (Layer 3).

Table 3. Summary of the water circulation balance in the Gałęzice–Bolechowice–Borków syncline, obtained on the basis of the model research.

	Inflow/Outflow (m^3/d)					
	Calibrated		Prognosed			
Component	Variant 0		Variant 1		Variant 2	
	In	Out	In	Out	In	Out
Effective infiltration of atmospheric precipitation	97,311	0	97,258	0	97,258	0
River infiltration/drainage	34,116	54,998	18,841	58,817	18,588	65,203
Exploitation of groundwater intakes	0	6037	0	6037	0	6037
Drainage of open-pit mines	0	70,400	0	51,254	7935 [1]	52,549
Total	131,427	131,435	116,099	116,108	123,781	123,789

[1] The amount of water from the drainage of the Kowala Mała mine, recirculated to the Radkowice mine.

The arrangement of the groundwater table in the Middle Devonian aquifer (Layer 3), obtained after model calibration, was determined by the operation of rock mine drainage systems. The exploitation and, simultaneously, drainage levels of the mines acted as drainage bases for groundwater inflow (Figure 5). The dewatering ordinates assumed in the model, corresponding to the actual state in mid-2017, were significantly lower compared to the original water table position before the dewatering started, which translates into the volume of inflows to the dewatered workings.

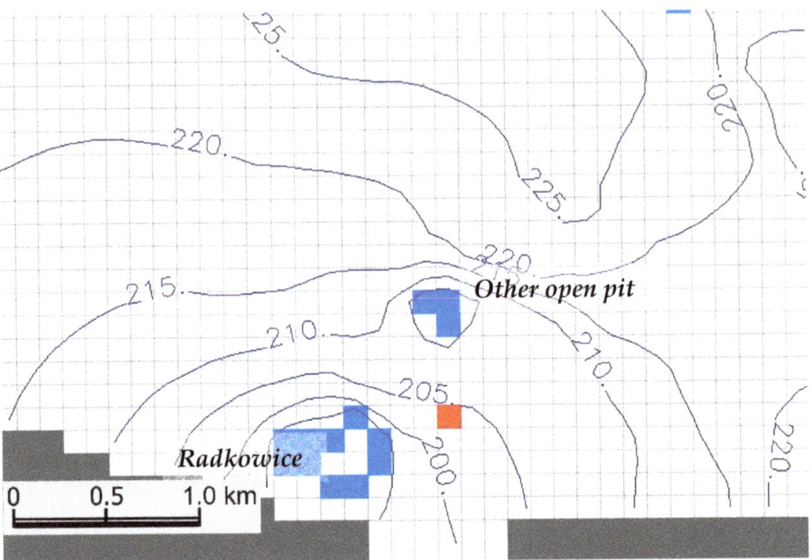

Figure 5. Hydrodynamic field distribution of the middle Devonian aquifer in the central part of Layer 3. Variant 0—calibrated state (mid-2017). Blue cells—contours of open-pits, red cell—exploitation well.

The water balance and hydroisohypse arrangement show that the magnitude of inflow to the Radkowice mine depended on the ordinates of exploitation and the extent of the dewatered mine workings, but infiltration of water from the nearby Bobrza River also had a significant impact. Exploitation and related dewatering operations carried out in mine workings of neighbouring mines situated in groundwater flow directions also produced a considerable effect on inflow to the Radkowice mine. The central and north-western parts of the syncline make a particular contribution to the inflow.

3.2. Assumptions of Prognostic Variants

Forecast simulations were conducted for two variants (numbered 1 and 2). The key objective of both variants was to predict the hydrodynamic field distribution and inflows to the mine workings assuming that the exploitation and dewatering of the Radkowice mine are abandoned and, at the same time, the Kowala Mała mine starts to be drained. A common assumption for all variants was that the remaining mines continue to operate within the horizontal extent of the workings and at dewatering ordinates consistent with the status quo (model calibration stage—mid-2017). The same assumption was also made for groundwater exploitation by intakes.

Both prognostic variants assume the discontinuation of the Radkowice pit dewatering, which means flooding the existing workings to a level corresponding to hydrodynamic equilibrium in the rock mass, taking into account the operation and drainage in other mines. The dewatering of the workings was also considered in view of the planned deepening of mining operations at the Kowala Mała mine to a level below the water table. The adopted dewatering contour, in the form of first-type boundary conditions, corresponded to the horizontal extent of the designed excavation. The ordinate of the lowered water table reflected the assumed ordinate of the lowest exploitation level (+200 m a.s.l.). The main difference between the variants was a different structure of the dewatering system of the

Kowala Mała mine and the course of flooding the Radkowice mine. Variant 1 assumes a typical solution, which is to discharge drainage water from the Kowala Mała mine into ditches and surface watercourses, while their further fate was not subject to analysis. In this solution, the Radkowice pit would be self-sinking with waters coming from natural inflow from the aquifer. Variant 2 includes the pumping of waters coming from the dewatering of the Kowala Mała mine to the abandoned pit of the Radkowice mine. In the Radkowice open pit, additional water accumulation would occur, which would result in an increase in the hydraulic gradient in the rock mass and, consequently, in an intensification of inflows to the Kowala Mała mine.

3.3. Results of Prognostic Variants

3.3.1. Variant 1

A major change in groundwater pressure distribution would occur in the vicinity of the abandoned Radkowice pit and in the neighbourhood of the deepened Kowala Mała mine (Figure 6). The discontinuation of the Radkowice pit dewatering would result in the damming of groundwater and flooding of the pit to the level of about +220 ÷ +221 m a.s.l. Quite a large difference in groundwater pressure would be created between the flooded Radkowice pit and the drained Kowala Mała pit at an ordinate of +200 m a.s.l. Due to a relatively short distance between the nearest edges of both pits (about 700 m), a significant hydraulic gradient could lead to the intensification of inflows to the Kowala Mała pit, making it a local drainage base. The abandonment of the Radkowice pit dewatering would result in a noticeable increase in the groundwater table in the Middle Devonian aquifer on the side of the previously dominant inflows (N and NW).

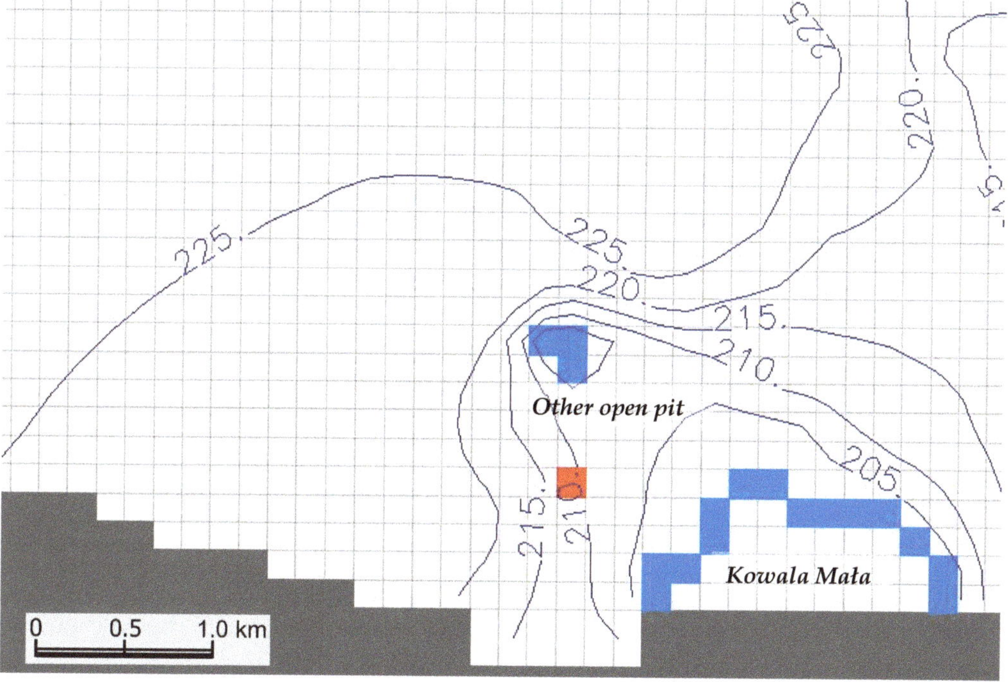

Figure 6. Hydrodynamic field distribution of the middle Devonian aquifer (Layer 3 in the model)—Variant 1. Blue cells—contours of open-pits, red cell—exploitation well.

The balance of inflows to particular mines (Table 2) revealed that the abandonment of mining and dewatering operations at the Radkowice mine would result in the intensification of inflows to almost all other mines. The highest increase in inflows would be seen for the Kowala Mała mine (6909 m^3/day). However, the total inflow to all mines would be considerably lower (about 51,000 m^3/day), meaning that the value of the decrease compared to the calibrated state would amount to 19,000 m^3/day (about 27%).

In the aggregated balance of water circulation in the Gałęzice–Bolechowice–Borków syncline obtained from the model for Variant 1 (Table 3), the total amount of water (about 116,000 m^3/day) would see a considerable decrease in comparison with the calibrated state—by about 15,000 m^3/day (11.6%). Recharge would originate mainly from unchanged effective infiltration of precipitation (about 97,000 m^3/day) and, to a lesser extent, from river infiltration (about 19,000 m^3/day). On the drainage side, apart from mine drainage systems (about 51,000 m^3/day), rivers would play an important role (about 59,000 m^3/day), with deep groundwater intakes playing a secondary role in the balance (about 6000 m^3/day). A significant decrease in inflows to the mines in the regional groundwater circulation system means that less water has to be pumped (and discharged), bringing measurable economic and environmental benefits. In relation to the summary of the water circulation balance for the reproduced present conditions (Variant 0), a significant reduction of water quantity from river infiltration (nearly halved) was noticeable. This would be a direct effect of abandoning the exploitation and dewatering of the Radkowice pit situated in a close vicinity of the Bobrza River.

3.3.2. Variant 2

As in the case of Variant 1, a substantial change in groundwater head distribution would be seen in the vicinity of the abandoned (flooded) Radkowice pit and in the vicinity of the deepened Kowala Mała mine (Figure 7). The impact of the assumed changes in dewatering systems on the distribution of the hydrodynamic field would be additionally intensified as a result of the transfer of waters from the Kowala Mała mine dewatering to the flooded pit of the Radkowice mine. As a result, the water table in the Radkowice reservoir would stabilise several meters higher in relation to the prognosed value for Variant 1—at the level of about +225 ÷ +226 m a.s.l. The Kowala Mała excavation would still be drained at the ordinate of +200 m a.s.l., as a result of the difference in hydrostatic pressures between the Radkowice reservoir and the Kowala Mała excavation that could reach approximately 25 m. Due to a relatively short distance between the nearest edges of both pits (about 700 m), a significant hydraulic gradient (3.5%) could contribute to the intensification of inflows to the Kowala Mała pit in the long run.

The analysis of the detailed water balance of particular mines (Table 2) points to a further intensification of inflows to almost all remaining pits as a result of the discontinued exploitation and dewatering of the Radkowice mine. In the most distant mines, the increase in inflows would be small. The largest increase in inflows would occur at the Kowala Mała mine, up to about 8000 m^3/day, i.e., 1000 m^3/day more than for Variant 1. The simulation for Variant 2 assumes that the total amount of water flowing from the rock mass into the Kowala Mała workings is to be treated as a supply (inflow to the rock mass) in cells mapping the contour of the water reservoir in the abandoned Radkowice pit. This approach corresponds to the inflow of waters coming from the dewatering of the Kowala Mała mine into the water reservoir forming in the flooded Radkowice pit. The total inflow to the workings of all mines would be considerably smaller than at the model calibration stage, reaching about 53,000 m^3/day. As a result of the water transfer to the Radkowice reservoir, the total inflow to all mine workings simulated for Variant 2 would be slightly higher than for Variant 1.

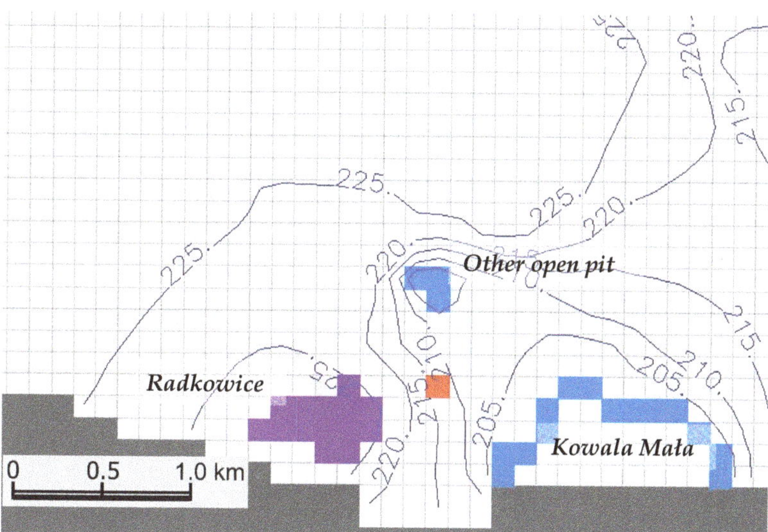

Figure 7. Hydrodynamic field distribution of the middle Devonian aquifer (Layer 3 in the model)—Variant 2. Blue cells—contours of open-pits, red—exploitation well, purple cells—flooded pit of the Radkowice mine.

The total amount of circulating waters in the Gałęzice–Bolechowice–Borków syncline (Table 3) obtained from model tests under Variant 2 would reach about 124,000 m^3/day. In relation to the calibrated state, this would mean a decrease in the amount of circulating waters by approximately 7600 m^3/d, while in relation to Variant 1, the total amount of circulating waters would be higher by 7700 m^3/day. The inflow side would be quantitatively dominated by effective infiltration of precipitation (approximately 97,000 m^3/day), with the rest coming from river infiltration (nearly 19,000 m^3/day). On this side of the balance, the amount of water discharged into the flooded Radkowice pit from the dewatering of the Kowala Mała mine should also be considered (almost 8000 m^3/day). On the side of the rock mass drainage, the outflow of groundwater to rivers would dominate—about 65,000 m^3/day, but a very significant contribution to the predicted drainage levels would be made by the dewatering of mine workings—amounting to almost 53,000 m^3/day in total. As in the case of Variant 1, a significant decrease in the amount of waters originating from river infiltration is observed in the water circulation balance for Variant 2, which should be associated with the new arrangement of the hydrodynamic field in the vicinity of the flooded Radkowice pit.

4. Discussion

The obtained prognostic variants were based on assumptions—both target (Kowala Mała and Radkowice mines) or simulated at the model calibration stage—related to mine drainage (extents of workings and their ordinates, Radkowice mine—drainage discontinuation). The presented results of the predictions do not consider intermediate states, but only the final state resulting from the assumed stresses. In fact, until the new conditions are fully established, both inflows and groundwater levels could reach intermediate values between the reconstructed and forecasted ones for the target conditions. A precise determination of the times of reaching intermediate and target conditions would require model tests in transient flow simulation. Such predictive simulations are much more sophisticated and require knowledge of rock mass capacitance parameters (porosity and specific yield). The results of additional laboratory and tracer tests may form the basis for supplementing the model with information on rock mass capacitance parameters, making it possible to recalibrate the model. The model completed in this way is the only one that could be used for the prognostic evaluation of the course of the considered hydrodynamic processes, allowing for time schedules of mining works in individual mines to be taken into account.

The obtained results of the model tests support the arguments that led the investor to opt for the second variant, i.e., transferring water from the deepened pit at the Kowala Mała deposit to the flooded Radkowice pit. The self-sinking process began in March 2019. The relatively stable hydrodynamic field formed as a result of long-term mine exploitation was strongly disturbed as a result of the discontinuation of the Radkowice mine dewatering. Observations carried out since then point to a gradual recovery of the groundwater table in the neighbouring rock mass of the flooded pit (Figure 8). At the end of July 2019, the observed water table in the formed reservoir reached the ordinate of about 212 m above sea level, with the sinking process itself slightly slowing down (increase in the water table in the reservoir by about 0.07 m per day). In the last quarter of 2020, the water table in the flooded reservoir slightly exceeded +221 m a.s.l. The gradually conducted observations indicate that the state of hydrodynamic equilibrium in the rock mass is yet to be achieved and the process of the water table rise is still underway, with the rate of these processes being already very limited.

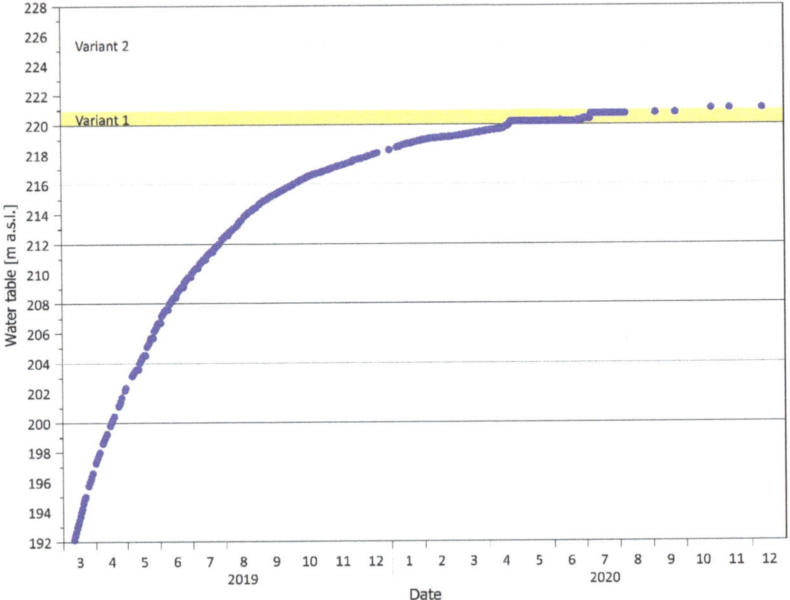

Figure 8. Water table elevation in the flooded Radkowice pit.

Currently (the first quarter of 2021), the water table elevation in the flooded Radkowice pit rose to the level corresponding to the height predicted for Variant 1. At present, the extraction of dolomite from the Kowala Mała deposit was conducted at level IV (+230 m a.s.l.). Works on an excavation to gain access to level V (+215 m a.s.l.) were commenced in mid-2020 and the opening pit was formed in the first quarter of 2021. This means that current dewatering operations were performed at a level higher than assumed in the predicted variants.

The implemented dewatering system used a gradually developed system of drainage ditches (both permanent and temporary) within the excavation level to feed, by gravity, the sump with water from the rock mass. Outside the excavation area, water is transported through two 400 mm diameter pumping pipelines, one of which has a backup function. The water is then transported to the neighbouring Radkowice pit, which is in the process of flooding, via a 700 m long transfer pipeline. Works on the transmission system and the sump at a level of +215 m a.s.l were completed recently. The target sump will be created for the exploitation of the next level—+200 m a.s.l. Only then, after the conditions have stabilised, the actual state will correspond to that modelled for Variant 2 and the comparison of the actual ordinates in the flooded excavation to the predicted ones will be justified.

The completion of installation works in recent months did not make it possible to collect reliable data on actual inflows. However, no significant volumes are expected in the current phase. The exploitation level of +230 m a.s.l. is above the present water table ordinate, originally formed at 226.7 ÷ 228.6 m a.s.l. The opening pit of small dimensions, made to the level of +215 m a.s.l., will cause only slight inflows whose recirculation through redirecting to the reservoir in the recultivated Radkowice excavation will not significantly affect the rate of flooding and the ordinate of the water surface.

The actual inflow to the Kowala Mała pit may slightly differ from the numerical prognoses. It will be determined by pressure differences between the flooded Radkowice pit and the dewatered Kowala Mała mine and by filtration parameters of the rock mass (mainly in the zone between the two pits). In reality, a significant hydraulic gradient between the workings (3.5%; approximately 25 m difference in the water table position at a distance of about 700 m) would lead to the saturation of the vadose zone that reactivated preferential flow paths. The prognostic simulations were carried out on the basis of the calibrated variant, without making any modifications to rock mass filtration parameters. It is not possible to precisely and clearly determine to what extent the intensification of the flow between the workings will affect the reactivation of formerly inactive flow paths. Working simulations, which hypothetically assume the intensification of water flow through additional migration routes to a different extent, estimate that the actual inflows may increase by 20–25% in comparison with the quantities obtained in the prognoses (Variants 1 and 2). This applies mainly to the Kowala Mała open pit, where care should be taken to ensure an adequate drainage system reserve. The possibility of an increase in inflows to the mine workings also results from the possible occurrence of karst phenomena in the rock mass, which cannot be excluded in the case of geological data from the studied area.

The results of the predictive calculations were obtained using a model calibrated to assume that inflows to the drainage systems averaged over the actual 2016 inflow. However, the effective recharge from precipitation infiltration was determined based on multiyear average precipitation. In reality, inflows to the pits considered over short time intervals may differ from the projected average inflows. This is due to the relatively fast response of the aquifer system to not exactly predictable changes in precipitation. The predicted inflows to the mine workings should be considered the average inflows for a longer time interval, corresponding to the average meteorological conditions, generally for medium states. In periods of increased precipitation, the actual inflows to the mine workings may temporarily exceed the predictions as a result of both direct surface runoff and increased infiltration into aquifers. Due to the uneven spatial and temporal intensity of precipitation and the probability of short-term heavy rainfall, which is possible in the long run, the necessity to ensure appropriate reserve capacity for draining or retaining excessive water inflow should be taken into account when designing the drainage system of the mine workings.

The fact that actual inflows to individual mines will, to a large extent, depend on the exploitation and drainage conditions of the neighbouring mines should also be considered. With this in mind, the presented results obtained with the use of model research should be treated as representative for a longer time horizon assuming that the mining and operation conditions of the neighbouring open-pit mines do not change.

The presented results of the prognoses apply to the state of full hydrodynamic equilibrium in the entire aquifer after complete water table stabilisation. In reality, the state of this type will most probably not be fully achieved within a certain period of time as a result of further changes in the exploitation and drainage of individual mines and changes in the volumes of exploited groundwater for municipal purposes. It can be assumed that the actual water table will reach a quasi-steady state close to the steady state obtained in the prediction only after some time from the occurrence of the stresses provided that there are no considerable changes in the range of exploitation and dewatering in other mines.

5. Conclusions

The variant analysis of changes in hydrogeological conditions in the vicinity of the Kowala Mała deposit in the perspective of deepening the exploitation and flooding of the neighbouring depleted Radkowice-Podwole deposit contributed to the selection of the drainage concept. Among the considered factors, the impact of the forecast mining drainage on the surrounding groundwater environment was of particular importance.

As an extensive and productive aquifer, the Gałęzice–Bolechowice–Borków syncline has, for several decades, been subject to intensive drainage, primarily due to the development of the exploitation of rock materials and, to a lesser extent, the intake of water for municipal purposes. MGR No. 418, covering most of the syncline, is yet to be provided with hydrogeological documentation establishing available resources. The calculations from a decade ago [27] estimate the available resources to be 63,816 m^3/day. In the light of further research [34], this value seems to be underestimated. The total consumption resulting from both mine drainage systems and groundwater intakes (e.g., from 2016—Table 3) seem to indicate that the available resources of the aquifer were being overexploited. In light of this knowledge, the environmental decision-making process of relevant offices and institutions pays particular attention to further plans for the operation of mines in the context of their impact on the aquatic environment.

The consequences of water discharge from dewatering carbonate resource deposits are not particularly threatening to the environment. Discharged waters originate from natural inflow with neutral physicochemical parameters (having no negative impact on the quality of receiving waters). The practicality of the proposed changes comes down to the possibility of using existing drainage infrastructure (Variant 2 vs. Variant 1) in order to reduce investment costs. Fundamental environmental problems result from drainage. The combined depression cones formed around the neighbouring mines created a zone of the groundwater table that was lowered more than the local range. Diverting pumped waters to the neighbouring flooded pit, within the same aquifer, limited the spread of the depression cone by increasing the disposable resources of the aquifer and decreasing river infiltration. On a regional scale, the amount of circulating water decreased significantly, resulting in an overall improvement to the aquatic environment of the overexploited reservoir. The mining entity, to a less measurable extent, became better perceived by state institutions dealing with environmental protection issues, which, in the future, should translate into a better negotiating position when obtaining subsequent environmental decisions.

Through the termination of the Radkowice-Podwole deposit exploitation and the decision to flood the existing excavation, both prognostic variants pointed to a significant improvement in water balance on a local scale, which was a significant benefit for the natural environment. With a considerable reduction in the total mining water consumption from about 70,000 m^3/day to just over 50,000 m^3/day, the total amount of water involved in the circulation in the aquifer was reduced, and the amount of river infiltration almost halved. Due to the more favourable environmental effect, the second variant was selected for implementation, which assumes the transfer of water from the deepened excavation in the Kowala Mała deposit to a water reservoir formed as a result of self-sinking in the Radkowice excavation. The cessation of discharging waters from drainage to the Bobrza River by directing them to the adjacent water reservoir will result in an increase in the water table level in the reservoir and in the surrounding rock mass in line with the predictions. The resulting depression cone will be limited in the west and north-west directions, as a result of which the Bobrza River will regain its drainage function to a considerable extent. At the same time, with the increased hydraulic gradient, the inflows to the Kowala Mała mine will rise due to the increased seepage of recirculated waters through the pillar between the workings. The projected growth in inflows of about 1000 m^3/day will further burden the mine drainage system and directly contribute to higher operating costs. Still, the assumed environmental benefits related to the reduction of the negative drainage effects and the simplification of the drainage system combined with a simultaneous reduction in investment outlays (shorter transmission pipelines as water is discharged into the reservoir rather than the river) were prioritised and ultimately prevailed in the analysis of benefits and losses of the proposed solutions.

Author Contributions: Conceptualisation, K.R. and R.Z.; methodology, K.R. and R.Z.; software, R.Z.; validation, K.R. and R.Z.; formal analysis, K.R. and R.Z.; investigation, K.R. and R.Z.; resources, W.C.; data curation, K.R., R.Z. and W.C.; writing—original draft preparation, K.R. and R.Z.; writing—review and editing, K.R. and R.Z.; visualisation, K.R. and R.Z.; supervision, K.R. and R.Z. All authors have read and agreed to the published version of the manuscript.

Funding: The APC was funded by the AGH University of Science and Technology in Poland (scientific subsidy number: 16.16.100.215, 16.16.140.315) and by the "Excellence Initiative–Research University" programme for the AGH University of Science and Technology.

Institutional Review Board Statement: Not applicable.

Informed Consent Statement: Not applicable.

Conflicts of Interest: The authors declare no conflict of interest.

References

1. El Idrysy, H.; Connelly, R. Water-the Other Resource a Mine Needs to Estimate. *Procedia Eng.* **2012**, *46*, 206–212. [CrossRef]
2. Hawkins, J.W.; Dunn, M. Hydrologic Characteristics of a 35-Year-Old Underground Mine Pool. *Mine Water Environ.* **2007**, *26*, 150–159. [CrossRef]
3. Unsal, B.; Yazicigil, H. Assessment of Open Pit Dewatering Requirements and Pit Lake Formation for the Kışladağ Gold Mine, Uşak, Turkey. *Mine Water Environ.* **2016**, *35*, 180–198. [CrossRef]
4. Motyka, J.; d'Obyrn, K.; Kasprzak, A.; Szymkiewicza, A. Sources of groundwater inflows into the "Czatkowice" limestone quarry in southern Poland. *Arch. Min. Sci.* **2018**, *63*, 417–424.
5. Chunhu, Z.; Dewu, J.; Qiangmin, W.; Hao, W.; Zhixue, L.; Xiaolong, S.; Mingpei, L.; Shaofeng, W. Water inflow characteristics of coal seam mining aquifer in Yushen mining area, China. *Arab. J. Geosci.* **2021**, *14*, 1–12. [CrossRef]
6. Spitz, K.; Moreno, J. *A Practical Guide to Groundwater and Solute Transport Modelling*; John Wiley & Sons Inc.: New York, NY, USA, 1996; p. 480.
7. Kresic, N. *Hydrogeology and Groundwater Modeling*, 2nd ed.; CRC Press: Boca Raton, FL, USA, 2006. [CrossRef]
8. Anderson, M.P.; Woessner, W.W.; Hunt, R.J. *Applied Groundwater Modeling: Simulation of Flow and Advective Transport*, 2nd ed.; Elsevier: Amsterdam, The Netherlands, 2015.
9. Zdechlik, R. A Review of Applications for Numerical Groundwater Flow Modeling. In Proceedings of the 16th International Multidisciplinary Scientific GeoConference SGEM, Vienna, Austria, 2–5 November 2016; Book 3; Volume 3, pp. 11–18, ISBN 978-619-7105-81-0. [CrossRef]
10. Bukowski, P.; Haładus, A.; Zdechlik, R. *The Process of Mine Workings Flooding on the Example of Selected Hard Coal Mines in the Upper Silesian Coal Basin*; [In Polish: Zatapianie wyrobisk górniczych na przykładzie wybranych kopalń węgla kamiennego w Górnośląskim Zagłębiu Węglowym]; Sadurski, A., Krawiec, A., Eds.; Współczesne Problemy Hydrogeologii, t.XII, Wydawnictwo Uniwersytetu Mikołaja Kopernika: Toruń, Poland, 2005; pp. 85–90.
11. Rapantova, N.; Grmela, A.; Vojtek, D.; Halir, J.; Michalek, B. Ground Water Flow Modelling Applications in Mining Hydrogeology. *Mine Water Environ.* **2007**, *26*, 264–270. [CrossRef]
12. Haładus, A.; Zdechlik, R.; Bukowski, P. *Numerical Modeling of Flooding Process in the Part of the Janina Hard Coal Mine*; [In Polish: Modelowanie przebiegu zatapiania Ruchu II ZG Janina]; Szczepański, A., Kmiecik, E., Żurek, A., Eds.; XIII Sympozjum Współczesne Problemy Hydrogeologii. Krakow-Krynica: Krakow, Poland, 21–23 June 2007; pp. 789–795.
13. Hu, L.; Zhang, M.; Yang, Z.; Fan, Y.; Li, J.; Wang, H.; Lubale, C. Estimating dewatering in an underground mine by using a 3D finite element model. *PLoS ONE* **2020**, *15*, e0239682. [CrossRef] [PubMed]
14. Surinaidu, L.; Rao, V.G.; Rao, N.S.; Srinu, S. Hydrogeological and groundwater modeling studies to estimate the groundwater inflows into the coal Mines at different mine development stages using MODFLOW, Andhra Pradesh, India. *Water Resour. Ind.* **2014**, *7–8*, 49–65. [CrossRef]
15. Szczepiński, J. The Significance of Groundwater Flow Modeling Study for Simulation of Opencast Mine Dewatering, Flooding, and the Environmental Impact. *Water* **2019**, *11*, 848. [CrossRef]
16. Brown, K.; Trott, S. Groundwater Flow Models in Open Pit Mining: Can We Do Better? *Mine Water Environ.* **2014**, *33*, 187–190. [CrossRef]

17. Niedbalska, K.; Haładus, A.; Bukowski, P.; Augustyniak, I.; Kubica, J. Modelling of Changes of Hydrodynamic Conditions in the Aquatic Environment of the Maczki-Bor Sand Pit due to the Fact of Planned Closure of Mining Operations (NE part of Upper Silesian Coal Basin -Poland). In Proceedings of the International Mine Water Association Congress, Aachen, Germany, 4–11 September 2011; pp. 231–234.

18. Niedbalska, K.; Bukowski, P.; Haładus, A. Using of Groundwater Flow Modeling to Optimize the Methods of Liquidation of Open Pit Mine Reclaimed by Post-Mining Wastes. In Proceedings of the 14th International Multidisciplinary Scientific Geoconference SGEM, Geoconference on Science and Technologies in Geology, Exploration and Mining, Albena, Bulgaria, 17–26 June 2014; Volume II, pp. 1035–1042.

19. Zhao, L.; Ren, T.; Wang, N. Groundwater impact of open cut coal mine and an assessment methodology: A case study in NSW. *Int. J. Min. Sci. Technol.* **2017**, *27*, 861–866. [CrossRef]

20. Aryafar, A.; Ardejani, F.D.; Singh, R.; Shokri, B.J. Prediction of Groundwater Inflow and Height of the Seepage Face in a Deep Open Pit Mine Using Numerical Finite Element Model and Analytical Solutions. In *IMWA Symposium 2007*; Cidu, R., Frau, F., Eds.; Water in Mining Environments, International Mine Water Association: Cagliari, Italy, 2007.

21. Sayit, A.P.; Cankara-Kadioglu, C.; Yazicigil, H. Assessment of Dewatering Requirements and their Anticipated Effects on Groundwater Resources: A Case Study from the Caldag Nickel Mine, Western Turkey. *Mine Water Environ.* **2015**, *34*, 122–135. [CrossRef]

22. Bukowski, P.; Augustyniak, I.; Niedbalska, K. Some Methods of Hydrogeological Properties Evaluation and Their Use in Mine Flooding Forecasts. In Proceedings of the 15th International Multidisciplinary Scientific Geoconference SGEM, Science and technologies in geology, exploration and mining, Albena, Bulgaria, 18–24 June 2015; Volume II, pp. 717–724.

23. Wu, Q.; Hu, B.X.; Wan, L.; Zheng, C. Coal mine water management: Optimization models and field application in North China. *Hydrol. Sci. J.* **2010**, *55*, 609–623. [CrossRef]

24. Treichel, W.; Haładus, A.; Zdechlik, R. Simulation and optimization of groundwater exploitation for the water supply of Tarnów agglomeration (southern Poland). *Bull. Geogr. Phys. Geogr. Ser.* **2015**, *9*, 21–29. [CrossRef]

25. MODFLOW and Related Programs. Available online: https://water.usgs.gov/ogw/modflow (accessed on 16 April 2021).

26. Diersch, H.-J. *FEFLOW Finite Element Modeling of Flow, Mass and Heat Transport in Porous and Fractured Media*; Springer: Berlin/Heidelberg, Germany, 2014; p. 996. [CrossRef]

27. Prażak, J. *Position of Hydrodynamic and Economic Significance of Devonian Groundwater Reservoirs in the Holy Cross Mountains*; [In Polish: Pozycja hydrodynamiczna i znaczenie gospodarcze dewońskich zbiorników wód podziemnych w Górach Świętokrzyskich]; Prace PIG, Polish Geological Institute—National Research Institute: Krakow, Poland, 2012; Volume 198, p. 72.

28. Available online: https://www.pgi.gov.pl/en/phs/tasks/8878-gzwp-major-groundwater-reservoirs.html (accessed on 3 June 2021).

29. Chiang, W.-H. *3D-Groundwater Modeling with PMWIN*; Springer: Berlin/Heidelberg, Germany, 2005. [CrossRef]

30. Kulma, R.; Zdechlik, R. *Groundwater Flow Modeling*; [In Polish: Modelowanie Procesów Filtracji]; The AGH University of Science and Technology Press: Krakow, Poland, 2009.

31. Herman, G. *Hydrogeological Map of Poland, Scale 1:25 000, No. 851 Morawica*; Polish Geological Institute: Warsaw, Poland, 1997.

32. Herman, G. *Hydrogeological Map of Poland, Scale 1:25 000, No. 852 Daleszyce*; Polish Geological Institute: Warsaw, Poland, 1997.

33. Prażak, J. *Hydrogeological Map of Poland, Scale 1:25 000, No. 814 Piekoszów*; Polish Geological Institute: Warsaw, Poland, 1997.

34. Różkowski, K.; Zdechlik, R.; Polak, K.; Pawlecka, K.; Bielec, B. *Hydrogeological Documentation Specifying the Hydrogeological Conditions in Connection with the Drainage of the Kowala Mała Deposit to +200 m a.s.l*; [In Polish: Dokumentacja hydrogeologiczna określająca warunki hydrogeologiczne w związku z odwodnieniem złoża „Kowala Mała" do rzędnej +200 m n.p.m.]; Fundacja Nauka i Tradycje Górnicze, z siedzibą Wydział Górnictwa i Geoinżynierii AGH w Krakowie: Krakow, Poland, 2016; not published.

35. Kielce Climate. Available online: https://pl.climate-data.org/location/764743/ (accessed on 1 April 2021).

Article

Gate Road Support Deformation Forecasting Based on Multivariate Singular Spectrum Analysis and Fuzzy Time Series

Luka Crnogorac *, Rade Tokalić, Zoran Gligorić , Aleksandar Milutinović, Suzana Lutovac and Aleksandar Ganić

Faculty of Mining and Geology, University of Belgrade, Đušina 7, 11000 Belgrade, Serbia;
rade.tokalic@rgf.bg.ac.rs (R.T.); zoran.gligoric@rgf.bg.ac.rs (Z.G.); aleksandar.milutinovic@rgf.bg.ac.rs (A.M.);
suzana.lutovac@rgf.bg.ac.rs (S.L.); aleksandar.ganic@rgf.bg.ac.rs (A.G.)
* Correspondence: luka.crnogorac@rgf.bg.ac.rs; Tel.: +381-011-3219-202

Abstract: Underground mining engineers and planners in our country are faced with extremely difficult working conditions and a continuous shortage of money. Production disruptions are frequent and can sometimes last more than a week. During this time, gate road support is additionally exposed to rock stress and the result is its progressive deformation and the loss of functionality of gate roads. In such an environment, it is necessary to develop a low-cost methodology to maintain a gate road support system. For this purpose, we have developed a model consisting of two main phases. The first phase is related to support deformation monitoring, while the second phase is related to data analysis. To record support deformations over a defined time horizon we use laser scanning technology together with multivariate singular spectrum analysis to conduct data processing and forecasting. Fuzzy time series is applied to classify the intensity of displacements into several independent groups (clusters).

Keywords: support deformation; laser scanning; multivariate singular spectrum analysis; forecasting; fuzzy time series clusters

Citation: Crnogorac, L.; Tokalić, R.; Gligorić, Z.; Milutinović, A.; Lutovac, S.; Ganić, A. Gate Road Support Deformation Forecasting Based on Multivariate Singular Spectrum Analysis and Fuzzy Time Series. *Energies* **2021**, *14*, 3710. https://doi.org/10.3390/en14123710

Academic Editor: Krzysztof Skrzypkowski

Received: 19 May 2021
Accepted: 14 June 2021
Published: 21 June 2021

Publisher's Note: MDPI stays neutral with regard to jurisdictional claims in published maps and institutional affiliations.

Copyright: © 2021 by the authors. Licensee MDPI, Basel, Switzerland. This article is an open access article distributed under the terms and conditions of the Creative Commons Attribution (CC BY) license (https://creativecommons.org/licenses/by/4.0/).

1. Introduction

Due to hard working conditions and a chronic shortage of money, the underground mining in our country is very a difficult task. Most underground mines are characterized by low mechanized mining methods and production disruptions can happen often. Such mining methods are characterized by a long production cycle, and if we add disruptions then it is obvious that the gate road support is exposed to the rock stress much longer than is the case with the application of modern mechanized methods. In such a production environment, the maintenance of the gate road stability is almost impossible and is performed in an unsystematic way. In order to improve such a difficult situation, we use a laser scanner to obtain high accuracy deformation data, and multivariate singular spectrum analysis to process them and forecast the future state of deformations. Fuzzy time series is used to classify deformation data into several independent groups with respect to the magnitude of deformation intensity. In this way, we are enabling underground mining engineers to develop a plan to support maintenance. They can plan activities related to the identification of support states in the future, time of reconstruction, and type of support that can be used to decrease the progressive deformation. An effective deformation forecast can result in a decrease of costs and delays.

There are many different approaches to obtaining future states of support deformation. Ding-Ping Xu et al. [1] compared the predictions from the rock–soil composite material model, where an analytic method, along with existing data from physical model tests, was applied to acquire results. Good consistency, both in terms of strength and failure mode, was provided when a comparison of these three types of information was done [1].

Danial Jahed Armaghani et al. [2] applied three-nonlinear forecasting methods, namely a non-linear multiple regression, an artificial neural network, and an adaptive neuro-fuzzy inference system, in order to estimate the uniaxial compressive strength of rock.

Grey system prediction models were developed by Xiaobo Xiong [3] for forecasting rock tunnel displacement considering the non-linear parameters of the deformation of the tunnel. Ping Tang [4] forecasted the working face underground pressure by applying a grey system theory and genetic algorithm. Sun Yu et al. [5] proposed a time series prediction model based on a generalized regression neural network to predict the long-term potential deformation trends of surrounding rocks in a soft rock roadway tunnel. Haiming Chen et al. [6] built an intelligent displacement back analysis network of a deep mine roadway surrounding rock that is based on MATLAB NN toolbox. Francesca Bozzano et al. [7] used terrestrial SAR interferometry to support the management of each phase of tunneling. Qian Zhang et al. [8] made a numerical simulation for advanced displacement of tunnel with weak and broken surrounding rock in order to reveal the adjusting process of displacement in the main parts of a tunnel.

Grossauer et al. [9] developed an efficient way to predict displacements caused by tunnel excavation based on the use of analytical functions that describe the movements as a function of time and the face advance. A procedure based on analytical functions has been developed to support the prediction of the tunnel performance and surface movements. Merlini and Falanesca [10] illustrated the comparison between the prognosis and the crosscheck of the support methods and the replacement solutions. Several back-analyses were carried out in order to achieve the correct validation of the interventions and an optimal understanding of the deformation behavior of the rock mass. Bao-Zhen Yao et al. [11] has presented a multi-step-ahead prediction model, based on a support vector machine, for tunnel surrounding rock displacement forecasting.

This paper is divided into five sections. In the section Materials and Methods, we describe the model of forecasting based on observed data and multivariate singular spectrum analysis. We introduce the fuzzy time series concept to make clusters describing the gate road support configuration deformation for every time point over observed and forecasted periods. A computational process is performed in the Numerical Example section to represent the possibilities of model and conclusions are provided in the Results and Conclusion section. Obtained results show a high level of correlation of original and reconstructed series of the support deformations. Based on these results, it can be concluded that the model is reliable and applicable for solving real-time problems in terms of predicting gate road support deformations.

2. Materials and Methods

2.1. Dynamic of Support Deformations

The dynamic behavior of support is induced by the rock stress surrounding the gate road. Deformation dynamics is the study of time varying response of support under dynamic loads. These loads are primarily considered a change of rock stress intensity induced by mining activities. As a stope approaches the gate road, the rock stress intensity increases.

In order to describe deformations of a gate road support, we consider the time dependent displacements only in the vertical cross-section (planar problem). Let $S(t)$ be a surface of the support bounded by $B(t) = [B_l(t), B_u(t)]$ at the current time t, where $B_l(t)$ represents the lower edge and $B_u(t)$ represents the upper edge of the support. Without loss of generality, we equal the upper edge with the lower and transform the surface of the support into one line, i.e., we reduce a real two-dimensional support to an artificial one-dimensional support; $S(t = 0) \in R^2 \to B_l(t = 0) \in R$, (Figure 1).

The initial configuration of $B_l(t)$, $t = 0$ is undeformed and known (recorded by laser scanning). The motion of each point on $B_l(t)$, $t = 0$, from the initial to the current configuration, is completely defined by a time dependent mapping function. Since the equation of this function is unknown and it is necessary to make monitoring of support deformations at equal or nonequal time intervals. Monitoring at equal intervals is much more applicable for forecasting deformations of a gate road. It is very important to use the same coordinate system to describe deformations over the time of monitoring.

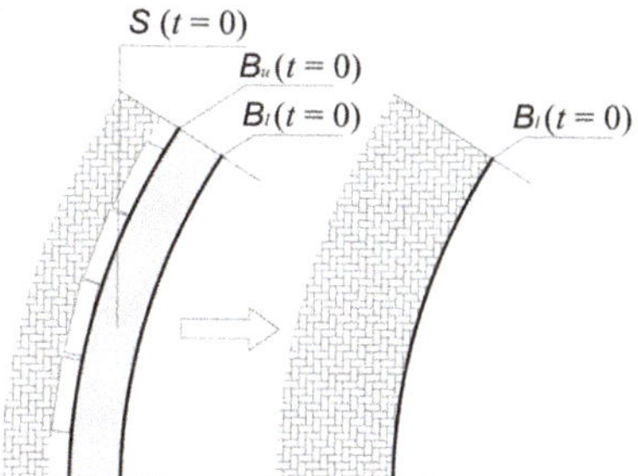

Figure 1. Transformation of real two-dimensional support to an artificial one-dimensional support.

The position of the marker $m\left(\overrightarrow{v},t\right)$, induced by rock stress, changes according to the following vector equation (Figure 2):

$$\overrightarrow{v}(t) = \overrightarrow{v}(t-1) + \overrightarrow{u}(t),\ t = 1, 2, \ldots, N,\tag{1}$$

where
$\overrightarrow{v}(t)$—the position vector of the marker m in the current support configuration.
$\overrightarrow{v}(t-1)$—the position vector of the marker m in the previous support configuration.
$\overrightarrow{u}(t)$—the vector of displacement.
N—time of monitoring.

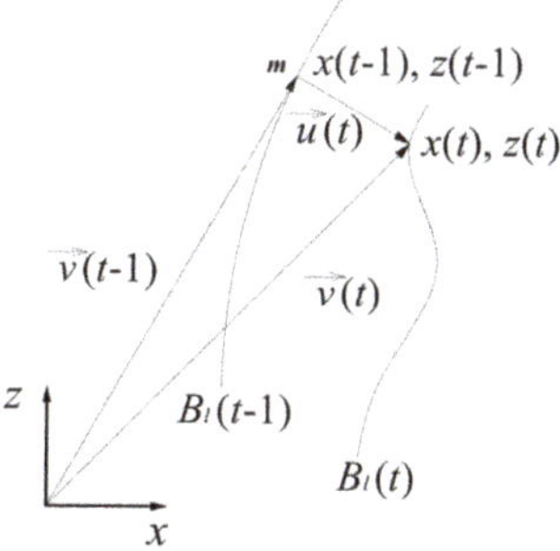

Figure 2. Change of the marker position over time.

In the xz-coordinate plane, the position of the marker $m(x, z, t)$ is defined as follows:

$$\begin{aligned} x(t) &= x(t-1) + \Delta x(t) \\ z(t) &= z(t-1) + \Delta z(t), \end{aligned}\tag{2}$$

The gate road support domain $B_l(t)$ is divided into a finite number of segments "connected" at the marker's position. Support configuration at current time t is defined by the set $B_l(t) = \{m_{t,i}\}, i = 1, 2, \ldots, M$, where M is the total number of markers. Let us define a section of the initial support configuration by the set $D_l(t = 0) \subset B_l(t = 0); D_l(t = 0) = \{m_{0,1}, m_{0,2}, m_{0,3}\}$. At time $t = 1$, the section of initial support configuration will be transformed and defined by the set $D_l(t = 1) = \{m_{1,1}, m_{1,2}, m_{1,3}\}$. At time $t = 2$, the topology of the section is defined by the set $D_l(t = 2) = \{m_{2,1}, m_{2,2}, m_{2,3}\}$. The time-dependent path of the section topology transformation for $t = 0, 1, 2$ is defined as follows: $D_l(0, 1, 2) = \{m_{0,1}, m_{0,2}, m_{0,3}\} \rightarrow \{m_{1,1}, m_{1,2}, m_{1,3}\} \rightarrow \{m_{2,1}, m_{2,2}, m_{2,3}\}$ (Figure 3).

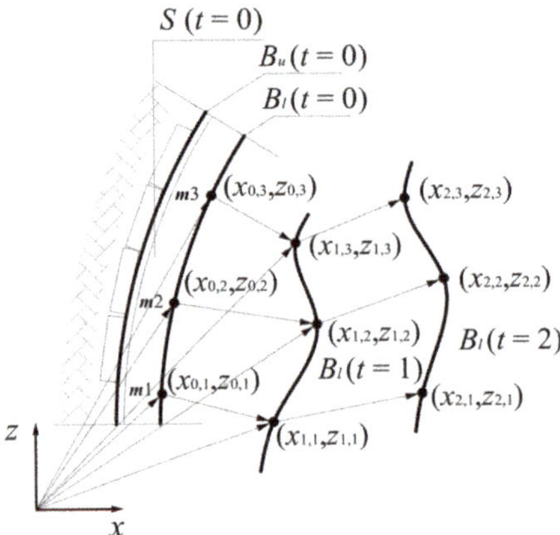

Figure 3. Change of the section of support configuration over time interval [0, 2].

In conventional matrix form, the data of position of all markers obtained by scanning for $1 \leq t \leq N$ can be expressed by the time-dependent position vectors as follows:

$$\vec{V}(t) = \left[\vec{v}_{ij}\right]_{M \times N} = \begin{bmatrix} \vec{v}_{11} & \vec{v}_{12} & \vec{v}_{13} & \cdots & \vec{v}_{1N} \\ \vec{v}_{21} & \vec{v}_{22} & \vec{v}_{23} & \cdots & \vec{v}_{2N} \\ \vdots & \vdots & \vdots & \ddots & \vdots \\ \vec{v}_{M1} & \vec{v}_{M2} & \vec{v}_{M3} & \cdots & \vec{v}_{MN} \end{bmatrix}, \tag{3}$$

The position of each marker is represented by each row of the previous matrix, while each column represents the configuration of the gate road support over time. According to Equation (2), x and z coordinates of observed markers are defined as follows:

$$X(t) = \left[x_{ij}\right]_{M \times N} = \begin{bmatrix} x_{11} & x_{12} & x_{13} & \cdots & x_{1N} \\ x_{21} & x_{22} & x_{23} & \cdots & x_{2N} \\ \vdots & \vdots & \vdots & \ddots & \vdots \\ x_{M1} & x_{M2} & x_{M3} & \cdots & x_{MN} \end{bmatrix}, \tag{4}$$

$$Z(t) = \left[z_{ij}\right]_{M \times N} = \begin{bmatrix} z_{11} & z_{12} & z_{13} & \cdots & z_{1N} \\ z_{21} & z_{22} & z_{23} & \cdots & z_{2N} \\ \vdots & \vdots & \vdots & \ddots & \vdots \\ z_{M1} & z_{M2} & z_{M3} & \cdots & z_{MN} \end{bmatrix}, \tag{5}$$

As can be seen, the obtained data can be treated as a multichannel time series. Multivariate singular spectrum analysis is a very useful tool to process such datasets.

2.2. Gateroad Support Deformation Forecasting Algorithm

Our forecasting algorithm is based on the methodology of the multivariate singular spectrum analysis (MSSA). In the paper by Harris and Yuan [12], the basic SSA (singular spectrum analysis) algorithm was presented. Hassani and Zhigljavsky [13] showed that SSA is a powerful method for time-series analysis and forecasting. Hassani and Mahmoudvand [14] pointed out that this method can be applied both to single series and jointly for several series (MSSA). When it is a case of a two or more series, we are talking about MSSA. The application of the SSA technique was found to be very useful for time series analysis in many different fields, such as medicine, biology, genetics, finance, engineering and many others [15]. The basic concept of the singular spectrum analysis is well presented in the paper by Hassani et al. [16].

According to Equation (2), x and z coordinates of the marker m over time correspond to one channel time series. If we take into consideration the total number of markers is M, then we are faced with an M-channel time series.

Consider an M-channel time series with a series length of N_i; $V_{N_i}^{(i)} = \left(v_1^{(i)}, \ldots, v_{N_i}^{(i)} \right)$, $(i = 1, \ldots, M)$, where N is the time of monitoring. Here, we provide the procedure for the x coordinates of the markers. The procedure for the z coordinates is completely the same.

At the first stage of the algorithm, the decomposition, includes the following two steps: embedding and singular value decomposition (SVD) [14,16].

Embedding is a mapping that transfers a one dimensional time series of coordinates $X_{N_i}^{(i)} = \left(x_1^{(i)}, \ldots, x_{N_i}^{(i)} \right)$ into a multidimensional matrix $\left[Y_1^{(i)}, \ldots, Y_{K_i}^{(i)} \right]$ with vectors $Y_j^{(i)} = \left(x_j^{(i)}, \ldots, x_{j+L_i+1}^{(i)} \right)^T \in R^{L_i}$, where $L_i (1 < L_i < N_i)$ is the window length for each series with length N_i and $K_i = N_i - L_i + 1$. The result of this step is the trajectory matrix $Y^{(i)} = \left[Y_1^{(i)}, \ldots, Y_{K_i}^{(i)} \right] = (y_{mn})_{m,n=1}^{L_i, K_i}$. The trajectory matrix $Y^{(i)}$ is known as a Hankel matrix. Thus, the above procedure for each series separately provides M different $L_i \times K_i$ trajectory matrices $Y^{(i)} (i = 1, \ldots, M)$. To form a new block Hankel matrix in a vertical form, we need to have $K_1 = \ldots = K_M = K$. The result of this step is the following block Hankel trajectory matrix:

$$Y_V = \begin{bmatrix} Y^{(1)} \\ \vdots \\ Y^{(M)} \end{bmatrix}, \tag{6}$$

where Y_V indicates that the output of the first step is a block Hankel trajectory matrix formed in a vertical form; index V means vertical.

In the second step, we perform the SVD of Y_V. Denote $\lambda_{V_1}, \ldots, \lambda_{V_{L_{sum}}}$ as the eigenvalues of $Y_V Y_V^T$, arranged in decreasing order $\left(\lambda_{V_1} \geq \ldots \lambda_{V_{L_{sum}}} \geq 0 \right)$ and $U_{V_1}, \ldots, U_{V_{L_{sum}}}$, the corresponding eigenvectors, where $L_{sum} = \sum_{i=1}^{M} L_i$. Note also that the structure of the matrix $Y_V Y_V^T$, is as follows:

$$Y_V Y_V^T = \begin{bmatrix} Y^{(1)}Y^{(1)T} & Y^{(1)}Y^{(2)T} & \cdots & Y^{(1)}Y^{(M)T} \\ Y^{(2)}Y^{(1)T} & Y^{(2)}Y^{(2)T} & \cdots & Y^{(2)}Y^{(M)T} \\ \vdots & \vdots & \ddots & \vdots \\ Y^{(3)}Y^{(1)T} & Y^{(3)}Y^{(2)T} & \cdots & Y^{(3)}Y^{(M)T} \end{bmatrix}, \tag{7}$$

The structure of the matrix $Y_V Y_V^T$ is similar to the variance-covariance matrix in the classical multivariate statistical analysis literature. The matrix $Y^{(i)}Y^{(i)T}$, which is used in the univariate SSA, for the series $X_{N_i}^{(i)}$ appears along the main diagonal, and the products of two Hankel matrices $Y^{(i)}Y^{(i)T}$ $(i \neq j)$ that are related to the series $X_{N_i}^{(i)}$ and $X_{N_j}^{(j)}$ appear in the off-diagonal. The SVD of Y_V can be written as $Y_V = Y_{V_1} + \ldots + Y_{V_{L_{sum}}}$, where $Y_{V_i} = \sqrt{\lambda_i} U_{V_i} U_{V_i}^T$ and $V_{V_i} = Y_V^T U_{V_i} / \sqrt{\lambda_{V_i}}, (Y_{V_i} = 0 \, if \, \lambda_{V_i} = 0)$. U_{V_i} are called factor empirical orthogonal functions and V_{V_i} are the left and right eigenvectors of the trajectory matrix, often called principal components.

The second stage of the algorithm, called reconstruction, includes the following two steps: grouping and diagonal averaging or Hankelization [14,16].

The grouping step corresponds to splitting the matrices $Y_{V_1}, \ldots, Y_{V_{L_{sum}}}$ into several disjointed groups and summing the matrices within each group. The split of the set of indices $\{1, \ldots, L_{sum}\}$ into disjointed subsets $I_1, \ldots, I_k$ corresponds to the representation of $Y_V = Y_{I_1} + \ldots + Y_{I_k}$. The procedure of choosing the sets $I_1, \ldots, I_k$ is called grouping. For a given group I, the contribution of the component Y_{V_I} is measured by the share of the corresponding eigenvalues: $\sum_{i \in I} \lambda_{V_i} / \sum_{i=1}^{d_V} \lambda_{V_i}$, where d_V is the rank of Y_{V_I} and $I \subset \{1, \ldots, L_{sum}\}$. In a simple case where we have only signal and noise components, we use two groups of indices, $I_1 = \{1, \ldots, r\}$ and $I_2 = \{r+1, \ldots, L_{\text{sum}}\}$ and associate the group I_1 with the signal component and the group I_2 with the noise.

The purpose of diagonal averaging is to transform the reconstructed matrix $\hat{Y}_{V_i}$ into a Hankel matrix, which can subsequently be converted into a time series. Let $\tilde{Y}^{(h)}$ be the approximation of $Y^{(i)}$ obtained from the diagonal averaging step. If $\tilde{y}_{mn}^{(i)}$ stands for an element of a matrix $\tilde{Y}^{(i)}$, then the jth term of the reconstructed series $\tilde{X}_{N_i}^{(i)} = \left(\tilde{x}_1^{(i)}, \ldots, \tilde{x}_j^{(i)}, \ldots, \tilde{x}_{N_i}^{(i)} \right)$ is achieved by arithmetic averaging $\tilde{y}_{mn}^{(i)}$ over all (m, n) such that $m + n - 1 = j$.

The third stage of the algorithm concerns the future positions of the markers and is based on the vertical multivariate singular spectrum analysis recurrent procedure (VMSSA-R).

Let us have M-channel series $X_{N_i}^{(i)} = \left(x_1^{(i)}, \ldots, x_{N_i}^{(i)} \right)$ and corresponding window length L_i, $1 < L_i \leq N_i$, $i = 1, \ldots, M$. Optimal values of the window length are discussed in chapter 4 of the paper by Hassani and Mahmoudvand [14].

The VMSSA-R forecasting algorithm for the h-step ahead forecast is as follows:

For a fixed value of K, construct the trajectory matrix $Y^{(i)} = \left[Y_1^{(i)}, \ldots, Y_K^{(i)} \right] = (y_{mn})_{m,n=1}^{L_i, K}$ for each single series separately; construct the block trajectory matrix Y_V as follows:

$$Y_V = \begin{bmatrix} Y^{(1)} \\ \vdots \\ Y^{(M)} \end{bmatrix}, \tag{8}$$

let $U_{V_j} = \left(U_j^{(1)}, \ldots, U_j^{(M)} \right)^T$ be the jth eigenvector of the $Y_V Y_V^T$, where $U_j^{(i)}$ with length L_i corresponds to the series $X_{N_i}^{(i)}$ $(i = 1, \ldots, M)$; consider $\hat{Y}_V = \left[\hat{Y}_1 : \ldots : \hat{Y}_K \right] = \sum_{i=1}^{r} U_{V_i} U_{V_i}^T Y_V$ the reconstructed matrix achieved from r eigentriples:

$$\hat{Y}_V = \begin{bmatrix} \hat{Y}^{(1)} \\ \vdots \\ \hat{Y}^{(M)} \end{bmatrix}, \tag{9}$$

consider matrix $\tilde{Y}^{(i)} = H \times \hat{Y}^{(i)} = \mathcal{H} \, \hat{Y}^{(i)}$ $(i = 1, \ldots, M)$ as the result of the Hankelization procedure of the matrix $\tilde{Y}^{(i)}$ obtained from the previous step, where $\mathcal{H}$ is a Hankel operator; assume $U_j^{(i)\nabla}$ denotes the vector of the first $L_i - 1$ components of the vector $U_j^{(i)}$ and $\pi_j^{(i)}$ is the last component of the vector $U_j^{(i)}$ $(i = 1, \ldots, M)$; select the number of r eigentriples for the reconstruction stage that can also be used for forecasting purposes; define matrix $U^{\nabla M} = \left(U_1^{\nabla M}, \ldots, U_r^{\nabla M} \right)$, where $U_j^{\nabla M}$ is as follows:

$$U_j^{\nabla M} = \begin{bmatrix} U_j^{(1)\nabla} \\ \vdots \\ U_j^{(M)\nabla} \end{bmatrix}, \tag{10}$$

define matrix W as follows:

$$W = \begin{bmatrix} \pi_1^{(1)} & \pi_2^{(1)} & \cdots & \pi_r^{(1)} \\ \pi_1^{(2)} & \pi_2^{(2)} & \cdots & \pi_r^{(2)} \\ \vdots & \vdots & \ddots & \vdots \\ \pi_1^{(M)} & \pi_2^{(M)} & \cdots & \pi_r^{(M)} \end{bmatrix}, \tag{11}$$

if the matrix $\left(I_{M \times M} - WW^T\right)^{-1}$ exists and $r \leq L_{sum} - M$, then the h-step ahead VMSSA forecasts exist and is achieved by the following formula:

$$\left[\hat{x}_{j_1}^{(1)}, \ldots, \hat{x}_{j_M}^{(M)}\right]^T = \begin{cases} \left[\tilde{x}_{j_1}^{(1)}, \ldots, \tilde{x}_{j_M}^{(M)}\right], & j_i = 1, \ldots, N_i \\ \left(I_{M \times M} - WW^T\right)^{-1} WU^{\nabla M^T} Z_h, & j_i = N_i + 1, \ldots, N_i + h \end{cases}, \tag{12}$$

where $Z_h = \left[Z_h^{(1)}, \ldots, Z_h^{(M)}\right]^T$ and $Z_h^{(1)} = \left[\hat{x}_{N_i - L_i + h + 1}^{(i)}, \ldots, \hat{x}_{N_i + h + 1}^{(i)}\right]$ $(i = 1, \ldots, M)$. It should be noted that Equation (12) indicates that the h-step ahead forecasts of the refined series are obtained by a multi-dimensional linear recurrent formula (LRF).

2.3. Displacement Time Series Clustering

The time displacement intensity vector of marker m is defined as follows:

$$\Delta v(t) = \sqrt{\Delta x^2(t) + \Delta z^2(t)}, \; t = 1, 2, \ldots, N, \tag{13}$$

The universe of displacement (clustering) is defined as the union of the two following sets:

$$\Delta V = \Delta V_{N_i}^{(i)} \cup \Delta \hat{V}_{N_i + h}^{(i)}, \tag{14}$$

where $\Delta V_{N_i}^{(i)}$ is related to the observed displacement series and $\Delta \hat{V}_{N_i + h}^{(i)}$ to the the h-step ahead forecasts of the refined displacement series. Note that clustering is performed for each set separately.

To create the clusters within a defined universe of clustering, we apply the concept of fuzzy time series. Such series, which were first proposed by Song and Chissom, ref. [17] are based on fuzzy set theory proposed by Zadeh [18,19]. The most important advantage of fuzzy time series approaches is to be able to work with a very small set of data and not to require the linearity assumption. In the process of measuring the cross-section of the underground roadways, certain measurement errors can occur (conditioned by the error of the measurement method, instrumental error, personal error by the operator of the instrument, or the error due to the working environment itself), the nature of fuzzy time series is a good choice for a credible presentation of displacements of the markers and its clustering.

Let $\Delta V_{M \times N}$ be the universe of observed displacement. Data obtained by monitoring can be clustered in different ways, for example, a cluster composed of marker displacement over time separately $(\Delta v_{12}, \Delta v_{32}, \Delta v_{24}, \ldots)$, a cluster composed of each row or column data of the matrix ΔV. We focus on each column data in order to see how displacement intensity of configuration of the gate road support changes over time. For that purpose, we apply the concept of fuzzy time series with multiple observations [20–22].

A fuzzy set A of $\Delta V_{M \times N}$ is defined as follows:

$$A = \frac{f_A(\Delta v_{11})}{\Delta v_{11}} + \frac{f_A(\Delta v_{12})}{\Delta v_{12}} + \ldots + \frac{f_A(\Delta \hat{v}_{MN})}{\Delta \hat{v}_{MN}}, \tag{15}$$

where f_A is the membership function of A, and $f_A : \Delta V \to [0, 1]$. The symbol "+" denotes the union operator. Membership function represents the degree of membership of Δv_i in A, where $f_A(\Delta v_i) \in [0, 1]$, $1 \leq i \leq M \times N$. Let $\Delta V(t)$ be the universe of displacement on which fuzzy sets $f_i(t), (i = 1, 2, \ldots)$ are defined and $F(t)$ is a collection of $f_1(t), f_2(t), \ldots, F(t)$ that is referred to as a fuzzy time series of $\Delta V(t)$. Here, $F(t)$ is viewed as a linguistic variable and $f_i(t)$ represents possible linguistic values of $F(t)$. If $F(t)$ is caused by $F(t-1)$ only, the relationship can be expressed as [23]. If the maximum degree of membership of $F(t)$ belongs to A_i then $F(t)$ is considered to be A_i. Hence, $F(t-1) \to F(t)$ becomes $A_i(t-1) \to A_j(t)$.

The algorithm of a fuzzy time series model for multiple observations is composed of the following steps [23]:

Step 1: define the fuzzy sets for the fuzzy time series. We set the beginning and the end of the universe of displacement as $l = \min\Delta V_{M\times N}$ and $u = \max\Delta V_{M\times N}$, respectively. The observed data are sorted in ascending order. The distance of any two consecutive values of displacement is calculated as follows:

$$d = |\Delta v_{i+1} - \Delta v_i|, \; i = 1, 2, \dots, (M \times N) - 1, \tag{16}$$

Now it is necessary to compute the corresponding average value $\bar{d}$ and standard deviation σ_d. The intention is to eliminate the outliers from the sorted data and obtain an average distance value free of distortion. Outliers are values that are either abnormally high or abnormally low in the sorted dataset. The process of elimination of outliers is performed as follows:

$$\bar{d} - \sigma_d \leq d \leq \bar{d} + \sigma_d, \tag{17}$$

Since the process of elimination is completed, a revised average distance value $\bar{d}_R$ is calculated for the remaining values in the sorted data set. Accordingly, the universe of displacement is also revised and defined as follows:

$$\Delta V_R = [l_R, u_R] = \left[l - \bar{d}_R, u + \bar{d}_R\right], \tag{18}$$

The number of equal intervals n is given by the user and each interval is characterized by an adequate linguistic variable. The variable whose values are words or sentences in a natural or artificial language is called a linguistic variable. We use five linguistic variables $n = 5$ to describe displacement intensity of configuration of the gate road support; $A_1 =$(very small), $A_2 =$(small), $A_3 =$(medium), $A_4 =$(high), and $A_5 =$(very high). The length of interval can be calculated by the equation:

$$L = \frac{u_R - l_R}{n + 1}, \tag{19}$$

Each interval is obtained as: $\Delta v_1 = [lR, lR + L]$, $\Delta v_2 = [lR + L, lR + 2L]$, $\dots, \Delta v_n = [lR + (n-1)L, lR + nL]$. Linguistic variable A_i $(i = 1, 2, \dots, 5)$ can now be defined according to defined intervals as follows:

$$
\begin{aligned}
A_1 &= \left\{ \frac{1}{\Delta v_1}, \frac{0.5}{\Delta v_2}, \frac{0}{\Delta v_3}, \frac{0}{\Delta v_4}, \frac{0}{\Delta v_5} \right\} \\
A_2 &= \left\{ \frac{0.5}{\Delta v_1}, \frac{1}{\Delta v_2}, \frac{0.5}{\Delta v_3}, \frac{0}{\Delta v_4}, \frac{0}{\Delta v_5} \right\} \\
A_3 &= \left\{ \frac{0}{\Delta v_1}, \frac{0.5}{\Delta v_2}, \frac{1}{\Delta v_3}, \frac{0.5}{\Delta v_4}, \frac{0}{\Delta v_5} \right\} \\
A_4 &= \left\{ \frac{0}{\Delta v_1}, \frac{0}{\Delta v_2}, \frac{0.5}{\Delta v_3}, \frac{1}{\Delta v_4}, \frac{0.5}{\Delta v_5} \right\} \\
A_5 &= \left\{ \frac{0}{\Delta v_1}, \frac{0}{\Delta v_2}, \frac{0}{\Delta v_3}, \frac{0.5}{\Delta v_4}, \frac{1}{\Delta v_5} \right\}
\end{aligned}
\tag{20}
$$

The triangular fuzzy number is used to quantify the linguistic variable. It can be defined as a triplet (a, b, c). The corresponding membership function is defined as [24]:

$$
\mu_A(\Delta v) = \begin{cases}
0, & \Delta v < a \\
\frac{\Delta v - a}{b - a}, & a \leq \Delta v \leq b \\
\frac{c - \Delta v}{c - b}, & b \leq \Delta v \leq c' \\
0, & \Delta v > c
\end{cases}
\tag{21}
$$

Step 2: determine a fuzzy observation of displacement. Triangular fuzzy number $O(t) = (a(t), b(t), c(t))$ is also used to represent displacement of all markers at time point t, where $a(t)$, $b(t)$, $c(t)$ are the left, middle and right values of the triangular fuzzy number, respectively. These values are defined in the following way:

$$a(t) = min(\Delta v_{1t}, \Delta v_{2t}, \ldots, \Delta v_{Mt})$$
$$b(t) = \frac{\Delta v_{1t} + \Delta v_{2t} + \ldots + \Delta v_{Mt}}{M}$$
$$c(t) = max(\Delta v_{1t}, \Delta v_{2t}, \ldots, \Delta v_{Mt}), \ t = 1, 2, \ldots, N, \tag{22}$$

In other words, for every observation (column of matrix), it is necessary to separate minimum, maximum, and average values of displacement for all markers.

Step 3: calculate the fuzzy relationship, $Y(t)$, between each fuzzified observation and defined fuzzy time series as in Step 1.

$$Y(t) = (\mu_1(t), \mu_2(t), \ldots, \mu_n(t)) = \sum_{i=1}^{n} max \bullet min(O(t), A_i), \ \mu_i(t) \geq 0, \tag{23}$$

Fuzzified observation of displacement at time point t belongs to cluster A_i if and only if their intersection has the highest value of membership function; $\mu_i(t) \rightarrow max$.

Each $\mu_i(t)$ is calculated according to Figure 4, representing the intersection between a fuzzy observation of displacement $O(t)$ and fuzzy time series consisting of $A_1, A_2, \ldots, A_i$.

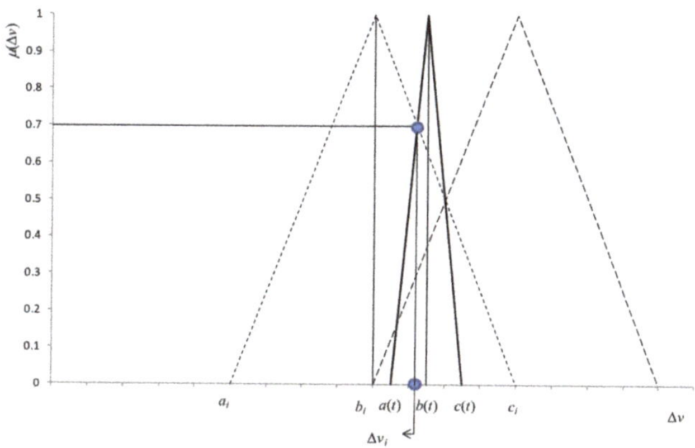

Figure 4. Relationship between fuzzy observation and fuzzy time series.

The calculation of the membership function value of intersection is based on the following approach:

$$\frac{1}{b(t) - a(t)} = \frac{\mu_i(\Delta v_i)}{\Delta v_i - a(t)}, \tag{24}$$

$$\frac{1}{c_i - b_i} = \frac{\mu_i(\Delta v_i)}{c_i - \Delta v_i}, \tag{25}$$

From Equations (24) and (25), we can calculate the value of $\mu_i(t)$ as follows:

$$\mu_i(\Delta v_i) = \frac{c_i - a(t)}{b(t) - a(t) + c_i - b_i}, \tag{26}$$

3. Numerical Example

Artificial data related to displacement of markers are generated in order to verify the validity of the proposed forecast model. A cross section of the gate road support and positions of the markers are represented by Figure 5. The coordinate system origin is located at the left lower corner of the gate road support (marker 1). The time resolution of observation is five days.

Coordinate data obtained after six observations are represented in Tables 1 and 2 and Figure 6.

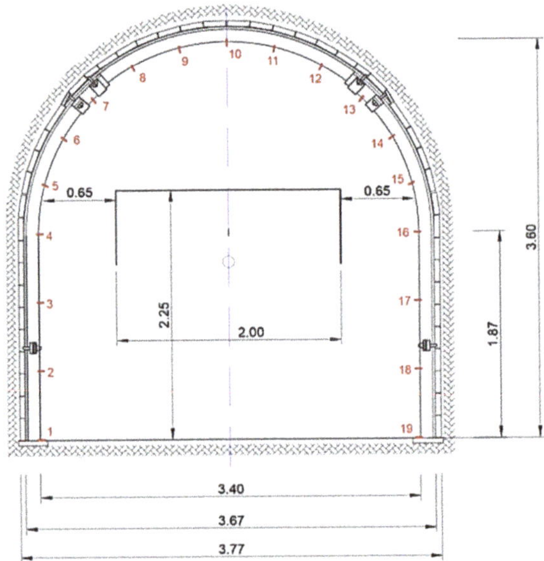

Figure 5. Gate road cross section with marker positions.

Table 1. X coordinates of the markers (m).

Marker	$t = 0$	$t = 1$	$t = 2$	$t = 3$	$t = 4$	$t = 5$
1 X(1)	0.000	0.011	0.025	0.039	0.055	0.068
2 X(2)	0.000	0.019	0.035	0.053	0.071	0.085
3 X(3)	0.000	0.022	0.047	0.079	0.104	0.135
4 X(4)	0.000	0.030	0.072	0.111	0.153	0.207
5 X(5)	0.058	0.095	0.132	0.192	0.230	0.281
6 X(6)	0.233	0.258	0.340	0.384	0.473	0.517
7 X(7)	0.502	0.549	0.630	0.646	0.720	0.732
8 X(8)	0.859	0.876	0.927	0.934	1.010	1.038
9 X(9)	1.278	1.299	1.360	1.351	1.308	1.343
10 X(10)	1.700	1.700	1.730	1.700	1.741	1.729
11 X(11)	2.122	2.089	2.119	2.079	2.054	2.027
12 X(12)	2.541	2.521	2.468	2.449	2.447	2.408
13 X(13)	2.898	2.830	2.814	2.726	2.721	2.665
14 X(14)	3.167	3.108	3.080	3.031	2.995	2.903
15 X(15)	3.342	3.300	3.260	3.184	3.120	3.081
16 X(16)	3.400	3.372	3.339	3.287	3.257	3.226
17 X(17)	3.400	3.387	3.364	3.341	3.313	3.286
18 X(18)	3.400	3.394	3.376	3.357	3.341	3.326
19 X(19)	3.400	3.391	3.379	3.365	3.351	3.341

Table 2. Z coordinates of the markers (m).

Marker	$t = 0$	$t = 1$	$t = 2$	$t = 3$	$t = 4$	$t = 5$
1 Z(1)	0.000	0.000	0.000	0.000	0.000	0.000
2 Z(2)	0.624	0.625	0.622	0.621	0.617	0.614
3 Z(3)	1.245	1.231	1.239	1.221	1.235	1.219
4 Z(4)	1.860	1.846	1.853	1.835	1.826	1.823
5 Z(5)	2.305	2.284	2.266	2.266	2.263	2.269
6 Z(6)	2.723	2.669	2.610	2.615	2.579	2.549
7 Z(7)	3.072	2.998	2.916	2.899	2.851	2.843
8 Z(8)	3.371	3.261	3.188	3.133	3.057	2.982

Table 2. *Cont.*

Marker	$t = 0$	$t = 1$	$t = 2$	$t = 3$	$t = 4$	$t = 5$
9 Z(9)	3.538	3.386	3.251	3.163	3.101	3.044
10 Z(10)	3.590	3.388	3.308	3.207	3.126	3.026
11 Z(11)	3.537	3.344	3.258	3.124	3.031	2.930
12 Z(12)	3.371	3.219	3.146	3.079	3.017	2.936
13 Z(13)	3.072	2.939	2.890	2.815	2.778	2.730
14 Z(14)	2.723	2.662	2.634	2.606	2.587	2.552
15 Z(15)	2.305	2.271	2.241	2.239	2.219	2.209
16 Z(16)	1.860	1.853	1.844	1.839	1.820	1.812
17 Z(17)	1.245	1.241	1.238	1.249	1.244	1.242
18 Z(18)	0.624	0.622	0.621	0.620	0.620	0.622
19 Z(19)	0.000	0.000	0.000	0.000	0.000	0.000

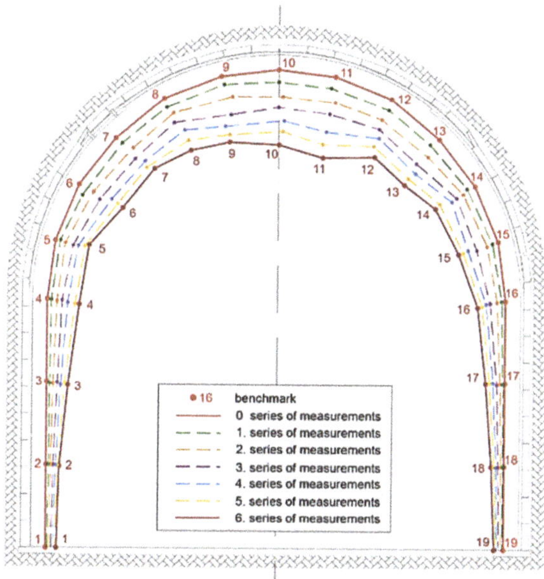

Figure 6. Configuration of the gate road support over the time of observation.

Measurements for $t = 0, 1, \ldots , 5$ are used to create a forecast model, while the measurement for $t = 6$ is used as a control series. Detail calculation is provided for the x coordinates. Figure 7 represents x coordinates of the marker 10 obtained by observation.

The inputs needed to perform multivariate singular spectrum analysis are:

- the number of markers (series) M = 19,
- the time series length needs to be equal to each marker (series) N = 6,
- the window length for each marker (series) L = 3,
- parameter K also needs to be equal to each marker (series) K = 4.

The trajectory matrix of marker 1 is obtained as follows: from the original series $X^{(1)} = (0.000\ 0.011\ 0.025\ 0.039\ 0.055\ 0.068)$, we create the following trajectory matrix of dimension $L \times K = 3 \times 4$.

$$Y^{(1)} = \begin{bmatrix} 0.000 & 0.011 & 0.025 & 0.039 \\ 0.011 & 0.025 & 0.039 & 0.055 \\ 0.025 & 0.039 & 0.055 & 0.068 \end{bmatrix}, \tag{27}$$

The above procedure is performed for each series separately and provides nineteen different $L \times K = 3 \times 4$ trajectory matrices $Y^{(i)}$ $(i = 1, 2, \ldots, 19)$. The result of this step is the following vertical formed block Hankel trajectory matrix:

$$
Y_V = \begin{bmatrix} Y^{(1)} \\ Y^{(2)} \\ \vdots \\ Y^{(19)} \end{bmatrix} = \begin{bmatrix} 0.000 & 0.011 & 0.025 & 0.039 \\ 0.011 & 0.025 & 0.039 & 0.055 \\ 0.025 & 0.039 & 0.055 & 0.068 \\ 0.000 & 0.019 & 0.035 & 0.053 \\ 0.019 & 0.035 & 0.053 & 0.071 \\ 0.035 & 0.053 & 0.071 & 0.085 \\ \vdots & \vdots & \vdots & \vdots \\ 3.400 & 3.391 & 3.379 & 3.365 \\ 3.391 & 3.379 & 3.365 & 3.351 \\ 3.379 & 3.365 & 3.351 & 3.341 \end{bmatrix}, \tag{28}
$$

The eigenvalues of $Y_V Y_V^T$, arranged in decreasing order are $\lambda_{V_1} = 1066.7105$; $\lambda_{V_2} = 0.2704$; $\lambda_{V_3} = 0.0135$; $\lambda_{V_4} = 0.0080$. The values of the corresponding eigenvectors $U_{V_1}, U_{V_2}, U_{V_3}, U_{V_4}$ are represented in Table 3.

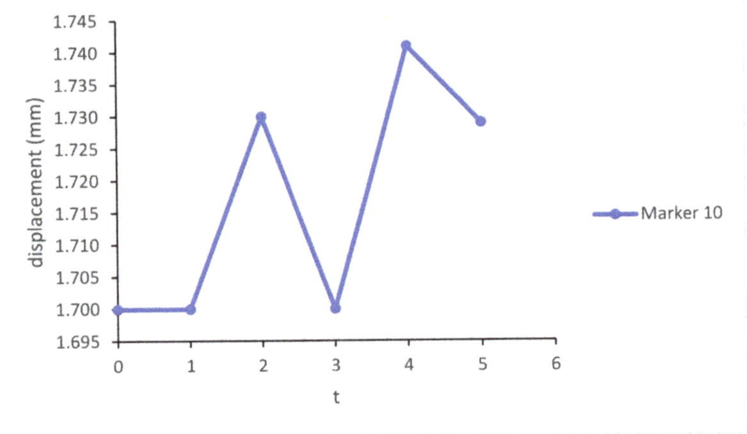

Figure 7. x coordinates of marker 10.

Table 3. Corresponding eigenvectors of $Y_V Y_V^T$.

Marker	U_{V1}	U_{V2}	U_{V3}	U_{V_4}
	0.0011	−0.0570	−0.0017	−0.0151
1	0.0020	−0.0639	−0.0113	−0.0075
	0.0029	−0.0640	0.0038	0.0064
	0.0016	−0.0761	−0.0163	0.0108
2	0.0027	−0.0764	−0.0050	−0.0085
	0.0037	−0.0744	0.0027	0.0238
	...	...	...	...
	...	...	...	...
	...	...	...	...
	0.2072	−0.0705	−0.0268	0.0102
19	0.2065	−0.0628	−0.0298	−0.0064
	0.2057	−0.0649	−0.0375	−0.0386

The rank of the matrix Y_V is $d = 4$. The contribution of the component Y_{V_I} is measured by the share of the corresponding eigenvalues (Table 4).

Table 4. Share of the eigenvalues.

Eigenvalue	$\sum_{i \in I} \lambda_{V_i} / \sum_{i=1}^{4} \lambda_{V_i}$	Share (%)
$\lambda_{V_1} = 1066.7105$	1066.7105/1067.0024	99.9726
$\lambda_{V_2} = 0.2704$	0.2704/1067.0024	0.02534
$\lambda_{V_3} = 0.0135$	0.0135/1067.0024	0.00126
$\lambda_{V_4} = 0.0080$	0.0080/1067.0024	0.00075
Sum: 1067.0024		

To perform the reconstruction stage, it is necessary to calculate the principal components $V_{V_i} = Y_V^T U_{V_i} / \sqrt{\lambda_{V_i}}$. The results of this calculation are represented in Table 5.

Table 5. Values of V_{V_i}.

V_{V_1}	V_{V_2}	V_{V_3}	V_{V_4}
0.5061209	0.6596638	0.1650227	−0.5318409
0.5020078	0.2372012	−0.5817402	0.5924250
0.4982835	−0.2396131	0.7359443	0.3984461
0.4935079	−0.6718575	−0.3059158	−0.4548699

The reconstructed trajectory matrix of all markers achieved from four eigentriples is presented in Table 6.

The reconstructed series of marker M_1 expressed in the form of principal components achieved by arithmetic averaging of $\hat{Y}_{V_1}^{(1)}$, $\hat{Y}_{V_2}^{(1)}$, $\hat{Y}_{V_3}^{(1)}$, $\hat{Y}_{V_4}^{(1)}$ are represented in Table 7 and Figure 8 separately.

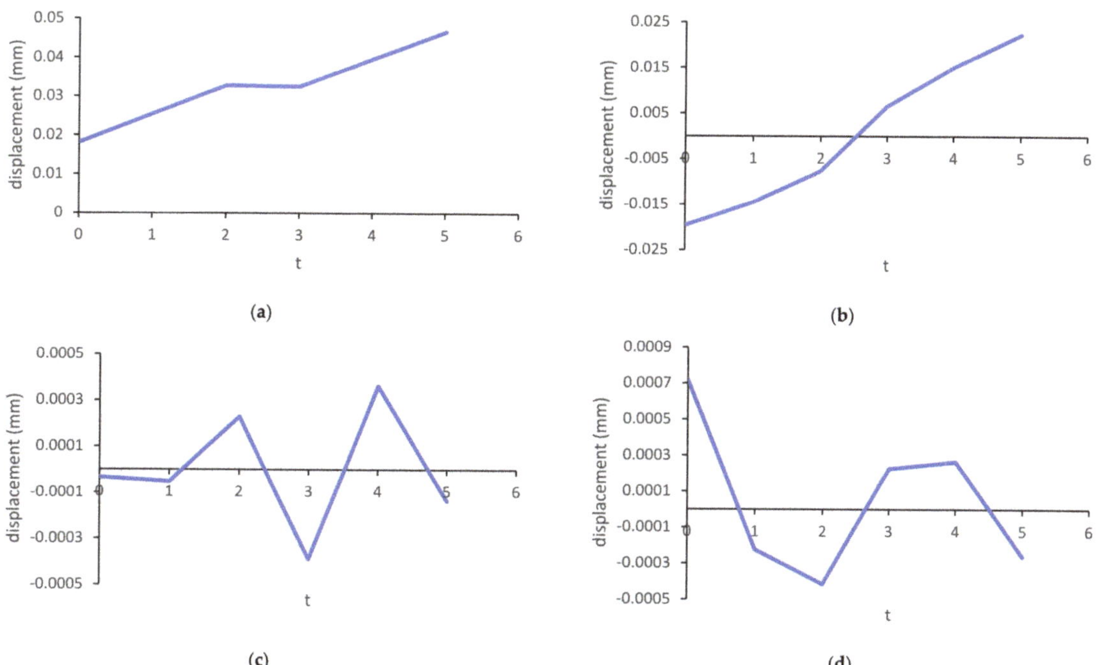

Figure 8. The reconstructed series of x coordinates of the marker M_1 decomposed on principal components: (**a**) the first principal component 99.9726%; (**b**) the second principal component 0.02534%; (**c**) the third principal component 0.00126%; (**d**) the fourth principal component 0.00075%.

Table 6. Reconstructed trajectory matrix of $\hat{Y}_V$.

$\hat{Y}_{V_1}^{(i)}, i = 1,2,\ldots,19$				$\hat{Y}_{V_2}^{(i)}, i = 1,2,\ldots,19$				$\hat{Y}_{V_3}^{(i)}, i = 1,2,\ldots,19$				$\hat{Y}_{V_4}^{(i)}, i = 1,2,\ldots,19$			
0.01818	0.01804	0.01790	0.01773	−0.01955	−0.00703	0.00710	0.01991	−0.00003	0.00011	−0.00015	0.00006	0.00072	−0.00080	−0.00054	0.00061
0.03306	0.03279	0.03255	0.03224	−0.02192	−0.00788	0.00796	0.02232	−0.00022	0.00076	−0.00097	0.00040	0.00036	−0.00040	−0.00027	0.00031
0.04794	0.04755	0.04720	0.04674	−0.02195	−0.00789	0.00797	0.02236	0.00007	−0.00026	0.00032	−0.00014	−0.00030	0.00034	0.00023	−0.00026
0.02645	0.02623	0.02604	0.02579	−0.02610	−0.00939	0.00948	0.02659	−0.00031	0.00110	−0.00139	0.00058	−0.00051	0.00057	0.00038	−0.00044
0.04463	0.04427	0.04394	0.04352	−0.02621	−0.00942	0.00952	0.02669	−0.00010	0.00034	−0.00043	0.00018	0.00040	−0.00045	−0.00030	0.00035
0.06116	0.06066	0.06021	0.05964	−0.02552	−0.00918	0.00927	0.02599	0.00005	−0.00018	0.00023	−0.00010	−0.00113	0.00126	0.00085	−0.00097
...	...	...	...	...	...	...	...	...	...	...	...	...	...	...	...
...	...	...	...	...	...	...	...	...	...	...	...	...	...	...	...
...	...	...	...	...	...	...	...	...	...	...	...	...	...	...	...
3.42505	3.39722	3.37202	3.33970	−0.02418	−0.00870	0.00878	0.02463	−0.00051	0.00181	−0.00229	0.00095	−0.00049	0.00054	0.00036	−0.00041
3.41348	3.38574	3.36062	3.32842	−0.02154	−0.00775	0.00782	0.02194	−0.00057	0.00201	−0.00255	0.00106	0.00030	−0.00034	−0.00023	0.00026
3.40026	3.37263	3.34760	3.31552	−0.02226	−0.00801	0.00809	0.02267	−0.00072	0.00253	−0.00321	0.00133	0.00184	−0.00205	−0.00138	0.00157

Table 7. Reconstruction stage of principal components of marker M_1.

	$\tilde{x}_0^{(1)}$	$\tilde{x}_1^{(1)}$	$\tilde{x}_2^{(1)}$	$\tilde{x}_3^{(1)}$	$\tilde{x}_4^{(1)}$	$\tilde{x}_5^{(1)}$
$\tilde{X}_{V_1}^{(1)}$	0.0181832	0.0255479	0.0328769	0.0326088	0.0397158	0.0467429
$\tilde{X}_{V_2}^{(1)}$	−0.0195524	−0.0144750	−0.0075777	0.0066606	0.0151494	0.0223594
$\tilde{X}_{V_3}^{(1)}$	−0.0000326	−0.0000509	0.0002304	−0.0003876	0.0003633	−0.0001351
$\tilde{X}_{V_4}^{(1)}$	0.0007183	−0.0002217	−0.0004133	0.0002287	0.0002666	−0.0002604

Within the grouping step, we only have signal and noise components and we use two groups of indices, $I_1 = \{\lambda_{V_1}, \lambda_{V_2}\}$ and $I_2 = \{\lambda_{V_3}, \lambda_{V_4}\}$ and associate the group I_1 with signal component and the group I_2 with noise. The reconstructed series of x coordinates of the marker M_1 composed of the signal components are represented in Table 8.

Table 8. The reconstructed series of the marker M_1 composed of the signal components.

	$\tilde{x}_0^{(1)}$	$\tilde{x}_1^{(1)}$	$\tilde{x}_2^{(1)}$	$\tilde{x}_3^{(1)}$	$\tilde{x}_4^{(1)}$	$\tilde{x}_5^{(1)}$
$\tilde{X}_{V_1}^{(1)} + \tilde{X}_{V_2}^{(1)}$	-0.001369	0.011073	0.025299	0.039269	0.054865	0.069102

Figure 9 represents the original and reconstructed series of x coordinates of the marker M_1 using only signal group (I_1).

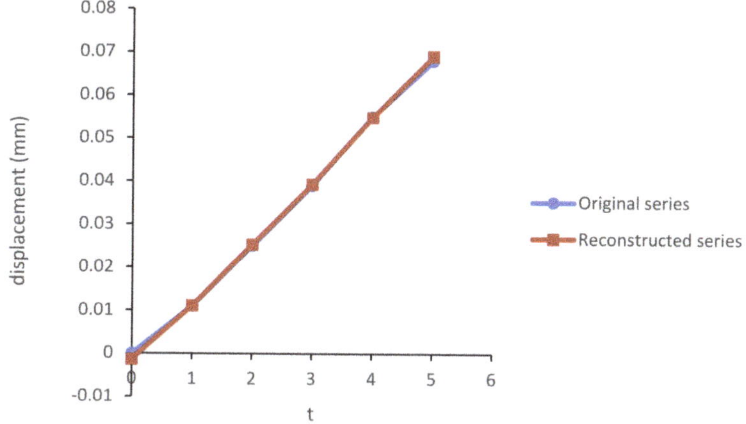

Figure 9. Original and reconstructed series of x coordinates of M_1.

The reconstructed series $\tilde{X}_{N_i}^{(i)}$, $i = 1, 2, \ldots, 19$ achieved by arithmetic averaging of $Y_V(I_1)$ are represented in Table 9.

Table 9. Reconstructed series of x coordinates of markers based on the signal components.

Marker	$t = 0$	$t = 1$	$t = 2$	$t = 3$	$t = 4$	$t = 5$
1 X(1)	-0.001	0.011	0.025	0.039	0.055	0.069
2 X(2)	0.000	0.018	0.035	0.052	0.070	0.086
. . .	. . .	. . .	. . .	. . .	. . .	. . .
19 X(19)	3.401	3.390	3.379	3.366	3.353	3.338

The error of the reconstruction stage is calculated according to the following equation of mean absolute percentage error (MAPE):

$$MAPE_i = \frac{1}{N} \sum_{N_i=1}^{N} \left| \frac{\tilde{X}_{N_i}^{(i)} - X_{N_i}^{(i)}}{X_{N_i}^{(i)}} \right| \times 100\%, \tag{29}$$

The aggregated mean absolute percentage error (AMAPE) of the reconstruction stage is calculated as follows:

$$AMAPE = \frac{1}{M} \sum_{i=1}^{M} MAPE_i, \tag{30}$$

MAPEi and AMAPE are represented in Table 10.

Table 10. Error of reconstruction stage.

Marker	$MAPE_i^{(0)}$	$MAPE_i^{(1)}$	$MAPE_i^{(2)}$	$MAPE_i^{(3)}$.	$MAPE_i^{(4)}$	$MAPE_i^{(5)}$	$MAPE_i(\%)$
1	0.14	0.66	1.20	0.69	0.25	1.62	0.76
2	0.03	7.18	0.96	1.05	1.62	0.74	1.93
...	...	...	...	...	...	...	...
19	0.03	0.02	0.00	0.02	0.06	0.08	0.04
AMAPE							0.95

Signal components used for the reconstruction stage were also used for the forecasting purpose. Vector $U_j^{(i)\nabla}$, $j = 1, 2$, composed of the first $L_i - 1$ components of the vector $U_j^{(i)}$, $j = 1, 2, 3, 4$, is represented in Table 11.

Table 11. Components of the vector $U_j^{(i)}$, $j = 1, 2$.

Marker	U_{V1}	U_{V2}
1	0.0011	−0.0570
	0.0020	−0.0639
2	0.0016	−0.0761
	0.0027	−0.0764
...	...	...
	...	...
19	0.2072	−0.0705
	0.2065	−0.0628

Vector $\pi_j^{(i)}$, $j = 3$, composed of the last component of the vector $U_j^{(i)}$, $(i = 1, \ldots, M)$, is represented in Table 12.

Table 12. Components of the vector $\pi_j^{(i)}$, $j = 3$.

Marker	U_{V1}	U_{V2}
1	0.0029	−0.0640
2	0.0037	−0.0744
...	...	...
19	0.2057	−0.0649

Forecasted x coordinates of the markers for $h = 1$ step ahead are obtained according to Equation (12) and represented in Table 13.

Table 13. Forecasted x coordinates for $h = 1$.

Marker	$t = 6$	$t = 7$	$t = 8$
1	0.083	0.096	0.108
2	0.102	0.117	0.131
...	...	...	...
19	3.323	3.303	3.281

The analogous procedure is performed for z coordinates. The reconstructed series $\tilde{Z}_{N_i}^{(i)}$, $i = 1, 2, \ldots, 19$ are represented in Table 14.

Table 14. Reconstructed series of Z coordinates of markers based on the signal components.

Marker	$t = 0$	$t = 1$	$t = 2$	$t = 3$	$t = 4$	$t = 5$
1 $Z^{(1)}$	0.000	0.000	0.000	0.000	0.000	0.000
2 $Z^{(2)}$	0.624	0.624	0.622	0.620	0.617	0.615
...	...	...	...	...	...	...
19 $Z^{(19)}$	0.000	0.000	0.000	0.000	0.000	0.000

The difference between the original and reconstructed series for marker 2 is represented in Figure 10.

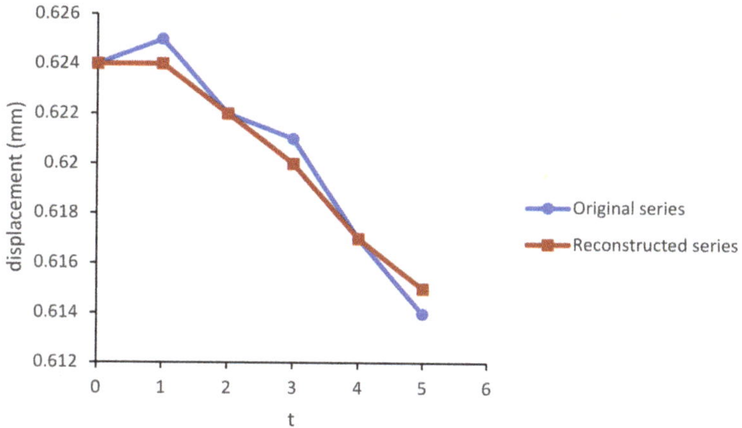

Figure 10. Original and reconstructed series of z coordinates of M_2.

Aggregated mean absolute percentage error (AMAPE) of reconstruction stage of z coordinates is 0.18%. Forecasted z coordinates of the markers for $h = 1$ step ahead are obtained according to Equation (12) and represented in Table 15.

Table 15. Forecasted z coordinates for $h = 1$.

Marker	$t = 6$	$t = 7$	$t = 8$
1	0.000	0.000	0.000
2	0.612	0.611	0.609
.	.	.	.
19	0.000	0.000	0.000

The time displacement intensity vector of marker m for $t = 0, 1, \ldots, 5$ and $t = 6, 7, 8$ is represented in Table 16.

Table 16. Displacement intensity vector ΔV.

Marker	$\Delta v(t = 0)$	$\Delta v(t = 1)$	$\Delta v(t = 2)$	$\Delta v(t = 3)$	$\Delta v(t = 4)$	$\Delta v(t = 5)$	$\Delta v(t = 6)$	$\Delta v(t = 7)$	$\Delta v(t = 8)$
1	0.000	0.011	0.014	0.014	0.016	0.013	0.015	0.013	0.012
2	0.000	0.019	0.016	0.018	0.018	0.014	0.017	0.015	0.015
...	...	...	...	...	...	...	...	...	...
19	0.000	0.009	0.012	0.014	0.014	0.010	0.018	0.020	0.022

The universe of displacement is defined as follows: $D = [l, u]$, where $l = \min\Delta V_{19 \times 19} = 0$ and $u = \max\Delta V_{19 \times 19} = 0.2020$. The observed and forecasted data are sorted in the following ascending order: $0, 0, \ldots, 0.0063, 0.0090, \ldots, 0.1958$, and 0.2020. The distance of any two consecutive values is: $0, 0, \ldots, 0.0027, \ldots, 0.0062$. The corresponding average value $\bar{d}$ and standard deviation σ_d of obtained distances are: $\bar{d} = 0.0011$, $\sigma_d = 0.0038$. The process of elimination of outliers is performed with respect to the following condition: $-0.00265 \le d \le 0.00503$. A revised average distance is $\bar{d}_R = 0.000695$. A revised universe of displacement is $\Delta V_R = [-0.0007, 0.2027]$. According to Equation (19) the length of interval is 0.0338.

Quantification of the linguistic variables by the triangular fuzzy numbers is represented in Table 17.

Table 17. Variables obtained by quantification.

Variable	a_i	b_i	c_i
A_1	−0.0007	0.0332	0.0671
A_2	0.0332	0.0671	0.1010
A_3	0.0671	0.1010	0.1349
A_4	0.1010	0.1349	0.1688
A_5	0.1349	0.1688	0.2027

According to Equation (22), displacement intensity of configuration of the gate road support for $t = 1$ is: $O(1) = (a(1), b(1), c(1)) = (0.0632, 0.0758, 0.2020)$ (Figure 11).

Fuzzy relationships between fuzzified observation $O(1) = (a(1), b(1), c(1))$ for $t = 1$ and defined fuzzy time series are: $\mu_{A_1}(1) = 0.587$; $\mu_{A_2}(1) = 0.754$; $\mu_{A_2}(1) = 0.915$; $\mu_{A_3}(1) = 0.842$; $\mu_{A_3}(1) = 0.727$; $\mu_{A_4}(1) = 0.631$; $\mu_{A_4}(1) = 0.359$; $\mu_{A_5}(1) = 0.419$; $\mu_{A_5}(1) = 0$. Fuzzified observation of displacement intensity of configuration of the gate road support for $t = 1$ belongs to the cluster A_2 because the intersection between $O(1)$ and A_2 has the highest value of membership function, $\mu_{A_2}(1) = 0.915$, (Figure 11).

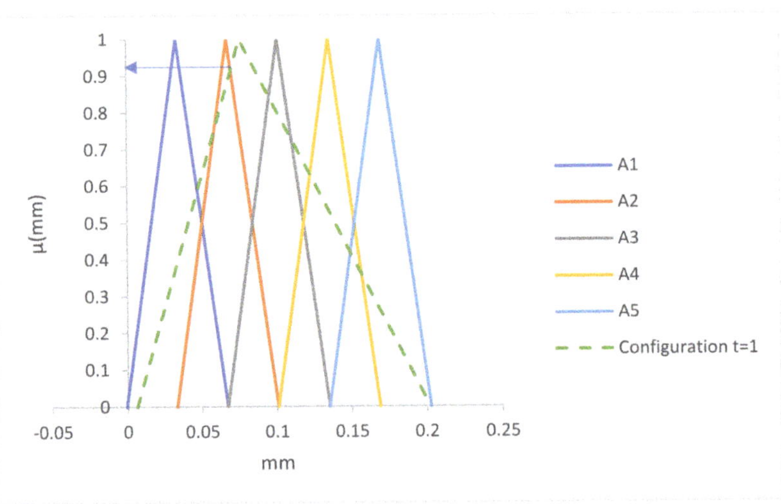

Figure 11. Fuzzy relationships between $O(1)$ and A_1, A_2, A_3, A_4, and A_5.

4. Results

Finally, the obtained sequences of displacement intensity of configuration of the gate road support over observed and forecasted time is as follows: A2→A2→A2→A2→A2→A1 →A1→A1, see Figure 12.

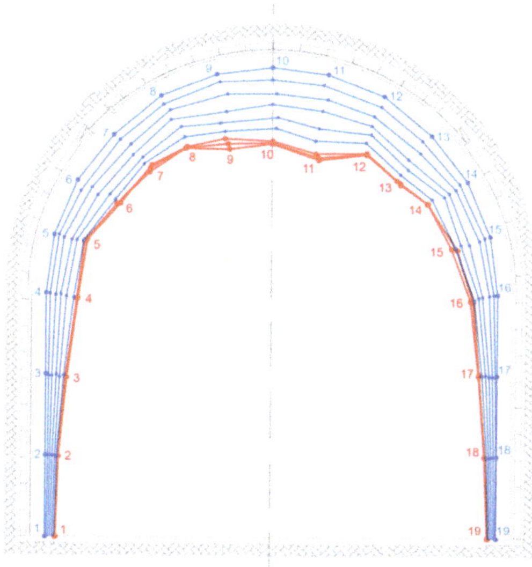

Figure 12. Topology of the gate road support over time, blue color-observed and red color-forecasted configuration.

5. Conclusions

Efficiency of underground mine production directly depends on functionality of the main gate roads. Due to hard geological conditions and mining activities, deformations of gate roads are very intensive and happen frequently. Having the ability to monitor and forecast deformations is recognized as a key role to maintaining the stability of gate roads, i.e., to maintain their functionality. The signal and noise of deformations are isolated by multivariate singular spectrum analysis and it is also used to forecast future deformations with respect to different combinations of obtained principal components. Clustering based on fuzzy time series enables us to see how the configuration of the gate road support has generally changed over time and defines the intervals of deformation by adequate triangular fuzzy numbers. Making decisions and plans based on the intervals is much more reliable and suitable than performing such activities based on crisp values.

The developed model enables mine planners to create an effective plan to support repairs, including duration time and costs of reparation with respect to forecasted deformations. Future research will be focused on carrying an in situ measuring of the deformations in an underground mine to quantify the quality of the method. Furthermore, the focus will be on the making of a 3-D model of gate road support deformation forecasting.

Author Contributions: Conceptualization, R.T.; software, A.G.; formal analysis, S.L.; writing—original draft preparation, L.C.; visualization, A.M.; supervision, Z.G. All authors have read and agreed to the published version of the manuscript.

Funding: This research received no external funding.

Institutional Review Board Statement: Not applicable.

Informed Consent Statement: Not applicable.

Data Availability Statement: Not applicable.

Conflicts of Interest: The authors declare no conflict of interest.

References

1. Xu, D.P.; Feng, X.T.; Cui, Y.J.; Jiang, Q. Use of the equivalent continuum approach to model the behaviour of a rock mass containing an interlayer shear weakness zone in an underground cavern excavation. *Tunn. Undergr. Space Technol.* **2015**, *47*, 35–51. [CrossRef]
2. Armaghani, D.J.; Mohamad, E.T.; Hajihassani, M.; Yagiz, S.; Motaghedi, H. Application of several non-linear prediction tools for estimating uniaxial compressive strength of granitic rocks and comparison of their performances. *Eng. Comput.* **2016**, *32*, 189–206. [CrossRef]
3. Xiong, X. Research on Grey System Model and its Application on Displacement Prediction in Tunnel Surrounding Rock. *Open Mech. Eng. J.* **2014**, *8*, 514–518. [CrossRef]
4. Tang, P. Forecast the Working Face Underground Pressure Using Grey Model Improved by the Genetic Algorithm. *Electron. J. Geotech. Eng.* **2011**, *16*, 1215–1226.
5. Yu, S.; Hongzhen, Z.; Yanna, C. Application of GRNN in Time Series Prediction for Deformation of Surrounding Rocks in Soft Rock Roadway. In Proceedings of the 2011 Fourth International Conference on Intelligent Computation Technology and Automation, Shenzhen, China, 28–29 March 2011; Volume 1, pp. 63–66.
6. Chen, H.; Wang, R. Artificial Neural Network's Application in Intelligent Displacement Back Analysis of Deep Mine Roadway Surrounding Rock. In Proceedings of the 2010 International Conference on Intelligent Computation Technology and Automation, Changsha, China, 11–12 May 2010; Volume 1, pp. 808–811.
7. Bozzano, F.; Mazzanti, P.; Prestininzi, A. Supporting Tunnelling Excavation of an Unstable Slope by Long term Displacement Monitoring. In Proceedings of the International Conference on Case Histories in Geotechnical Engineering, Chicago, IL, USA, 29 April–4 May 2013; p. 42.
8. Zhang, Q.; Li, S.C.; Li, L.P.; Zhang, Q.Q.; Shi, S.S.; Song, S.G. Numerical Analysis of Advanced Displacement in Construction Progress of Tunnel Excavation with Weak Surrounding Rock. *Res. J. Appl. Sci. Eng. Technol.* **2013**, *6*, 1497–1503. [CrossRef]
9. Grossauer, K.; Leitner, R.; Schubert, W.; Sellner, P. Prediction of subsidence during tunnel construction. In *Underground Space—The 4th Dimension of Metropolises, Proceedings of the 33rd ITA-AITES World Tunnel Congress*; Prague, Czech Republic, 5–10 May 2007, Barták, J., Hrdina, I., Romancov, G., Zlámal, J., Eds.; Taylor & Francis Group: London, UK, 2007; pp. 857–862.
10. Merlini, D.; Falanesca, M. Ceneri Base Tunnel advancement in difficult rock conditions: Tunnel design and construction optimization through back-analysis of the geomechanical parameters. In Proceedings of the World Tunnel Congress 2013 Geneva Underground—The Way to the Future, Geneva, Switzerland, 31 May–7 June 2013; Anagnostou, G., Ehrbar, H., Eds.; Taylor & Francis Group: London, UK, 2013; pp. 1722–1729.
11. Yao, B.Z.; Yang, C.Y.; Yao, J.B.; Sun, J. Tunnel Surrounding Rock Displacement Prediction Using Support Vector Machine. *Int. J. Comput. Intell. Syst.* **2010**, *3*, 843–852. [CrossRef]
12. Harris, T.J.; Yuan, H. Filtering and frequency interpretations of Singular Spectrum Analysis. *Phys. D* **2010**, *239*, 1958–1967. [CrossRef]
13. Hassani, H.; Zhigljavsky, A. Singular Spectrum Analysis: Methodology and Application to Economics Data. *J. Syst. Sci. Complex.* **2009**, *22*, 372–394. [CrossRef]
14. Hassani, H.; Mahmoudvand, R. Multivariate Singular Spectrum Analysis: A General View and New Vector Forecasting Approach. *Int. J. Energy Stat.* **2013**, *1*, 55–83. [CrossRef]
15. Kalantari, M.; Hassani, H. Automatic Grouping in Singular Spectrum Analysis. *Forecasting* **2019**, *1*, 189–204. [CrossRef]
16. Hassani, H.; Yeganegi, M.R.; Silva, E.S. A New Signal Processing Approach for Discrimination of EEG Recordings. *Stats* **2018**, *1*, 155–168. [CrossRef]
17. Song, Q.; Chissom, B.S. Fuzzy time series and its models. *Fuzzy Sets Syst.* **1993**, *54*, 269–277. [CrossRef]
18. Zadeh, L.A. Fuzzy sets. *Inform. Control* **1965**, *8*, 338–353. [CrossRef]
19. Zadeh, L.A. The Concept of a Linguistic Variable and its Application to Approximate Reasoning. In *Learning Systems and Intelligent Robots*; Fu, K.S., Tou, J.T., Eds.; Springer: Boston, MA, USA, 1974. [CrossRef]
20. Huarng, K.H.; Hui, T.; Yu, K. Modelling fuzzy time series with multiple observations. *Int. J. Innov. Comput. Inf. Control* **2012**, *8*, 7415–7426.
21. Chen, S.M. Forecasting enrolments based on fuzzy time series. *Fuzzy Sets Syst.* **1996**, *81*, 311–319. [CrossRef]
22. Liang, C.T.; Hsue, C.C.; Ching, C.Y. Using Extracted Fuzzy Rules Based on Multi-Technical Indicators for Forecasting TAIEX. In Proceedings of the International Conference on Artificial Intelligence (ICAI'10), Las Vegas, NV, USA, 12–15 July 2010; pp. 30–34.
23. Su, C.H.; Chen, T.L.; Cheng, C.H.; Chen, Y.C. Forecasting the Stock Market with Linguistic Rules Generated from Minimize Entropy Principle and the Cumulative Probability Distribution Approaches. *Entropy* **2010**, *12*, 2397–2417. [CrossRef]
24. Kaufmann, A.; Gupta, M.M. *Introduction to Fuzzy Arithmetic: Theory and Applications*; Van Nostrand Reinhold: New York, NY, USA, 1985.

MDPI

St. Alban-Anlage 66

4052 Basel

Switzerland

Tel. +41 61 683 77 34

Fax +41 61 302 89 18

www.mdpi.com

Energies Editorial Office

E-mail: energies@mdpi.com

www.mdpi.com/journal/energies